Bones, Clones, and Biomes

Bones, Clones, and Biomes

The History and Geography of Recent Neotropical Mammals

Edited by Bruce D. Patterson and Leonora P. Costa

THE UNIVERSITY OF CHICAGO PRESS CHICAGO & LONDON

Bruce D. Patterson is the MacArthur Curator of Mammals at the Field Museum of Natural History in Chicago. **Leonora P. Costa** is associate professor in the Departamento de Ciências biológicas at Universidade federal do espírito santo, Vitória, Brazil.

The University of Chicago Press, Chicago 60637
The University of Chicago Press, Ltd., London

Printed in the United States of America

21 20 19 18 17 16 15 14 13 12   1 2 3 4 5

ISBN-13: 978-0-226-64919-1 (cloth)
ISBN-10: 0-226-64919-9 (cloth)

Library of Congress Cataloging-in-Publication Data
Bones, clones, and biomes : the history and geography of recent neotropical mammals / edited by Bruce D. Patterson and Leonora P. Costa.
pages ; cm
Includes bibliographical references and index.
ISBN-13: 978-0-226-64919-1 (cloth : alkaline paper)
ISBN-10: 0-226-64919-9 (cloth : alkaline paper) 1. Mammals—Latin America—History. 2. Mammals—Latin America—Geographical distribution. 3. Biogeography—Latin America. 4. Paleobiogeography—Latin America. I. Patterson, Bruce D. II. Costa, Leonora Pires.
QL721.8.B66 2012
599.098—dc23

2011029614

♾ This paper meets the requirements of ANSI/NISO Z39.48-1992 (Permanence of Paper).

Contents

PART 2. REGIONAL PATTERNS

1 Introduction to the History and Geography of Neotropical Mammals

Bruce D. Patterson and Leonora P. Costa

The Neotropics are home to more than 1500 species of living mammals, almost 30% of all extant species. These charismatic organisms include remarkable animals found nowhere else on earth: armadillos and anteaters, capybaras and capuchins, maned wolves and mouse opossums, sloths and sakis. Endemic groups include the platyrrhine monkeys (136 species), the phyllostomid bats (58 genera and 192 species, exploiting almost every dietary strategy used by bats worldwide), the caviomorph rodents (12 families and 288 species of guinea-pig relatives), and the sigmodontines (a subfamily of mice that explosively radiated into at least 88 genera and 399 species over the past 6–8 million years). Local diversity of mammals in the Neotropics is so great that biologists can capture representatives of 5–6 families and 30–35 species of bats in a single night by using mist nets. Individual reserves in the Neotropics can support more than 220 species of mammals, twice the number known from the eastern third of North America!

This biological bounty can be partly attributed to the striking diversity of Neotropical landscapes and biomes. The region supports lush rainforests (both temperate and tropical), parched deserts, sprawling savannas, thorny scrublands, alpine steppes, and towering peaks. A plethora of climatic and edaphic conditions accompany and shape this biotic diversity. Another agency fostering its endemism has been South America's isolation as an island continent for much of the last 65 million years (Ma). However, unlike its Gondwana neighbor Australia, South America's history of isolation has been interrupted by a succession of continental connections that permitted intermittent faunal exchanges with Africa, Antarctica and Australia, and North America at different times. These episodes of exchange, each involving distinctive sets of immigrants, appear as discrete "strata" in South America's extensive fossil record. To comprehend the development of modern Neotropical mammal faunas requires not only mastery of the Neotropics' substantial mammalian diversity, but also knowledge of lineages and landscapes stretching back to the Mesozoic. A synthesis of this information requires expertise that spans the disciplines of geology, paleontology, physical geography, systematics, and biogeography. Few modern workers claim this breadth.

As a result, the leading works on the evolution and biogeography of Neotropical mammals, books like *Mammals on Southern Continents*, *Splendid Isolation*, and the *Great American Biotic Interchange* (Keast, Erk, and Glass 1972; Simpson 1980; Stehli and Webb 1985) are grossly outdated. Although there are now excellent field guides (Emmons and Feer 1997; Reid 1997) and some regional treatises on the fauna (Eisenberg and Redford 1989, 1999; Redford and Eisenberg 1992), no recent work has treated the origins, interrelationships, and biogeography of living Neotropical mammals. Even the taxonomic foundations for this undertaking lack refinement. Angel Cabrera's classic *Catálogo* (Cabrera 1958, 1961) is a half-century old, and the three-volume series now destined to replace it, *Mammals of South America*, is only partly completed (Gardner 2008). In any event, that work has been organized around taxa (orders, families, genera, and species), not around time periods and regions. For this reason, a synthesis focused on regional and historical features of the modern Neotropical mammal fauna nicely complements that taxonomic series.

Since the last attempt at a synthesis, a host of practical and conceptual advances have been made in disciplines flanking historical biogeography. Continental drift has replaced hypothetical land bridges as a means of intercontinental dispersal (cf. Simpson 1980). Absolute chronometric methods have been developed and applied to precisely date faunal horizons (Flynn and Swisher 1995). Additional land mammal ages have been described (Flynn et al. 2003), while existing ones have been better delimited (Kay et al. 1999). Nearly a quarter of living Neotropical mammal species has been newly discovered or rediscovered (Patterson 1996, 2000), offering taxonomic and geographic refinement to biogeographic reconstructions. Cladistics has replaced earlier paradigms in systematics (Cracraft and Donoghue 2004; Felsenstein 2004), and explicit analyses such as phylogeography and dispersal-vicariance approaches have refined historical biogeographic reconstructions (Crisci, Katinas, and Posadas 2003; Ree et al. 2005). All of these advances contribute substantial new understanding to discussions on the origin, diversity, and distribution of Neotropical mammals, and a comprehensive synthesis of their findings is long overdue.

We addressed this need by assembling expert paleontologists and neontologists and asking them to detail their complementary spatiotemporal perspectives by focusing on a central question: *How did changing climates and landscapes, intercontinental connections, and newly evolved lineages interact to populate Central and South America and the Antilles with almost 30% of the world's living mammals?* The paleontologists were tasked with exploring the historical context of the modern fauna. Their descriptions and characterizations of Mesozoic, Paleogene, and Neogene landscapes and inhabitants offer depth and context to understanding of modern landscapes and faunas.

The neontologists were challenged to account for modern faunas region by region. Each of the regional chapters explores the historical development, diversification, and endemism of the mammal groups living there. We make no pretense of documenting modern diversity or its distribution, but rather try to summarize patterns reflected in those distributions and relationships, and do so to the extent that regional studies permit. Our treatment does not extend to aquatic and marine mammals, because very different physical and geographic issues are involved with those groups and, for most, even less information is available. Restrictions on manuscript length precluded a fuller exploration of Neotropical paleontology—although some intervals remain poorly known and studied, this decision was largely owing to the richness and complexity of modern faunas and the desire to treat them as comprehensively as possible. Even so, some distinctive biomes, such as the Gran Chaco and Valdivian Forest, are only touched upon by chapters devoted to neighboring biomes.

Earlier attempts to chronicle the development of Neotropical mammal faunas were plagued by labored arguments and controversy over plate tectonics and phylogenetic methods (Keast, Erk, and Glass 1972; Simpson 1980). Here, chapters employ these well-worked methods and conceptual frameworks to generate new and refined interpretations of how current faunas came to be. We devote two of the volume's chapters to Neotropical carnivore radiations: one is paleontological and documents spatial and temporal records, while the other chapter uses molecular trees to date key events in the radiation, especially those in the tropics and during their initial radiation, when fossils are generally lacking.

Chapters articulate research visions that are scattered widely across physical geography, geology, paleontology, systematics, mammalogy, and biogeography. Each of the contributors is an acknowledged authority in his or her field, and we were fortunate to secure the participation of virtually all the scientists that we approached with our plans. From the start, their participation was to be both oral and written. These contributions were first presented to the scientific community in two back-to-back symposia at the 10th International Mammalogical Congress in Mendoza, Argentina, on August 14, 2009. The symposia gave each contributor an overview of the ensemble and the approaches, analyses and results of other participants. We hope that this has helped us to integrate these papers into a coherent volume whose scope exceeds the expertise and conceptual and analytical range of any single person.

The various chapters were submitted to us by January 31, 2010, and were sent out for anonymous peer review; revised versions of the chapters were submitted to the publisher on July 15, 2010; updated citations were possible in the proof stage.

A number of people aided us in this project. We thank the organizers of IMC-10, especially Ricardo Ojeda, for the invitation to organize a symposium and subsequently giving us the latitude (and schedule) to make it last all day. A host of peer reviewers offered helpful criticisms and helped us shape and rework the contributed manuscripts. In alphabetical order, we thank Sergio Ticul Alvarez, Robert Anderson, Mariano Bond, François Catzeflis, Guillermo D'Elia, Albert Ditchfield, Analia Forasiepi, Alfred Gardner, Chris Himes, Christopher Johnson, Douglas Kelt, Thomas Lacher, Lucia Luna, Bruce MacFadden, Ross MacPhee, Kevin Murray, James Patton, Alexandre Percequillo, Bruce Shockey, Maria Nazareth da Silva, Richard Tedford, Robert Voss, Marcelo Weksler, Lars Werdelin, Michael Woodburne, and Neal Woodman. Our contributors all bore in good cheer (but variable timeliness) our efforts to orchestrate their scholarship. We thank three anonymous reviewers of the volume as a whole for their excellent feedback and suggestions; depth of knowledge and generous editorial assistance unmasked one of these (special thanks to Alfred Gardner). Finally, of course, we thank Christie Henry and Amy Krynak at the University of Chicago Press for their expertise in bringing this project forward in its current form and Rachel Cabaniss for her careful copy editing.

Bruce D. Patterson
Chicago, USA

Leonora P. Costa
Vitoria, Brazil

Literature Cited

Cabrera, A. 1958. "Catálogo de los Mamíferos de América del Sur." *Revista del Museo Argentino de Ciencias Naturales. "Bernardo Rivadivia" Instituto Nacional de Investigación Ciencias Naturales, Ciencias Zoológicas* 4 (1):1–307.

———. 1961. "Catálogo de los Mamíferos de America del Sur. Vol. 2." *Revista del Museo Argentino de Ciencias Naturales, "Bernardo Rivadivia" Instituto Nacional de Investigación Ciencias Naturales, Ciencias Zoológicas* 4 (2):308–732.

Cracraft, J., and M. J. Donoghue. 2004. *Assembling the Tree of Life*. New York: Oxford University Press.

Crisci, J. V., L. Katinas, and P. Posadas. 2003. *Historical Biogeography: An Introduction*. Cambridge: Harvard University Press.

Eisenberg, J. F., and K. H. Redford. 1989. *Mammals of the Neotropics*. Vol. 1, *The Northern Neotropics: Panama, Colombia, Venezuela, Guyana, Suriname, French Guiana*. Chicago: University of Chicago Press.

———. 1999. *Mammals of the Neotropics*. Vol. 3, *The Central Neotropics: Ecuador, Peru, Bolivia, Brazil*. Chicago: University of Chicago Press.

Emmons, L. H., and F. Feer. 1997. *Neotropical Rainforest Mammals: A Field Guide*. Chicago: University of Chicago Press.

Felsenstein, J. 2004. *Inferring Phylogenies*. Sunderland, MA: Sinauer Associates.

Flynn, J. J., and C. C. Swisher, III. 1995. "Cenozoic South American Land Mammal Ages: Correlation to Global Geochronologies." In *Geochronology Time Scales and Global Stratigraphic Correlation*, edited by W. Berggren et al., 317–34. Tulsa: Society for Sedimentary Geology (SEPM).

Flynn, J. J., A. R. Wyss, D. A. Croft, and R. Charrier. 2003. "The Tinguiririca Fauna, Chile: Biochronology, Paleoecology, Biogeography, and a new Earliest Oligocene South American Land Mammal 'Age'." *Palaeogeography, Palaeoclimatology, Palaeoecology* 195:229–59.

Gardner, A. L., ed. 2008. *Mammals of South America*. Vol. 1, *Marsupials, Xenarthrans, Shrews, and Bats*. Chicago: University of Chicago Press.

Kay, R. F., R. H. Madden, M. G. Vucetich, A. A. Carlini, M. M. Mazzoni, G. H. Re, M. Heizler et al. 1999. "Revised Geochronology of the Casamayoran South American Land Mammal Age: Climatic and Biotic Implications." *Proceedings of the National Academy of Sciences USA* 96:13235–40.

Keast, A. F., F. C. Erk, and B. Glass, eds. 1972. *Evolution, Mammals, and Southern Continents*. Albany: State University of New York Press.

Patterson, B. D. 1996. "The 'Species Alias' Problem." *Nature* 380:589.

———. 2000. "Patterns and Trends in the Discovery of New Neotropical Mammals." *Diversity and Distributions* 6:145–51.

Redford, K. H., and J. F. Eisenberg. 1992. *Mammals of the Neotropics*. Vol. 2, *The Southern Cone: Chile, Argentina, Uruguay, Paraguay*. Chicago: University of Chicago Press.

Ree, R. H., B. R. Moore, C. O. Webb, and M. J. Donoghue. 2005. "A Likelihood Framework for Inferring the Evolution of Geographic Range on Phylogenetic Trees." *Evolution* 59:2299–311.

Reid, F. A. 1997. *A Field Guide to the Mammals of Central America and Southeast Mexico*. New York: Oxford University Press.

Simpson, G. G. 1980. *Splendid Isolation: The Curious History of South American Mammals*. New Haven: Yale University Press.

Stehli, F. G., and S. D. Webb, eds. 1985. *The Great American Biotic Interchange*. New York: Plenum Press.

Part 1
The Geological Setting

2

Punctuated Isolation
The Making and Mixing of South America's Mammals

Darin A. Croft

South America's fossil record of mammals from the Cenozoic Era (the past 65 million years or so) is far from perfect. No fossils have yet been discovered from some intervals of this time span, whereas only one or a few poorly sampled localities represent others. Moreover, most intervals are known only from a relatively restricted portion of the continent. In many respects, South America's mammalian record pales in comparison to that of northern continents.

On the other hand, South America overall probably has the best fossil record of mammals among southern hemisphere continents. Thanks to a handful of spectacular localities and an abundance of others, the broad picture of mammal evolution on the continent has been known for more than half a century (e.g., Simpson 1950). As detailed in the chapters that follow, the mammals of this continent evolved largely within the context of isolation, punctuated by rare episodes of waif dispersal and faunal interchange. This contrasts starkly with the history of mammal evolution on northern continents, in which episodes of faunal interchange were commonplace (Janis 1993). Whereas northern continents have acted as thoroughfares for numerous mammal groups, South America and other southern continents have acted more as cul-de-sacs: areas off the beaten path with limited access and little large-scale turnover (Keast, Erk, and Glass 1972). They have not been evolutionary dead ends for mammals (with the exception of Antarctica), but rather have produced communities of thriving, endemic species exhibiting mixtures of novel and convergent traits.

Although the general context of evolving mammal communities in South America has been known for many years, remarkable advances have taken place in recent decades in virtually every aspect of mammalian paleontology. These advances have been made by scores of researchers from South America and elsewhere using a variety of methods ranging from traditional to cutting edge. Some of these experts, representing many of the main areas of scientific investigation, have been brought together in the first part of this volume to summarize these recent advances and to provide a framework for the origins and biogeography of modern mammal communities.

The next chapter, by Goin and collaborators, offers a broad-stroke perspective on changing faunas and habitats of South America over the past 250 Ma. Of necessity, their observations are based primarily on the Patagonian fossil record, since it is the only region of the continent where nearly all major time intervals are represented. The excellent record in Patagonia is the result of fortuitous geological and climatic factors. For much of the Cenozoic, parts of Patagonia gradually accumulated water- and wind-borne sediments, conditions highly suitable for the preservation of teeth, bones, tracks, and other evidence of past life. Some of these sediments were sands and silts derived from the erosion of topographically higher areas, whereas others were ashes from volcanic eruptions. Occasionally, transgressions of the ocean onto the continent added marine sediments over the continental record. With the rise of the Andes and changes in global climate patterns, Patagonia developed the dry, windy, temperate conditions that characterize it today (Barreda and Palazzesi 2007). Because these conditions support little vegetation, the sediments that accumulated over millions of years, and the fossils within them, have been exposed to the elements, as well as to the eyes of paleontologists. Without such exposures, fossils would remain inaccessible and undiscovered, an observation that partly explains the scarcity of fossil localities in tropical lowlands, which are typified by abundant vegetation. Water and wind erosion in Patagonia continuously expose new fossils and, given the vast expanse of Patagonia, many new fossil localities undoubtedly remain to be discovered.

Based on the record from Patagonia, Goin and collaborators recognize five main phases in the evolution of South American mammal faunas. The transitions between these phases are recognized by more or less drastic changes in faunal composition. Among these transitions, the most significant might have been the one that occurred near the end of the Age of Dinosaurs (the Mesozoic Era, ending ~65 Ma). It is almost as if the list of mammals living in South America at the time had been wiped clean; the main constituents of Mesozoic ecosystems, the nontribosphenic mammals (which mostly lack living descendants) disappeared and were replaced by therians (marsupials and placentals), likely of North American origin. The fossil record documenting this remarkable change is relatively sparse, and the exact sequence of events and their causes remain unknown. Nevertheless, current evidence suggests that the results of the interchange between the Americas at this time were just as dramatic as those that occurred only a few million years ago (as a result of the Great American Biotic Interchange [GABI], discussed later). Goin and collaborators refer to this early event as the FABI (First American Biotic Interchange). The brief period following the FABI appears to have been the heyday for South American metatherians (marsupials) in terms of taxonomic diversity and abundance. It came to an end about 50 Ma as placentals definitively gained the upper hand, having differentiated into the main groups so characteristic of later South American paleocommunities.

Another major shift in South American mammal faunas (as well as those in many parts of the world) occurred ~35 Ma; Antarctica's land connections to South America and Australia were finally severed, isolating all three continents and initiating a cascade of events that, in combination with other factors, resulted in a substantial drop in global temperatures (Zanazzi et al. 2007). The immediate effects of the so-called Eocene-Oligocene transition in South America were unrecognized in the mammal fossil record until the discovery of a new locality in central Chile in the late 1980s (Wyss et al. 1993, 1994). The many taxonomic, paleogeographic, and paleoecological insights that have come from this fauna are highlighted in the chapter on Cenozoic Andean faunas by Flynn and collaborators (chapter 4, this volume). The story of the serendipitous discovery of this locality—and of additional localities in the region—is nearly as interesting as the science (see Flynn, Wyss, and Charier 2007). At last tally, the rock formation whose unusual conditions preserved these fossils had produced more than 1400 specimens, mostly skulls, mandibles, and teeth, ranging in age from 40–15 Ma. Several hundred other specimens have been collected from a remarkable succession of strata in a similar formation slightly farther south in Chile (Flynn et al. 2008). These assemblages represent an increasingly important and expanding record of extra-Patagonian mammal evolution.

Just as Patagonia has the right conditions for exposing vast thicknesses of fossiliferous rocks at the Earth's surface, so too do the Andes. It is therefore little wonder that, as detailed by Flynn and collaborators, other areas of the Andes (Colombia and Bolivia in particular) have also produced significant collections of Cenozoic mammals (for overviews, see Kay et al. 1997; Marshall and Sempere 1991). Smaller collections have come from the Andes of Ecuador and Peru (e.g., Madden et al. 1994; Shockey et al. 2003). Most of these localities are north of the Tropic of Capricorn (~23°S), well outside the traditionally well-sampled southern part of the continent. The only other mammal faunas known from this vast area of South America are a handful of lowland faunas from Brazil, Peru, and Venezuela (e.g., Antoine et al. 2006, 2007; Campbell 2004; Cozzuol 2006; Sánchez-Villagra and Aguilera 2006, 2008; Vucetich and Ribeiro 2003).

Together, these low and middle latitude assemblages have provided critical insights into the development of Neotropical mammal communities (reviewed in MacFadden 2006). Considering that they are spread over tens of millions of years, however, they provide mere glimpses of habitats and communities rather than an integrated picture of evolving mammal faunas. Sampling thus remains a key issue in Neotropical paleomammalogy, especially in light of biogeographic hypotheses that recognize Patagonia as a region distinct from the rest of South America (a point further elaborated by Goin and collaborators). Documenting new fossil mammal localities in low and intermediate latitudes is a priority for future research.

In addition to recent progress documenting past mammal distributions, major advancements have been made in the field of paleobiology, the form and function of extinct organisms. Paleobiological analyses of many South American mammals have been hindered by the lack of living representatives and close living analogues: diverse groups of hoofed herbivores such as notoungulates and litopterns are extinct, and modern faunas lack anything closely similar to the large ground sloths and armored glyptodonts common in Neotropical paleofaunas. Circumventing this issue has required a combination of large modern comparative data sets and mechanical modeling. A leading research group in this regard has been that of Vizcaíno and collaborators, whose analysis (chapter 5, this volume) examines how mammal body mass maxima and ranges have changed over the past 40 Ma. Their study demonstrates that different taxonomic groups have occupied the role of largest herbivore during this interval, and indicates that the greatest diversity of large species was present in South America only 10,000 BP. This latter observation has important implications for the functioning of modern ecosystems (see later) and suggests that these paleocommunities were structured quite differently than those of today, a finding that echoes results of similar studies of other time intervals (Croft 2001; Croft, Flynn, and Wyss 2008). Their chapter adds to an expanding body of literature on diet (e.g., Bargo and Vizcaíno 2008; Croft and Weinstein 2008; MacFadden 2005; Townsend and Croft 2008; Vizcaíno et al. 2004, Vizcaíno, Bargo, and Cassini 2006), locomotion (e.g., Elissamburu 2004; Elissamburu and Vizcaíno 2004, 2005; Shockey 1999, 2001; Shockey and Anaya 2008; Shockey, Croft, and Anaya 2007; Vizcaíno and Milne 2002), body mass (e.g., Croft 2001; Fariña, Vizcaíno, and Bargo 1998; Sánchez-Villagra, Aguilera, and Horovitz 2003), and community structure (e.g., Croft 2006; Croft and Townsend 2005; Flynn et al. 2003; Kay and Madden 1997; MacFadden and Shockey 1997; Vizcaíno et al. 2006) that together have created a much more detailed picture of the life and times of extinct South American mammals.

The final great perturbation in South American mammal faunas noted by Goin and collaborators, the Great American Biotic Interchange (GABI), is familiar to anyone who has taken a college biology class: the intermingling of terrestrial mammals that resulted from the gradual reduction and eventual elimination of the water barrier separating North and South America (a full connection having been established between two and three million years ago). As has been the case for other episodes in the history of South American mammals, the details of this classic tale have been refined in recent years (Woodburne 2010) and different aspects of the GABI are discussed in many chapters of this volume. The final two chapters of part 1 focus exclusively on a clade intimately linked to this intercontinental exchange, the Carnivora.

The absence of carnivorans (i.e., members of the order Carnivora) is a characteristic and intriguing feature of most South American paleocommunities. In their absence, the role of large, terrestrial, warm-blooded meat-eater (carnivore) was filled by metatherian mammals—specifically sparassodonts (borhyaenids and relatives)—as well as phorusrhacids, also known as terror birds, which were large to giant flightless birds with oversized heads and hooked beaks (Alvarenga and Höfling 2003; Chiappe and Bertelli 2006; Marshall 1977). Although sparassodonts included highly specialized forms such as the saber-toothed *Thylacosmilus atrox* (Riggs 1934) and the bear-sized *Paraborhyaena boliviana* (Hoffstetter and Petter 1983), as a group they were apparently both taxonomically depauperate and extremely rare (Croft 2006). Whether such attributes resulted from competition with phorusrhacids, or whether phorusrhacids simply exploited a vacant niche, is not known, though certain aspects of metatherian craniodental anatomy favor the latter hypothesis (Werdelin 1987, 1988). Regardless, these two groups coexisted for most of the Cenozoic, suggesting effective niche partitioning by habitat and diet. This arrangement came to a fairly abrupt end about the time that carnivorans were added to the mix.

The timing and pattern of carnivoran arrival and diversification in South America, as evidenced by the paleontological record, are detailed in Prevosti and Soibelzon's chapter (chapter 6, this volume). Although sampling deficiencies exist, the overall trend is one of staggered immigrations to South America followed by virtually unrestrained diversification (dampened only by a Quaternary extinction event, discussed later). The last sparassodonts coexisted with the earliest South American carnivorans (Marshall and Patterson 1981) and there is no consensus regarding the role of carnivorans, if any, in their extinction (Forasiepi, Martinelli, and Goin 2007). By any standard, however, the evolutionary success of carnivorans in the continent has been remarkable. The number of species presently inhabiting South America rivals the number of sparassodonts known from the entire Cenozoic (A. Forasiepi, pers. comm.; Wozencraft 2005).

The chapters by Vizcaíno and collaborators (chapter 5, this volume) and Prevosti and Soibelzon (chapter 6, this volume) address the effect that Quaternary megafaunal extinctions had on herbivores and carnivorans, respectively. This event, which took place ~10,000 BP, marked the last appearance of many large mammal species in the Americas including mammoths, saber-toothed cats, and ground sloths. The causes of this extinction are still being debated, but it seems likely that humans were a contributing factor (Johnson 2002). While receiving much more attention in North America than in South America, the effects of this extinction phase on South American mammals were more profound. Among northern mammals, no extinctions represented the loss of an ordinal group at a global scale, and only two represented global losses of families (Mammutidae and Gomphotheriidae). In the south, however, the extinction spelled the end for the two great orders of endemic South American herbivores, the notoungulates and the litopterns, as well as for four families of xenarthrans (mylodontid and megatheriid sloths, and pampatheriid and glyptodontid cingulates). Representatives of these xenarthran groups had been fairly recent immigrants to North America, and their loss might not be expected to have had much impact on ecosystem functioning there. In contrast, Vizcaíno and collaborators contend that modern South American ecosystems might still reflect the effects of the loss of these and other megaherbivores, and that this should be an area of further investigation. An argument could thus be made for adding a sixth phase to the scheme of Goin and collaborators that might appropriately be called the Hypoamerican phase, given its diminished diversity of higher clades. It would be marked by the arrival of humans to the continent, as well as the extinction of the last representatives of many lineages of native South American mammals. Among these, notoungulates and litopterns have the unfortunate distinction of being the only well-documented mammalian ordinal groups to have gone extinct at least partly due to human activity (the monogeneric Bibymalagasia of Madagascar may also fall into this category).

That said, immigration and speciation can compensate for losses due to extinction; although modern South American mammal faunas are in some ways depauperate versions of relatively recent paleofaunas, in other ways they are richer owing to these factors. Classic examples of recent (post-GABI) mammalian radiations in South America include cricetid rodents (especially sigmodontines) and carnivorans, both of which include many species filling varied ecological niches throughout the continent. The patterns of diversification of carnivorans are the subject of Eizirik's chapter, the final one in part 1. In contrast to the preceding chapters, Eizirik focuses on molecular rather than morphological data and begins rather than ends with the Recent. Combined with paleontological data, he is able to infer patterns and times of divergence for lineages unrecorded in the fossil record, thus providing a scale of resolution greater than that typically permitted by paleontological data alone. Molecular analyses such as these provide a vital bridge between the coarser, deep-time perspective of paleontology and the finer-scale, more temporally limited neontological perspective. The addition of data from ancient DNA to such analyses will provide an even greater level of detail about the GABI, Quaternary extinctions, and speciation events than would be possible relying exclusively on DNA from extant species (e.g., Orlando et al. 2009). New molecular data and techniques such as these, combined with new fossils and types of paleontological analyses, will undoubtedly make the ensuing decades an exciting time to study the development of Neotropical mammal faunas from both perspectives.

Literature Cited

Alvarenga, H. M. F., and E. Höfling. 2003. "Systematic Revision of the Phorusrhacidae (Aves: Ralliformes)." *Papéis Avulsos de Zoologia* 43:55–91.

Antoine, P.-O., D. De Franceschi, J. J. Flynn, A. Nel, P. Baby, M. Benammi, Y. Calderón et al. 2006. "Amber from Western Amazonia Reveals Neotropical Diversity During the Middle Miocene." *Proceedings of the National Academy of Sciences USA* 103:13595–600.

Antoine, P.-O., R. Salas-Gismondi, P. Baby, M. Benammi, S. Brusset, D. De Franceschi, N. Espurt et al. 2007. "The Middle Miocene (Laventan) Fitzcarrald Fauna, Amazonian Peru." In *4th European Meeting on the Paleontology and Stratigraphy of Latin America, Cuadernos del Museo Geominero*, No. 8, edited by E. Díaz-Martínez and I. Rábano, 19–24. Madrid: Instituto Geológico y Minero de España.

Bargo, M. S., and S. F. Vizcaíno. 2008. "Paleobiology of Pleistocene Ground Sloths (Xenarthra, Tardigrada): Biomechanics, Morphogeometry and Ecomorphology Applied to the Masticatory Apparatus." *Ameghiniana* 45:175–96.

Barreda, V., and L. Palazzesi. 2007. "Patagonian Vegetation Turnovers During the Paleogene-Early Neogene: Origin of Arid-Adapted Floras." *Botanical Review* 73:31–50.

Campbell, K. E., Jr., ed. 2004. *The Paleogene Mammalian Fauna of Santa Rosa, Amazonian Peru*, *Science Series*, 40. Los Angeles: Natural History Museum of Los Angeles County.

Chiappe, L. M., and S. Bertelli. 2006. "Skull Morphology of Giant Terror Birds." *Nature* 443:929.

Cozzuol, M. A. 2006. "The Acre Vertebrate Fauna: Age, Diversity, and Geography." *Journal of South American Earth Sciences* 21:185–203.

Croft, D. A. 2001. "Cenozoic Environmental Change in South American as Indicated by Mammalian Body Size Distributions (Cenograms)." *Diversity and Distributions* 7:271–87.

———. 2006. "Do Marsupials Make Good Predators? Insights from Predator-Prey Diversity Ratios." *Evolutionary Ecology Research* 8:1193–214.

Croft, D. A., J. J. Flynn, and A. R. Wyss. 2008. "The Tinguiririca Fauna of Chile and the Early Stages of 'Modernization' of South American Mammal Faunas." *Arquivos do Museu Nacional, Rio de Janeiro* 66:191–211.

Croft, D. A., and K. E. Townsend. 2005. "Inferring Habitat for the Late Early Miocene Santa Cruz Fauna (Santa Cruz Province, Argentina) Using Ecological Diversity Analysis." *Journal of Vertebrate Paleontology* 25:48A.

Croft, D. A., and D. Weinstein. 2008. "The First Application of the Mesowear Method to Endemic South American Ungulates (Notoungulata)." *Palaeogeography, Palaeoclimatology, Palaeoecology* 269:103–14.

Elissamburu, A. 2004. "Análisis Morfométrico y Morfofunctional del Esqueleto Appendicular de *Paedotherium* (Mammalia, Notoungulata)." *Ameghiniana* 41:363–80.

Elissamburu, A., and S. F. Vizcaíno. 2004. "Limb Proportions and Adaptations in Caviomorph Rodents (Rodentia: Caviomorpha)." *Journal of Zoology* 262:145–59.

———. 2005. "Diferenciación Morfométrica del Húmero y Femur de las Especies de *Paedotherium* (Mammalia, Notoungulata) del Plioceno y Pleistoceno Temporano." *Ameghiniana* 42:159–66.

Fariña, R. A., S. F. Vizcaíno, and M. S. Bargo. 1998. "Body Mass Estimations in Lujanian (Late Pleistocene–Early Holocene of South America) Mammal Megafauna." *Mastozoología Neotropical* 5:87–108.

Flynn, J. J., R. Charrier, D. A. Croft, P. B. Gans, T. M. Herriott, J. A. Wertheim, and A. R. Wyss. 2008. "Chronologic Implications of New Miocene Mammals from the Cura-Mallín and Trapa Trapa Formations, Laguna del Laja Area, South Central Chile." *Journal of South American Earth Sciences* 26:412–23.

Flynn, J. J., A. R. Wyss, and R. Charrier. 2007. "South America's Missing Mammals." *Scientific American* 296 (5):68–74.

Flynn, J. J., A. R. Wyss, D. A. Croft, and R. Charrier. 2003. "The Tinguiririca Fauna, Chile: Biochronology, Paleoecology, Biogeography, and a New Earliest Oligocene South American Land Mammal 'Age'." *Palaeogeography, Palaeoclimatology, Palaeoecology* 195:229–59.

Forasiepi, A. M., A. G. Martinelli, and F. J. Goin. 2007."Revisión Taxonómica de *Parahyaenodon argentinus* Ameghino y Sus Implicancias en el Conocimiento de los Grandes Mamíferos Carnívoros del Mio-Plioceno de América de Sur." *Ameghiniana* 44:143–59.

Hoffstetter, R., and G. Petter. 1983. "*Paraborhyaena boliviana* et *Andinogale sallensis*, Deux Marsupiaux (Borhyaenidae) Nouveaux de Déséadien (Oligocène Inférieur) de Salla (Bolivie)." Comptes Rendus des Seances de l'Académie des Sciences, Serie 3. *Sciences de la Vie* 296:143–46.

Janis, C. M. 1993. "Tertiary Mammal Evolution in the Context of Changing Climates, Vegetation, and Tectonic Events." *Annual Reviews in Ecology and Systematics* 24:467–500.

Johnson, C. N. 2002. "Determinants of Loss of Mammal Species During the Late Quaternary 'Megafauna' Extinctions: Life History and Ecology, but not Body Size." *Proceedings of the Royal Society of London, B, Biological Sciences* 269:2221–27.

Kay, R. F., and R. H. Madden. 1997. "Paleogeography and Paleoecology." In *Vertebrate Paleontology in the Neotropics: The Miocene Fauna of La Venta, Colombia*, edited by R. F. Kay, R. H. Madden, R. L. Cifelli, and J. J. Flynn, 520–50. Washington, DC: Smithsonian Institution Press.

Kay, R. F., R. H. Madden, R. L. Cifelli, and J. J. Flynn, eds. 1997. *Vertebrate Paleontology in the Neotropics: The Miocene Fauna of La Venta, Colombia*. Washington, DC: Smithsonian Institution Press.

Keast, A., F. C. Erk, and B. Glass, eds. 1972. *Evolution, Mammals, and Southern Continents*. Albany: State University of New York Press.

MacFadden, B. J. 2005. "Diet and Habitat of Toxodont Megaherbivores (Mammalia, Notoungulata) from the Late Quaternary of South and Central America." *Quaternary Research* 64:113–24.

———. 2006. "Extinct Mammalian Biodiversity of the Ancient New World Tropics." *Trends In Ecology & Evolution* 21:157–65.

MacFadden, B. J., and B. J. Shockey. 1997. "Ancient Feeding Ecology and Niche Differentiation of Pleistocene Mammalian Herbivores from Tarija, Bolivia: Morphological and Isotopic Evidence." *Paleobiology* 23:77–100.

Madden, R. H., R. Burnham, A. A. Carlini, C. C. Swisher, III, and A. H. Walton. 1994. "Mammalian Paleontology, Paleobotany and Geochronology of the Miocene Intermontain Basins of Southern Ecuador." *Journal of Vertebrate Paleontology* 14:35A.

Marshall, L. G. 1977. "Evolution of the Carnivorous Adaptive Zone in South America." In *Major Patterns in Vertebrate Evolution*, edited by M. K. Hecht, P. C. Goody, and B. M. Hecht, 709–21). New York: Plenum Press.

Marshall, L. G., and B. Patterson. 1981. "Geology and Geochronology of the Mammal-Bearing Tertiary of the Valle de Santa María and Río Corral Quemado, Catamarca Province, Argentina." *Fieldiana: Geology*, n.s., 9:1–80.

Marshall, L. G., and T. Sempere. 1991. "The Eocene to Pleistocene Vertebrates of Bolivia and Their Stratigraphic Context: A Review." In *Fósiles y Facies de Bolivia, Vol. 1. Vertebrados*, edited by R. Suárez-Soruco, 631–52. *Revista Técnica de Yacimientos Petrolíferos Fiscales de Bolivia*, 12(3–4), Santa Cruz, Bolivia.

Orlando, L., J. L. Metcalf, M. T. Alberdi, M. Telles-Antunes, D. Bonjean, M. Otte, F. Martin et al. 2009. "Revising the Recent Evolutionary History of Equids Using Ancient DNA." *Proceedings of the National Academy of Sciences USA* 106:21754–59.

Riggs, E. S. 1934. "A New Marsupial Saber-Tooth from the Pliocene of Argentina and its Relationships to Other South American Predacious Marsupials." *Transactions of the American Philosophical Society* 24:1–32.

Sánchez-Villagra, M. R., and O. A. Aguilera. 2006. "Neogene Vertebrates from Urumaco, Falcón State, Venezuela: Diversity and Significance." *Journal of Systematic Palaeontology* 4:213–20.

———. 2008. "Contributions on Vertebrate Paleontology in Venezuela." Preface. *Paläontologische Zeitschrift* 82:103–04.

Sánchez-Villagra, M. R., O. Aguilera, and I. Horovitz. 2003. "The Anatomy of the World's Largest Extinct Rodent." *Science* 301:1708–10.

Shockey, B. J. 1999. "Postcranial Osteology and Functional Morphology of the Litopterna of Salla, Bolivia (late Oligocene)." *Journal of Vertebrate Paleontology* 19:383–90.

———. 2001. "Specialized Knee Joints in Some Extinct, Endemic, South American Herbivores." *Acta Palaeontologica Polonica* 46:277–88.

Shockey, B. J., and F. Anaya. 2008. "Postcranial Osteology of Mammals from Salla, Bolivia (late Oligocene): Form, Function, and Phylogenetic Implications." In *Mammalian Evolutionary Morphology: A Tribute to Frederick S. Szalay*, edited by E. J. Sargis and M. Dagosto, 135–57. New York: Springer.

Shockey, B. J., D. A. Croft, and F. Anaya. 2007. "Analysis of Function in the Absence of Extant Functional Analogs: A Case Study of Mesotheriid Notoungulates." *Paleobiology* 33:227–47.

Shockey, B. J., R. Salas, E. J. Sargis, R. Quispe, A. Flores, and J. Acosta. 2003. "Moquegua: The First Deseadan SALMA (Late Oligocene) Local Fauna of Peru." *Journal of Vertebrate Paleontology* 23:97A.

Simpson, G. G. 1950. "History of the Faunas of Latin America." *American Scientist* 38:361–89.

Townsend, K. E., and D. A. Croft. 2008. "Diets of Notoungulates from the Santa Cruz Formation, Argentina: New evidence From Enamel Microwear." *Journal of Vertebrate Paleontology* 28:217–30.

Vizcaíno, S. F., M. S. Bargo, and G. H. Cassini. 2006. "Dental Occlusal Surface Area in Relation to Body Mass, Food Habits and Other Biological Features in Fossil Xenarthrans." *Ameghiniana* 43:11–26.

Vizcaíno, S. F., M. S. Bargo, R. F. Kay, and N. Milne. 2006. "The Armadillos (Mammalia, Xenarthra, Dasypodidae) of the Santa Cruz Formation (Early-Middle Miocene): An Approach to Their Paleobiology." *Palaeogeography, Palaeoclimatology, Palaeoecology* 237:255–69.

Vizcaíno, S. F., R. A. Fariña, M. S. Bargo, and G. De Iuliis. 2004. "Functional and Phylogenetic Assessment of the Masticatory Adaptations in Cingulata (Mammalia, Xenarthra)." *Ameghiniana* 41:651–64.

Vizcaíno, S. F., and N. Milne. 2002. "Structure and Function in Armadillo Limbs (Mammalia: Xenarthra: Dasypodidae)." *Journal of Zoology* 257:117–27.

Vucetich, M. G., and A. M. Ribeiro. 2003. "A New and Primitive Rodent from the Tremembé Formation (Late Oligocene) of Brazil, with Comments on the Morphology of the Lower Premolars of Caviomorph Rodents." *Revista Brasileira de Paleontologia* 5:73–82.

Werdelin, L. 1987. "Jaw Geometry and Molar Morphology in Marsupial Carnivores: Analysis of a Constraint and its Macroevolutionary Consequences." *Paleobiology* 13:342–50.

———. 1988. "Circumventing a Constraint—The Case of *Thylacoleo* (Marsupialia, Thylacoleonidae)." *Australian Journal of Zoology* 36:565–71.

Woodburne, M. O. 2010. "The Great American Biotic Interchange: Dispersals, Tectonics, Climate, Sea Level and Holding Pens." *Journal of Mammalian Evolution* 17:245–64.

Wozencraft, W. C. 2005. "Order Carnivora." In *Mammal Species of the World*, 3rd ed., edited by D. E. Wilson and D. M. Reeder, 532–628. Baltimore: Johns Hopkins University Press.

Wyss, A. R., J. J. Flynn, M. A. Norell, C. C. Swisher, III, R. Charrier, M. J. Novacek, and M. C. McKenna. 1993. "South America's Earliest Rodent and Recognition of a New Interval of Mammalian Evolution." *Nature* 365:434–7.

Wyss, A. R., J. J. Flynn, M. A. Norell, C. C. Swisher, III, M. J. Novacek, M. C. McKenna, and R. Charrier. 1994. "Paleogene Mammals from the Andes of Central Chile: A Preliminary Taxonomic, Biostratigraphic, and Geochronologic Assessment." *American Museum Novitates* 3098:1–31.

Zanazzi, A., M. J. Kohn, B. J. MacFadden, and D. O. Terry. 2007. "Large Temperature Drop Across the Eocene-Oligocene Transition in Central North America." *Nature* 445:639–42.

Origins, Radiations, and Distribution of South American Mammals
From Greenhouse to Icehouse Worlds

Francisco J. Goin, Javier N. Gelfo, Laura Chornogubsky, Michael O. Woodburne, and Thomas Martin

Abstract

At least five successive phases in South American mammalian evolution can be envisaged, the oldest one being largely hypothetical: (1) Early Gondwanan (?Late Triassic-Early Cretaceous), mammals of Pangaean (triconodontids with amphilestid affinities) and Gondwanan (autralosphenids) origin; (2) Late Gondwanan (Late Cretaceous), strong endemism in most lineages of Pangaean (Dryolestida) and Gondwanan (Gondwanatheria) origin; (3) Early South American (?latest Cretaceous-latest Eocene), major radiations within Metatheria and Eutheria; (4) Late South American (early Oligocene-middle Pliocene), standardization of relatively few lineages among metatherians and marked radiation of hypsodont types among South American native ungulates; and (5) Inter-American (late Pliocene-Recent), mixture of North and South American therian lineages, with progressive decline of native faunas. The biotic and abiotic events that triggered these phases include the last global warming event at the beginning of the Late Cretaceous (between phases 1 and 2); an intermittent connection between North and South America by the Late Cretaceous, enabling the First American Biotic Interchange, and the decline of native nontherians (between 2 and 3); global cooling and full development of the Circumpolar Antarctic Current, and arrival of platyrrhines and caviomorphs (between 3 and 4); and finally, the Panamanian connection between the Americas and the beginning of the Great American Biotic Interchange (between 4 and 5).

3.1 Introduction

Several reviews in recent years have highlighted the historical keystones in our knowledge of South American mammalian successions (e.g., Pascual 2006 and literature cited therein). Others include the works of Ameghino (1906), Ortiz-Jaureguizar (1986), Pascual and Ortiz-Jaureguizar (1990a, 1990b), Pascual et al. (1985), Pascual, Ortiz-Jaureguizar, and Prado (1996), Patterson and Pascual (1972), Reig (1981), and Simpson (1950). Most of these studies have focused on the better known Cenozoic evolution of South American mammals, even though in the past few years their Mesozoic history has also been considered (Pascual 2006; Pascual and Ortiz-Jaureguizar 2007). These works, especially those following Simpson (1950, 1980), have stressed the episodic nature of the mammalian successions. This is partly a consequence of several important hiatuses in our biochronological knowledge of these successions, which exaggerate the differences between each "stratum," "stage," "cycle," "horofauna," or "episode," as they have been variously named. Nevertheless, when the biochronological sequence is relatively well known, as with the Neogene mammalian associations of southern South America, it seems obvious that there are such things as cycles or phases among them, each being characterized by faunal turnovers and a distinctive taxonomic composition. In addition to the intrinsic evolutionary patterns of each mammalian lineage, a few major driving forces of these phases can be recognized, with paleoclimate and tectonics being the most relevant.

Two aspects preclude a better understanding of the early history of South American land mammals (for alternative views of the concepts of Mammaliamorpha and Mammalia, see Kielan-Jaworowska, Cifelli, and Luo 2004; Luo 2007; Rowe et al. 2008 [whom we follow]; regarding cynodont evolution in South America, see Bonaparte, Martinelli, and Schultz 2005; Martinelli and Rougier 2007). First, the Mesozoic fossil record is extremely fragmentary (see Rougier, Chornogubsky, et al. 2009). Most known Mesozoic taxa come from Late Cretaceous (Campanian-Maastrichtian) sites, in that the pre-Campanian Cretaceous and Jurassic mammal record still is largely unknown. Second, almost all of the Mesozoic and most of the early Cenozoic (Paleocene-Eocene) fossil record in South America comes from its southernmost tip, Patagonia (e.g., Marshall, Hoffstetter, and Pascual 1983; Rougier, Chornogubsky, et al. 2009). Given this limitation, and for reasons discussed more fully later, we need to be cautious in making inferences about the whole South American continent on the basis of Patagonian mammalian sequences. South America and its biotas are and were neither isolated nor homogeneous.

Two recent studies (Pascual 2006; Pascual and Ortiz-Jaureguizar 2007) have considered the evolution of South American land mammals throughout Mesozoic and Cenozoic times. Both recognize two major episodes, a Gondwanan Episode and a South American Episode. "The Gondwanan Episode is so termed because it is exclusively represented by endemic mammals of Gondwanan origin, i.e., Mesozoic lineages. In contrast, the South American Episode is almost exclusively distinguished by endemic therian mammals whose ancestors emigrated from the Laurasian North American Continent" (Pascual 2006, 209). As striking as the difference between therian and nontherian assemblages is, we feel that this primary distinction in the mammalian succession obscures other equally important patterns and processes that affected, and are reflected by, South American land mammals. Dispersal events were not the only factors that triggered the significant turnovers that occurred during the continent's history; a two-episode portrayal of this period simplifies what actually occurred. Moreover, it is not clear that the earliest mammalian phase in South America was strictly of Gondwanan origin. Finally, consideration of South America as a single biogeographical unit hides a more complex, richer evolutionary history whose early phases in tropical and intertropical parts of the continent remain largely unknown.

Here we propose a new model for the successive phases reflected by southern South American mammals throughout history. Our proposal is preliminary and subjective, as we have not attempted a formal analysis of faunal turnovers (e.g., Alroy 2004)—our current knowledge of Mesozoic South American mammals is too incomplete for such an attempt. Instead, we have tried to emphasize the interactions between global climate, tectonics, and South American mammal assemblages, in some cases in a highly speculative manner. In short, we offer a narrative framework for considering the mammalian successions in South America, hoping that future tests of our model will confirm, clarify, and extend its applicability, causes, and consequences. We view it as an initial step in a general, multidisciplinary attempt to understand the evolution of South America's biota as a whole over the last 250 million years of its history.

3.2 Biogeographical Context

A traditional approach to the biogeography of South America is that it can be referred to as a single, Neotropical unit (e.g., Cox 2001). Following studies by Crisci et al. (1991), and on the basis of panbiogeographic and cladistic biogeographic studies, Morrone (2001a, 2001b, 2002) argued in favor of the composite nature of this continent, ". . . with southern South America closely related to the southern temperate areas (Australia, Tasmania, New Zealand, New Guinea, and New Caledonia), and tropical South America closely related to Africa and North America" (Morrone 2002, 149). Thus, he formally proposed the Andean Region (southern South America and a strip of the Andean Cordillera) as part of an Austral Kingdom. The concept of an Austral biogeographic kingdom (or Austral realm, fig. 3.1) is not new, and can be traced back

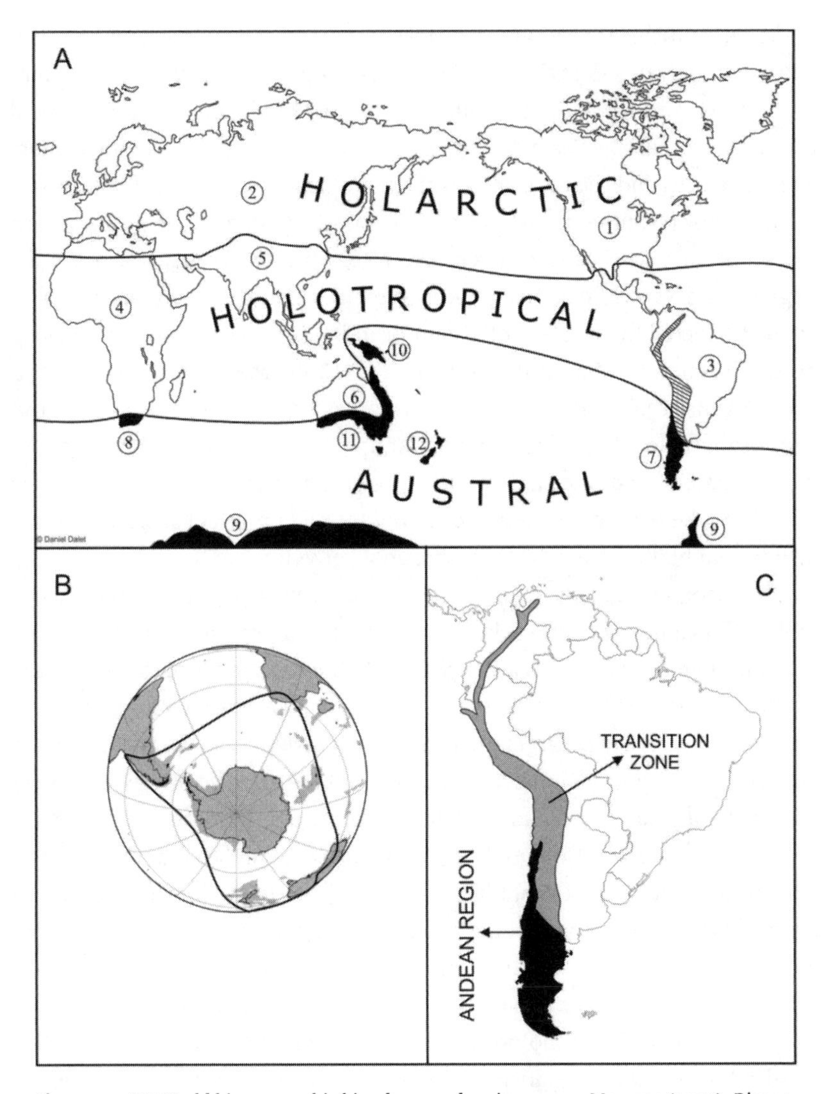

Figure 3.1 (A) World biogeographic kingdoms and regions sensu Morrone (2002). Biogeographic regions are, for the Holarctic Kingdom (=Laurasia): (1) Nearctic, (2) Palearctic; for the Holotropical Kingdom: (3) Neotropical, (4) Afrotropical, (5) Oriental, (6) Australotropical; for the Austral Kingdom: (7) Andean, (8) Cape or Afrotemperate, (9) Antarctic, (10) Neoguinean, (11) Australotemperate, and (12) Neozelandic. (B) The Austral biogeographic kingdom in polar projection (based in the ODSN Plate Tectonic Reconstruction Service, http://www.odsn.de/odsn/services/paleomap/ paleomap.html). (C) map of South America showing the Andean Region and the Transition Zone (from Morrone 2006). Sources for maps in (A) and (C) are: http://dmaps.com/carte.php?lib=mundo_centrado_oceano_pacifico_mapa&num_car=3227&lang=es and http://d-maps.com/carte.php?lib=america_del_sur_mapa&num_car=2313&lang=es, respectively.

to the early nineteenth century (see Moreira-Muñoz 2007). Morrone's (2001a, 2001b) model considers three global biotic realms: The Holarctic Kingdom, the Holotropical Kingdom (essentially Western Gondwana), and the Austral Kingdom (essentially Eastern Gondwana). The latter included southernmost South America and the Andean Range (the Andean Region), Antarctica (Antarctic Region), southernmost Africa (Cape or Afrotemperate Region), eastern and southern Australia (Australotemperate Region), New Zealand (Neozelandic Region), and New Guinea (Neoguinean Region). Later, Morrone (2004a, 2004b, 2006) restricted the concept of the Andean Region to the southernmost tip of South America (i.e., central and southern Chile, the southern Andes, and Patagonia in Argentina). In turn (Morrone 2006, fig.1), he regarded the central and northern sections of the Andean Range as a transitional zone between the Neotropical and Andean regions. "Transition zones . . . are located at the boundaries between biogeographic regions . . . and represent areas of mixture of different biotic elements" (Morrone 2006, 469; fig. 3.1C).

The recognition of the dual nature of South America's historical biogeography offers a new insight to the understanding of mammalian evolution in this continent, especially its earlier phases. The general tendency to extrapolate results to a continental scale, based on empirical evidence coming from Patagonia, may prove to be largely inadequate. For instance, analyzing the middle Miocene (Laventan) mammalian fauna of Quebrada Honda, in southernmost Bolivia, Croft (2007, 277) stated that it is "more similar to the slightly older high-latitude fauna of Collón-Curá [Patagonia] than to the contemporaneous low-latitude fauna of La Venta [Colombia], suggesting that isolating mechanisms between the low and middle latitudes were in place during the early and/or middle Miocene." Owing to few comparable taxa and lacking isotopic dates for the faunas at La Venta and Quebrada Honda, it would be difficult to assign both to the same biochronological unit! Interestingly, Quebrada Honda is located at the heart of the Transition Zone between the Neotropical and Andean regions, something that could explain its mix of earlier southern taxa with more modern, middle/low latitude forms. Another example of mammal provinciality is represented by the Cerdas fauna from southwestern Bolivia (Croft et al. 2009). This fauna ranges in age from 16.3–15.1 Ma, but seems to be faunally more like that of Quebrada Honda (13.0–12.7 Ma) than to comparably old faunas of nearby Chile (Chucal; 18.8–17.5 Ma). This indicates that faunas typically found during the Laventan South American Land Mammal Age (SALMA) have an earlier occurrence in Bolivia than would be expected for faunas of Cerdas age based on traditional faunal compositions derived from sites farther south.

In that context, Croft et al. (2009, 193) noted that "significant provinciality was present in South America at least by the early Miocene." We interpret this to indicate that elements of the modern Holotropical Realm in South America have considerable antiquity.

We elaborate on the historical development of the Austral Kingdom elsewhere. Briefly, we suggest that the initial stages in the development of an Austral realm could date from the Late Triassic. Artabe, Morel, and Spalletti (2003) summarized the relatively well-known paleobotanical evidence in southwestern Gondwana, claiming that the extensive floral extinctions that occurred at the end of the Permian, together with the expansion of northern lineages into southern Gondwana, characterized the main components of the Gondwanan Triassic turnover. The Gondwanan Triassic (floral) Kingdom (defined by the presence of *Dicroidium* floras) can be divided in two main dominions, or areas: the Tropical Area and the "Tethys Wet Corridor" on one side (northern Gondwana), and the Extratropical Area on the other, south of 30° S paleolatitude (the latter is defined by the presence of the Ipswich microfloras; see Zavattieri 2002). In turn, the Extratropical Area includes two major regions, or floral provinces: Southwestern Gondwana, mostly including central and southern South America and Africa, and Southeastern Gondwana, for the remaining landmasses of South Gondwana. Most of Southwestern Gondwana is included within the dry subtropical climatic zone (from 30° to 60° S paleolatitude), while Patagonia in South America, the Karoo Basin in southernmost Africa, and most of Southeastern Gondwana belong to the warm-temperate climatic belt (south of 60° S; Artabe, Morel, and Spalletti 2003; fig. 3.2). Because of their similar latitudinal position and a similar climatic regime, several characteristic lineages were shared between Patagonia and Southeastern Gondwana (e.g., Marattiales; Artabe, Morel, and Spalletti 2003). It is probable that, under these paleolatitudinal and climatic constraints, the Austral realm began its differentiation in the Late Triassic.

3.3 Phases in the Evolution of Austral South American Mammals

We distinguish five successive phases in the evolution of South American southern mammals: (1) Early Gondwanan (?Late Triassic-Early Cretaceous), (2) Late Gondwanan (Late Cretaceous), (3) Early South American (?latest Cretaceous-latest Eocene), (4) Late South American (early Oligocene-Middle Pliocene), and (5) Inter-American (late Pliocene-Recent). Because of their scarcity in the fossil record, we provide a short overview of our knowledge regarding Mesozoic

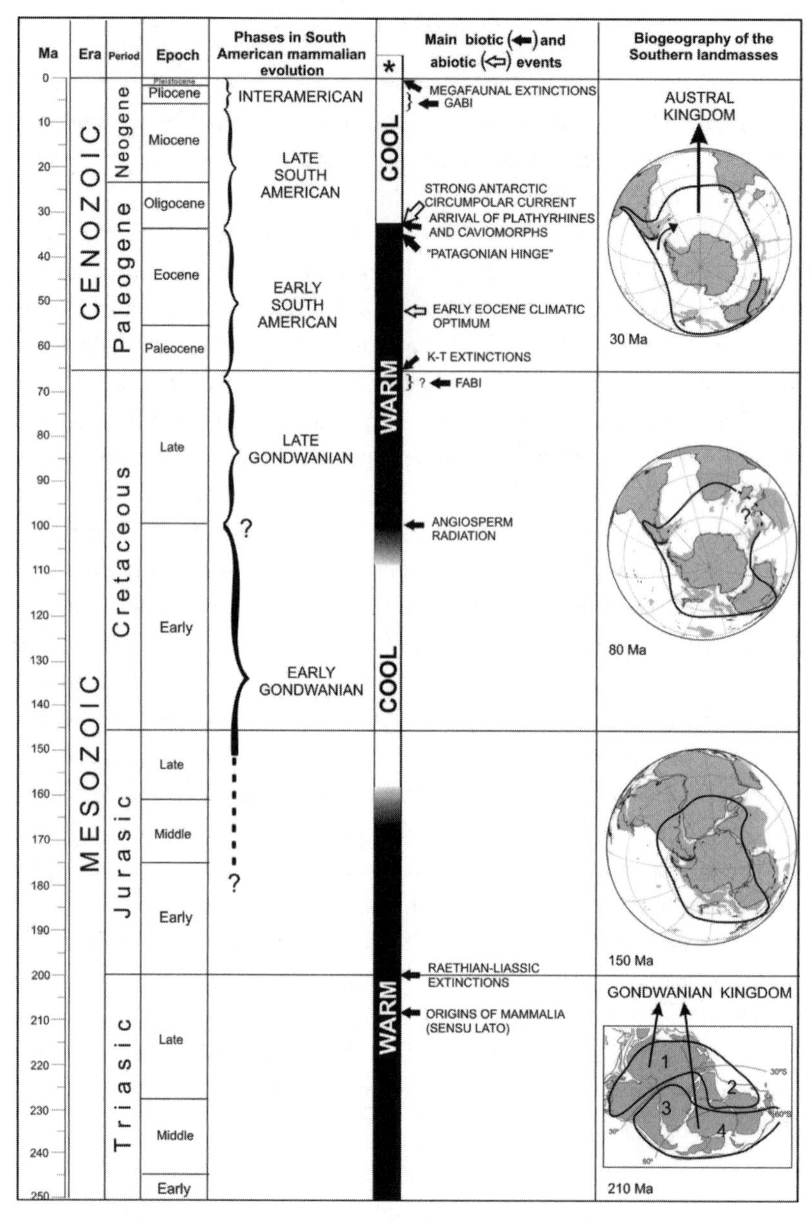

Figure 3.2 Phases in the evolution of South American mammals. Maps at the right show plate reconstructions at 30 Ma, 80 Ma, 150 Ma (based in the ODSN Plate Tectonic Reconstruction Service (http://www.odsn.de/odsn/services/paleomap/paleomap.html), and 210 Ma (*). Temperatures (long waves) modified after Scotese, Boucot, and McKerrow (1999). The arrow in the uppermost map indicates the full opening of the Drake Passage. Floral areas and provinces of the Gondwanan Floristic Kingdom in the lowermost map: (1) tropical area, (2) wet warm corridor of Tethys, (3) southwestern Gondwana, and (4) southeastern Gondwana (after Artabe, Morel, and Spalletti 2003, fig. 3).

mammals (phases 1 and 2). Regarding phases 3 through 5, we briefly mention only the main processes undergone by marsupials and native ungulates, which are better known than xenarthrans in the early Tertiary. Much more detailed reviews of these and other lineages of South American therians are cited in the introduction. Regarding the Inter-American phase, we detail relatively recent information on the timing of arrival of several mammalian lineages (see also fig. 3.3).

3.3.1 EARLY GONDWANAN PHASE

From the Late Triassic up to the late Early Jurassic, Pangaea persisted as a single supercontinent. By the Middle Jurassic, the early phases of Pangaean breakup had already begun, with several plates beginning to drift from Gondwana toward Eurasia. During the Early Cretaceous, Gondwana separated from Laurasia. Climate during the Late Triassic was warm (10° C warmer than present), polar ice caps were absent, and there was a gradual gradient from the Equator toward the poles (Scotese 2001). The Middle Jurassic climate of the Austral realm was warm-temperate. From the Late Jurassic up until the end of the Early Cretaceous, there was a steep decrease in global temperatures. However, these new icehouse conditions were not extreme, as extended polar ice caps and glacial systems did not develop. Most of the Austral realm during the Early Cretaceous was in the cool-temperate climate zone (Scotese, Boucot, and McKerrow 1999).

The earliest records of fossils probably belonging to a Mesozoic South American mammal are also the oldest: the ichnogenus *Ameghinichnus patagonicus* (Casamiquela 1961) from Middle Jurassic levels in Santa Cruz Province, southern Patagonia. Its mammalian status has been recently supported by De Valais (2009), who cited alternative views of *Ameghinichnus* as an advanced therapsid.

Middle Late Jurassic mammals from the Queso Rallado locality near Cerro Cóndor (Cañadón Asfalto Formation, Chubut Province, central Patagonia) are represented by the eutriconodont *Argentoconodon fariasorum* (Rougier, Garrido, et al. 2007) and the australosphenidans *Asfaltomylos patagonicus* and *Henosferus molus* (Rauhut et al. 2002; Rougier, Forasiepi, et al. 2007). The concept, contents, and phylogenetic position of the Gondwanan Australosphenida are still a matter of debate (contrast Luo, Kielan-Jaworowska, and Cifelli 2002; Martin and Rauhut 2005; Rauhut et al. 2002; and Rougier, Garrido, et al. 2007 with Rowe et al. 2008 and Woodburne, Rich, and Springer 2003).

Early Cretaceous (Hauterivian-Barremian) South American mammals are

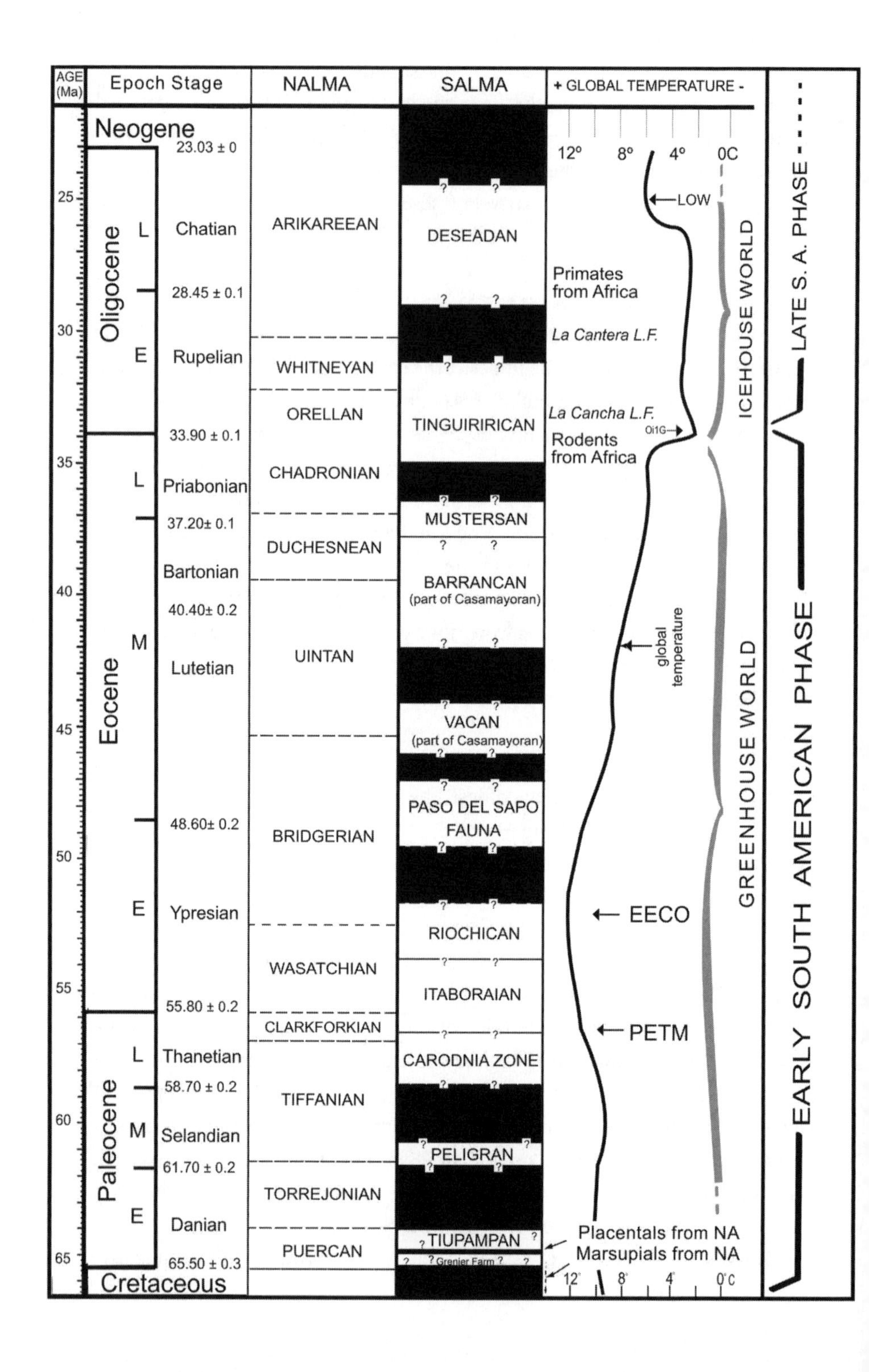

AGE (Ma)
Epoch Stage
NALMA
SALMA
+ GLOBAL TEMPERATURE -
Neogene
23.03 ± 0
Oligocene
L
Chatian
28.45 ± 0.1
E
Rupelian
33.90 ± 0.1
Eocene
L
Priabonian
37.20± 0.1
Bartonian
40.40± 0.2
M
Lutetian
48.60± 0.2
E
Ypresian
55.80 ± 0.2
Paleocene
L
Thanetian
58.70 ± 0.2
M
Selandian
61.70 ± 0.2
E
Danian
65.50 ± 0.3
Cretaceous
ARIKAREEAN
WHITNEYAN
ORELLAN
CHADRONIAN
DUCHESNEAN
UINTAN
BRIDGERIAN
WASATCHIAN
CLARKFORKIAN
TIFFANIAN
TORREJONIAN
PUERCAN
DESEADAN
TINGUIRIRICAN
MUSTERSAN
BARRANCAN
(part of Casamayoran)
VACAN
(part of Casamayoran)
PASO DEL SAPO FAUNA
RIOCHICAN
ITABORAIAN
CARODNIA ZONE
PELIGRAN
TIUPAMPAN
Grenier Farm
12°
8°
4°
0C
LOW
Primates from Africa
La Cantera L.F.
La Cancha L.F.
Oi1G
Rodents from Africa
global temperature
EECO
PETM
Placentals from NA
Marsupials from NA
ICEHOUSE WORLD
GREENHOUSE WORLD
LATE S. A. PHASE
EARLY SOUTH AMERICAN PHASE

solely represented by *Vincelestes neuquenianus* (La Amarga Formation, Neuquén Province, northern Patagonia; Bonaparte 1986a). *Vincelestes* was first related to the Zatheria (Rougier 1992), although more recently Bonaparte (2008) made a case for australosphenidan relationships.

In general terms, despite the still spotty record, Jurassic and Early Cretaceous mammals from Patagonia seem to represent a combination of cosmopolitan (eutriconodontids) and Austral (australosphenidan) taxa.

3.3.2 LATE GONDWANAN PHASE

Before the end of the Early Cretaceous, South America and Africa were already separated (see a discussion in Krause et al. 2006). However, paleogeographic bridges between South America and Antarctica persisted until the Paleogene (Woodburne and Case 1996), and Antarctica and Australasia remained connected throughout the Late Cretaceous. The timing of the separation of Antarctica and Indo-Madagascar is controversial. Case (2002) and Krause et al. (2006) suggested that links between these landmasses were maintained until 80 Ma (i.e., Campanian). As previously mentioned, global greenhouse conditions persisted during the whole period (up to the late Eocene). Burgoyne et al. (2005) suggested a slight cooling phase at the end of the Cretaceous and a transient warming at the Cretaceous/Tertiary boundary.

Several mammals from the early Late Cretaceous (Cenomanian-Coniacian) locality of La Buitrera (Candeleros Formation, Río Negro Province) remain undescribed (Rougier and Apesteguía 2004). One, represented by several partial skulls, belongs to a long-snouted dryolestoid superficially resembling living elephant shrews. "It is probably part of a Gondwanan dryolestoid lineage that [subsequently gave rise to] the latest Cretaceous and Paleocene South American forms" (Apesteguía, Gaetano, and Rougier 2009, 26).

Figure 3.3 (*facing page*) Paleogene time scale, global temperatures, North American Land Mammal Age (NALMA), and South American Land Mammal Age (SALMA). Marsupials immigrated to South America in the latest Cretaceous, with the earliest records preserved in the early Paleocene Grenier Farm site. Placental mammals immigrated in the Tiupampian. Rodents, presumably from Africa, are recorded in the Tinguiririan. The La Cancha and La Cantera Local Faunas are known from the Gran Barranca of Argentina (Goin, Abello, and Chornogubsky 2010). Primates are recorded from the Deseadan, but may have arrived in South America concurrently with rodents. The Paleogene time scale is after Luterbacher et al. (2004). The NALMA chronology is after Woodburne (2004) and Woodburne, Gunnell, and Stucky (2009a, 2009b), while SALMA chronology follows Gelfo et al. (2009), with Tinguirican after Goin, Abello, and Chornogubsky (2010) and Deseadan after Ré, Geuna, and Vilas (2010). Global temperature, PETM (Paleocene-Eocene Thermal Maximum), EECO (Early Eocene Climatic Optimum), and Oi1G (initiation of Antarctic glaciations) are after Zachos et al. (2001).

Bonaparte and Soria (1985) and Bonaparte (1986a, 1986b) executed a series of studies on one of the richest and diverse Mesozoic localities for mammals, the Late Cretaceous (Campanian-Maastrichtian, Alamitan SALMA) Los Alamitos site in Río Negro Province, northern Patagonia. These specimens provided completely new insights into the evolution of mammals in the Southern Hemisphere, as most previous ideas had been made exclusively from the Northern Hemisphere record. In the north, tribosphenic mammals dominated mammalian assemblages of the Late Cretaceous; the multituberculates and triconodonts being exceptions. Strikingly, in the Los Alamitos assemblage, only nontribosphenic mammals are represented, including: a variety of symmetrodonts, triconodonts, dryolestoids, and gondwanatheres (Bonaparte 1986a, 1986b, 1987, 1990, 1992; see alternative views of gondwanatherian affinites in Gurovich and Beck 2008; Kielan-Jaworowska, Cifelli, and Luo 2004; Pascual et al. 1999). Representatives of this locality include the gondwanatherians *Ferugliotherium* and *Gondwanatherium* and a number of dryolestoids including *Mesungulatum*, *Groebertherium*, *Brandonia*, *Leonardus*, and *Reigitherium* (the latter formerly regarded as a docodont; Pascual et al. 2000). At least one group of dryolestoids, the Mesungulatidae, had a dentition in some ways functionally convergent on that of tribosphenic mammals (Bonaparte 1996). Two additional faunas of similar (Late Cretaceous) age have been reported subsequently: La Colonia in Chubut Province, central Patagonia (Kielan-Jaworowska et al. 2007; Pascual et al. 2000; Rougier et al. 2000; Rougier, Forasiepi, et al. 2009), and Cerro Tortuga (Allen Formation, Río Negro Province, northern Patagonia; Rougier, Chornogubsky, et al. 2009). La Colonia produced the first records for multituberculates from South America (Kielan-Jaworowska, Cifelli, and Luo 2004; Kielan-Jaworowska et al. 2007). From the Late Cretaceous site of Paso Córdoba (Río Colorado Formation, Río Negro Province), an edentulous mandible probably belonging to a metatherian was reported by Goin, Carlini, and Pascual (1986). However, Martinelli and Forasiepi (2004) suggested this specimen could be a dryolestoid. Rougier, Forasiepi, et al. (2009) described several petrosals tentatively assigned to the mesungulatid *Coloniatherium cilinskii*, and suggested its phylogenetic position is close to *Vincelestes*.

Mesozoic mammals are known from some additional South American localities far north of Patagonia. A fragmentary mandible with one premolar was recovered from the Late Cretaceous Adamantina Formation at Santo Anastasio, Brazil. Bertini et al. (1993) suggested its affinities lay with eutherians, while its fragmentary nature led Candeiro et al. (2006) to treat it as Mammalia *incertae sedis*. Several isolated molars referable to tribosphenic and nontribosphenic mammals have been reported from the Late Cretaceous of Pajcha Pata, El Molino Formation, in western Bolivia; some dryolestoid teeth apparently show affinities with those of contemporary strata in Patagonia (Gayet et al. 2001; Rougier, Forasiepi, et al. 2009). The Tiupampa fauna, also in Bolivia, originally referred to the Late Cretaceous, is currently regarded as early Paleocene in age (Muizon 1991; reviewed in Gelfo et al. 2009). Sigé (1968, 1971) reported the Laguna Umayo site and fauna in southern Peru as late Cretaceous, but more recently, Sigé et al. (2004) suggested the fauna dates from the late Paleocene or early Eocene.

Late Cretaceous mammals from Patagonia (and hypothetically all the Austral kingdom) differ from those of Laurasia in two important respects. First, dryolestoids are mostly known from the Late Jurassic and Early Cretaceous of Laurasian landmasses, but southern South America's dryolestids apparently reached their climax during the Late Cretaceous (Cenomanian-Maastrichtian). Second, "To date, no unequivocal cranial or dental remains of therian mammals are known from the Late Cretaceous of South America" (Rougier, Forasiepi, et al. 2009, 207); the same can be said for all other Mesozoic localities and levels of this continent. In contrast, therian mammals were dominant during the Late Cretaceous in all Laurasian continents.

Although the roots of some lineages can be traced back to the Jurassic of Laurasia (e.g., dryolestoids; Bonaparte 1994), Late Cretaceous mammals from Patagonia show high endemism and the development of some remarkable adaptive morphological types (e.g., mesungulatids, reigitheriids, gondwanatherians). Mesungulatids ". . . achieve a dentition reflecting omnivorous and herbivorous habits, with a progressive reduction of the orthal component during mastication culminating finally with the acquisition of an almost lophodont dentition in [the Paleocene dryolestoid] *Peligrotherium*" (Rougier, Forasiepi, et al. 2009, 208; see also Gelfo and Pascual 2001). Gondwanatherians were one of the earliest mammals to develop hypsodont cheek-teeth having thick cementum surfaces (Koenigswald, Goin, and Pascual 1999). We suggest that the radiation of these endemic mammalian lineages coincided with the global warming trend that had begun by the early Late Cretaceous, roughly contemporaneous with the radiation of angiosperms (Gurovich and Beck 2008; Koenigswald, Goin, and Pascual 1999; Rougier, Forasiepi, et al. 2009). We also hypothesize that the radiation of several, if not most, of these Patagonian lineages was an event biogeographically restricted to the Austral Kingdom. Testing this last hypothesis will require more prospecting in other Austral fossiliferous levels of contemporary age, as well as in northern (Neotropical) South America. The discovery of other Austral elements of the mammalian fauna (e.g., monotremes) is expected in Late Cretaceous strata of Patagonia. Rowe et al. (2008) estimated that the platypus and echidna clades were already distinct by the Early Cretaceous. Paleocene monotremes have already been recorded in central Patagonia (e.g., Pascual and Ortiz-Jaureguizar 2007; Pascual et al. 1992).

3.3.3 EARLY SOUTH AMERICAN PHASE

At the end of the Cretaceous, there was an intermittent connection between North and South America. Case, Goin, and Woodburne (2005) summarized previous data on the Caribbean tectonics and discussed several alternatives regarding land bridges between the Americas. They concluded that the best candidate for a pathway for biotic dispersal was the Aves Ridge and adjacent Cuba during the Campanian and especially the Maastrichtian. A major consequence of this development was the arrival of the first therian mammals in South America (but see Rich 2008), as well as other vertebrate groups (e.g., hadrosaurs, neoceratopsians). We call this dispersal event between the Americas as the First American Biotic Interchange (FABI). Rage (1978) reported at least one group of snakes (Boidae) as having dispersed northward at this time. The South America and Antarctica connection was interrupted by the early Eocene (48.6–55.8 Ma), as happened between Antarctica and Australia with the opening of the Tasmanian Gate (Woodburne and Case 1996). There were widespread tropical conditions in the global climate, especially during the Early Eocene Climatic Optimum (~50–53 Ma; Zachos et al. 2001). Patagonian climates from the early Paleocene through the late Eocene were warm-temperate (Bijl et al. 2009).

As mentioned, no unequivocal therian remains have been discovered in the South American Mesozoic. The earliest known Cenozoic therian is, most probably, a polydolopimorphian marsupial represented by an isolated lower molar. The specimen comes from early Paleocene levels of the Lefipán Formation in Chubut Province, Argentina, just 5 m above a level referred to as the Cretaceous-Tertiary boundary (Goin et al. 2006). It is noteworthy that the specimen does not belong to a generalized "opossum-like" marsupial but instead shows several derived features common among bonapartheriiform polydolopimorphians. Its discovery strongly suggests that the diversification of metatherians in South America can be traced back to the Late (probably latest) Cretaceous. Alternatively, because of the dual nature of South America's biogeography, the first therians may have arrived on this continent earlier in the Neotropical Region (e.g., Campanian; Case, Goin, and Woodburne 2005), only arriving later (e.g.,

Late Maastrichtian) in the Andean Region. Present evidence suggests that the faunal turnover that marked the beginning of the Early South American Phase happened sometime between the Late Cretaceous (Alamitian Age; Bonaparte 1986a) and the earliest Paleocene (Tiupampian age, 65–64 Ma; Ortiz-Jaureguizar 1996; Ortiz-Jaureguizar and Pascual 2007; Pascual and Ortiz-Jaureguizar 1990a, 1990b; see Gelfo et al. 2009 on the age of the Tiupampian).

Three major events in the evolution of southern South American mammals took place during the Early South American Phase: (1) the rapid decline and extinction of the nontherian native lineages, probably by mid-Paleocene times (but see Goin et al. 2004); (2) the arrival of therians, probably in several waves (Gelfo et al. 2009), already by the Late Cretaceous; and (3) major radiations of therians during the late Paleocene-early Eocene, possibly coincident with the Early Eocene Climatic Optimum (see fig. 3.4).

During the Paleocene and early Eocene, marsupials were dominant in Patagonian faunas. Tiupampian marsupials (early Paleocene) reveal that they had already undergone a rapid diversification, probably in the Late Cretaceous (Muizon 1991). Mid-Paleocene marsupials of Peligran age are already quite derived. Between the late Paleocene and early middle Eocene, polydolopimorphians reached their climax. A wide variety of opossum-like marsupials (e.g., Derorhynchidae, Peradectidae, Protodidelphidae, Sternbergiidae), many still of uncertain affinities, developed by the Itaboraian (here regarded as latest Paleocene-early Eocene in age; Gelfo et al. 2009). The earliest microbiotherians and paucituberculates are also of Itaboraian age, and sparassodonts of this time are small-to-medium in size. Marsupials of this phase exploited a variety of adaptive zones and diets: insectivorous, omnivorous, frugivorous, carnivorous, or a combination of them.

The earliest record of eutherians dates from the early Paleocene of Tiupampa and pertains to North American groups (Pantodonta, Mioclaenidae). Native ungulates also appear at Tiupampa, in the form of an indeterminate notoungulate referred to henricosbornids or oldfieldthomasiids (Muizon and Cifelli 2000; Muizon and Marshall 1992; Muizon, Marshall, and Sigé 1984). Medial Paleocene to early Eocene ungulates developed a wide variety of forms that characterized the first radiation of native South American ungulates. They developed low-crowned dentitions of various types: strictly bunodont (i.e., Didolodontidae, Protolipternidae), bilophodont (i.e., Xenungulata), bunolophodont (e.g., Notonychopidae, Trigonostylopidae, Henricosborniidae, Oldfieldthomasiidae, notopithecine Interatheriidae), and more lophoselenodont (e.g., Sparnotheriodontidae). Middle and late Eocene ungulates show the first trend toward hypsodonty, particularly among archaeohyracid Typotheria from the Casamayoran (late Eocene; Bond and López 1993, 1995). About the time of the late Eocene-early Oligocene global cooling (see later), there was an increase in grass phytoliths and volcanic activity, which generated volcanic ash that deposited upon vegetation. Both generated positive selective pressures in favor of higher-crowned dentitions among ungulates (Madden et al. 2010; Scarano 2009).

A marked biogeographic distinction between northern (Neotropical) and southern (Andean) regions was also apparent during this phase. Among the best known Paleogene faunas are those of Itaboraí (Brazil) and Las Flores (southern Argentina), of contemporary age (Itaboraian SALMA; late Paleocene-early Eocene; see Bergqvist, Lima Moreira, and Ribeiro Pinto 2006). Marsupials are the best represented taxa in both local associations. While polydolopid polydolopimorphians are the most diverse and abundant marsupials throughout the Patagonian and Antarctic Paleogene, including the Las Flores association, they are completely absent in Itaboraí.

3.3.4 LATE SOUTH AMERICAN PHASE

By the Eocene-Oligocene boundary (33.9 Ma), climatic conditions changed again toward a new icehouse phase. The opening of the Southern Ocean with the Drake Passage (i.e., the disconnection of South America and Antarctica) led to the formation of the Antarctic Circumpolar Current (Livermore et al. 2004). The result was the first major expansion of Antarctic ice in the Cenozoic. The sharp decrease in global temperatures was the primary driving force causing generalized turnovers in Paleogene marine and terrestrial biota. Probably, soon after these shifts, a major dispersal event occurred: the arrival of caviomorph rodents and platyrrhine primates in South America.

Goin, Abello, and Chornogubsky (2010) called the major taxonomic and ecologic shift in land-mammal faunas of southern South America at the Eocene-Oligocene boundary the *Bisagra Patagónica* (Patagonian Hinge; see also Flynn et al. 2003). They regarded the Hinge as equally important as other contemporary, major biotic events as the European *Grand Coupure* or the Central Asian *Mongolian Remodeling*. Recalling a more general statement made by the late Stephen Gould, Pascual (2006, 221) summarized the major differences in the early-to-late Cenozoic mammalian successions as follows (brackets added): "At this point, the history of South American mammals passed from 'Early Experimentation' [our Early South American Phase] to 'Modern Standardization' [our Late South American Phase]."

The recently discovered mammalian faunas of La Cancha and La Cantera, in central Patagonia (~33 Ma and 31–29 Ma, respectively; Goin, Abello, and Chornogubsky 2010), offer a detailed panorama of faunal changes at the beginning of this phase. Marsupials, especially sensitive to low temperatures, had already experienced a major turnover by the earliest Oligocene, including the last records of Caroloameghiniid peradectoids (Didelphimorphia), Sternbergiidae opossum-likemarsupials (Didelphimorphia), Hatcheriformes (Polydolopimorphia), bonapartherioid Bonapartheriiformes (Polydolopimorphia), and all Polydolopiformes (Polydolopimorphia), as well as the beginning of a rapid diversification of the "shrew opossums" (Paucituberculata), the radiation of large, modern borhyaenoids (Sparassodonta), and the origins of the Argyrolagoidea (Polydolopimorphia; Goin, Abello, and Chornogubsky 2010). This reflects a modernization of the marsupial fauna and extinction of some of the earlier, archaic taxa. Records of the first modern opossums (Didelphidae sensu stricto, possibly including also a caluromyine), the first Thylacosmilidae sparassodonts, as well as major radiations within the Microbiotheria and Paucituberculata all appear by the early Miocene (Colhuehuapian, 19–20 Ma; Goin et al. 2007).

Among native ungulates, there was a noticeable increase in the diversity of hypsodont notoungulates (e.g., Archaeohyracidae, Interatheriinae, Hegetotheriidae, and Mesotheriidae) after the early Oligocene. Low-crowned, bunodont ungulate types (i.e., megadolodine Litopterna) became scarce and were completely restricted to the lower latitudes of the Miocene of Colombia and Venezuela (Carlini, Gelfo, and Sánchez 2006; Cifelli and Villarroel 1997). Astrapotheria and Pyrotheria developed their larger forms during the Oligocene-Miocene and disappeared by the middle late Miocene (see Vizcaíno et al., chapter 5, this volume). In some lineages, limb specialization was convergent on that of equids; lophoselenodont dentitions (e.g., in proterotheriid Litopterna) were well developed by the Miocene (Soria 2001).

The oldest record of a platyrrhine in South America is *Branisella* from the Deseadan of Salla, Bolivia (late Oligocene, ~26 Ma; Kay et al. 1998). The oldest records of caviomorphs come from Tinguiririca levels of central Chile (Wyss et al. 1993). The age of the Tinguirirican SALMA has been estimated as 33.4–31 Ma (latest Eocene-early Oligocene; Goin, Abello, and Chornogubsky 2010). Both lineages probably arrived from Africa, although there is still debate on

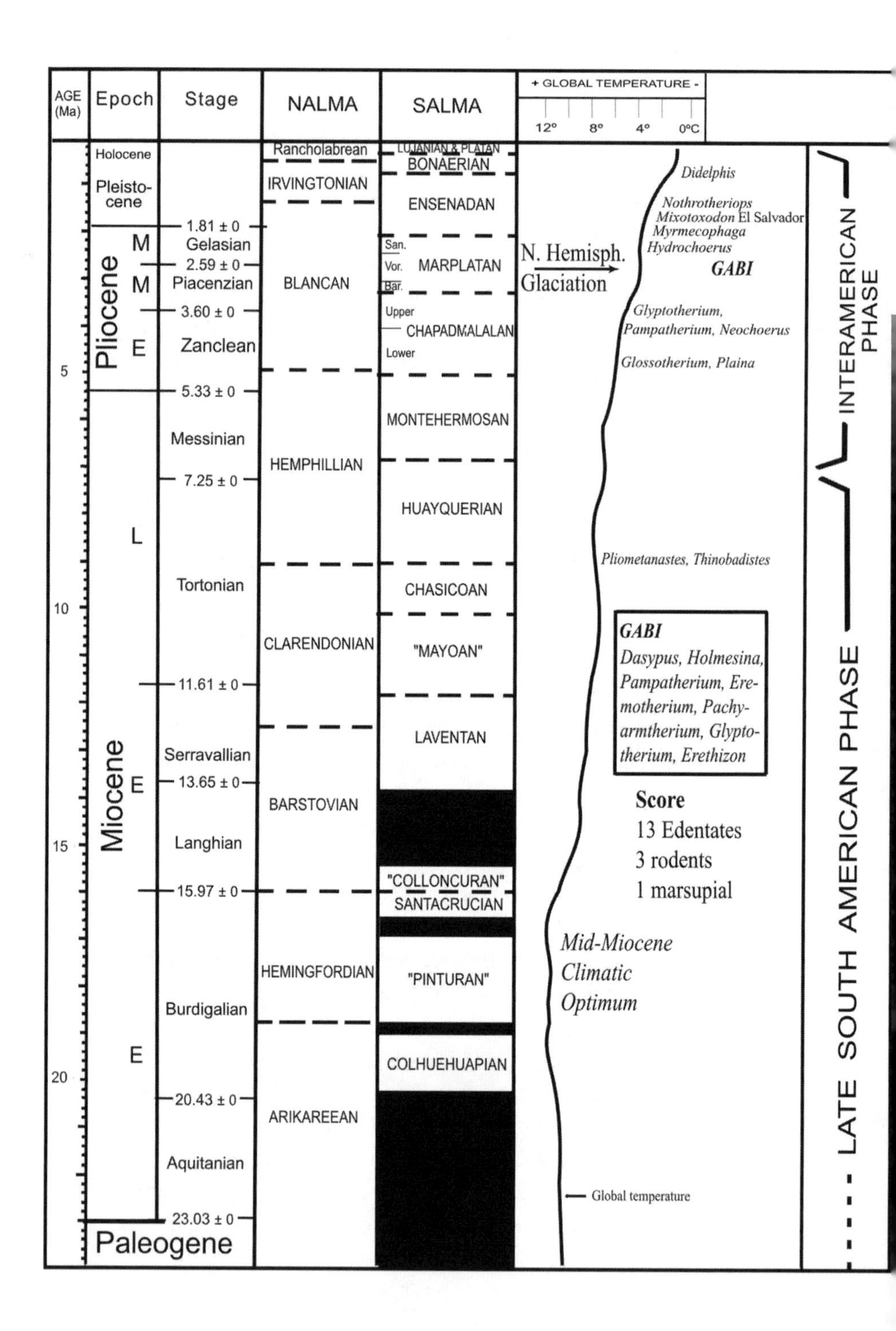
AGE (Ma)
Epoch
Stage
NALMA
SALMA
+ GLOBAL TEMPERATURE -
12°
8°
4°
0°C
Holocene
Pleisto-cene
Pliocene
Miocene
Paleogene
M
M
E
L
E
E
1.81 ± 0
Gelasian
2.59 ± 0
Piacenzian
3.60 ± 0
Zanclean
5.33 ± 0
Messinian
7.25 ± 0
Tortonian
11.61 ± 0
Serravallian
13.65 ± 0
Langhian
15.97 ± 0
Burdigalian
20.43 ± 0
Aquitanian
23.03 ± 0
5
10
15
20
Rancholabrean
IRVINGTONIAN
BLANCAN
HEMPHILLIAN
CLARENDONIAN
BARSTOVIAN
HEMINGFORDIAN
ARIKAREEAN
LUJANIAN & PLATAN
BONAERIAN
ENSENADAN
San.
Vor.
Bar.
MARPLATAN
Upper
CHAPADMALALAN
Lower
MONTEHERMOSAN
HUAYQUERIAN
CHASICOAN
"MAYOAN"
LAVENTAN
"COLLONCURAN"
SANTACRUCIAN
"PINTURAN"
COLHUEHUAPIAN
N. Hemisph. Glaciation
Didelphis
Nothrotheriops
Mixotoxodon El Salvador
Myrmecophaga
Hydrochoerus
GABI
Glyptotherium,
Pampatherium, Neochoerus
Glossotherium, Plaina
Pliometanastes, Thinobadistes
GABI
Dasypus, Holmesina, Pampatherium, Eremotherium, Pachyarmtherium, Glyptotherium, Erethizon
Score
13 Edentates
3 rodents
1 marsupial
Mid-Miocene Climatic Optimum
Global temperature
INTERAMERICAN PHASE
LATE SOUTH AMERICAN PHASE

whether this took place in synchrony or in asynchronous waves (e.g., Poux et al. 2006). The early records of caviomorphs are far more abundant than those of platyrrhines. Recently, Vucetich et al. (2010) described several taxa of relatively generalized, pre-Deseadan caviomorphs from the post-Tinguiririican (early Oligocene) levels of La Cantera, in central Patagonia (ca. 31–29 Ma), and estimated that the arrival of caviomorphs into South America "was not much before the early Oligocene" (Vucetich et al. 2010, 189). A recent molecular dating of early divergences within Hystricognathi, using a Bayesian "relaxed clock" approach and multiple fossil calibrations, suggested that the split between phiomorphs and caviomorphs occurred around 36 Ma (late Eocene; Sallam et al. 2009). Again, we propose an alternative hypothesis that takes into account the dual nature of South America's biogeography: rodents and primates arrived earlier in the Neotropical Region, by middle or late Eocene times (e.g., Frailey and Campbell 2004), only arriving later in the Andean Region.

3.3.5 INTER-AMERICAN PHASE

The Great American Biotic Interchange (GABI) (Stehli and Webb 1985) refers to the establishment of the strongest biogeographic linkage between North and South America ever (Marshall et al. 1982; Morgan 2008; Simpson 1953; Webb 1976, 1985; Woodburne, Cione, and Tonni 2006). Although the Panamanian region was an upland isthmus as early as 6 Ma, permitting limited dispersals from about that time, the main pulse of the interchange began about 2.6 Ma, essentially coeval with the onset of major Northern Hemisphere glaciation (Mudelsee and Raymo 2005), and persisted to ~2.4 Ma. Traditional views on the GABI regard this process as a response to the tectonic closure of the Panamanian seaway. It is now clear from geological evidence that this closure was under way by ~12 Ma and effectively completed by ~6 Ma (Coates et al. 2004), with final closure of the Central American seaway transpiring by ~2.8 Ma (Bartoli et al. 2005). By 6 Ma, a few mammalian lineages known as the "heralds" of the GABI (Webb 1985) dispersed throughout the Americas. However, the main thrust of the GABI was the multitaxon exchanges that began ~2.6 Ma when

---

Figure 3.4 (*facing page*) Neogene time scale, NALMA, SALMA, Great American Biotic Interchange (GABI), global temperatures, and taxa that dispersed from South America to North America during this time. The Neogene time scale is after Lourens et al. (2004), the NALMA chronology follows Woodburne (2004), while the SALMA chronology is based on Woodburne, Cione, and Tonni (2006) and Flynn and Swisher (1995). Taxa that dispersed to North America are from Woodburne, Cione, and Tonni (2006). Global temperature scale is after Zachos et al. (2001).

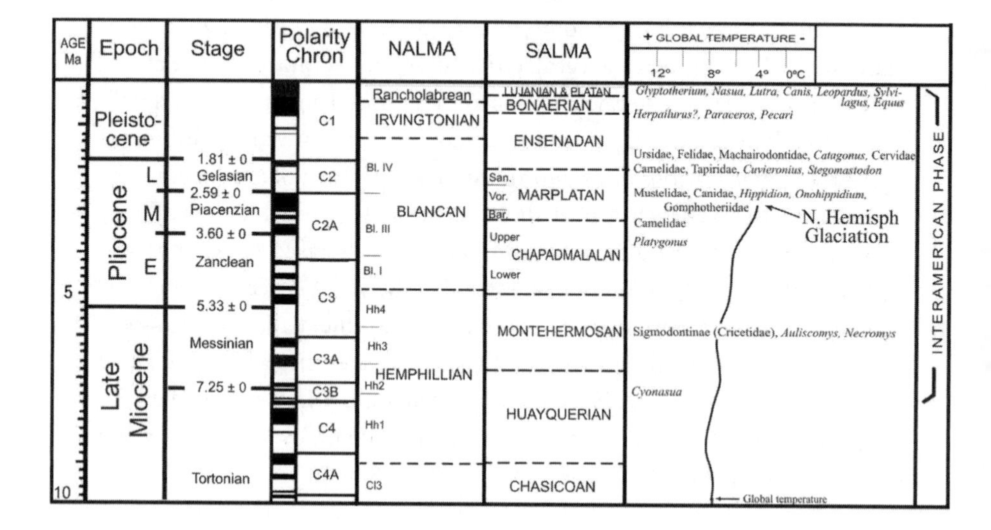

Figure 3.5 Great American Biotic Interchange to South America, NALMA and SALMA chronology, and global temperature. Taxa that dispersed to South America are from Woodburne, Cione, and Tonni (2006), Cione et al. (2007), and Woodburne (2010).

climate changed with Northern Hemisphere glaciations, permitting temperate-adapted taxa to cross the Panamanian isthmus (fig. 3.5).

The traditional viewpoint has been that GABI signaled the development of an overland corridor across the Panamanian region, and that this transpired ~3.0 Ma. A more persuasive interpretation now seems to be that these dispersal episodes reflect the onset and presence of glacial conditions in the Northern Hemisphere (Webb 1991). This stems not only from paleobotanical and other evidence, but also reflects the more temperate, versus tropical, adaptations of the taxa making the crossings. In fact, Coates et al. (2004) reconstructed the geological evolution of the Panamanian region and indicated that it was an upland isthmus as early as 6 Ma, with the former Central American Seaway being interrupted by an evolving volcanic arc as early as ~12 Ma.

Even under Isthmian tropical conditions, North America was a beneficiary of the tectonic process, with the first South American immigrants being the mylodontid sloth, *Thinobadistes*, and the megalonychid sloth, *Pliometanastes*, found in tropical (Florida) to more temperate sites (California) of early Hemphillian age (~9 Ma; Morgan 2008). Island-hopping across short marine barriers apparently did not deter these edentates from colonizing. Subsequently, at

~7 Ma, *Pliometanastes* was the sister taxon to *Megalonyx*, which became a relatively common constituent of late Miocene to Pleistocene faunas of southern North America (Bell et al. 2004). As summarized by Morgan (2008), a variety of other edentates entered North America and successfully inhabited temperate regions across the southern part of North America. Also at about 7.2 Ma, the procyonid carnivore *Cyonasua* dispersed to South America.

From about 6 Ma, the pre-GABI pace of dispersal began to accelerate (Cione et al. 2007; Woodburne 2010). Sigmodontine rodents apparently dispersed to South America by 6 Ma, and at about 4.7 Ma, pampathere (*Plaina*) and mylodont (*Glossotherium*) edentates are recorded in Mexico. Somewhat later (3.0 Ma), *Glossotherium* is recorded in Florida, which was still subtropical. This is one of a number of instances (Flynn et al. 2005) in which Mexico appears to have been a waiting room for taxa that subsequently completed their dispersal to the continental United States, despite the lack of known physical or ecological barriers (Woodburne, Cione, and Tonni 2006). Although Verzi and Montalvo (2008) presented a very interesting scenario regarding the early dispersal of rodents and carnivores in the Huayquerian SALMA, we follow the interpretation of Prevosti and Pardiñas (2009) and place the colonization of sigmodontines in this timeframe.

At 3.8 Ma, *Neochoerus* was a northward immigrant to Mexico and, until its discovery in North Carolina strata of the same age (Sanders 2002), was considered one of the Mexican laggards. This genus, a hydrochoerid rodent, is considered to have been an inhabitant of tropical to subtropical conditions. The peccary *Platygonus* is recorded in Argentina at 3.7 Ma, so it apparently dispersed about that time, and *Pampatherium* (Cingulata) is recorded then in Mexico. At 3.3 Ma, Camelidae dispersed to South America: A *Lama* precursor and the peccary apparently are the first savanna-adapted taxa to disperse southward, and the camel record appears coeval with the advent of Northern Hemisphere glaciation. Just prior to GABI, at 3.0 Ma, the mylodont *Glossotherium* appears in Florida (after an earlier presence in Mexico).

As documented by the GABI chronology, at least part of the tropical rain forests of the isthmus shifted toward savanna-like habitat (Webb, 1991); this is reflected in the ecologic diversity of the dispersing taxa. South American records of mustelid and canid carnivores, equid perissodactyls, and a gomphothere appear at about 2.6 Ma. An immigrant to North America at about this time (2.7 Ma), the porcupine *Erethizon* is now adapted to cool temperate conditions.

At 2.4 Ma, the pampathere *Pampatherium* enters from Mexico (a laggard) along with its through-moving relatives *Holmesina*, armadillos (*Dasypus*, *Pachyarmatherium*), and a megatheriid (*Eremotherium*). Of these, *Dasypus* still inhabits temperate regions in the Gulf Coast region and formerly occurred as far north as Iowa in the Pleistocene (Morgan 2008). Overall, this appears to be a savanna-adapted group of taxa.

An apparent gap in dispersals lasted from 2.4 to 1.8 Ma, which seems to have been a tropical holding pattern for hydrochoerid rodents. *Hydrochoerus* was present in Florida about 2.2 Ma in presumably wet conditions and might have lingered in local sites in the Isthmian region. But about 1.8 Ma, South America experienced a strong pulse of immigration that included ursids, canids, felids, machairodontids, peccaries, deer, camels, tapirs, and proboscideans. Together, they document the presence of (or perhaps a return to) savanna-like conditions in the Isthmian region and adjacent South America. Another apparent gap in dispersal lasted from just after 1.8 Ma to about 0.7 Ma, but immigrants to North America included a toxodont to El Salvador (1.5 Ma) and a megatheriid sloth, *Nothrotheriops* (1.3 Ma).

At about 0.8 Ma *Didelphis* entered North America, followed at ~0.7 Ma by the dispersal to South America by a mustelid, a deer, and a peccary. Finally, at ~0.125 Ma, a major southward dispersal is recorded by a xenarthran, procyonid, mustelid, canid, and felid carnivores, a sylvilagine rabbit, and the equid *Equus*, a diversified if not balanced assemblage.

In summary, the GABI can be resolved as a series of pulses, possibly reflecting glacial versus interglacial conditions in the Northern Hemisphere. A major exchange from 2.7–2.4 Ma resulted in a variety of edentates and a porcupine arriving in North America and a larger group of carnivores, ungulates, and a gomphothere colonizing South America coincident with major Northern Hemisphere glaciations (Mudelsee and Raymo 2005). A dispersal gap from ~2.4–1.8 Ma saw limited northward dispersal of a hydrochoerid rodent, possibly reflecting its earlier presence in the Isthmian tropics. Renewed dispersals to South America at ~1.8 Ma entailed a diversity of carnivores and various artiodactyls, perissodactyls, and proboscideans, with a myrmecophagid entering North America as well. Another gap from ~1.8–0.8 Ma saw limited dispersal to North America by a toxodont to El Salvador at 1.5 Ma and a megatheriid sloth at 1.3 Ma. A more limited exchange at 0.8–0.7 Ma is recorded by a marsupial going north and a mustelid, a deer, and a peccary to the south. The last major GABI at 0.125 Ma saw the southward dispersal of an edentate, a diversity of carnivores, a sylvilagine rabbit, and *Equus*. The foregoing indicates that more taxa went south than went north; Webb (1991) explained this numerical dominance of North American taxa as a reflection of the sixfold difference in the source area of open country habitats in North America versus those in the south.

Beginning in 2000, K. E. Campbell and colleagues have presented a series of articles that propose a modification of the early phase of the aforementioned scenario. In essence, the gomphotheriid proboscidean *Amahuacatherium peruvium* is considered to have entered South America in the late Miocene, where it is found in Peruvian sediments along with mammal fossils of Huayquerian age (Campbell, Frailey, and Romero-Pittman 2000, 2009). The paleontological age is supported by both radioisotopic and paleomagnetic evidence (Campbell et al. 2001, 2010), with the fossil site considered to be ~9.5 Ma old. Campbell, Frailey, and Romero-Pittman (2000) also suggest that camels, peccaries, and tapirs are part of the immigrant mammalian component at this time. Although the chronologic setting for these taxa seems established, a persistent difficulty is that there is as yet no subsequent record of their pre-GABI presence anywhere in South America (e.g., Pascual and Ortiz-Jaureguizar 1990b; Pascual 2006; Patterson and Pascual 1972), regardless of the potential biogeographic considerations mentioned earlier (Croft 2007; Croft et al. 2009). Thus, the intriguing anomaly suggested by *Amahuacatheriuim* and its faunal associates remains incompletely understood.

Acknowledgments

We thank Analía Artabe and Alba Zamuner for their help with the palobotanical literature, and Marcela Tomeo for the composition of the figures. The CONICET provided financial support to one of us (F. Goin; PIP 5621). Reviewers Analía Forasiepi and Mariano Bond helped greatly in improving the submitted manuscript. Whatever the merits of this contribution may be, the ideas presented here owe much to numerous talks and discussions over the last twenty years with Rosendo Pascual, to whom this work is dedicated.

Literature Cited

Alroy, J. 2004. "Are Sepkoski's Evolutionary Faunas Dynamically Coherent?" *Evolutionary Ecology Research* 6:1–32.

Ameghino, F. 1906. "Les Formations Sédimentarires du Crétacé Supérieur et du Tertiaire de Patagonie Avec un Parallèle Entre Leurs Faunes Mammalogiques et Celles de L'ancien Continent." *Anales del Museo Nacional de Historia Natural de Buenos Aires* 8:1568.

Apesteguía, S., L. Gaetano, and G. W. Rougier. 2009. "The Highly Unusual Skull of the Dryolestoid Mammals: Keys from the Other Side of the Atlantic." *I North African Congress on Vertebrate Paleontology, Marrakech* (Abstracts): 26.

Artabe, A. E., E. M. Morel, and L. A. Spalletti. 2003. "Caracterización de las Provincias Fitogeográficas Triásicas del Gondwana Extratropical." *Ameghiniana* 40:387–405.

Bartoli, G., M. Sarnthein, M. Weinelt, H. Erlenkeuser, D. Garbe-Schönberg, and D. W. Lea. 2005. "Final Closure of Panama and the Onset of Northern Hemisphere Glaciation." *Earth and Planetary Science Letters* 237:33–44.

Bell, C. J., E. L. Lundelius, Jr., A. D. Barnosky, R. W. Graham, E. H. Lindsay, D. R. Ruez, Jr., H. A. Semken et al. 2004. "The Blancan, Irvingtonian, and Rancholabrean Mammals Ages." In *Late Cretaceous and Cenozoic Mammals of North America: Biostratigraphy and Geochronology*, edited by M. O. Woodburne, 232–314. New York: Columbia University Press.

Bergqvist, L., A. Lima Moreira, and D. Ribeiro Pinto. 2006. *Bacia de Sao José de Itaboraí. 75 anos de História e Ciéncia*. Rio de Janeiro: Servico Geológico do Brasil.

Bertini, R. J., L. G. Marshall, M. Gayet, and P. Brito. 1993. "Vertebrate Faunas from the Adamantina and Marília Formations (Upper Baurú Group, Late Cretaceous, Brazil) in Their Stratigraphic and Paleobiogeographic Context." *Neues Jahrbuch für Geologie und Palaöntologie, Abhandlungen* 188:71–101.

Bijl, P. K., S. Shouten, A. Sluijs, G.-J. Reichart, J. C. Zachos, and H. Brinkhuis. 2009. "Early Palaeogene Temperature Evolution of the Southwest Pacific Ocean." *Nature* 461:776–79.

Bonaparte, J. F. 1986a. "History of the Terrestrial Cretaceous Vertebrates of Gondwana." *IV Congreso Argentino de Paleontología y Bioestratigrafía, Mendoza, Actas* 2:63–95.

———. 1986b."Sobre *Mesungulatum houssayi* y Nuevos Mamíferos Cretácicos de Patagonia, Argentina." *IV Congreso Argentino de Paleontología y Bioestratigrafía, Mendoza, Actas* 2:48–61.

———, ed. 1987. "The Late Cretaceous Fauna of Los Alamitos, Patagonia, Argentina." *Revista del Museo Argentino de Ciencias Naturales, Paleontología* 3:103–78.

———. 1990. "New Late Cretaceous Mammals from the Los Alamitos Formation, Northern Patagonia." *National Geographic Research* 6 (1):63–93.

———. 1992. "Una Nueva Especie de Triconodonta (Mammalia) de la Formación Los Alamitos, Provincia de Río Negro y Comentarios Sobre su Fauna de Mamíferos." *Ameghiniana* 29 (2):99–110.

———. 1994. "Approach to the Significance of the Late Cretaceous Mammals of South America." *Berliner Geowissenschaften Abhandlugen* E 13:31–44.

———.1996. "Cretaceous Tetrapods of Argentina." *Münchner Geowissenschaftliche Abhandlungen (A)* 30:73–130.

———. 2008. "On the Phylogenetic Relationships of *Vincelestes neuquenianus*." *Historical Biology* 20:81–6.

Bonaparte, J. F., A. G. Martinelli, and C. L. Schultz. 2005. "New Information on *Brasilodon* and *Brasilitherium* (Cynodontia, Probainognathia) from the Late Triassic of Southern Brazil." *Revista Brasileira de Paleontologia* 8:25–46.

Bonaparte, J. F. and M. F. Soria. 1985. "Nota Sobre el Primer Mamífero del Cretácico Argentino, Campaniano-Maastrichtiano (Condylarthra)." *Ameghiniana* 21:177–83.

Bond, M., and G. López. 1993. "El Primer Notohippidae (Mammalia, Notoungulata) de la Formación Lumbrera (Grupo Salta) del Noroeste Argentino: Consideraciones Sobre la Sistemática de la Familia Notohippidae." *Ameghiniana* 30:59–68.

———. 1995. "Los Mamíferos de la Formación Casa Grande (Eoceno) de la Provincia de Jujuy, Argentina." *Ameghiniana* 32:301–09.

Burgoyne, P. M., A. E. van Wyk, J. M. Anderson, and B. D. Schrire. 2005. "Phanerozoic Evolution of Plants on the African Plate." *Journal of African Earth Sciences* 43:13–52.

Campbell, K. E. Jr., C. D. Frailey, and L. Romero-Pittman. 2000. "The Late Miocene Gomphothere *Amahuacatherium peruvium* (Proboscidea: Gomphotheriidae) from Amazonian Peru: Implications for the Great American Faunal Interchange." *Instituto Geológico Minero y Metalúrgico, Serie D, Estudios Regionales, Boletín* 23:1–152.

———. 2009. "In Defense of *Amahuacatherium* (Proboscidea: Gomphotheriidae)." *Neues Jahrbuch für Geologie und Paläontologie, Abhandlungen* 252:113–28.

Campbell, K. E. Jr., Heizler, M., C. D. Frailey, L. Romero-Pittman, and D. R. Prothero. 2001. "Upper Cenozoic Chronostratigraphy of the Southwestern Amazon Basin." *Geology* 29:595–8.

Campbell, K. E. Jr., D. R. Prothero, L. Romero-Pittman, F. Hertel, and N. Rivera. 2010. "Amazonian Magnetostratigraphy: Dating the First Pulse of the Great American Faunal Interchange." *Journal of South American Earth Sciences* 29: 619–26.

Candeiro, C. R. A., A. R. Santos, T. H. Rich, T. S. Marinho, and E. C. Oliveira. 2006. "Vertebrate Fossils from the Adamantina Formation (Late Cretaceous), Prata Paleontological District, Minas Gerais State, Brazil." *Geobios* 39:319–27.

Carlini, A. A., J. N. Gelfo, and R. Sánchez. 2006. "First Record of the Strange Megadolodinae (Mammalia: Litopterna: Proterotheriidae) in the Urumaco Formation (Late Miocene), Venezuela." *Journal of Systematic Palaeontology* 4:279–84.

Casamiquela, R. M. 1961. "El Hallazgo del Primer Elenco (Icnológico) Jurásico de Vertebrados Terrestres de Latinoamérica (Noticia)." *Revista de la Asociación Geológica Argentina* 15:1–14.

Case, J. A. 2002. "A New Biogeographic Model for Dispersal of Late Cretaceous Vertebrates into Madagascar and India." *Journal of Vertebrate Paleontology* 22:42A.

Case, J. A., F. J. Goin, and M. O. Woodburne. 2005. "'South American' Marsupials from the Late Cretaceous of North America and the Origin of Marsupial Cohorts." *Journal of Mammalian Evolution* 12:461–94.

Cifelli, R. L., and C. Villarroel. 1997. "Paleobiology and Affinities of *Megadolodus*." In *A History of the Neotropical Fauna: Vertebrate Paleobiology of the Miocene of Tropical South America*, edited by R. F. Kay, R. L. Cifelli, and R. H. Madden, 265–88. New York: Springer-Verlag.

Cione, A. L., E. P. Tonni, S. Bargo, M. Bond, A. M. Candela, A. A. Carlini, C. M. Deschamps et al. 2007. "Mamíferos Continentales del Mioceno Tardío a la Actualidad en Argentina: Cincuenta Años de Estudios." In *Ameghiniana 50° Aniversario, Asociación Paleontológica Argentina, Publicación Especial* 11, 257–78. Buenos Aires: Asociación Paleontológica Argentina.

Coates, A. G., L. S. Collins, M.-P. Aubry, and W. A. Berggren. 2004. "The Geology of the Darien, Panama, and Miocene-Pliocene Collision of the Panama Arc with Northwestern South America." *Geologial Society of America Bulletin* 116:1327–44.

Cox, C. B. 2001. "The Biogeographic Regions Reconsidered." *Journal of Biogeography* 28:511–23.

Crisci, J. V., M. M. Cigliano, J. J. Morrone, and S. Roig-Nuñent. 1991. "Historical Biogeography of Southern South America." *Systematic Zoology* 40:152–71.

Croft, D. A. 2007. "The Middle Miocene (Laventan) Quebrada Honda Fauna, Southern Bolivia and a Description of its Notoungulates." *Palaeontology* 50:277–303.

Croft, D. A., F. Anaya, D. Auerbach, and C. Garzione. 2009. "New Data on Miocene Neotropical Provinciality From Cerdas, Bolivia." *Journal of Mammalian Evolution* 16:175–98.

De Valais, S. 2009. "Ichnotaxonomic Revision of *Ameghinichnus*, a Mammalian Ichnogenus from the Middle Jurassic La Matilde Formation, Santa Cruz Province, Argentina." *Zootaxa* 2203:1–21.

Flynn, J. J., B. J. Kowalis, C. Núñez, O. Carranza-Castañeda, W. E. Miller, C. C. Swisher, III, and E. H. Lindsay. 2005. "Geochronology of Hemphillian-Blancan Aged Strata, Guanajuato, Mexico, and Implications for Timing of the Great American Biotic Interchange." *Journal of Geology* 113:287–307.

Flynn, J. J., A. R. Wyss, D. A. Croft, and R. Charrier. 2003. "The Tinguiririca Fauna, Chile: Biochronology, Paleoecology, Biogeography, and a New Earliest Oligocene South American Land Mammal 'Age'." *Palaeogeography, Palaeoclimatology, Palaeoecology* 195:229–59.

Frailey, C. D., and K. E. Campbell, Jr. 2004. "Paleogene Rodents from Amazonian Perú: The Santa Rosa Local Fauna." In *The Paleogene Mammalian Fauna of Santa Rosa, Amazonian Peru, Science Series*, 40, edited by K. E. Campbell, Jr., 71–130. Los Angeles: Natural History Museum of Los Angeles County.

Gayet, M., L. G. Marshall, T. Sempere, F. J. Meunier, H. Cappetta, and J.-C. Rage. 2001. "Middle Maastrichtian Vertebrates (Fishes, Amphibians, Dinosaurs and Other Reptiles, Mammals) from Pajcha Pata (Bolivia). Biostratigraphic, Palaeoecologic and Palaeobiogeographic Implications." *Palaeogeography, Palaeoclimatology, Palaeoecology* 169:39–68.

Gelfo, J. N., F. J. Goin, M. O. Woodburne, and C. de Muizon. 2009. "Biochronological Relationships of the Earliest South American Paleogene Mammalian Faunas." *Palaeontology* 52:251–69.

Gelfo, J. N., and R. Pascual. 2001. "*Peligrotherium tropicalis* (Mammalia, Dryolestida) from the Early Paleocene of Patagonia, a Survival from a Mesozoic Gondwanan Radiation." *Geodiversitas* 23:369–79.

Goin, F. J., M. A. Abello, E. Bellosi, R. F. Kay, R. H. Madden, and A. A. Carlini. 2007. "Los Metatheria Sudamericanos de Comienzos del Neógeno (Mioceno Temprano, Edad-mamífero Colhuehuapense). Parte 1: Introducción, Didelphimorphia y Sparassodonta." *Ameghiniana* 44:29–71.

Goin, F. J., M. A. Abello, and L. Chornogubsky. 2010. "Middle Tertiary Marsupials from Central Patagonia (Early Oligocene of Gran Barranca): Understanding South America's Grande Coupure." In *The Paleontology of Gran Barranca: Evolution and Environmental Change Through the Middle Cenozoic of Patagonia*, edited by R. H. Madden, A. A. Carlini, M. G. Vucetich, and R. F. Kay, 71–107. New York: Cambridge University Press.

Goin, F. J., A. A. Carlini, and R. Pascual. 1986. "Un Probable Marsupial del Cretácico Tardío del Norte de Patagonia, Argentina." *IV Congreso Argentino de Paleontología y Bioestratigrafía, Mendoza, Actas* 2:43–47.

Goin, F. J., R. Pascual, M. F. Tejedor, J. N. Gelfo, M. O. Woodburne, J. A. Case, M. A. Reguero et al. 2006. "The Earliest Tertiary Therian Mammal from South America." *Journal of Vertebrate Paleontology* 26:505–10.

Goin, F. J., E. C. Vieytes, M. G. Vucetich, A. A. Carlini, and M. Bond. 2004. "Enigmatic Mammal from the Paleogene of Perú." In *The Paleogene Mammalian Fauna of Santa Rosa, Amazonian Perú, Science Series*, 40, edited by K. E. Campbell, Jr., 145–153. Los Angeles: Natural History Museum of Los Angeles County.

Gurovich, Y., and R. Beck. 2008. "The Phylogenetic Affinities of the Enigmatic Mammalian Clade Gondwanatheria." *Journal of Mammalian Evolution* 16:25–49.

Kay, R. F., B. J. MacFadden, R. H. Madden, H. Sandeman, and F. Anaya. 1998. "Revised Age of the Salla Beds, Bolivia, and its Bearing on the Age of the Deseadan South American Land Mammal 'Age'." *Journal of Vertebrate Paleontology* 18:189–99.

Kielan-Jaworowska, Z., R. L. Cifelli, and Z.-X. Luo. 2004. *Mammals from the Age of Dinosaurs. Origins, Evolution, and Structure*. New York: Columbia University Press.

Kielan-Jaworowska, Z., E. Ortiz-Jaureguizar, E. C. Vieytes, R. Pascual, and F. J. Goin. 2007. "First ?Cimolodontan Multituberculate Mammal from South America." *Acta Palaeontologica Polonica* 52:257–62.

Koenigswald, W. von, F. J. Goin, and R. Pascual. 1999. "Hypsodonty and Enamel Microstructure in the Paleocene Gondwanatherian Mammal *Sudamerica ameghinoi*." *Acta Palaeontologica Polonica*, 44:263–300.

Krause, D. W., P. M. O'Connor, C. C. Rogers, S. D. Sampson, G. A. Buckley, and R. R. Rogers. 2006. "Late Cretaceous Terrestrial Vertebrates from Madagascar: Implications for Latin American Biogeography." *Annals of the Missouri Botanical Garden* 93:178–208.

Livermore, R., G. Eagles, P. Morris, and A. Maldonado. 2004. "Shackleton Fracture Zone: No Barrier to Early Circumpolar Ocean Circulation." *Geology* 32:797–800.

Lourens, L., F. Hilgen, N. J. Shackleton, J. Laskar, and D. Wilson. 2004. "The Neogene Period." In *A Geologic Time Scale*, edited by F. Gradstein, J. Ogg, and A. Smith, 409–40. Cambridge: Cambridge University Press.

Luo, Z.-X. 2007. "Transformation and Diversification in Early Mammal Evolution." *Nature* 450:1011–19.

Luo, Z.-X., Z. Kielan-Jaworowska, and R. L. Cifelli. 2002. "In Quest for a Phylogeny of Mesozoic Mammals." *Acta Paleontologica Polonica* 47:1–78.

Luterbacher, H. P., J. R. Ali, H. Brinkhuis, F. M. Gradstein, J. J. Hooker, S. Monechi, J. G. Ogg et al. 2004. "The Paleogene Period." In *A Geologic Time Scale*, edited by F. Gradstein, J. Ogg, and A. Smith, 384–408. Cambridge: Cambridge University Press.

Madden, R. H., R. F. Kay, M. G. Vucetich, and A. A. Carlini. 2010. "Gran Barranca: A 23-Million-Year Record of Middle Cenozoic Faunal Evolution in Patagonia." In *The Paleontology of Gran Barranca: Evolution and Environmental Change Through the Middle Cenozoic of Patagonia*, edited by R. H. Madden, A. A. Carlini, M. G. Vucetich, and R. F. Kay, 419–35. New York: Cambridge University Press.

Marshall, L. G., R. Hoffstetter, and R. Pascual. 1983. "Mammals and Stratigraphy: Geochronology of the Continental Mammal-Bearing Tertiary of South America. *Palaoevertebrata, Montpellier, Mémoire Extraordinaire* 1983:1–93.

Marshall, L. G., S. D. Webb, J. J. Sepkoski, Jr., and D. M. Raup. 1982. "Mammalian Evolution and the Great American Interchange." *Science* 215:1351–57.

Martin, T., and O. W. M. Rauhut. 2005. "Mandible and Dentition of *Asfaltomylos patagonicus* (Australosphenida, Mammalia) and the Evolution of Tribosphenic Teeth." *Journal of Vertebrate Paleontology* 25:414–25.

Martinelli, A. G., and A. Forasiepi. 2004. "Late Cretaceous Vertebrates from Bajo de Santa Rosa (Allen Formation), Río Negro Province, Argentina, With the Description of a New Sauropod Dinosaur (Titanosauridae)." *Revista del Museo Argentino de Ciencias Naturales, Buenos Aires*, n.s., 6 (2):257–305.

Martinelli, A. G., and G. W. Rougier. 2007. "On *Chaliminia musteloides* (Eucynodontia: Tritheledontidae) from the Late Triassic of Argentina, and a Phylogeny of Ictidosauria." *Journal of Vertebrate Paleontology* 27:442–60.

Moreira-Muñoz, A. 2007. "The Austral Floristic Realm Revisited." *Journal of Biogeography* 34:1649–60.

Morgan, G. R. 2008. "Vertebrate Fauna and Geochronology of the Great American Biotic Interchange in North America." In *Neogene Mammals*, edited by S. G. Lucas, G. S. Morgan, J. A. Spielmann, and D. R. Prothero, 93–140. Albuquerque: New Mexico Museum of Natural History and Sciences Bulletin 44.

Morrone, J. J. 2001a. *Biogeografía de América Latina y el Caribe*, Vol. 3. Zaragoza: Manuales and Tesis SEA.

———. 2001b. "A Proposal Concerning Formal Definitions of the Neotropical and Andean Regions." *Biogeographica* 77:65–82.

———. 2002. "Biogeographical Regions Under Track and Cladistic Scrutiny." *Journal of Biogeography* 29:149–52.

———. 2004a. "La Zona de Transición Sudamericana: Caracterización y Relevancia Evolutiva." *Revista Entomológica Chilena* 28:41–50.

———. 2004b. "Panbiogeografía, Componentes Bióticos y Zonas de Transición." *Revista Brasileira de Entomologia* 48:149–62.

———. 2006. "Biogeographic Areas and Transition Zones of Latin America and the Caribbean Islands Based on Panbiogeographic and Cladistic Analyses of the Entomofauna." *Annual Review of Entomology* 51:467–94.

Mudelsee, M., and M. E. Raymo. 2005. "Slow Dynamics of the Northern Hemisphere Glaciation." *Paleoceanography* 20:PA4022.

Muizon, C. de. 1991. "La Fauna de Mamíferos de Tiupampa (Paleoceno Inferior, Formación Santa Lucía), Bolivia. In *Fósiles y Facies de Bolivia, Vol. 1. Vertebrados*, edited by R. Suárez-Soruco, 575–624. Santa Cruz: Revista Técnica de Yacimientos Petrolíferos Fiscales de Bolivia, 12(3–4).

Muizon, C. de, and R. Cifelli. 2000. "The 'Condylarths' (Archaic Ungulata, Mammalia) from the Early Palaeocene of Tiupampa (Bolivia): Implications on the Origin of the South American Ungulates." *Geodiversitas* 22:47–150.

Muizon, C. de, and L. G. Marshall. 1992. "*Alcidedorbignya inopinata* (Mammalia: Pantodonta) from the Early Paleocene of Bolivia: Phylogenetic and Paleobiogeographic Implications." *Journal of Palaeontology* 66:509–30.

Muizon, C. de, L. G. Marshall, and B. Sigé. 1984. "The Mammal Fauna from the El Molino Formation (Late Cretaceous-Maastrichtian) at Tiupampa, Southcentral Bolivia." *Bulletin du Muséum National d'Histoire Naturelle du Paris, 4e Série, Section C* 6:315–27.

Ortiz-Jaureguizar, E. 1986. "Evolución de las Comunidades de Mamíferos Cenozoicos Sudamericanos: Un Estudio Basado en Técnicas de Análisis Multivariado." *IV Congreso Argentino de Paleontología y Bioestratigrafía, Mendoza, Actas* 2:191–207.

———. 1996. "Paleobiogeografía y Paleoecología de los Mamíferos Continentales de América del Sur Durante el Cretácico Tardío-Paleoceno: Una Revisión." *Estudios Geológicos* 52:83–94.

Ortiz-Jaureguizar, E., and R. Pascual. 2007. "The Tectonic Setting of the Caribbean Region: Key in the Radical Late Cretaceous-Early Paleocene South American Land-Mammal Turnover. In *Fourth European Meeting on the Palaeontology and Stratigraphy of Latin America*, edited by E. Díaz-Martínez and I. Rábano, 301–07. Madrid: Instituto Geológico y Minero de España, Cuadernos del Museo Geominero n° 8.

Pascual, R. 2006. "Evolution and Geography: The Biogeographic History of South American Land Mammals." *Annals of the Missouri Botanical Gardens* 93:209–30.

Pascual, R., M. Archer, E. Ortiz-Jaureguizar, J. L. Prado, H. Godthelp, and S. J. Hand. 1992. "First Discovery of Monotremes in South America." *Nature* 356:704–5.

Pascual, R., F. J. Goin, P. González, A. Ardolino, and P. Puerta. 2000. "A Highly Derived Docodont from the Patagonian Late Cretaceous: Evolutionary Implications for Gondwana Mammals." *Geodiversitas* 22:395–414.

Pascual, R., F. J. Goin, D. W. Krause, E. Ortiz-Jaureguizar, and A. A. Carlini. 1999. "The First Gnathic Remains of *Sudamerica*: Implications for Gondwanathere Relationships." *Journal of Vertebrate Paleontology* 19:373–82.

Pascual, R., and E. Ortiz-Jaureguizar. 1990a. "Evolutionary Pattern of Land Mammal Faunas During the Late Cretaceous and Paleocene in South America: A Comparison with the North American Pattern." *Annales Zoologici Fennici* 28:245–52.

———. 1990b. "Evolving Climates and Mammal Faunas in Cenozoic South America." *Journal of Human Evolution* 19:23–60.

———. 2007. "The Gondwanan and South American Episodes: Two Major and Unrelated Moments in the History of the South American Mammals." *Journal of Mammalian Evolution* 14:75–137.

Pascual, R., E. Ortiz-Jaureguizar, and J. L. Prado. 1996. "Land Mammals: Paradigm for Cenozoic South American Geobiotic Evolution." In *Contributions of Southern South America to Vertebrate Paleontology*, edited by G. Arratia, 265–319. Munich: Münchner Geowissenschaften Abh. 30.

Pascual, R., M. G. Vucetich, G. J. Scillato-Yané, and M. Bond. 1985. "Main Pathways of Mammalian Diversification in South America." In *The Great American Biotic Interchange* (Topics in Geobiology, Vol. 4), edited by F. G. Stehli and S. D. Webb, 219–47. New York: Plenum Press.

Patterson, B., and R. Pascual. 1972. "The Fossil Mammal Fauna of South America." In *Evolution, Mammals, and Southern Continents*, edited by A. Keast, F. C. Erk, and B. Glass, 247–309. Albany: State University of New York Press.

Poux, C., P. Chevret, D. Huchon, W. W. De Jong, and E. J. P. Douzery. 2006. "Arrival and Diversification of Caviomorph Rodents and Platyrrhine Primates in South America." *Systematic Biology* 55:228–44.

Prevosti F., and U. F. J. Pardiñas. 2009. "Comment on 'The Oldest South American Cricetidae (Rodentia) and Mustelidae (Carnivora): Late Miocene Faunal Turnover in Central Argentina and the Great American Biotic Interchange', by D. H. Verzi and C. I. Montalvo [*Palaeogeography, Palaeoclimatology, Palaeoecology* 267 (2008), 284–91]." *Palaeogeography, Palaeoclimatology, Palaeoecology* 280:543–47.

Rage, J. C. 1978. "Une Connexion Continentale Entre Amérique du Nord et Amérique du Sud au Crétacé Superieur? L'exemple des Vertébrés Continentaux." *Compte rendu sommaire de Scéances de la Société Géologique de France* 6:281–85.

Rauhut, O. W. M., T. Martin, E. Ortiz-Jaureguizar, and P. F. Puerta. 2002. "A Jurassic Mammal from South America." *Nature* 416:165–8.

Ré, G., S. E. Geuna, and J. F. A. Vilas. 2010. "Paleomagnetism and Magnetostratigraphy of the Sarmiento Formation (Eocene-Miocene) at Gran Barranca, Chubut, Argentina." In *The Paleontology of Gran Barranca: Evolution and Environmental Change Through the Middle Cenozoic of Patagonia*, edited by R. H. Madden, A. A. Carlini, M. G. Vucetich, and R. F. Kay, 32–45. New York: Cambridge University Press.

Reig, O. A. 1981. "Teoría del Origen y Desarrollo de la Fauna de Mamíferos de América del Sur." *Monographiae Naturae (Museo Municipal de Ciencias Naturales de Mar del Plata)* 1:1–162.

Rich, T. H. 2008. "The Palaeobiogeography of Mesozoic Mammals: A Review." *Arquivos do Museu Nacional do Rio de Janeiro* 66:231–49.

Rougier, G. W. 1992. "*Vincelestes neuquenianus* Bonaparte (Mammalia, Theria), un Primitivo Mamífero del Cretácico Inferior de la Cuenca Neuquina." Unpublished PhD diss., Facultad de Ciencias de Ciencias Exactas y Naturales, Universidad de Buenos Aires.

Rougier, G. W., and S. Apesteguía. 2004. "The Mesozoic Radiation of Dryolestoids in South America: Dental and Cranial Evidence." *Journal of Vertebrate Paleontology* 24:106A.

Rougier, G. W., L. Chornogubsky, S. Casadío, N. Páez Arango, and A. Giallombardo. 2009. "New Mammals from the Allen Formation, Late Cretaceous, Argentina." *Cretaceous Research* 30:223–38.

Rougier, G. W., A. M. Forasiepi, R. V. Hill, and M. J. Novacek. 2009. "New Mammalian Remains from the Late Cretaceous La Colonia Formation, Patagonia, Argentina." *Acta Palaeontologica Polonica* 54:195–212.

Rougier, G. W., A. M. Forasiepi, A. G. Martinelli, and M. J. Novacek. 2007. "New Jurassic Mammals from Patagonia, Argentina: A Reappraisal of Australosphenidan Morphology and Interrelationships." *American Museum Novitates* 3566:1–54.

Rougier, G. W., A. Garrido, L. Gaetano, P. Puerta, C. Corbitt, and M. J. Novacek. 2007. "First Jurassic Triconodont from South America." *American Museum Novitates* 3580: 1–33.

Rougier, G. W., M. J. Novacek, R. Pascual, J. N. Gelfo, and G. Cladera. 2000. "New Late Cretaceous Mammals from Argentina and the Survival of Mesozoic Lineages in the Patagonian Early Tertiary." *Journal of Vertebrate Paleontology* 20:65A.

Rowe, T. B., T. H. Rich, P. Vickers-Rich, M. Springer, and M. O. Woodburne. 2008. "The

Oldest Platypus and its Bearing on Divergence Timing of the Platypus and Echidna Clades." *Proceedings of the National Academy of Sciences, USA* 105:1238–42.

Sallam, H. M., E. R. Seiffert, M. E. Steiperc, and E. L. Simons. 2009. "Fossil and Molecular Evidence Constrain Scenarios for the Early Evolutionary and Biogeographic History of Hystricognathous Rodents." *Proceedings of the National Academy of Sciences, USA* 106:16722–7.

Sanders, A. E. 2002. "Additions to the Pleistocene Mammal Faunas of South Carolina, North Carolina, and Georgia." *Transactions of the American Philosophical Society* 92:1–152.

Scarano, A. C. 2009. "El Proceso de Desarrollo de la Hipsodoncia Durante la Transición Eoceno-Oligoceno. El caso de los Ungulados Autóctonos del Orden Notoungulata [Mammalia]." Unpublished PhD diss., Facultad de Ciencias Naturales y Museo de La Plata, La Plata.

Scotese, C. R. 2001. "Atlas of Earth History, vol. 1, Paleogeography, PALEOMAP Project, Arlington, Texas." Available at http://www.scotese.com.

Scotese, C. R., A. J. Boucot, and W. S. McKerrow. 1999. "Gondwanan Palaeogeography and Palaeoclimatology." *Journal of African Earth Sciences* 28:99–114.

Sigé, B. 1968. "Dents de Micromammifères et Fragments de Coquilles d´oeufs de Dinosauriens Dans la Faune de Vertébrés du Crétacé Supérieur de Laguna Umayo (Andes Péruviennes)." *Comptes Rendus Hebdomadaires des Séances de l´Académie des Sciences, Paris* 273:2479–81.

———. 1971. "Les Didelphoidea de Laguna Umayo (Formation Vilquechico, Crétacé Supérieur, Pérou), et le Peuplement Marsupial d´Amérique du Sud." *Comptes Rendus de l'Académie des Sciences, Paris* 273:2479–81.

Sigé, B., T. Sempere, R. F. Butler, L. G. Marshall, and J.-Y. Crochet. 2004. "Age and Stratigraphic Reassessment of the Fossil-Bearing Laguna Umayo Red Mudstone Unit, SE Peru, from Regional Stratigraphy, Fossil Record, and Paleomagnetism." *Geobios* 37: 771–94.

Simpson, G. G. 1950. "History of the Fauna of Latin America." *American Scientist* 38:361–89.

———. 1953. *Evolution and Geography: An Essay on Historical Biogeography with Special Reference to Mammals*. Corvallis: Condon Lectures, Oregon State System of Higher Education.

———. 1980. *Splendid Isolation: The Curious History of South American Mammals*. New Haven: Yale University Press.

Soria, M. F. 2001. "Los Proterotheriidae (Litopterna, Mammalia): Sistemática, Origen y Filogenia." *Monografías del Museo Argentino de Ciencias Naturales de Buenos Aires* 1:1–167.

Stehli, F. G., and S. D. Webb, eds. 1985. *The Great American Biotic Interchange*. New York: Plenum Press.

Verzi, D. H., and C. I. Montalvo. 2008. "The Oldest South American Cricetidae (Rodentia) and Mustelidae (Carnivora): Late Miocene Faunal Turnover in Central Argentina and the Great American Biotic Interchange." *Palaeogeography, Palaeoclimatology, Palaeoecology* 267:284–91.

Vucetich, M. G., E. C. Vieytes, M. E. Pérez, and A. A. Carlini. 2010. "The Rodents from La Cantera and the Early Evolution of Caviomorphs in South America." In *The Paleontology* of *Gran Barranca: Evolution and Environmental Change Through the Middle Cenozoic of Patagonia*, edited by R. H. Madden, A. A. Carlini, M. G. Vucetich, and R. F. Kay, 189–201. New York: Cambridge University Press.

Webb, S. D. 1976. "Mammalian Faunal Dynamics of the Great American Biotic Interchange." *Paleobiology* 2:230–4.

———. 1985. "Late Cenozoic Mammal Dispersals Between the Americas." In *The Great American Biotic Interchange*, edited by F. G. Stehli and S. D. Webb, 357–86. New York: Plenum Press.

———. "Ecogeography and the Great American Interchange." *Paleobiology* 17:266–80.

Woodburne, M. O. 2004. "Global Events and the North American Mammalian Biochronology." In *Late Cretaceous and Cenozoic Mammals of North America: Biostratigraphy and Geochronology*, edited by M. O. Woodburne, 315–44. New York: Columbia University Press.

———. 2010. "The Great American Biotic Interchange: Dispersals, Tectonics, Climate, Sea-Level and Holding Pens." *Journal of Mammalian Evolution* 17:245–64.

Woodburne, M. O., and J. A. Case. 1996. "Dispersal, Vicariance, and the Late Cretaceous to Early Tertiary Land Mammal Biogeography from South America to Australia." *Journal of Mammalian Evolution* 3:121–62.

Woodburne, M. O., A. L. Cione, and E. P. Tonni. 2006. "Central American Provincialism and the Great American Biotic Interchange." In *Advances in Late Tertiary Vertebrate Paleontology in Mexico and the Great American Biotic Interchange*, edited by O. Carranza-Castaneda and E. H. Lindsay, 73–101. Mexico City: Universidad Nacional Autonoma de México, Instituto de Geología y Centro de Geociencias, Publicación Especial 4.

Woodburne, M. O., G. F. Gunnell, and R. S. Stucky. 2009a. "Climate Directly Influences Early Eocene Mammal Faunal Dynamics in North America." *Proceedings of the National Academy of Sciences,USA* 106:13399–403.

———. 2009b. "Land Mammal Faunas of North America Rise and Fall During the Early Eocene Climatic Optimum." *Annals of the Denver Museum of Natural History* 1:1–74.

Woodburne, M. O., T. H. Rich, and M. S. Springer. 2003. "The Evolution of Tribospheny and the Antiquity of Mammalian Clades." *Molecular Phylogeny and Evolution* 28:360–85.

Wyss, A. R., J. J. Flynn, M. A. Norell, C. C. Swisher, III, R. Charrier, M. J. Novacek, and M. C. McKenna. 1993. "South America's Earliest Rodent and Recognition of a New Interval of Mammalian Evolution." *Nature* 365:434–7.

Zachos, J., M. Pagani, L. Sloan, E. Thomas, and K. Billups. 2001. "Trends, Rhythms, and Aberrations in Global Climate 65 Ma to present." *Science* 292:686–93.

Zavattieri, A. M. 2002. "Aspectos Biogeográficos y Paleoclimáticos de las Sucesiones Triásicas de Argentina, en Base a Registros Palinológicos." IANIGLA, 1973:203–7.

Cenozoic Andean Faunas
Shedding New Light on South American Mammal Evolution, Biogeography, Environments, and Tectonics

John J. Flynn, Reynaldo Charrier, Darin A. Croft, and Andre R. Wyss

Abstract

For most of the past 200 years, knowledge of South American fossil mammals was derived largely from Argentine lowland, high-latitude sites. The continent's mammalian record, therefore, was not only highly skewed geographically, but it also contained several important temporal gaps. A few dramatic faunal changes were traditionally seen as punctuating an otherwise steady series of evolutionary and environmental transitions (G. G. Simpson's "Three-Stratum" concept). There is growing evidence, however, that the actual pattern is far more complex, with Cenozoic mammal faunas responding to biogeographic, climatic, tectonic, sea level, ecologic, and environmental changes. Data from previously unsampled regions have clarified regional and continental patterns of faunal change. Notable Tertiary mammal faunas uncovered in the Andes in recent decades include: Tiupampa (early Paleocene, Bolivia); several from central Chile (Eocene-middle Miocene); Salla and Moquegua (early Oligocene, Bolivia); others from diverse sites in Bolivia (Miocene-Pliocene); two from Chilean Patagonia and Altiplano (early Miocene); and La Venta (middle Miocene, Colombia). In contrast to many lowland temperate sequences, Andean faunas are often precisely dated because of associated volcanics. Here we review the late Mesozoic to middle Cenozoic mammal record, including: (1) new Andean faunas (and other key assemblages); (2) major taxonomic groups of South American mammals (early nontribosphenic and nontherian forms, as well as Cenozoic monotremes, marsupials, xenarthrans, the endemic "ungulates" [basal forms plus Xenungulata, Astrapotheria, Pyrotheria, Litopterna, and Notoungulata], platyrrhine primates, and caviomorph rodents); and (3) their broad continental paleoenvironmental context. In addition we discuss biogeographic, paleoenvironmental, and tectonic implications of early to middle Cenozoic Andean mammal faunas, emphasizing new assemblages across the length of Chile, including a series of midlatitude sites spanning considerable time (Eocene to mid-Miocene) and space (more than 5° of latitude). For example, the ~31.5 Ma Tinguiririca Fauna, representing a new South American Land Mammal Age, contains the oldest known caviomorph rodents and provides the earliest global evidence of open grassland habitats. Evidence from the Tinguirirican caviomorphs and the 20.1 Ma platyrrhine primate *Chilecebus* support African origins for these immigrant clades.

4.1 Introduction

South American Cenozoic mammals have long figured prominently in studies of evolutionary patterns and biogeography. Yet for almost 200 years, since the startling discoveries of giant ground sloths and other unusual fossils by early naturalists and explorers, including Charles Darwin (e.g., Fernicola, Vizcaíno, and De Iuliis 2009), knowledge of South American fossil mammals has been dominated by the remarkable, but hiatus-riddled record from Argentine lowland, temperate, and high-latitude sites. Although some aspects of South American mammal history are known in considerable detail, more than a century of intensive study has failed to close a number of critical temporal and geographic gaps in the paleontological record—leaving us with an overall picture highly biased toward one small portion of the continent. In addition, parts of the South American Cenozoic land mammal sequence (particularly its earlier portion) remain poorly controlled chronologically. Recent recovery of extensive assemblages spanning much of the Andean Cordillera, from the modern tropics to the high latitudes (in Chile, Peru, Bolivia, Colombia, and Ecuador), often in association with other chronologic data, have broadened the geographic, temporal, elevational, and paleoenvironmental scope of this record (figs. 4.1–4.2). Here we review this emerging Andean record, with an emphasis on the many new early middle Cenozoic faunas from the central Chilean Main Range, as these assemblages are of growing importance in studies of the evolution of South American biotas and environments as well as the tectonics of the Andean Cordillera (figs. 4.3–4.4).

Geodynamic and paleontological evidence indicates that plate tectonic movements entirely, or at least largely, isolated continental South America as an island for most of the past 80 My, since its split from Africa (with a complete deep-water barrier as early as 100 Ma [Eagles 2007] and later separation from Antarctica) during the fragmentation of Gondwana, until its reconnection to North America through final emergence of the Central American isthmus approximately 3.5 Ma. Some 70 years ago, George Gaylord Simpson (1940) suggested that the history of South American mammals is divisible into three phases (or "strata"). Simpson's concept envisioned the emplacement of archaic faunas and floras in South America prior to what we now know was the complete fragmentation of Gondwana (Simpson formulated this concept within a stabilist framework that predated plate tectonics). Subsequent isolation fostered the evolution of endemic archaic forms during the early Cenozoic (Simpson's Stratum 1, between roughly 65–40 Ma) characterized by marsupials,

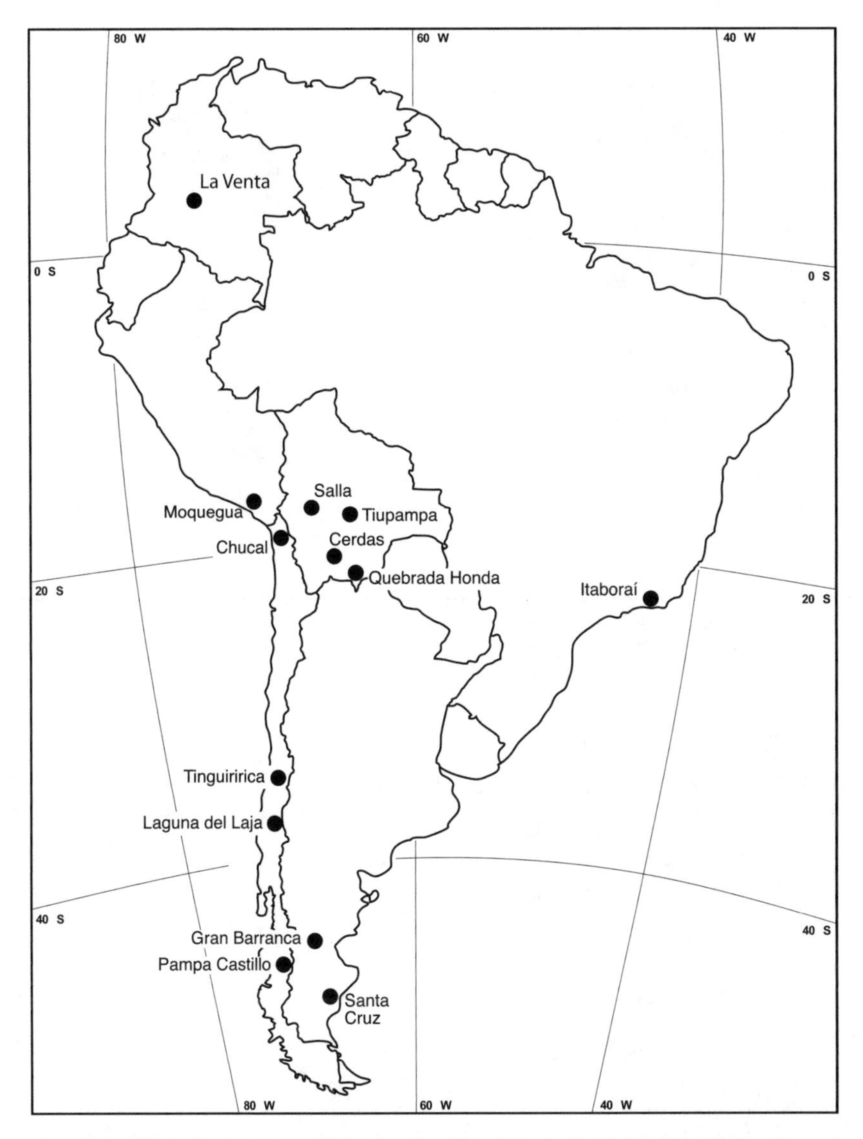

Figure 4.1 Map indicating mammal-bearing fossil localities in the Andes of South America and other key localities referred to in the text.

placental ungulates, and edentates. Stratum 2 is marked by the modernization of many of these endemic lineages, particularly notoungulates, and the appearance of two immigrant groups, caviomorph rodents and platyrrhine primates by overwater "sweepstakes" dispersal. Finally, beginning about 10 Ma and accelerating after 3.5 Ma, Stratum 3 heralded the establishment of South

MA	Epoch	SALMA
0	PLEIS	Lujanian Ensenadan Marplatan Chapadmalalan
	PLIO	
5	MIOCENE	Montehermosan
		Huayquerian
10		? Chasicoan ?
		Mayoan
		Laventan
15		Colloncuran
		?Friasian
		Santacrucian
		????
20		Colhuehuapian
		????
25	OLIGOCENE	Deseadan
30		Tinguiririca
		????

STRATUM 3

Northern invaders and Great American Interchange

- Many non-native groups enter S. America: deer, camels, tapirs, bears, cats, raccoons horses, gomphothere elephants
- Native groups enter Central & N. America
- Extinction of notoungulates and litopterns

STRATUM 2

Primates and rodents arrive, modernization of ancient lineages

- Increased hypsodont (high-crowned) teeth in many mammal lineages; earliest evidence for open, grassland habitats in South America, 15-20 million years earlier than on other continents
- Peak diversity of notoungulates in the late Oligocene
- Hypselodont notoungulates common and widespread during Miocene
- First appearance of large notoungulates
- Extinction of many notoungulate groups, like notostylopids, isotemnids, archaeohyracids, homalodotheriids, and leontiniids
- Extinction of astrapotheres and pyrotheres

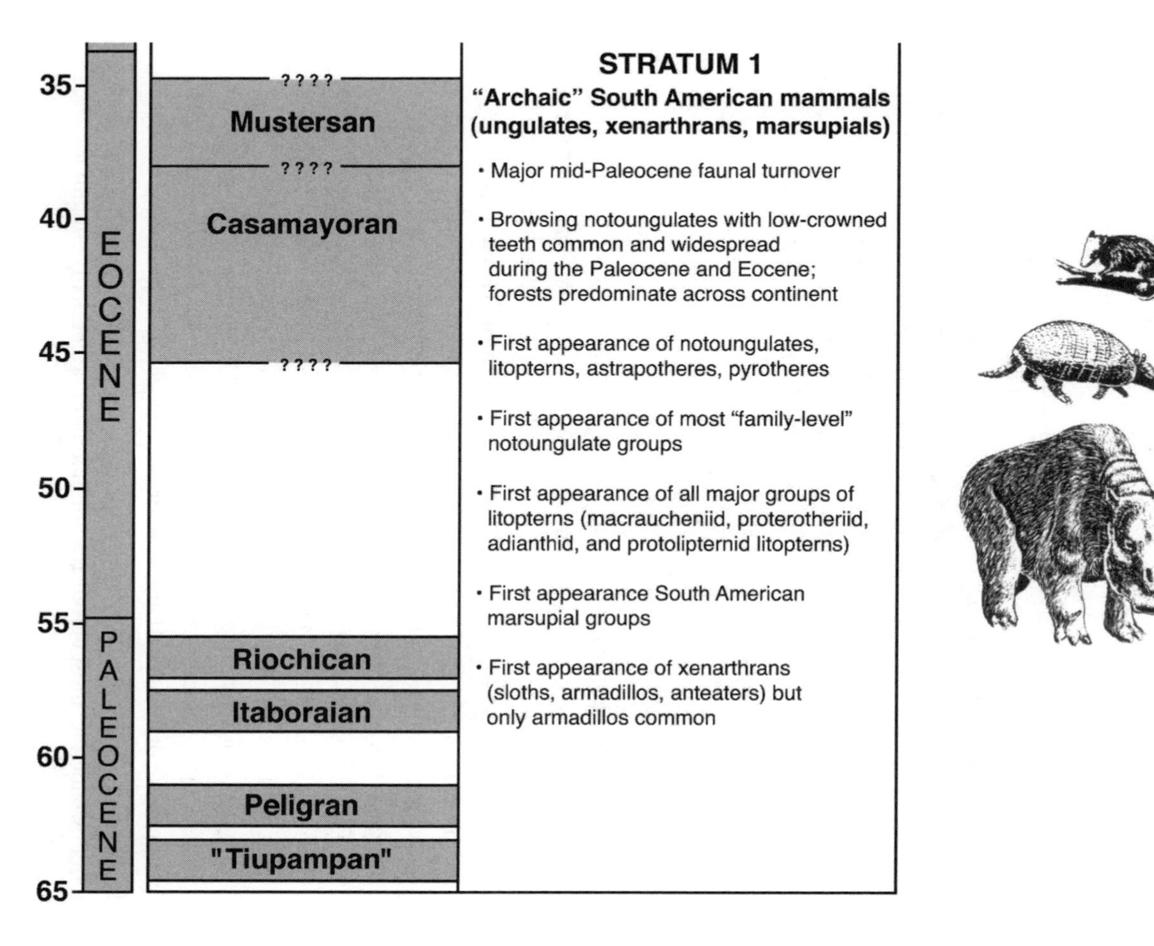

35
40
45
50
55
60
65
EOCENE
PALEOCENE
????
Mustersan
????
Casamayoran
????
Riochican
Itaboraian
Peligran
"Tiupampan"
STRATUM 1
"Archaic" South American mammals (ungulates, xenarthrans, marsupials)
• Major mid-Paleocene faunal turnover
• Browsing notoungulates with low-crowned teeth common and widespread during the Paleocene and Eocene; forests predominate across continent
• First appearance of notoungulates, litopterns, astrapotheres, pyrotheres
• First appearance of most "family-level" notoungulate groups
• First appearance of all major groups of litopterns (macraucheniid, proterotheriid, adianthid, and protolipternid litopterns)
• First appearance South American marsupial groups
• First appearance of xenarthrans (sloths, armadillos, anteaters) but only armadillos common

Figure 4.2 *(previous pages)* Timescale of South American Land Mammal Ages (SALMA) summarizing the timespans, major events, and representative faunal composition of the "3-stratum" model of Cenozoic mammalian faunal succession in South America. Strata 1–3 are calibrated to the timescale on the left, but the events listed within each stratum are not (representing only a summary of events within the interval). The FA and LA represent First Appearance and Last Appearance, respectively. The SALMA biochronology was modified from Flynn and Swisher (1995), Flynn et al. (2003), and Croft, Flynn, and Wyss (2008); the four youngest SALMAs have been consolidated to enhance legibility. (Animals illustrated in each stratum may be living or fossil forms, and only represent examples of the groups that first appeared during that time interval.)

America's modern mammal fauna, culminating with a continuous overland connection to North America marking the beginning of the Great American Biotic Interchange (GABI). Each of these episodes contributed key components of the modern biota, and while some ancient lineages have modern descendants, others are entirely extinct.

Although Simpson's three-stratum scenario remains a reasonable first-order summary of key events, it is incomplete. As detailed later, an intervening half-century of discoveries from throughout South America—and especially the Andes—have clarified details about the timing of evolutionary and geological events, and dramatically altered our understanding of the patterns and causes of biotic and geologic change in South America.

4.2 The Setting: Temporal Sampling

The Andes have yet to provide much insight into the evolution of South American Mesozoic mammals, as all but one South American Mesozoic site are located in the lowlands of Patagonia and Brazil. The oldest sites have yielded few mammal species to date—two tribosphenic taxa from the Jurassic/Cretaceous of Argentina (*Asfaltomylos patagonicus*, *Henosferus molus*) and *Vincelestes* from the Early Cretaceous (~130 Ma) La Amarga Formation (see Rauhut et al. 2002; Rougier et al. 2007; and summary in Rougier et al. 2009). In the latest Cretaceous, a few isolated mammal teeth of uncertain affinities have been reported from localities outside Argentina, such as in the Bolivian Andes (Gayet et al. 2001) and Brazil (Bertini et al. 1993). The Late Cretaceous (~75 Ma) Los Alamitos Fauna is the most diverse South American Mesozoic assemblage known, with teeth of some 14 mammal species reported (Bonaparte 1990). Surprisingly, none of the Los Alamitos mammals appear to be therians (only gondwanatheres, triconodonts, symmetrodonts, and dryolestoids) and the fauna is highly endemic, most closely resembling older rather than contemporaneous faunas from Laurasia. Two other small Argentine Late Cretaceous mammal assemblages—La Colonia (which has produced a docodont [previously considered a dryolestoid], Pascual et al. 2000) and Cerro Tortuga (producing small dryolestoids, including mesungulatids, and ferugliotheriid gondwanatheres; Rougier et al. 2009)—also resemble the fauna from Los Alamitos.

Pre-latest Paleocene mammalian assemblages are rare; the three most important come from the Bolivian Andes (Tiupampa; Muizon 1991), southern Argentina (Punta Peligro; Bonaparte, Van Valen, and Kramartz 1993), and Brazil (Itaboraí; Bergqvist, Abrantes, and Avilla 2004; Paula Couto 1952). Collectively these sites provide insights into the early Cenozoic diversity of mammals in the mid-high latitudes, even though debate continues about their numerical and relative ages. Noteworthy occurrences include a diverse array of early marsupials, a somewhat unexpected assemblage of placental ungulates, the persistence of gondwanatheres (otherwise reported only from Cretaceous [India, Madagascar] or Eocene [Antarctica] deposits in other Gondwanan landmasses), and substantial differences in early Paleocene mammalian communities between tropical and temperate regions as well as possibly montane versus lowland regions (Pascual 2006; Pascual and Ortiz-Jaureguizar 1990, 2007; Pascual, Ortiz-Jaureguizar, and Prado 1996; see section The Setting: Taxa later for details). The marked diversity of marsupials (Muizon 1991, 1998) and near absence of most characteristic native South American ungulate groups (Muizon and Cifelli 2000) in the early Cenozoic may suggest that marsupials arrived and diversified in South America before most of the placental groups that characterized younger time intervals. It is also noteworthy that the most recently discovered early Cenozoic faunas (Tiupampan and Peligran SALMAs) have yet to record xenarthrans (Gelfo et al. 2009).

An exceptional series of Andean faunas has been recovered over the past few decades from Eocene to late Miocene deposits in Chile, the Bolivian Altiplano, and in other parts of the Andes (summarized in Flynn 2002). In contrast to most lowland temperate sequences, many of these can be precisely dated because of their Andean setting and consequent abundance of volcaniclastics. Perhaps the best-sampled temporal interval in the Andes is the early to late Miocene (~20–10 Ma); important Andean assemblages complementing the classical record from foreland and coastal Argentina are now known from southern Chile (47°S, Pampa Castillo Fauna; Flynn, Novacek, et al. 2002) to north-central Colombia (3°N, La Venta Fauna; Kay et al. 1997), with intervening Andean Miocene sequences throughout southern-central Chile (Abanico Formation of the Maipo, Las Leñas, Teno, and Upeo rivers, and equivalents at

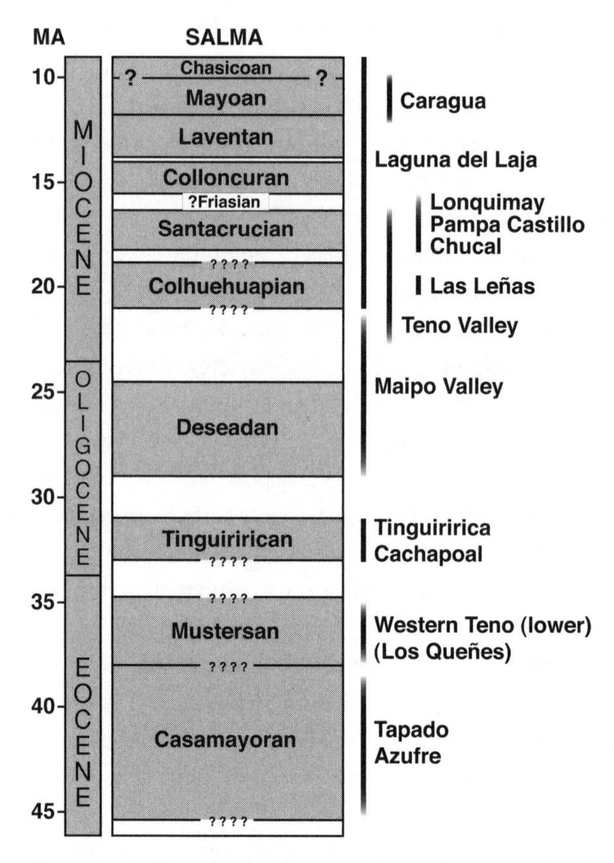

Figure 4.3 Estimated ages of mammal faunas from the Andes of central Chile, correlated to the South American Land Mammal Age (SALMA) biochronology. Sources as in fig. 4.2.

Lonquimay and Laguna del Laja [see references in Flynn et al. 2008]) and the Chilean Altiplano (Chucal, Santacrucian SALMA [Croft, Flynn, and Wyss 2007; Flynn, Croft, et al. 2002]). The La Venta and Pampa Castillo and Chucal faunas are briefly summarized here, while the new series of Andean assemblages from central Chile are detailed in a subsequent section.

The northernmost well-represented Andean fauna, the middle Miocene (13.5–11.8 Ma) La Venta Fauna from Colombia, includes more than 70 mammal species (and roughly the same number of other vertebrates) and documents marked biotic provinciality between tropical and temperate assemblages comparable to that among modern faunas (Kay et al. 1997). La Venta also marks the oldest definitive presence of many modern mammal groups, including true opossums (didelphids), dasypodin and tolypeutin dasypodids (armadillos), most major New World primate subgroups, and several bat lineages. The fauna and associated geochronologic studies document a number of tectonic events, including pronounced basin subsidence shortly before 13.5 Ma, a short and rapid phase of uplift and volcanism in the Central Cordillera at ~12.9 Ma coincident with initiation of uplift in the Eastern Cordillera (continuous by 11.9 Ma), and major depositional changes in the intervening Magdalena River basin. Paleoecological analyses of the La Venta sequence indicate a mix of varied forest and more open biotopes throughout, rather than a purely closed canopy habitat, as might be expected for an equatorial fauna of this age (Kay and Madden 1997; Kay et al. 1997).

At the southern extreme, the diverse late early Miocene Pampa Castillo Fauna of Chile contains more than 40 taxa and is the westernmost high-latitude mammal fauna known from South America (Flynn, Novacek, et al. 2002; assignable to the Santacrucian SALMA, ~16–19 Ma [16.3–17.5 Ma in Flynn and Swisher 1995, extended to 19 Ma based on the Andean Chucal Fauna of northern Chile by Croft, Flynn, and Wyss 2007]). Similarity of the Pampa Castillo assemblage to coastal Santacrucian faunas (Tauber 1999) suggests similar conditions (increasing aridity, extensive grasslands, but continued presence of significant riparian forests and woodlands) across much of Patagonia. The mammal-bearing unit transitionally overlies fossiliferous marine strata, documenting the northwestern-most extent of the last significant incursion of southern Atlantic epicontinental seaways. Uplift in this region is thus constrained to have begun by 16 Ma, likely in response to migration of the nearby Chile Margin Triple Junction.

The Chucal Fauna (Croft, Flynn, and Wyss 2007; Flynn, Croft, et al. 2002), recovered from a series of sites at ~4500 m present-day elevation, clarifies the age of the hosting unit, timing of deformation, and initiation of major uplift in the Chilean Altiplano. These deposits likely accumulated in an intermontane, but not yet extremely elevated basin, with the spectrum of mammal taxa, a frog, and leaf macrofossils indicating a mixed habitat of grasslands and significant lake and riverine habitats flanked by forests and woodlands. A distinctive hoofed herbivore group (mesotheres) is abundant and diverse here and in slightly younger faunas from the Bolivian Altiplano, but is unrecorded elsewhere during this interval (Patagonia, La Venta). Chinchilline rodents (a mostly montane group today) are also abundant at Chucal, predating the next-youngest occurrence of the group by some 15 Ma. Miocene faunas from the Altiplano of Chile and Bolivia are similar compositionally, but differ markedly from contemporaneous faunas in both the low-latitude tropics (Colombia) and the high latitudes (Patagonia)—these patterns almost certainly reflect pronounced regional provinciality at this time, which in turn may have resulted from broader scale climate changes, Andean tectonics (initiation of substantial uplift), and associated paleoenvironmental changes.

No early Miocene sites contemporaneous with Pampa Castillo and Chucal have yet been recognized from the Bolivian Altiplano, but the middle Miocene is represented there by several well-characterized localities (Croft 2007; Croft et al. 2009; MacFadden et al. 1994; Marshall and Sempere 1991). Cerdas, a locality at 4000 m near the eastern edge of the Altiplano, and Nazareno, a locality at 3200 m from the Eastern Cordillera, likely pertain to the early middle Miocene. They differ faunistically both from contemporaneous sites in Patagonia and from Chucal, likely the combined result of geographic provinciality and sampling distinct temporal intervals. Quebrada Honda, the best-sampled Bolivian Miocene locality, lies at 3500 m in the far southern tip of the Eastern Cordillera. This late middle Miocene site is contemporaneous with La Venta, but has almost no species in common, a testament to the significant differences between low- and middle-latitude faunas by this time.

4.3 The Setting: Taxa

The summary by Pascual, Ortiz-Jaureguizar, and Prado (1996) provides the most recent comprehensive compendium of temporal distributions for South American Cenozoic mammal genera, and is the source for the ranges cited later, except where noted. Pascual and Ortiz-Jaureguizar (2007) provided additional discussions of temporal, biogeographic, and paleoecological implications of Mesozoic-Cenozoic South American mammals, while Gelfo et al. (2009) detailed important updates to knowledge of early Cenozoic temporal distributions. Here we briefly review the major clades of South American mammals from the Mesozoic and Cenozoic Strata 1 and 2, emphasizing new insights derived from the study of Andean sequences.

Cretaceous mammals: As noted earlier, the few known South American Cretaceous faunas are quite distinctive in their taxonomic composition, some resembling older Laurasian faunas, others sharing some higher-level taxa with other Gondwanan assemblages, and with many taxa being endemic to South America. Although they cautioned that small pre-Late Cretaceous samples and temporal gaps may exaggerate faunal distinctiveness, Rougier and colleagues (2009, 223) concluded that: "The Late Cretaceous nontribosphenic mammals have no clear link with the Jurassic and Early Cretaceous South American mammals, emphasizing the distinctiveness and episodic nature of the Mesozoic South American mammalian assemblages." Gelfo and Pascual (2001) suggested that dryolestoids persisted into the early Paleocene in Patagonia, based on reinterpretation of the affinities of *Peligrotherium* (previously considered a condylarth).

Monotremata: A monotreme, *Monotrematum sudamericanum*, from the early Paleocene of Patagonia, is one of the most unexpected fossil mammal discoveries in South America in recent years (Pascual et al. 1992). Resembling the Australian Miocene platypus *Obdurodon*, *Monotrematum* demonstrates that monotremes ranged more broadly across Gondwana than previously supposed.

Marsupialia: The oldest known marsupial from South America is the recently described *Cocatherium* of Danian-equivalent (early Paleocene) age from Chubut, Argentina (Goin et al. 2006). The growing early Cenozoic record from Bolivia, Brazil, and Patagonia documents an exceptional diversity of marsupials, including early representatives of endemic groups, as well as microbiotheres, considered related to Australian forms (Case, Goin, and Woodburne 2005). Tiupampa in the Bolivian Andes has yielded remarkably complete fossils, including the oldest known skulls and skeletons for any South American marsupial (e.g., Muizon 1991, 1998; Muizon, Cifelli, and Paz 1997). Material from Tiupampa includes predatory borhyaenoids (two taxa), an array of the somewhat more anatomically generalized didelphoids (five taxa), and *Andinodelphys* (Ameridelphia *incertae sedis* in Case, Goin, and Woodburne 2005).

Xenarthra: Relatively complete material of South American xenarthrans is first known from the Eocene (teeth, jaws, and a partial skull and skeleton of a Casamayoran armadillo; Simpson 1932). Less informative but still diagnostic armadillo specimens (isolated osteoderms) occur in the middle late Paleocene of Brazil and Patagonia (Itaboraian, Riochican; Bergqvist, Abrantes, and Avilla 2004). The occurrence of middle late Paleocene armadillos indicates that the other branch of Xenarthra (Pilosa: anteaters and sloths) must have existed by then as well, although their fossils are not known until much later. Crown-clade sloths are unknown before the Deseadan (~25 Ma; Carlini and Scillato-Yané 2004; Pujos and De Iuliis 2007; Shockey and Anaya 2008). The range of *Pseudoglyptodon*, the nearest outgroup to sloths, now extends to the Tinguiririican (~32 Ma) based on a skull from Chile, and possibly to the Mustersan (perhaps 34–36 Ma; McKenna, Wyss, and Flynn 2006). Among anteaters, only the living genus *Myrmecophaga* has an extensive fossil record, extending to the Montehermosan (4–6.8 Ma), and representatives of the anteater clade are unknown prior to 17 Ma. The scarcity of early cingulates and their current lack of a pre-Itaboraian record, as well as the lack of pre-Mustersan record of pilosans, are curious. Early representatives of these groups are either unrecorded in presently sampled regions due to some preservational bias, or inhabited regions not yet yielding fossils. The first explanation becomes increasingly less likely as more fossils from multiple time slices are recovered from well-sampled areas like Patagonia. The second explanation, widely accepted by paleontologists, is consistent with the high diversity of xenarthrans in the modern tropics and the correspondingly poor early Cenozoic record of that region. The second explanation remains plausible, although good early Cenozoic samples from several other continents (North America, Europe, and Asia) lack definitive occurrences of xenarthrans (Rose et al. 2005). It is also possible that xenarthrans originated after the earliest Cenozoic, contradicting traditional views (and some molecular phylogenies; cf. Murphy et al. 2001) that place them among the earliest diverging placental lineages.

Endemic "ungulates:" Early South American "ungulates" anatomically ranged across very generalized, small-bodied forms (didolodontid condylarths), the earliest litopterns (later differentiating into several anatomically specialized subgroups), several lineages of Notoungulata (later to become the most diverse of the South American "ungulates"), and two of the three early "ungulate" giants — the Astrapotheria and Xenungulata (the third group, Pyrotheria, does not appear until the Casamayoran). Neither "ungulates" nor condylarths appear to represent monophyletic groups, hence the use of quotation marks initially (but for simplicity's sake, not hereafter).

Xenungulata: The enigmatic Xenungulata, a small, short-lived group of ungulates, includes just three genera. *Carodnia* is known only from the Itaboraian in both Patagonia, where it was first discovered, and Brazil, where it is better represented. This large animal, with simple, ridged teeth, was likely amphibious in habits. Villarroel (1987) reported a second poorly known form, *Etayoa*, of uncertain age (questionably middle Paleocene) from the tropics of Colombia, and Gelfo, López, and Bond (2008) described *Notoetayoa* from the same beds as *Carodnia* in Patagonia.

Astrapotheria: Astrapotheres, with upper and lower tusks and high-crowned but simple teeth, are usually reconstructed with an elephant-like, albeit smaller, trunk because of features of the skull that are common in living mammals with a proboscis. Their body shape and simple teeth indicate that they too may have been semiaquatic or amphibious, although their high-crowned teeth provide a confusing counterindication. Appearing first in the Itaboraian (Paleocene or early Eocene; Gelfo et al. 2009), astrapotheres ranged widely in space, but became extinct by the mid-late Miocene.

Pyrotheria: Pyrotheres, the third large-bodied group of native South American mammals, remain among the least understood of the early native ungulates. They appeared in the Casamayoran and persisted at low species diversity and individual abundance through the late Oligocene (Deseadan, ~24–29 Ma). Their large size and distinctive dental attributes, readily identifiable in the field, account for them being the name bearer for the Deseadan, which Ameghino originally called the "*Pyrotherium* beds." Pyrotheres are known from Venezuela, Colombia, Peru, Bolivia, and Argentina (Salas, Sánchez, and Chacaltana 2006).

Litopterna: Litopterns were a group of native ungulates second only to the notoungulates in numbers of species and major clades. The two major lineages (proterotheres and macraucheniids) include some remarkably specialized forms. Litopterns possess a distinctive ankle joint, and while they also later developed characteristic dental specializations, these were not evident in the earliest members of the lineage (Cifelli 1983). Litopterns have not yet been recorded from the oldest known Cenozoic deposits (Tiupampa), but are well represented in the slightly younger Itaboraian assemblages from Brazil; they persisted until the Lujanian (0.01–0.8 Ma). Although the earliest litopterns include both proterotheres and macraucheniids, the group did not diversify in numbers of species or distinctive anatomical specializations until much later, peaking during the Miocene and being especially prominent in Miocene Santacrucian assemblages. Proterotheres included several forms that became almost completely monodactyl; toe reduction is typically associated with cursorial, open habitat animals such as horses. Macraucheniids typically were larger than proterotheres, had more complex teeth, and retained a less specialized three-toed foot with robust side toes. Retracted nasal bones suggest the presence of a proboscis.

Notoungulata: By far the best-known group of South American native ungulates was the Notoungulata. Present from virtually the beginning of the Cenozoic record, they radiated rapidly and extensively from the Riochican onward. Interestingly, even though notoungulates became the most diverse of native ungulates, with more than 150 genera and hundreds of species recognized, they were rare in the earliest assemblages, being represented by only a single species at Tiupampa (Muizon 1991; Muizon and Cifelli 2000) and by just two forms in the younger Itaboraian. While they already had declined tremendously in diversity by the end of their temporal range, a few notoungulate lineages

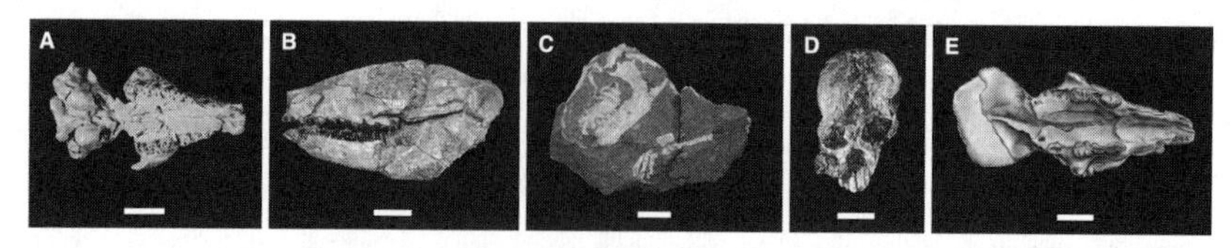

Figure 4.4 Representative mammal specimens from the Abanico Formation, Andes of central Chile. (A) cast of SGOPV 2900, skull of *Archaeotypotherium tinguiriricaense* (Archaeohyracidae, Notoungulata), Tinguiririca Fauna, in occlusal view; (B) SGOPV 2914, skull and jaws of *Santiagorothia chiliensis* (Interatheriidae, Notoungulata), Tinguiririca Fauna, in left lateral view; (C) SGOPV 3736, unprepared skull and forelimb of a typothere notoungulate, Upeo Fauna in left lateral view; (D) SGOPV 3213, skull of *Chilecebus carrascoensis* (Platyrrhini, Primates), Las Leñas Fauna, in anterior view; and (E) SGOPV 3476, skull of *Polydolops mckennai* (Polydolopidae, Marsupialia), Cachapoal Fauna, in occlusal view.

persisted through the beginnings of the Great American Biotic Interchange (GABI), with the last surviving lineage becoming extinct shortly before the end of the Pleistocene. One of these late surviving lineages, the rhino-like Toxodontidae, participated in the GABI, expanding its range into Central America. Although most early notoungulate species were quite generalized, each of the subsequent major lineages was already established by the Eocene. Notoungulates ranged from tiny to immense, evolved a wide spectrum of anatomical specializations (especially in the dentition and skull), and occupied virtually every habitat and geographic region of South America with a fossil record. Notoungulates are traditionally divided into about a dozen families. Assessment of monophyly of these groups and higher-level interrelationships are areas of active study.

Mid-Cenozoic Immigrants—Platyrrhini (Primates)and Caviomorpha (Rodentia): Primates first appear in the South American record near the beginning of Stratum 2. Following Ameghino's nineteenth century report of *Homunculus* from the Santacrucian of Patagonia, and discoveries of additional primates in the mid-1900s in Patagonia (*Tremacebus*, *Dolichocebus*) and La Venta in the Colombian Andes, additional Santacrucian as well as Colloncuran and older primates were discovered in the Bolivian Andes (*Branisella* [probably including *Szalatavus*]), Patagonia (*Carlocebus*, *Soriacebus*, *Proteropithecia*, *Killikaike*), and again in the Andes at La Venta (~one dozen taxa in total) and in Chile (*Chilecebus*; see Rosenberger et al. 2009 for summary of ages and localities for these platyrrhines and others cited below). Three taxa have also been recovered recently in the lowland tropics of Amazonia in Brazil (Acre region; estimated to be 9–6.8 Ma by Cozzuol 2006), while a number of other interesting new forms have been recovered

curred during the middle and late Cenozoic. The first, the Eocene-Oligocene transition, was characterized by the appearance of widespread open grassland/woodland habitats in higher latitudes—evidently the earliest occurrence of such biomes globally (Flynn et al. 2003)—possibly related to uplift that began during the middle Eocene Incaic tectonic phase (Charrier, Pinto, and Rodríguez 2007). The second major shift, occurring during the Pliocene and Pleistocene, included glacial-interglacial oscillations, increased aridity, pulsed shifts of grassland-forest biomes, and major faunal reorganization and extinctions associated with the GABI. Both of these phases appear to have involved extensive and relatively rapid biotic transformations. The large-scale faunal provinciality observed today, reflecting latitudinal climatic and elevational gradients, was well established by at least the middle to late Miocene (Laventan; Kay et al. 1997). New data from Andean and foreland sequences suggest that provinciality may have developed even earlier: by the early middle Miocene or late Oligocene (Bolivian-Chilean Altiplano faunas versus Patagonia, Andean-Patagonian faunas versus northeastern tropical faunas; Candela and Morrone 2003), perhaps by the Eocene, and possibly even by the early Paleocene. For example, Tinguirirican faunas of central Chile and Patagonia differ in taxonomic representation (e.g., central Chile lacks hegetotheres but possesses rodents, and a number of species are distinct between the two areas; Croft, Flynn, and Wyss 2008; Flynn et al. 2003), and Eocene faunas from northwest and west-central Argentina (e.g., Antofagasta de la Sierra, Lumbrera, and others; see Pascual and Ortiz-Jaureguizar 2007) differ from those at higher latitudes. A lengthy hiatus was long assumed to have separated the base of Simpson's Stratum 2 from the youngest faunal interval of Stratum 1 (Mustersan), thereby obscuring the rate at which this transition occurred. Discovery of the Tinguiririca Fauna (which predates the Deseadan and marks the current base of Stratum 2), coupled with the unexpected youth—late rather than early or middle Eocene—of the younger end of the Casamayoran (and hence also of the subsequent Mustersan; Kay et al. 1999), indicate that this transition was quite abrupt. It thus seems plausible to attribute this faunal shift to global tectonic (e.g., final separation of Antarctica-South America, eustatic sea level fall) and related oceanographic, atmospheric, and climatic changes across the Eocene/Oligocene transition around 34 Ma, including the well-known change from greenhouse to icehouse conditions (see Goin et al., chapter 3, this volume).

Periodically, smaller-scale changes were superimposed on the large-scale transformations. The presence of obligate humid forest-dwelling primates as far as 50°S in southern Patagonia indicates that humid forested habitats expanded poleward, perhaps several times during the middle Cenozoic and as late as the middle Miocene. Major epicontinental seaways along the Andean foreland from the north (Colombia-Venezuela), south (Argentina-Chile-Bolivia), and possibly east (southern Brazil) expanded and contracted during the Miocene (Webb 1995), and covered substantial parts of South America earlier in the Cenozoic.

4.5 Andean Insights: Fossils from Chile

Almost 2000 specimens have been recovered in almost two dozen new assemblages from volcaniclastic sediments of the Andean Main Range in central Chile over the past two decades. Spanning considerable time (Eocene–mid-Miocene) and space (nearly 5° of latitude), fossils are so ubiquitous that these deposits now represent a premier archive of mammal evolution in South America (see references in Flynn et al. 2008). The unusual depositional setting of these faunas (intermontane basins within an active volcanic arc) contributes to the remarkably high proportion of well-preserved specimens and permits the first precise radioisotopic dates for various South American Land Mammal Ages. The fossil mammals and associated dates have shown that the Abanico Formation is much younger than (early middle Cenozoic rather than late Mesozoic) and spans only a fraction of its previously assumed age (~20–40 million years, versus ~60–120 million years), and that unexpectedly complicated tectonic factors influenced Cenozoic extensional basin development in the Main Range (e.g., Charrier et al. 2002, Charrier, Flynn, Wyss, and Croft, forthcoming; Wyss, Charrier, and Flynn 1996).

The Tinguiririca Fauna, from near Termas del Flaco (35°S, 70°W), was the first mammal assemblage discovered in this sequence, and it remains one of the most important and certainly the best known (Flynn et al. 2003; Wyss et al. 1990). Some 400 specimens have been collected from the locality, representing at least 25 species of native ungulates, marsupials, xenarthrans, and rodents, the bulk of which represent new genera or species. Documenting the earliest Oligocene biochronologic interval (Tinguirirican), this assemblage includes the globally oldest mammalian herbivore fauna dominated by hypsodont taxa and suggests some faunal provinciality in southern South America by the early Oligocene. It dramatically alters understanding of the age and tectonic history of a large Andean lithosome (Abanico Formation; part of Darwin's "Porfiritica," plus its lateral equivalents). The fauna also provides key insights into environmental change in South America around the Eocene-Oligocene boundary. The

Tinguiririca Fauna includes the oldest South American rodents, a dasyproctid and chinchillid, the former of which supports an African origin for the Caviomorpha. Together, these two taxa suggest that the group's initial diversification was already well under way by 32 Ma. The fauna is also unique in documenting two groups of gliriform marsupials (groeberiids and polydolopids) co-occurring with true rodents. Paleoenvironmental analyses (hypsodonty, cenogram, macroniche rainfall estimation, ecological diversity analysis, and ankle index [relating to cursoriality]; Croft 2001; Croft, Flynn, and Wyss 2008; Flynn et al. 2003; Shockey and Flynn 2007) reveal nonanalog aspects of middle Cenozoic South American localities. Nevertheless, a moderately dry habitat was present at Tinguiririca, with relatively few trees and abundant open areas. These results also suggest that the most pronounced shift in Cenozoic South American paleoecology and paleoenvironments occurred by the earliest Oligocene (contrary to isotopic evidence indicating a lack of substantial regional climatic shift across the Eocene-Oligocene transition at the Gran Barranca in Argentine Patagonia; Kohn et al. 2004). The data from Tinguiririca provide compelling indications that open habitat, grassland-woodland environments flourished 15–20 million years earlier in South America than on other continents, likely related to climatic deterioration and associated paleoenvironmental events over the Eocene-Oligocene transition. Correlative faunas from elsewhere in Chile and in Argentina (Croft, Flynn, and Wyss 2008; Flynn and Wyss 2004; Flynn et al. 2003) document that Tinguiririca faunas were geographically widespread across southern South America. The younger (Deseadan) fauna from Salla, Bolivia (Kay et al. 1998), and a Deseadan assemblage recently reported from Moquegua, Peru (Shockey et al. 2006, 2009), expand upon previously available late Oligocene data from Patagonia and document some degree of intracontinental provinciality by the Oligocene.

The ?middle Eocene Tapado Fauna, the second mammalian assemblage discovered in the Tinguiririca drainage, as well as in the central Chilean Andes, consists of several dozen specimens, all native ungulates (Flynn et al. 2005). These include cf. *Ernestokokenia* (a didolodontid condylarth), cf. *Notonychops* (potentially a basal notoungulate), and four notoungulates: *Notostylops* sp. (Notostylopidae), *Ignigena minisculus*, (Interatheriidae), *Eohyrax* sp. (Hegetotheria), and *Pleurostylodon* sp. (Isotemnidae). The fauna will be extremely useful for understanding mammal distributions and biogeographic patterns during this interval, given the scarcity of well-sampled extra-Patagonia localities of this age—the fauna from the Lumbrera Formation of northwest Argentina being a notable exception. A faunule on the north side of the Río Tinguiririca may be as old as Paleocene. Sparse ?Eocene fossils have been recovered along the Río Azufre, just north of the Río Tinguiririca valley, including an excellent specimen of the interatheriid *Antepithecus brachystephanus* (Hitz, Flynn, and Wyss 2006). New localities discovered over the past two years along this river and an adjacent drainage (Río Los Helados) have provided additional Paleogene specimens of roughly similar age.

Four late Oligocene to Miocene assemblages have been recovered from the upper Río Maipo drainage ~100 km north of Tinguiririca (Boca Toma, Los Sapos, Estero San Francisco, and El Volcán). These, along with fragmentary bones from the Río Aconcagua drainage farther north, represent the northernmost localities within the formation. These are generally not as richly fossiliferous as some of the more southerly Abanico Formation localities, but nonetheless have produced excellent specimens of notoungulates and rodents.

Slightly further south (~60 km north of Tinguiririca), at least three new Oligocene-to-earliest-Miocene faunal assemblages were discovered along the upper drainages of the Cachapoal and Las Leñas rivers. The former may be correlative with the Tinguiririca Fauna (as noted earlier) and is important in documenting the early Oligocene presence of some groups not sampled at Tinguiririca (e.g., mesotheriid notoungulates). Las Leñas is noteworthy for producing the earliest well-preserved New World anthropoid primate skull (Flynn et al. 1995). This taxon, the 20.1 Ma *Chilecebus carrascoensis*, helped clarify the ancestral morphology of platyrrhines, pointing to an African rather than a North American ancestry for the group (paralleling the story told by Tinguirirican rodents relative to the origin of Caviomorpha). Recent morphometric analyses of *Chilecebus* indicated that its brain was unusually small, implying that modern platyrrhines and catarrhines developed large brains independently (Sears et al. 2008).

South of the Río Tinguiririca, important Eocene to early Miocene assemblages have been discovered at Lagunas de Teno, Río Vergara, Río Upeo, and Los Queñes (all within an approximately 20–40 km radius south and west of Termas del Flaco). Preliminary investigations of the first three of these faunas have identified small marsupials, several notoungulates, a litoptern, several species of rodents, various armadillos, and sloths, probably late Oligocene to early Miocene in age. These should provide an extremely useful point of comparison with early Miocene faunas from elsewhere in Chile (noted earlier). The stratigraphic section at Los Queñes is extraordinarily thick and fossiliferous, containing at least three stratigraphically superposed faunas. These were collected only recently and are not yet studied; nevertheless, the oldest of these faunas clearly predates the Tinguiririan and is probably Mustersan in age.

A lateral correlative of the Abanico Formation, the Curá-Mallín Formation contains a remarkable series of mid-late Miocene fossil mammal sites near Laguna del Laja (~300 km S of Tinguiririca; Flynn et al. 2008). These are associated with numerous high-precision 40Ar/39Ar dates indicating that the sequence spans at least 10 million years, and as many as 6 directly superposed SALMAs are represented (Colhuehuapian to Colloncuran or Laventan, with sparse occurrences in the overlying Trapa Trapa Formation perhaps extending to the Chasicoan). The Laguna del Laja succession thus rivals some of the most extensive sequences of fossil mammals in South America, including the Gran Barranca of Argentina. Notoungulates and rodents are by far the most common taxa at Laguna del Laja. The rodents have been studied in detail (Wertheim 2007; summary in Flynn et al. 2008), revealing a surprisingly high level of endemism: all 20 species, and some genera, are unique to the region. Moreover, some taxa that are abundant in nearby contemporaneous Patagonian localities, such as the dasyproctid rodent *Neoreomys*, are rare at correlative levels at Laguna del Laja. The rodents also generally appear to differ from those recovered from Miocene faunas in the Abanico Formation, only 2–4° farther north. This pronounced local endemism likely reflects a number of factors (e.g., differences in age, geography, paleoenvironment, and depositional setting).

Literature Cited

Bergqvist, L. P., É. A. L. Abrantes, and L. D. S. Avilla. 2004. "The Xenarthra (Mammalia) of São José de Itaboraí Basin (upper Paleocene, Itaboraian), Rio de Janeiro, Brazil." *Geodiversitas* 26:323–37.

Bertini, R. J., L. G. Marshall, M. Gayet, and P. Brito. 1993. "Vertebrate Faunas From the Adamantina and Marília Formations (Upper Baurú Group, late Cretaceous, Brazil) in Their Stratigraphic and Paleobiogeographic Context." *Neues Jahrbuch für Geologie und Paläontologie, Abhandlungen* 188:71–101.

Bonaparte, J. F. 1990. "New Late Cretaceous Mammals From the Los Alamitos Formation, Northern Patagonia." *National Geographic Research* 6:63–93.

Bonaparte, J. F., L. Van Valen, and A. Kramartz. 1993. "La Fauna Local de Punta Peligro, Paleoceno Inferior, de la Provincia del Chubut, Patagonia, Argentina." *Evolutionary Monographs* 14:1–61.

Campbell, K. E., Jr., ed. 2004. *The Paleogene Mammalian Fauna of Santa Rosa, Amazonian Perú.* Los Angeles: Natural History Museum of Los Angeles County, *Science Series* 40.

Candela, A. M., and J. J. Morrone. 2003. "Biogeografía de Puercoespines Neotropicales

(Rodentia: Hystricognathi): Integrando Datos Fósiles y Actuales a Través de un Enfoque Panbiogeográfico." *Ameghiniana* 40:361–8.

Carlini, A. A., and G. J. Scillato-Yané. 2004. "The Oldest Megalonychidae (Xenarthra: Tardigrada): Phylogenetic Relationships and an Emended Diagnosis of the Family." *Neues Jarburch für Geologie und Paläontologie Abhandlungen* 233:423–43.

Case, J. A., F. J. Goin, and M. O. Woodburne. 2005. "'South American' Marsupials from the Late Cretaceous of North America and the Origin of Marsupial Cohorts." *Journal of Mammalian Evolution* 12:461–94.

Charrier, R., O. Baeza, S. Elgueta, J. J. Flynn, P. Gans, S. M. Kay, N. Muñoz et al. 2002. "Evidence for Cenozoic Extensional Basin Development and Tectonic Inversion in the Southern Central Andes, Chile (33°–36° S.L.)." *Journal of South American Earth Sciences* 15:117–39.

Charrier, R., J. J. Flynn, A. R. Wyss, and D. A. Croft. Forthcoming. "Marco Geológico-Tectónico, Contenido Fosilífero y Cronología de los Yacimientos Cenozoicos Pre-Pleistocénicos de Mamíferos Terrestres Fósiles de Chile." In *Vertebrados Fósiles de Chile*, edited by D. Rubilar-Rogers and M. Sallaberry.

Charrier, R., L. Pinto, and M. P. Rodríguez. 2007. "Tectono-Stratigraphic Evolution of the Andean Orogen in Chile." In *Geology of Chile*, edited by W. Gibbons and T. Moreno, 21–116. London: Geological Society, Special Publication.

Cifelli, R. L. 1983. "The Origin and Affinities of the South American Condylarthra and Early Tertiary Litopterna (Mammalia)." *American Museum Novitates* 2772:1–49.

Cozzuol, M. A. 2006. "The Acre Vertebrate Fauna: Age, Diversity, and Geography." *Journal of South American Earth Sciences* 21:185–203.

Croft, D. A. 2001. "Cenozoic Environmental Change in South America as Indicated by Mammalian Body Size Distributions (Cenograms)." *Diversity and Distributions* 7:271–87.

———. 2007. "The Middle Miocene (Laventan) Quebrada Honda Fauna, Southern Bolivia and a Description of its Notoungulates." *Palaeontology* 50:277–303.

Croft, D. A., F. Anaya, D. Auerbach, C. Garzione, and B. J. McFadden. 2009. "New Data on Miocene Neotropical Provinciality from Cerdas, Bolivia." *Journal of Mammalian Evolution* 16:175–98.

Croft, D. A., J. J. Flynn, and A. R. Wyss. 2007. "A New Basal Glyptodontid and Other Xenarthra of the Early Miocene Chucal Fauna, Northern Chile." *Journal of Vertebrate Paleontology* 27:781–97.

———. 2008. "The Tinguiririca Fauna of Chile and the Early Stages of 'Modernization' of South American Mammal Fauna." *Arquivos do Museu Nacional (Rio de Janeiro)* 66:1–21.

Eagles, G. 2007. "New Angles on South Atlantic Opening." *Geophysical Journal International* 168:353–61.

Fernicola, J. C., S. F. Vizcaíno, and G. De Iuliis. 2009. "The Fossil Mammals Collected by Charles Darwin in South America During His Travels on Board the HMS Beagle." *Revista de la Asociación Geológica Argentina* 64:14–159.

Flynn, J. J. 2002. "Cenozoic Andean Paleoenvironments and Tectonic History: Evidence from Fossil Mammals." *5th International Symposium on Andean Geodynamics. Toulouse, France.* Editions IRD:215–18.

Flynn, J. J., R. Charrier, D. A. Croft, P. B. Gans, T. M. Herriott, J. A. Wertheim, and A. R. Wyss. 2008. "Chronologic Implications of New Miocene Mammals from the Cura-Mallín and Trapa Trapa Formations, Laguna del Laja area, South Central Chile." *Journal of South American Earth Sciences* 26:412–23.

Flynn, J. J., D. A. Croft, R. Charrier, G. Hérail, and A. R. Wyss. 2002. "The First Cenozoic Mammal Fauna from the Chilean Altiplano." *Journal of Vertebrate Paleontology* 22:200–6.

Flynn, J. J., D. A. Croft, R. B. Hitz, and A. R. Wyss. 2005. "The Tapado Fauna (?Casamayoran SALMA), Abanico Formation, Tinguiririca Valley, Central Chile." *Journal of Vertebrate Paleontology* 25 (suppl. to no. 3):57A–8A.

Flynn, J. J., M. J. Novacek, H. E. Dodson, D. Frassinetti, M. C. McKenna, M. A. Norell, K. E. Sears, C. C. Swisher, III, and A. R. Wyss. 2002. "A New Fossil Mammal Assemblage from the Southern Chilean Andes: Implications for Geology, Geochronology, and Tectonics." *Journal of South American Earth Sciences* 15:285–302.

Flynn, J. J., and Swisher, C. C., III. 1995. "Chronology of the Cenozoic South American Land Mammal Ages." In *Geochronology, Time-Scales, and Global Stratigraphic Correlation*, edited by W. A. Berggren, D. V. Kent, and J. Hardenbol, 317–33. Tulsa: Society for Sedimentary Geology (SEPM Special Publication 54).

Flynn, J. J., and A. R. Wyss. 2004. "A Polydolopine Marsupial Skull from the Cachapoal Valley, Andean Main Range, Chile." *Bulletin of the American Museum of Natural History* 285:80–92.

Flynn, J. J., A. R. Wyss, R. Charrier, and C. C. Swisher, III. 1995. "An Early Miocene Anthropoid Skull from the Chilean Andes." *Nature* 373:603–7.

Flynn, J. J., A. R. Wyss, D. A. Croft, and R. Charrier. 2003. "The Tinguiririca Fauna, Chile: Biochronology, Paleoecology, Biogeography, and a New Earliest Oligocene South American Land Mammal 'Age'." *Palaeogeography, Palaeoclimatology, Palaeoecology* 195:229–59.

Gayet, M., L. G. Marshall, T. Sempere, F. J. Meunier, H. Cappeta, and J.-C. Rage. 2001. "Middle Maastrichtian Vertebrates (Fishes, Amphibians, Dinosaurs and Other Reptiles, Mammals) from Pajcha Pata (Bolivia). Biostratigraphic, Palaeoecologic and Palaeobiogeographic Implications." *Palaeogeography, Palaeoclimatology, Palaeoecology* 169:39–68.

Gelfo, J. N., F. J. Goin, M. O. Woodburne, and C. de Muizon. 2009. "Biochronological Relationships of the Earliest South American Paleogene Mammalian Faunas." *Palaeontology* 52:251–69.

Gelfo, J. N., G. M. López, and M. Bond. 2008. "A New Xenungulata (Mammalia) from the Paleocene of Patagonia, Argentina." *Journal of Paleontology* 82:329–35.

Gelfo, J. N., and R. Pascual. 2001. "*Peligrotherium tropicalis* (Mammalia, Dryolestida) from the Early Paleocene of Patagonia, a Survival from a Mesozoic Gondwanan radiation." *Geodiversitas* 23:369–79.

Goin, F. J., R. Pascual, M. F. Tejedor, J. N. Gelfo, M. O. Woodburne, J. A. Case, M. A. Reguero et al. 2006. "The Earliest Tertiary Therian Mammal from South America." *Journal of Vertebrate Paleontology* 26:505–10.

Hitz, R. B., J. J. Flynn, and A. R. Wyss. 2006. "New Basal Interatheriidae (Typotheria,

Notoungulata, Mammalia) from the Paleogene of Central Chile." *American Museum Novitates* 3520:1–32.

Kay, R. F., B. J. MacFadden, R. H. Madden, H. Sandeman, and F. Anaya. 1998. "Revised Age of the Salla Beds, Bolivia, and its Bearing on the Age of the Deseadan South American Land Mammal 'Age'." *Journal of Vertebrate Paleontology* 18:189–99.

Kay, R. F., and R. H. Madden. 1997. "Mammals and Rainfall: Paleoecology of the Middle Miocene at La Venta (Colombia, South America)." *Journal of Human Evolution* 32:161–99.

Kay, R. F., R. H. Madden, R. L. Cifelli, and J. J. Flynn, eds. 1997. *Vertebrate Paleontology in the Neotropics*. Washington, DC: Smithsonian Institution Press.

Kay, R. F., R. H. Madden, M. G. Vucetich, A. A. Carlini, M. M. Mazzoni, G. H. Re, M. Heizler, and H. Sandeman. 1999. "Revised Geochronology of the Casamayoran South American Land Mammal Age: Climatic and Biotic Implications." *Proceedings of the National Academy of Sciences USA* 96:13235–40.

Kohn, M. J., A. A. Carlini, J. A. Josef, R. F. Kay, R. H. Madden, and M. G. Vucetich. 2004. "Climate Stability Across the Eocene-Oligocene Transition, Southern Argentina." *Geology* 32:621–4.

MacFadden, B. J., Y. Wang, T. E. Cerling, and F. Anaya. 1994. "South American Fossil Mammals and Stable Isotopes: A 25 Million-Year Sequence from the Bolivian Andes." *Palaeogeography, Palaeoclimatology, Palaeoecology* 107:257–68.

Marshall, L. G., and T. Sempere. 1991. "The Eocene to Pleistocene Vertebrates of Bolivia and Their Stratigraphic Context: A Review." In *Fósiles y Facies de Bolivia, Vol. 1, Vertebrados*, edited by R. Suárez-Soruco, 631–52. Santa Cruz: Revista Técnica de Yacimientos Petrolíferos Fiscales de Bolivia, 12(3–4).

McKenna, M. C., A. R. Wyss, and J. J. Flynn. 2006. "Paleogene Pseudoglyptodont Xenarthrans from Central Chile and Argentine Patagonia." *American Museum Novitates* 3536:1–18.

Muizon, C. de. 1991. "La Fauna de Mamíferos de Tiupampa (Paleoceno Inferior, Formación Santa Lucía), Bolivia." In *Fósiles y Facies de Bolivia, Vol. 1, Vertebrados*, edited by R. Suárez-Soruco, 575–624. Santa Cruz: Revista Técnica de Yacimientos Petrolíferos Fiscales Bolivianos 12.

———. 1998. "*Mayulestes ferox*, a Borhyaenoid (Metatheria, Mammalia) from the Early Palaeocene of Bolivia. Phylogenetic and Palaeobiologic Implications." *Geodiversitas* 20:19–142.

Muizon, C. de, and R. L. Cifelli. 2000. "The 'Condylarths' (Archaic Ungulata, Mammalia) from the Early Palaeocene of Tiupampa (Bolivia): Implications on the Origin of the South American Ungulates." *Geodiversitas* 22:47–150.

Muizon, C. de, R. L. Cifelli, and R. C. Paz. 1997. "The Origin of the Dog-Like Borhyaenoid Marsupials of South America." *Nature* 389:486–9.

Murphy, W. J., E. Eizirik, S. J. O'Brien, O. Madsen, M. Scally, C. J. Douady, E. Teeling et al. 2001. "Resolution of the Early Placental Mammal Radiation Using Bayesian Phylogenetics." *Science* 294:2348–51.

Pascual, R. 2006. "Evolution and Geography: The Biogeographic History of South American Land Mammals." *Annals of the Missouri Botanical Garden* 93:209–30.

Pascual, R., M. Archer, E. Ortiz-Jaureguizar, J. L. Prado, H. Godthelp, and S. J. Hand. 1992. "First Discovery of Monotremes in South America." *Nature* 356:704–5.

Pascual, R., F. J. Goin, P. González, A. Ardolino, and P. F. Puerta. 2000. "A Highly Derived Docodont from the Patagonian Late Cretaceous: Evolutionary Implications for Gondwanan Mammals." *Geodiversitas* 22:395–414.

Pascual, R., and E. Ortiz-Jaureguizar. 1990. "Evolving Climates and Mammal Faunas in Cenozoic South America." *Journal of Human Evolution* 19:23–60.

———. 2007. "The Gondwanan and South American Episodes: Two Major Moments in the History of South American Mammals." *Journal of Mammalian Evolution* 14:75–137.

Pascual, R., E. Ortiz-Jaureguizar, and J. L. Prado. 1996. "Land Mammals: Paradigm for Cenozoic South American Geobiotic Evolution." In *Contributions of Southern South America to Vertebrate Paleontology*, edited by G. Arratia, 265–319. Munich: Münchner Geowissenschaftliche abhandlungen Verlag Dr. F. Pfeil (A) 30.

Paula Couto, C. 1952. "Fossil Mammals from the Beginning of the Cenozoic in Brazil. Condylarthra, Litopterna, Xenungulata, and Astrapotheria." *Bulletin of the American Museum of Natural History* 99 (6):1–394.

Pujos, F., and G. De Iuliis. 2007. "Late Oligocene Megatherioidea Fauna (Mammalia: Xenarthra) from Salla-Luribay (Bolivia): New Data on Basal Sloth Radiation and Cingulata-Tardigrada Split." *Journal of Vertebrate Paleontology* 27:132–44.

Rauhut, O. W. M., T. Martin, E. Ortiz-Jaureguizar, and P. Puerta. 2002. "A Jurassic Mammal from South America." *Nature* 416:165–8.

Rose, K. D., R. J. Emry, T. J. Gaudin, and G. Storch. 2005. "Xenarthra and Pholidota." In *The Rise of Placental Mammals*, edited by K. D. Rose and J. D. Archibald, 106–26. Baltimore: John Hopkins University Press.

Rosenberger, A. L., M. F. Tejedor, S. B. Cooke, and S. Pekar. 2009. "Platyrrhine Ecophylogenetics in Space and Time." In *South American Primates: Comprehensive Perspectives in the Study of Behavior, Ecology and Conservation*. Developments in Primatology: Progress and Prospects, edited by P. A. Garber, A. Estrada, J. C. Bicca-Marques, E. W. Heymann, and K. B. Strier, 69–113. New York: Springer.

Rougier, G. W., L. Chornogubsky, S. Casadio, N. Paéz Arango, and A. Giallombardo. 2009. "Mammals from the Allen Formation, Late Cretaceous, Argentina." *Cretaceous Research* 30:223–38.

Rougier, G. W., A. G. Martinelli, A. M. Forasiepi, and M. J. Novacek. 2007. "New Jurassic Mammals from Patagonia, Argentina: A Reappraisal of Australosphenidan Morphology and Interrelationships." *American Museum Novitates* 3566:1–56.

Salas, R., J. Sánchez, and C. Chacaltana. 2006. "A New Pre-Deseadan Pyrothere (Mammalia) from Northern Perú and the Wear Facets of Molariform Teeth of Pyrotheria." *Journal of Vertebrate Paleontology* 26:760–9.

Sears, K. A., J. A. Finarelli, J. J. Flynn, and A. R. Wyss. 2008. "Estimating Body Mass in New World 'Monkeys' (Platyrrhini, Primates), with a Consideration of the Miocene Platyrrhine *Chilecebus carrascoensis*." *American Museum Novitates* 3617:1–29.

Shockey, B. J., and F. Anaya. 2008. "Postcranial Osteology of Mammals of Salla, Bolivia (Late Oligocene): Form, Function, and Phylogeny." In *Mammalian Evolutionary Morphology: A Tribute to Frederick S. Szalay*, edited by E. J. Sargis and M. Dagosto, 135–57. Dordrecht: Springer.

Shockey, B. J., and J. J. Flynn. 2007. "Morphological Diversity in the Postcranial Skeleton of Casamayoran (?Middle to Late Eocene) Notoungulata and Foot Posture in Notoungulates." *American Museum Novitates* 3601:1–26.

Shockey, B. J., R. Hitz, and M. Bond. 2004. "Paleogene Notoungulates from the Amazon Basin of Peru." In *The Paleogene Mammalian Fauna of Santa Rosa, Amazonian Peru, Science Series* 40, edited by K. E. Campbell, Jr., 61–69. Los Angeles: Natural History Museum of Los Angeles County.

Shockey, B. J., R. Salas Gismondi, P. Gans, A. Jeong, and Flynn, J.J. 2009. "Paleontology and Geochronology of the Deseadan (Late Oligocene) of Moquegua, Perú." *American Museum Novitates* 3668:1–24.

Shockey, B. J., R. Salas, R. Quispe, A. Flores, E. J. Sargis, J. Acosta, A. Pino et al. 2006. "Discovery of Deseadan Fossils in the Upper Moquegua Formation (Late Oligocene-?Early Miocene) of Southern Peru." *Journal of Vertebrate Paleontology* 26:205–8.

Simpson, G. G. 1932. "Enamel on the Teeth of an Eocene Edentate." *American Museum Novitates* 567:1–4.

———. 1940. "Review of the Mammal-Bearing Tertiary of South America." *Proceedings of the American Philosophical Society* 83:649–709.

Tauber, A. A. 1999. "Los Vertebrados de la Formación Santa Cruz (Mioceno Inferior-Medio) en el Extremo Sureste de la Patagonia y su Significado Paleoecológico." *Revista Española de Paleontología* 14:173–82.

Villarroel, C. A. 1987. "Características y Afinidades de *Etayoa* n. gen., Tipo de una Nueva Familia de Xenungulata (Mammalia) del Paleoceno Medio (?) de Colombia." *Comunicaciones Paleontológicas del Museo de Historia Natural de Montevideo* 19:241–53.

Webb, S. D. 1995. "Biological Implications of the Middle Miocene Amazon Seaway." *Science* 269:361–2.

Wertheim, J. A. 2007. "Fossil Rodents from Laguna del Laja, Chile: A Systematic, Phylogenetic, and Biochronologic Study." Unpublished PhD diss., University of California, Santa Barbara.

Willis, K. J., and J. C. McElwain. 2002. *The Evolution of Plants*. Oxford: Oxford University Press.

Wyss, A. R., R. Charrier, and J. J. Flynn. 1996. "Fossil Mammals as a Tool in Andean Stratigraphy: Dwindling Evidence of Late Cretaceous Volcanism in the South Central Main Range." *PaleoBios* 17:13–27.

Wyss, A. R., M. A. Norell, J. J. Flynn, M. J. Novacek, R. Charrier, M. C. McKenna, C. C. Swisher, III et al. 1990. "A New Early Tertiary Mammal Fauna from Central Chile: Implications for Andean Stratigraphy and Tectonics." *Journal of Vertebrate Paleontology* 10:518–22.

On the Evolution of Large Size in Mammalian Herbivores of Cenozoic Faunas of Southern South America

Sergio F. Vizcaíno, Guillermo H. Cassini,
Néstor Toledo, and M. Susana Bargo

Abstract

One of the major features of the continental Cenozoic faunas of South America is the presence of native lineages of herbivorous mammals, and among them the largest representatives of each fauna. They include a diversity of taxa within the Xenarthra, Pyrotheria, Astrapotheria, Notoungulata, Litopterna, Cetartiodactyla, Perissodactyla, Proboscidea and Rodentia. We analyze the evolution of the large body size of these mammals in relation to their taxonomic richness. As the South American mammalian fossil record is largely restricted to southern parts of the continent, with comparatively few Tertiary land mammal-bearing localities outside of Argentina, we limit our samples to faunas from that country. Faunal lists from 6 different ages were selected. Genera were classified in 4 body mass categories: (I) less than 100 kg, (II) 100 to 500 kg, (III) 500 to 1000 kg, and (IV) more than 1000 kg. The Pleistocene represents the spectacular climax in terms of body size. In general, but particularly for those faunas in which xenarthrans are dominant, the number of mega-mammals only distantly related to living counterparts raises problems in interpreting their paleobiology. Particularly for the Pleistocene, communities dominated by mega-mammals of very low metabolism (xenarthrans) have no counterpart in living faunas. Large size explains most of their vulnerability to extinction at the end of the Pleistocene in South America. Although the impact of the mega-mammal extinction on the post-Pleistocene evolution of plant communities has not been studied for South America, it is clear that it produced an enormous ecological gap in the herbivorous guild.

5.1 Introduction

The faunas of South America's Cenozoic (fig. 5.1) were dominated until the late Pliocene by endemic lineages of placental mammals (Marshall and Cifelli 1989; Simpson 1950, 1980; Webb 1991) in addition to the New World marsupials (Ameridelphia). Even the two placental groups (rodents and primates) that arrived from other continents in the late Eocene-Oligocene evolved as endemic

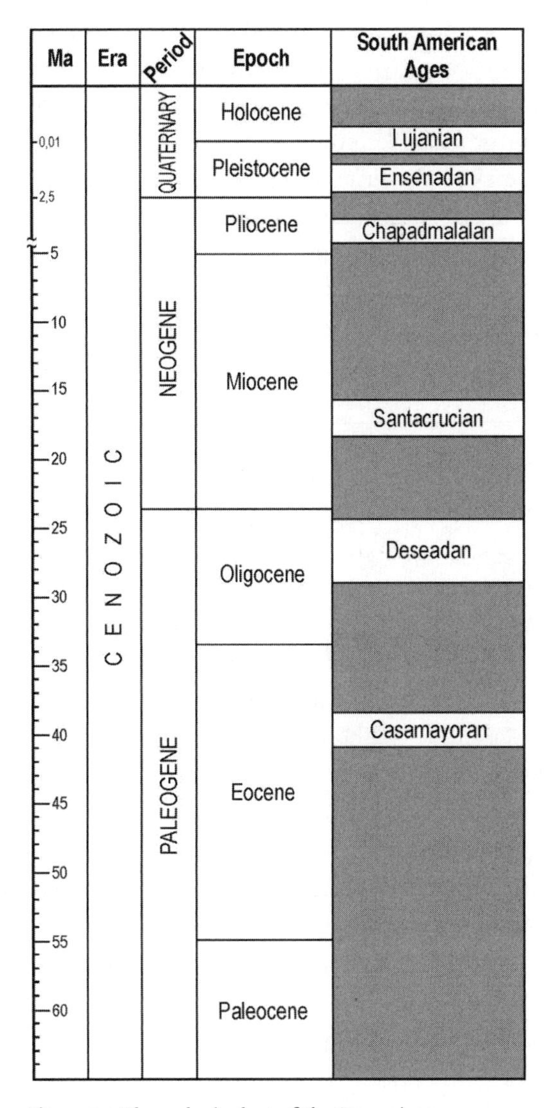

Figure 5.1 Chronologic chart of the Cenozoic.

lineages (caviomorphs and platyrrhines, respectively) in South America. Large marsupials occupied the mammalian predator guild in most faunas before placental carnivores arrived from the North during the Neogene, while the noncarnivorous forms were usually small sized. Most specialists recognize at least six major groups of predominantly herbivorous placentals that have been mostly or entirely restricted to South America: the xenarthrans and five main groups of ungulates.

Xenarthrans include the Cingulata and the Pilosa. The Cingulata are exemplified by armadillos and glyptodonts, noted for the bony armor covering the head, body, and tail. The Pilosa are composed of two groups: the Vermilingua and the Folivora. The Vermilingua, or anteaters, are characterized by an elongate, tubular skull, the absence of teeth, and a prehensile tongue used in acquiring prey. Among xenarthrans, only the armadillos have been recognized as exploiting a dietary spectrum from omnivory to carnivory (see Vizcaíno 2009; Vizcaíno, Fariña, Bargo, et al. 2004, Vizcaíno, Bargo, and Fariña 2008 and references therein). The Folivora, or sloths, include many fossil forms and are also known as Tardigrada or Phyllophaga (Delsuc and Douzery 2008; Fariña and Vizcaíno 2003; Gardner 2005). Modern sloths are almost exclusively arboreal herbivores; however, many extinct forms were much larger, more terrestrial, and are generally considered to have been exclusively herbivorous, although some degree of carnivory has been proposed for certain species (see Fariña and Blanco 1996, and discussions in Bargo 2001; Bargo and Vizcaíno 2008; Prevosti and Vizcaíno 2006).

Over the past several decades, morphology-based studies of xenarthrans and their phylogenetic position have tended to place them in a remote position within Placentalia (Gaudin and McDonald 2008), for example, as the sister taxon to other placentals (Epitheria), although perhaps allied with pangolins (Order Pholidota) and the extinct Palaeanodonta (Novacek 1992; Novacek and Wyss 1986). Placement of the Xenarthra as the sister group to the other placentals was first advocated by early cladistic studies of morphology (McKenna 1975). The Xenarthra/Epitheria dichotomy has been criticized by other studies (Gaudin and McDonald 2008; Gaudin et al. 1996; Rose and Emry 1993), and both morphological and molecular work have failed to support the xenarthran/pholidotan clade (Delsuc et al. 2002; Rose and Emry 1993; Rose et al. 2005) or have supported a close relationship between Xenarthra and Afrotheria (Prasad et al. 2008; Wible et al. 2007).

South American ungulates have always been compared to living herbivorous analogues (see Croft 1999 for a summary). South America's native ungulates include five groups: the astrapotheres, pyrotheres, notoungulates, litopterns, and xenungulates (Marshall and de Muizon 1988; Simpson 1980). The phylogenetic relationships of these groups are unclear (Cifelli 1985, 1993). They were all once united in a single taxon, Meridiungulata, on the idea that all endemic South American ungulates were monophyletic (McKenna 1975). However, the term "ungulate" here does not imply that the ungulate groups endemic to South

America and modern Ungulata (Perissodactyla and Cetartiodactyla) share a most recent common ancestor or even form a single clade themselves.

Astrapotheres (including trigonostylopids) were rhinoceros-like mammals (Cifelli 1985) found in deposits Paleocene to Miocene in age. They attained maximum diversity during the early Miocene Colhuehuapian and Santacrucian ages (Johnson and Madden 1997; Marshall and Cifelli 1989).

Pyrotheres were elephant-like, with tusks and bilophodont cheek teeth, and were never as diverse nor did they cover as great a time span as the astrapotheres. They are known only from the middle Eocene (Casamayoran) through the late Oligocene (Deseadan).

Notoungulates are by far the most diverse and abundant lineage of South American ungulates (nearly 140 species in 13 families; Croft 1999). This group includes animals similar to rhinoceroses, hippopotamuses, rabbits, and rodents. Other notoungulates do not closely resemble any living mammal. Many had high-crowned cheek teeth.

Litopterns were the second most successful group of South American ungulates in terms of diversity and longevity, spanning from the late Paleocene (Itaboraían) to the late Pleistocene (Marshall and Cifelli 1989). They include forms similar to antelopes, horses, and camels, all with relatively low-crowned cheek teeth.

Xenungulates are primitive, poorly known, tapir-like mammals, restricted to Paleocene deposits of Brazil and Argentina (Gelfo, López, and Bond 2008).

Based on both the fossil record and molecular dates, primates and rodents appeared in South America by the late Eocene and early Oligocene. They probably arrived as a result of trans-Atlantic migrations (see Pascual 2006; Poux et al. 2006). Each of these lineages of herbivorous placentals had a different fate in South America. Fossil South American primates are rare, and the present-day taxa are the result of a relatively recent (early Miocene or later) diversification that resulted in arboreal forms no larger than 10 kg. By contrast, caviomorph rodents are among the most abundant taxa in every post-Eocene fauna. Caviomorphs radiated soon after their arrival into a series of lineages that persist to the present, and some families evolved forms that weighed in excess of 100 kg.

"About 2.5 Ma tectonic activities along the Pacific margin caused the American continents to be sutured, and thus began one of the great biogeographic experiments, known to paleobiologists as the Great American Biotic Interchange" (Webb 1991, 266). The first North American forms to arrive in

South America included cricetid rodents, procyonid carnivorans, mustelid carnivorans, and tayassuid artiodactyls. The early Pleistocene marks the first appearance in South America of five major herbivore lineages: gomphotheriids among proboscideans, camelids and cervids among artiodactyls, and equids and tapirids among perissodactyls, in addition to several carnivoran lineages (felids, canids, and ursids).

One feature of these Cenozoic faunas is the tendency to increase maximum size in many lineages. Some remarkable gigantic forms are the mammalian Astrapotheria and Pyrotheria (Oligocene) and the avian phororhacoids or "terror birds" (Miocene). But the climax of body size evolution was reached during the Quaternary (the last 2 Ma). Among the more than 120 genera of mammals known, the estimated adult masses of about 40 genera exceeded 100 kg, of which about 20 were megaherbivores. Although a 2004 symposium on vertebrate giants revealed that no other fossil mammalian fauna is known to have contained such a diversity of megaherbivores (Bargo 2004), so far no analysis has considered the evolution of large body size in relation to taxonomic richness. The goal of this study is to provide a preliminary analysis of the evolution of large size in mammalian herbivores of the Cenozoic of South America. For this purpose, we followed a definition of "large mammal" that is commonly used in archaeology and paleontology (see Cione, Tonni, and Soibelzon 2003, 2009; Johnson 2002; Martin and Steadman 1999), considering only taxa >44 kg (100 pounds).

The South American mammalian fossil record is largely restricted to southern parts of the continent, with comparatively few Tertiary land mammal-bearing localities outside of Argentina (Ortiz-Jaureguizar and Cladera 2006, and references therein). Therefore, we limit our sample to faunas from Argentina. In this way, we are representing almost the same area that Ortiz- Jaureguizar and Cladera defined as Southern South America (i.e., the area south of 15° S) to reconstruct historic and ecologic relationships of the Patagonian biota with that of the rest of South America. Our goal is to identify new working hypotheses to test in a future long-term project on the paleobiology of the South American Cenozoic faunas. Our greatest interest is in the giants of the Pleistocene, which are dominated by ancient South American descendants: ground sloths and glyptodonts (Xenarthra), macrauchenians (Litopterna), toxodonts (Notoungulata), and mastodonts (Proboscidea). Some of these forms are characterized by peculiar adaptations of the masticatory and locomotory systems, which lack clear ecological equivalents among living mammals. Understanding the basis for their very large size is our research goal.

5.2 Evolution of the Paleofloras and Climates During the Cenozoic

The succession of environments in which the large herbivore mammals (and all the fauna) occurred can be envisioned through a summary of four major paleofloras described for the southern part of South America (Frenguelli 1953; Hinojosa and Villagrán 1997; Menéndez 1971; Romero 1978, 1986; Troncoso and Romero 1998). The spatial and temporal successions of these paleofloras appear to have been closely related to tectonic and climatic events that occurred during the Cenozoic (Hinojosa and Villagrán 1997). For extensive reviews on this topic, see recent articles by Hinojosa (2005) and Barreda et al. (2007).

First, the Neotropical Paleoflora developed mainly during the Paleocene when South America, Antarctica, and Australia maintained geographic connections, and warm climatic conditions extended to at least 50°S. The mean annual temperatures and rainfall ranged between 2 and 25°C, and 1500 and 2000 mm respectively (Hinojosa 2005). This flora was replaced during the Eocene and early Oligocene by the Mixed Paleoflora, coincident with a decrease of temperature levels and annual mean precipitation. According to Hinojosa (2005), the mean annual temperature was 17–20°C, and the mean annual precipitation was ~570 mm. The Mixed Paleoflora was characterized by taxa distributed today in the tropics and the Austral-Antarctic territories, as well as forms now endemic to tropical and subtropical regions of the continent.

From the late Eocene/early Oligocene up to early Miocene, global temperatures fell as glaciations developed in eastern Antarctica after the separation of Australia from Antarctica-South America. Accordingly the Mixed Paleoflora was mostly replaced by an Antarctic Paleoflora, characterized by the dominance of taxa from temperate-cold environments. The notable increase of the annual thermal amplitude would have favored the development of taxa adapted to colder conditions (mean annual ~15°C) and extreme temperatures. Annual rainfall averages increased to 870 mm by the late Oligocene and to 1120 mm by the Early Miocene (Hinojosa 2005).

Finally, the Subtropical Flora appeared as a consequence of the global warming that characterized the middle Miocene. The mean annual temperatures ranged 21–26°C (Hinojosa 2005). According to Zachos et al. (2001), the Climatic Optimum occurred around 15–17 Ma. After the late early Miocene Climatic Optimum, temperature and rainfall decreased (430 mm), related to an increase of the thermal difference between extreme temperatures, producing conditions favoring the development of xeric subtropical floras.

From the middle Miocene/Pliocene, with the culmination of the separation process between Antarctica and South America, the establishment of the Circumpolar and Humboldt oceanic currents in their present trajectories produced higher temperatures and increased aridity in the subtropics. The final elevation of the Andes produced a rain shadow effect that, by the Plio/Pleistocene, caused the fragmentation of the Subtropical Paleoflora and the spread of taxa of arid environments along the so-called "arid diagonal" that extends from the southeastern tip of the continent, across the Andes in Central Chile, and continuing along the Pacific coast to near the Equator (Hinojosa and Villagrán 1997; Villagrán and Hinojosa 1997).

The periodic climatic alternation of glacial and interglacial epochs during the middle late Pleistocene dramatically modified the distribution, composition, and biomass of plant and animal communities in South America (Cione, Tonni, and Soibelzon 2003).

5.3 Methods

Faunal lists from six different ages were selected, based on the quality of the information available: the middle Eocene Casamayoran from Patagonia and northern Argentina (Pascual and Ortiz-Jaureguizar 2007), the late Oligocene Deseadan, from Cabeza Blanca (Chubut) and La Flecha (Santa Cruz) in Patagonia (Reguero 1999), the early Miocene Santacrucian at the Atlantic coast (Santa Cruz) in Patagonia (Tauber 1997; Vizcaíno and colleagues, unpublished data), the late Pliocene Chapadmalalan at the Atlantic coast (Buenos Aires Province), Pampean region (Vizcaíno, Fariña, Zárate, et al. 2004), the early Pleistocene Ensenadan, and the late Pleistocene-early Holocene Lujanian (Buenos Aires Province), Pampean region (Cione and Tonni 2005; fig. 5.1).

A database of body mass estimates was built using different methods: from previously published estimates, using regression equations generated from modern relatives, from geometric similarity with a phylogenetically close relative (of known mass for living ones or with an appropriate estimation for fossil ones), or using gross anatomical similarity with living analogues (see table 5.1). Orders and genera were chosen as the working taxonomic levels. We chose to use genera rather than species because they are discrete taxonomic units accepted by most paleontologists, and they are less affected by the problems of evaluating intraspecific variation in fossils. Species of the same genus were considered only in those cases in which they attained body masses that fell into different categories, as described later.

Table 5.1 List of taxa represented for each age with their body mass estimates.

Order	Family	Genus	BM (Kg)	Source
CASAMAYORAN				
Astrapotheria	Astrapotheriidae	*Albertogaudrya*	60.28	Croft 2000
		Scaglia	200	CLA-*Tapirus terrestris*
	Trigonostylopidae	*Tetragonostylops*	200	CLA-*Tapirus terrestris*
		Trigonostylops	200	CLA-*Tapirus terrestris*
Litopterna	Sparnotheriodontidae	*Sparnotheriodon*	400	Vizcaíno et al. 1998
		Victorlemoinea	400	Vizcaíno et al. 1998
Notoungulata	Henricosborniidae	*Othnielmarshia*	50	CLA-*Ovis aries*
		Peripantostylops	50	CLA-*Ovis aries*
	Isotemnidae	*Anisotemnus*	48.68	Croft 2000
		Pampatemnus	50	CLA-*Ovis aries*
		Thomashuxleya	113.04	Croft 2000
	Notohippidae	*Pampahippus*	50	CLA-*Ovis aries*
		Plexotemnus	50	CLA-*Ovis aries*
	Notostylopidae	*Boreastylops*	50	CLA-*Ovis aries*
		Edvardotrouessartia	50	CLA-*Ovis aries*
		Homalostylops	50	CLA-*Ovis aries*
		Notostylops	50	CLA-*Ovis aries*
	Oldfieldthomasiidae	*Oldfieldthomasia*	50	CLA-*Ovis aries*
		Paginula	50	CLA-*Ovis aries*
Pyrotheria	Pyrotheriidae	*Carolozittelia*	200	CLA-*Tapirus terrestris*
DESEADAN				
Astrapotheria	Astrapotheriidae	*Parastrapotherium*	1900	GS-*Astrapotherium* (Kramarz and Bond 2008)
Notoungulata	Homalodotheriidae	*Asmodeus*	330	GS- *Homalodotherium*
	Leontiniidae	*Ancylocoelus*	288	GS-*Leontinia*
		Leontinia	288	RE
	Mesotheriidae	*Trachytherus*	45	Croft and Weinstein 2008
	Notohippidae	*Argyrohippus*	60	CLA-*Ovis aries*
		Eurygenium	60	CLA-*Ovis aries*
		Morphippus	60	CLA-*Ovis aries*
		Rhynchippus	120	CLA-*Ovis aries*
	Toxodontidae	*Proadinotherium*	90	GS-*Adinotherium*
Pyrotheria	Pyrotheriidae	*Pyrotherium*	3500	Shockey and Anaya 2004
Cingulata	Paleopelthidae	*Paleopeltis*	600	GS-*Neosclerocalyptus*
	Peltephilidae	*Peltephilus*	50	GS-*Priodontes*
Folivora	Orophodontidae	*Octodontotherium*	700	GS-*Glossotherium*

(continues)

Table 5.1 (*continued*)

Order	Family	Genus	BM (Kg)	Source
SANTACRUCIAN				
Astrapotheria	Astrapotheriidae	*Astrapotherium*	1021.63	RE
Litopterna	Macraucheniidae	*Theosodon*	95.61	RE
	Proterotheriidae	*Diadiaphorus*	70.25	RE
Notoungulata	Homalodotheriidae	*Homalodotherium*	340	RE
	Toxodontidae	*Adinotherium*	121.26	RE
	Toxodontidae	*Nesodon*	554.61	RE
Cingulata	Glyptodontidae	*Cochlops*	80	GS - *Propalaehoplophorus*
		Eucinepeltus	115	Croft 2001
		Propalaehoplophorus	73.40	Vizcaíno, Bargo, and Cassini 2006
Folivora	Megalonychidae	*Eucholoeops*	80	Bargo, Vizcaíno, and Kay 2009
	Megatherioidea	*Analcimorphus*	160	GS-*Hapalops*
		Hapalops	70	Bargo, Vizcaíno, and Kay 2009
		Planops	300	GS-*Hapalops*
		Prepotherium	525	GS-*Scelidotherium*
	Mylodontidae	*Nematherium*	423.98	GS-*Scelidotherium leptocephalum*
CHAPADMALALAN				
Cetartiodactyla	Tayassuidae	*Argyrohyus*	50	CLA-*Catagonus wagneri*
Litopterna	Macraucheniidae	*Promacrauchenia*	400	GS-*Macrauchenia patachonica*
Notoungulata	Toxodontidae	*Toxodon*	1642	Fariña, Vizcaíno, and Bargo 1998
		Xotodon	300	GS-*Toxodon platensis*
Rodentia	Dinomyidae	*Telicomys*	600	GS-*Hydrochoerus*
	Hydrochoeridae	*Chapalmatherium*	200	GS-*Hydrochaerus*
Cingulata	Glyptodontidae	*Paraglyptodon*	800	CF-*Glyptodon reticulatus*
		Plesiomegatherium	3000	GS-*Pyramiodontherium scillatoyanei* (De Iuliis et al. 2004)
		Plohophoroides	263	CF-*Plohophorus*
		Plohophorus	263	Vizcaíno, Bargo, and Cassini 2006
		Trachycalyptus	600	CF-*Neosclerocalyptus*
		Urotherium	600	CF-*Neosclerocalyptus*

Table 5.1 (*continued*)

Order	Family	Genus	BM (Kg)	Source
	Pampatheriidae	*Pampatherium*	200	GS-*Holmesina* (Vizcaíno, Bargo, and Cassini 2006).
Folivora	Mylodontidae	*Glossotheridium*	580	GS-*Glossotherium robustum*
		Proscelidodon	430	GS-*Scelidotherium leptocephalum*
		Scelidotheridium	430	GS-*Scelidotherium leptocephalum*
ENSENADAN				
Cetartiodactyla	Camelidae	*Hemiauchenia*	400	GS-*Lama guanicoe*
		Lama	120	CLA- *Lama guanicoe*
	Cervidae	*Antifer*	120	CLA- *Blastoceros dichotomus*
		Epieuricerus	120	CLA-*Blastoceros dichotomus*
	Tayassuidae	*Catagonus*	50	CLA-*Catagonus wagneri*
		Platygonus	50	CLA-*Catagonus wagneri*
Litopterna	Macrauchenidae	*Macraucheniopsis*	1200	GS-*Macrauchenia patachonica*
Notoungulata	Mesotheriidae	*Mesotherium*	60	GS-*Pseudotypotherium*
	Toxodontidae	*Toxodon*	1642	Fariña, Vizcaíno, and Bargo 1998
Perissodactyla	Equidae	*Equus*	379	Alberdi and Prado 2004
		Hippidion	460	Alberdi and Prado 2004
Proboscidea	Gomphotheriidae	*Stegomastodon*	7580	Fariña, Vizcaíno, and Bargo 1998
Rodentia	Caviidae	*Neochoerus*	110	GS-*Hydrochoerus hydrochaeris*
Cingulata	Dasypodidae	*Propraopus*	50	Fariña and Vizcaíno 1997
	Glyptodontidae	*Daedicuroides*	1100	CF- *Doedicurus clavicaudatus*
		Doedicurus	1468	Fariña, Vizcaíno, and Bargo 1998
		Glyptodon clavipes	2000	Fariña 1995

(*continues*)

Table 5.1 *(continued)*

Order	Family	Genus	BM (Kg)	Source
		Glyptodon reticulatus	862	Fariña, Vizcaíno, and Bargo 1998
		Lomaphorus	600	CF-*Neosclerocalyptus ornatus*
		Neosclerocalyptus	598	Vizcaíno, Bargo, and Cassini 2006
		Neothoracophorus	600	CF-*Neosclerocalyptus ornatus*
		Neuryurus	311	Vizcaíno, Bargo, and Cassini 2006
		Panochthus	1061	Fariña, Vizcaíno, and Bargo 1998
		Plaxhaplous	1100	CF- *Doedicurus clavicaudatus*
	Pampatheriidae	*Pampatherium*	200	GS-*Holmesina* (Vizcaíno, Bargo, and Cassini 2006)
Folivora	Megatheriidae	*Megatherium*	3950	Fariña, Vizcaíno, and Bargo 1998
	Mylodontidae	*Glossotherium*	1344	Fariña, Vizcaíno, and Bargo 1998; Christiansen and Fariña 2003
		Mylodon	1986	Christiansen and Fariña 2003
		Scelidotherium	1057	Fariña, Vizcaíno, and Bargo 1998
LUJANIAN				
Cetartiodactyla	Camelidae	*Eulamaops*	150	Prevosti and Vizcaíno 2006
		Hemiauchenia	400	GS-*Lama guanicoe*
		Lama	120	CLA-*Lama guanicoe*
	Tayassuidae	*Catagonus*	50	CLA-*Catagonus wagneri*
		Tayassus	50	CLA-*Tayassus tajacu*
Litopterna	Macraucheniidae	*Macrauchenia*	988	Fariña, Vizcaíno, and Bargo 1998
Notoungulata	Toxodontidae	*Toxodon*	1642	Fariña, Vizcaíno, and Bargo 1998
Perissodactyla	Equidae	*Equus (Amerhippus)*	379	Alberdi and Prado 2004

Table 5.1 *(continued)*

Order	Family	Genus	BM (Kg)	Source
		Hippidion	460	Alberdi and Prado 2004
Proboscidea	Gomphotheriidae	*Stegomastodon*	7580	Fariña, Vizcaíno, and Bargo 1998
Rodentia	Caviidae	*Neochoerus*	110	GS-*Hydrochoerus hydrochaeris*
Cingulata	Dasypodidae	*Eutatus*	50	Vizcaíno, Milne, and Bargo 2003
		Propraopus	50	Fariña and Vizcaíno 1997
	Glyptodontidae	*Doedicurus*	1468	Fariña, Vizcaíno, and Bargo 1998
		Glyptodon clavipes	2000	Fariña, 1995
		Glyptodon reticulatus	862	Fariña, Vizcaíno, and Bargo 1998
		Lomaphorus	600	CF-*Neosclerocalyptus ornatus*
		Neosclerocalyptus	598	Vizcaíno, Bargo, and Cassini 2006
		Neothoracophorus	600	CF-*Neosclerocalyptus ornatus*
		Panochthus	1061	Fariña, Vizcaíno, and Bargo 1998
		Plaxhaplous	1100	CF- *Doedicurus clavicaudatus*
	Pampatheriidae	*Pampatherium*	200	GS-*Holmesina* (Vizcaíno, Bargo, and Cassini 2006)
Folivora	Megatheriidae	*Megatherium*	3950	Fariña, Vizcaíno, and Bargo 1998
	Mylodontidae	*Glossotherium*	1344	Fariña, Vizcaíno, and Bargo 1998; Christiansen and Fariña 2003
		Lestodon	3397	Fariña, Vizcaíno, and Bargo 1998
		Mylodon	1986	Christiansen and Fariña 2003
		Scelidotherium	1057	Fariña, Vizcaíno, and Bargo 1998

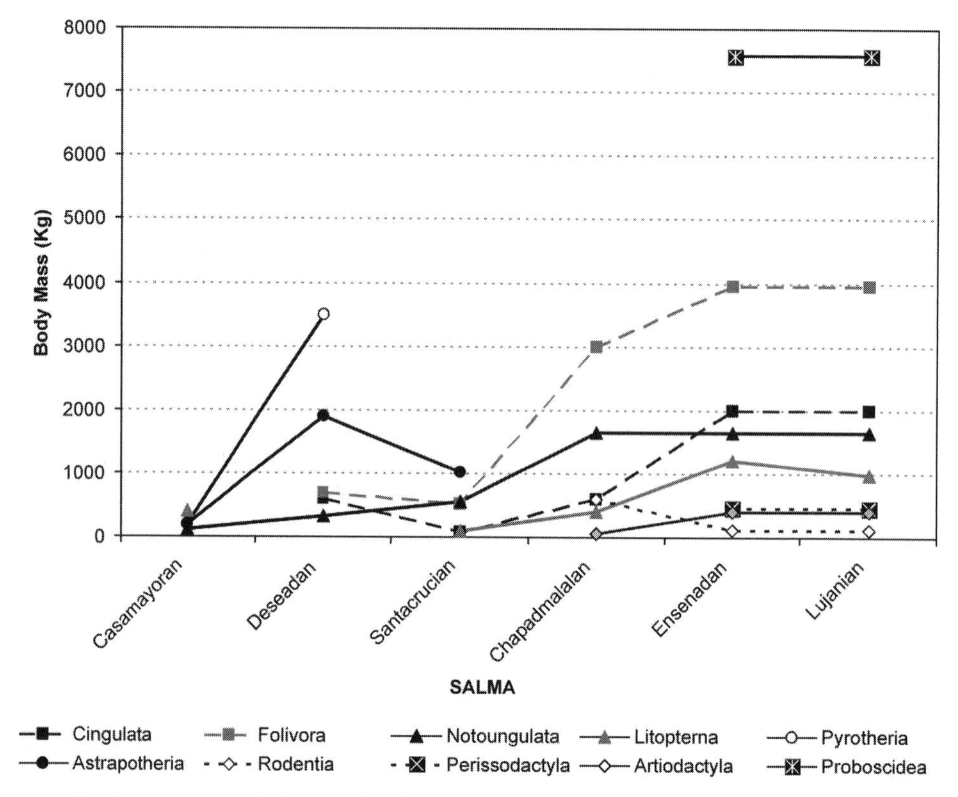

Figure 5.2 Maximum body mass of herbivorous mammals throughout the Cenozoic.

For each genus, the body mass estimate of the largest species was selected. As mentioned earlier, we considered large mammals as only those taxa >44 kg (100 pounds).

Two analyses were performed. First, in each age the genus with the maximum body mass (MBM) was selected to represent each order. The results were plotted on an XY graph, where the X-axis represents time (in mammal ages) and the Y-axis represents the body mass estimate in kg (fig. 5.2). For the second analysis, genera were classified in four body mass categories: (I) <100 kg; (II) 100–500 kg; (III) 500–1000 kg; and (IV) >1000 kg. The number of genera per order in each category and age was plotted in histograms (fig. 5.3).

5.4 Results

Maximum body mass of herbivorous mammals throughout the Cenozoic are depicted in table 5.1 and fig. 5.2. In general, maximum body mass (MBM) for most of the native orders of Cenozoic South American herbivorous mammals (notoungulates, litopterns, cingulates, and sloths) increases over time, especially after the Chapadmalalan (Pliocene). Pyrotheres and astrapotheres reach their largest sizes in the Deseadan (Oligocene). While pyrotheres are not recorded in the Santacrucian (Miocene) and following ages, astrapotheres, cingulates, and sloths attain smaller MBM during that age than in the previous Deseadan, and astrapotheres disappear after the Miocene. Among rodents, cow-sized forms are recorded during the Chapadmalalan (although maximal sizes were attained earlier; see Discussion section) and MBM decreases by the Ensenadan and Lujanian (Pleistocene-early Holocene).

Taxonomic representation of body mass categories during the Cenozoic is shown in table 5.2, which summarizes the number of taxa (and percentages) of each body mass category (I to IV) represented in each age. Figure 5.3 shows the representation of groups in each category through the different ages.

During the Eocene Casamayoran, category I is represented by mostly notoungulates, with only one astrapothere. Category II includes mainly astrapotheres and litopterns, with minor representation of pyrotheres and notoungulates. There are no representatives of categories III and IV.

During the Oligocene Deseadan, category I includes mostly notoungulates and one cingulate. Category II is represented only by notoungulates. Category III is equally represented by cingulates and sloths. Category IV is equally represented by astrapotheres and pyrotheres.

During the Miocene Santacrucian, category I is represented by similar proportions of litopterns, cingulates, and sloths. Category II includes mainly sloths and a minor proportion of notoungulates, while category III is represented by equal proportion of these orders. Category IV is made up of one astrapothere, with a mass estimation slightly above 1000 kg.

During the Pliocene Chapadmalalan, category I is represented by a single artiodactyl, although one mesotheriid notoungulate has mass estimates close to the lower limit. Category II includes 3 cingulates, 2 sloths, 1 notoungulate, 1 litoptern, and 1 rodent. Category III is represented by 3 cingulates, 1 sloth, and 1 rodent, while category IV is represented by only 1 sloth and 1 notoungulate.

During the early Pleistocene Ensenadan, category I is represented mainly by 2 artiodactyls, 1 cingulate, and 1 notoungulate. In Category II, artiodactyls are dominant while perissodactyls, cingulates, and rodents are represented in lower proportions. Category III is entirely composed of cingulates. Cingulates and sloths largely dominate category IV, which also includes 1 notoungulate, 1 litoptern, and 1 proboscidean.

During the late Pleistocene-early Holocene Lujanian, artiodactyls dominate

Table 5.2 Number (and percentages) of taxa in body mass categories I to IV represented in each age.

BMC	TAXA	CAS	DES	SAN	CHA	ENS	LU
Category I (44 to 100 kg)	Cingulata		1 (17%)	2 (34%)		1 (25%)	2 (33%)
	Folivora			2 (34%)			
	Notoungulata	12 (92%)	5 (83%)			1 (25%)	
	Litopterna			2 (34%)			
	Astrapotheria	1 (8%)					
	Artiodactyla				1 (100%)	2 (50%)	4 (67%)
Cat. II (100 to 500 kg)	Cingulata				3 (37.5%)	2 (22%)	1 (17%)
	Folivora			4 (67%)	2 (25%)		
	Notoungulata	1 (14%)	4 (100%)	2 (33%)	1 (12.5%)		
	Litopterna	2 (29%)			1 (12.5%)		
	Pyrotheria	1 (14%)					
	Astrapotheria	3 (43%)					
	Rodentia				1 (12.5%)	1 (11%)	1 (17%)
	Perissodactyla					2 (22%)	2 (33%)
	Cetartiodactyla					4 (45%)	2 (33%)
Cat. III (500 to 1000 kg)	Cingulata		1 (50%)		2 (40%)	4 (100%)	3 (75%)
	Folivora		1 (50%)	1 (50%)	2 (40%)		
	Notoungulata			1 (50%)			
	Litopterna						1 (25%)
	Rodentia				1 (20%)		
Cat. IV (more than 1000 kg)	Cingulata					5 (43%)	4 (36%)
	Folivora				1 (50%)	4 (33%)	5 (46%)
	Notoungulata				1 (50%)	1 (8%)	1 (9%)
	Litopterna					1 (8%)	
	Pyrotheria		1 (50%)				
	Astrapotheria		1 (50%)	1 (100%)			
	Proboscidea					1 (8%)	1 (9%)

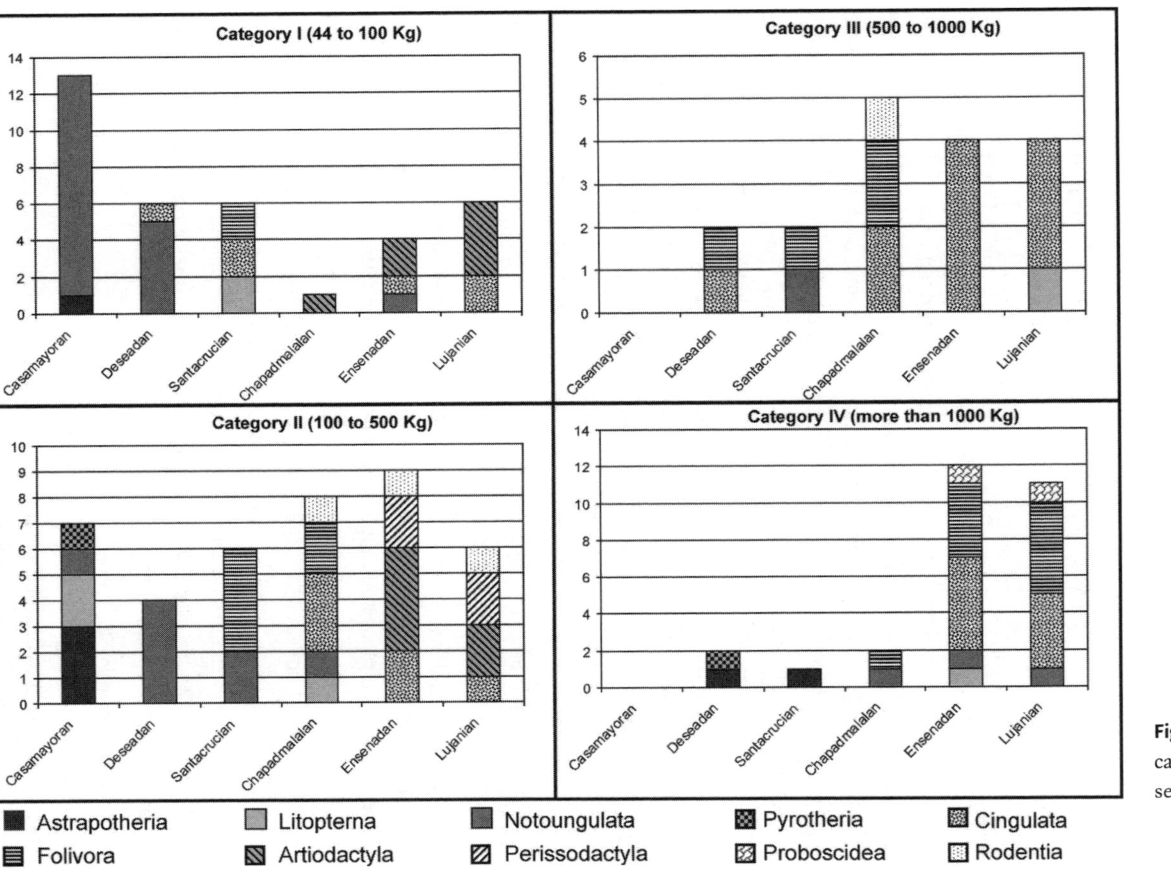

Figure 5.3 Body mass categories and their representation through time.

category I, the remaining members being cingulates. Category II is dominated by artiodactyls and perissodactyls in equal proportions, and also includes 1 rodent and 1 cingulate. Category III is mostly represented by cingulates and 1 litoptern. Sloths and cingulates largely dominate category IV, which also includes 1 notoungulate and 1 proboscidean.

5.5 Discussion

The evolution of large size in Cenozoic mammals in South America can be divided into two main periods. The first one unites what Pascual et al. (1985, 230–234) distinguished as the "most autochthonous part of the history," and the subsequent "first major change toward a modernization." During this lapse, the basic diversification of the South American mammals occurred, as did a faunal turnover that appears to be regionally connected to climatic and environmental changes related to various geotectonic events, like the Andean diastrophic phases and the opening of the Drake Passage that separated South America from Antarctica (see Pascual 2006).

During this period, the only mega-mammals sensu stricto (i.e., those >1000 kg) were pyrotheres and astrapotheres, archaic lineages of brachydont ungulates that peaked in size in the Oligocene Deseadan Age. Two exceptionally large genera were the astrapothere *Parastrapotherium*, at an estimated 1900 kg, and the pyrothere *Pyrotherium*, at 3500 kg.

By the Miocene Santacrucian Age, pyrotheres were extinct, and astrapotheres remained the only mega-mammals of the fauna (*Astrapotherium magnum*, ~1000 kg). By this time, the archaic lineages of brachydont ungulates (including litopterns and notoungulates) had gradually changed from protohypsodont to almost euhypsodont types (Pascual 2006). Eventually, these were replaced by euhypsodont native ungulates (notoungulates, and to a lesser extent litopterns), which characterized most of the native ungulates of all subsequent Cenozoic faunas. While large litopterns reached only category I (i.e., >44 kg but <100 kg, e.g., *Diadiaphorus*, 70 kg; *Theosodon*, 95 kg), large notoungulates rose to categories II and III (i.e., 100–1000 kg: *Adinotherium*, 121 kg; *Homalodotherium*, 340 kg; *Nesodon*, 554 kg). At the same time, large xenarthrans are more diverse but smaller than in the previous age, although sloths rivaled notoungulates in size.

Culminating in the Pliocene, the volcanic Isthmus of Panama rose up from the sea floor and bridged the formerly separated continents of North and South America. This geologic event had an extraordinary impact on South American faunas, including the evolution of large size. It allowed a bidirectional migration of fauna and flora between the continents, with its most dramatic effects on the mammals. This is the start of the second period considered here.

During the Pliocene-Pleistocene, and probably during the early Holocene, large native ungulates, now represented only by notoungulates and litopterns, increased in MBM, attaining only a modest representation in the categories >500 kg. By contrast, immigrant ungulates clearly dominated the categories <500 kg.

Interestingly, the xenarthrans did not suffer apparent decrease in diversity or abundance, at least not until the widespread extinction event at the very end of the Pleistocene. Furthermore, several xenarthrans crossed the bridge and became abundant in North America, and xenarthrans are the only mega-mammals that seem to have survived (albeit only briefly) the aforementioned extinction (Gutierrez et al. 2010). The persistent success of herbivorous xenarthrans subsequent to the Interchange might be explained by their ability to avoid competition with northern placental lineages by evolving increased size. Beginning in the Pliocene, sloths, pampatheres, and glyptodonts in particular, evolved a large diversity of gigantic forms unparalleled by analogous eutherian taxa in South America. The evolutionary success of xenarthrans has been considered in terms of the taxonomic diversity of the clade through the Cenozoic and related to the unusual set of features that characterize their dentitions. These features made them particularly well suited to dealing with an increase in the abundance of dentally abrasive particles in their environment (Vizcaíno 2009). Pascual (2006) and Ortiz-Jaureguizar and Cladera (2006) reviewed environmental and ecological changes in South America during the Cenozoic.

In the Pliocene Chapadmalalan, rodents attained unusually large size with *Chapalmatherium* (~200 kg) and *Telicomys* (~600 kg). *Chapalmatherium* is a hydrochoerid, related to extant amphibious capybaras (~60 kg), whereas *Telicomys* is a dinomyid related to the living pacaranas (10–15 kg). Other rodents not included in our analysis reached even larger sizes. During the late Miocene and early Pliocene of Argentina, Brazil, and Venezuela, the neoepiblemid (a chinchilloid) *Phoberomys* reached 700 kg (Horowitz et al. 2006; Sánchez-Villagra, Aguilera, and Horovitz 2003). The Pliocene-early Pleistocene dinomyid *Josephoartigasia monesi* (Uruguay) was proposed as the largest known rodent, with a body mass ~1000 kg (Rinderknecht and Blanco 2008). Since then, MBM of rodents has decreased. By the end of the Pleistocene, only hydrochoerids maintained large sizes (*Neochoerus*, ~100 kg), but even this exceeds the size of living rodents.

Mammals reached their most spectacular sizes in the Pleistocene. For instance, since the description of the first specimen ever collected from the Lujanian fauna—the ground sloth *Megatherium americanum*—the large body size of its members has been remarkable (see Simpson, 1980). This fauna compares very favorably to that of present-day savannas of Africa, where four or five species have adult body masses exceeding a ton. The number of Lujanian mammal species larger than a ton is even more impressive than in present-day African faunas; as many as 19 such species occur in a single locality (Fariña and Vizcaíno 1999). This presents interesting intellectual challenges to a paleobiologist. The question of the trophic relationships of the Lujanian megafauna has been addressed by Fariña (1996), following the general ecological relationships between population density and body size, and between basal metabolic rate and body size. Fariña considered that there are 30 species of herbivorous mammals greater than 10 kg found in the Luján local fauna, more than half of them (16) are xenarthrans: 9 glyptodonts, 1 pampathere, and 6 ground sloths. The list is completed by 1 giant rodent, 4 notoungulates, 1 litoptern, 1 perissodactyl, 6 artiodactyls, and a gomphothere. The energy requirements for each species were obtained by multiplying its standing biomass by its basal metabolic rate. Introducing some assumptions, Fariña estimated their consumption of the habitat's primary productivity, and concluded that there was an excess of herbivores in relation to the plant resources available and in relation to estimates of carnivore biomass. By the process of elimination, Fariña reached the surprising conclusion that ground sloths may have had scavenging habits, which may have helped to redress this ecological imbalance.

During the last decade, new research has promoted alternative interpretations. For instance, Vizcaíno, Bargo, and Cassini (2006) emphasized that the Pleistocene communities were dominated by mega-mammals of very low metabolism (xenarthrans), which have no counterpart in living faunas. These authors found that xenarthrans have less dental occlusal surface area available for triturating food than other placental mammals of similar sizes, and related this to the low basal metabolic rates characteristic of living xenarthrans (between 40% and 60% of the rates expected from mass via Kleiber's relation for placental mammals; McNab 1985). This implies that xenarthrans have lower energetic requirements than epitherians and, therefore, for a specific type of food, required lower intake than other placental mammals of similar body masses. Other research has proposed that if high herbivore biomass occurred during the Lujanian, then a higher density of carnivores could be supported than was inferred from the power function of body size and population density (Prevosti and Vizcaíno 2006).

New evidence indicates that the large mammal and mega-mammal faunas were very well diversified at the end of the Pleistocene and not suffering from any kind of declining trend in prior time intervals (Cione, Tonni, and Soibelzon 2009). Large size explains most of their vulnerability to extinction at the end of the Pleistocene in South America over other factors such as the geographic origin of the lineages (South American vs. North American) and dietary habits (herbivory vs. carnivory; Lessa and Fariña 1996). All mega-mammals (37 species) and most large mammals (46 species) present during the late Lujanian became extinct in South America (Cione, Tonni, and Soibelzon 2003, 2009). However, an analysis of the causes of large mammal extinctions in Australia, Eurasia, the Americas, and Madagascar during the late Quaternary concluded that large size was not directly related to risk of extinction. Rather, species with slow reproductive rates were at high risk regardless of their body size (Johnson 2002). This provides a biological framework for analyzing the extinction of large South American mammals. Cione, Tonni, and Soibelzon (2003, 2009) suggested that females probably reached sexual maturity late, had a very long gestation period and prolonged parental care, implying only one offspring in two or three years, and had a low lifetime reproductive yield. This reinforces the idea that large size influences extinction risk through several biological traits.

Humans arrived in South America ca. 13,000–11,000 BP. Some mega-mammals became extinct in South America during the early Holocene, perhaps as late as 7000 BP (but see Steele and Politis 2009). This extinction event lasted several thousand years: there is little support for the blitzkrieg extinction model for South America. However, a relatively small number of human foragers with specialized weapons could have been indirectly responsible for these extinctions if they focused their hunting efforts on key individuals (females, juveniles, or infants). This could have provoked a cascade of extinctions in herbivores as well as in large carnivores that preyed on the herbivores (Cione, Tonni, and Soibelzon 2003, 2009).

Johnson (2009) summarized the possible ecological consequences of late Quaternary megafaunal extinctions. Johnson maintained that there were significant changes in plant communities following megafaunal extinctions, and that the ecological aftershocks of those extinctions persist today (Johnson 2009). According to this author, large herbivores maintained biodiverse and complex habitats in dry, lowland, wooded landscapes. Some of these habitats became impoverished as a result of herbivore extinctions, while others contain anachronistic plants that may be in long-term decline.

Although the impact of the megafaunal extinctions on the evolution of Holocene plant communities has not been studied for South America, it clearly produced an enormous ecological gap in the herbivorous guild. As Darwin stated at the very beginning of his career: "It is impossible to reflect without the deepest astonishment, on the changed state of this continent. Formerly it must have swarmed with great monsters, like the southern parts of Africa, but now we find only the tapir, guanaco, armadillo, capybara; mere pigmies compared to antecedent races" (Darwin, 1839, 210).

The conditions that allowed the establishment of the ecosystem of the Pampean Region, as it was known by the Spanish conquerors, appeared very recently (Cione, Tonni, and Soibelzon 2009), probably ca. 1000 years BP (Tonni and Cione 1997). At least in this region, the herbivorous gap mentioned earlier persisted for about 6000 years, until it was filled by herds of feral cattle whose ancestors were introduced by the Spanish in the second half of the sixteenth century. By the end of that century, cattle had become so numerous that the trade in cow hides became one of the main economic colonial activities for the next two centuries. As happened with bison in North America in the second half of the nineteenth century, cattle were hunted for their hides, with the carcass left behind to decay on the ground. It also affected the economy of the native people, who shifted from hunting camelids (*Lama guanicoe*), deer (*Ozotoceros bezoarticus*), and rheas (*Rhea americana*), each of which produce less than 60 kg of meat, to hunting cattle, which weighed 400–800 kg (Ramos et al. 2008).

In conclusion, we emphasize Johnson's (2009) view that to understand living plant communities, we need to imagine them with the full complement of Pleistocene megafauna that shaped their evolutionary histories. To do so, more paleobiological studies are needed. The great phylogenetic distance between some mega-mammals and their living counterparts raises serious problems in understanding their paleobiology. This is especially true for those faunas in which xenarthrans dominated. Particularly for the Pleistocene, communities dominated by hypometabolic mega-mammals (xenarthrans) have no counterpart in living faunas. Hence, paleoecological reconstructions lack strict analogues; therefore, alternatives to purely comparative actualistic approaches must be used.

Acknowledgments

We are grateful to the conveners of the Symposium on Historical Biogeography of Neotropical Mammals, Bruce Patterson and Leonora Costa, for inviting us to participate. A. Abello, A. Candela, M. Reguero, and M. G. Vucetich provided us with information on body mass estimates. J. M. Perry, M. G. Vucetich, and M. Reguero critically reviewed early versions of the manuscript. Two reviewers made suggestions that improved this article. This is a contribution to the projects PICT 26219, UNLP N 474, and CONICET-PIP 1054.

Literature Cited

Alberdi, M. T., and J. L. Prado. 2004. *Caballos Fósiles de América del Sur: Una Historia de Tres Millones de Años*. Olavarría: INCUAPA.

Bargo, M. S. 2001. "The Ground Sloth *Megatherium americanum*: Skull Shape, Bite Forces, and Diet." In *Biomechanics and Paleobiology of Vertebrates*, edited by S. F. Vizcaíno, R. A. Fariña, and C. Janis, 41–60. Warsaw: *Acta Paleontologica Polonica*, 46.

———. 2004. "Cenozoic Giants of South America." *Journal of Morphology* 260:276.

Bargo, M. S., and S. F. Vizcaíno. 2008. "Paleobiology of Pleistocene Ground Sloths (Xenarthra, Tardigrada): Biomechanics, Morphogeometry and Ecomorphology Applied to the Masticatory Apparatus." *Ameghiniana* 45:175–96.

Bargo, M. S., S. F. Vizcaíno, and R. F. Kay. 2009. "Predominance of Orthal Masticatory Movements in the Early Miocene *Eucholaeops* (Mammalia, Xenarthra, Tardigrada, Megalonychidae) and Other Megatherioid Sloths." *Journal of Vertebrate Paleontology* 29:870–80.

Barreda, V., L. M. Anzótegui, A. R. Prieto, P. Aceñolaza, M. M. Bianchi, A. M. Borromei, M. Brea et al. 2007. "Diversificación y Cambios de las Angiospermas Durante el Neógeno en Argentina." *Asociación Paleontológica Argentina, Publicación Especial* 11:173–91.

Christiansen, P., and R. A. Fariña. 2003. "Mass Estimation of Two Fossil Ground Sloths (Xenarthra; Mylodontidae)." *Senckenbergiana Biologica* 83:95–101.

Cifelli, R. L. 1985. "South American Ungulate Evolution and Extinction." In *The Great American Biotic Interchange*, edited by F. G. Stehli and S. D. Webb, 249–66. New York: Plenum.

———. 1993. "The Phylogeny of the Native South American Ungulates." In *Mammal Phylogeny*, Vol. 2, edited by F. S. Szalay, M. L. Novacek, and M. C. McKenna, 195–216. New York: Springer-Verlag.

Cione, A. L., and E. P. Tonni. 2005. "Bioestratigrafía Basada en Mamíferos del Cenozoico Superior de la Provincia de Buenos Aires, Argentina." In *Geología y Recursos Minerales de la Provincia de Buenos Aires*, Relatorio 11, edited by R. E. de Barrio, R. O. Etcheverry, M. F. Caballé, and E. Llambías, 183–200. La Plata: 16° Congreso Geológico Argentino.

Cione, A. L., E. P. Tonni, and L. H. Soibelzon. 2003. "The Broken Zig-Zag: Late Cenozoic Large Mammal and Turtle Extinction in South America." *Revista del Museo Argentino de Ciencias Naturales "Bernardino Rivadavia"* 5:1–19.

———. 2009. "Did Humans Cause the Late Pleistocene-Early Holocene Mammalian Extinctions in South America in a Context of Shrinking Open Areas?" In *American Megafaunal Extinctions at the End of the Pleistocene*, edited by G. Haynes, 125–44. Vertebrate Paleobiology and Paleoanthropology Series. New York: Springer.

Croft, D. A. 1999. "Placentals: Endemic South American Ungulates." In *The Encyclopedia of Paleontology*, edited by R. Singer, 890–906. Chicago: Fitzroy-Dearborn Publishers.

———. 2000. "Archaeohyracidae (Mammalia: Notoungulata) from the Tinguiririca Fauna, Central Chile, and the Evolution and Paleoecology of South American Mammalian Herbivores." Unpublished PhD diss., University of Chicago.

Croft, D. A., and D. Weinstein. 2008. "The First Application of the Mesowear Method to Endemic South American Ungulates (Notoungulata)." *Palaeogeography, Palaeoclimatology, Palaeoecology* 269:103–14.

Darwin, C. R. 1839. *Narrative of the Surveying Voyages of His Majesty's Ships Adventure and Beagle Between the Years 1826 and 1836, Describing Their Examination of the Southern Shores of South America, and the Beagle's Circumnavigation of the Globe, Volume 3. Journal and Remarks 1832–1836*. London: Henry Colburn Press.

Delsuc F., and E. J. P. Douzery. 2008. "Recent Advances and Future Prospects in Xenarthran Molecular Phylogenetics." In *The Biology of the Xenarthra*, edited by S. F. Vizcaíno and W. J. Loughry, 11–23. Gainesville: University Press of Florida.

Delsuc, F., M. Scally, O. Madsen, M. J. Stanhope, W. W. De Jong, F. M. Catzeflis, M. S. Springer, and E. J. P. Douzery. 2002. "Molecular Phylogeny of Living Xenarthrans and the Impact of Character and Taxon Sampling on the Placental Tree Rooting." *Molecular Biology and Evolution* 19:1656–71.

Fariña, R. A. 1995. "Limb Bone Strength and Habits in Large Glyptodonts." *Lethaia* 28: 189–96.

———. 1996. "Trophic Relationships Among Lujanian Mammals." *Evolutionary Theory* 11:125–34.

Fariña, R. A., and E. R. Blanco. 1996. "*Megatherium*, the Stabber." *Proceedings of the Royal Society of London, B, Biological Sciences* 263:1725–29.

Fariña R. A., and S. F. Vizcaíno. 1997. "Allometry of the Leg Bones in Armadillos (Mammalia, Dasypodidae): A Comparison with Other Mammals." *Zeitschrift für Säugetierkunde* 62:65–70.

———. 1999. "A Century After Ameghino: The Palaeobiology of the Large Quaternary Mammals of South America Revisited." *Quaternary of South America and the Antarctic Peninsula* 12:255–77.

———. 2003. "Slow Moving or Browsers? A Note on Nomenclature." In *Morphological Studies in Fossil and Extant Xenarthra (Mammalia)*, edited by R. A. Fariña, S. F. Vizcaíno, and G. Storch, 3–4. Frankfurt am Main: Senckenbergiana Biologica, 83.

Fariña, R. A., S. F. Vizcaíno, and M. S. Bargo. 1998. "Body Mass Estimations in Lujanian (Late Pleistocene-Early Holocene of South America) Mammal Megafauna." *Mastozoología Neotropical* 5:87–108.

Frenguelli, J. 1953. "La Flora Fósil de la Región del Alto Río Chalia en Santa Cruz (Patagonia)." *Notas Museo de La Plata (Paleontología)* 16:239–57.

Gardner, A. L. 2005. "Order Cingulata; Order Pilosa." In *Mammal Species of the World: A Taxonomic and Geographic Reference*, 3rd ed., edited by D. E. Wilson and D. M. Reeder, 94–103. Baltimore: Johns Hopkins University Press.

Gaudin, T. J., and H. G. McDonald. 2008. "Morphology-Based Investigations of the Phylogenetic Relationships Among Extant and Fossil Xenarthrans." In *The Biology of the Xenarthra*, edited by S. F. Vizcaíno and J. W. Loughry, 24–36. Gainesville: University of Florida Press.

Gaudin, T. J., J. R. Wible, J. A. Hopson, and W. D. Turnbull. 1996. "Reexamination of the Morphological Evidence for the Cohort Epitheria (Mammalia, Eutheria)." *Journal of Mammalian Evolution* 3:31–79.

Gelfo, J. N., G. M. López, and M. Bond. 2008. "A New Xenungulata (Mammalia) from the Paleocene of Patagonia, Argentina." *Journal of Paleontology* 82:329–35.

Gutierrez, M. A., G. A. Martinez, M. S. Bargo, and S. F. Vizcaíno. 2010. "Supervivencia Diferencial de Mamíferos de Gran Tamaño en la Región Pampeana en el Holoceno Temprano y su Relación con Aspectos Paleobiológicos." In *Zooarqueología a Principios del Siglo XXI: Aportes Teóricos, Metodológicos y Casos de Estudio*, edited by M. A. Gutiérrez, M. De Nigris, P. M. Fernández, M. Giardina, A. F. Gil, A. Izeta, G. Neme, and H. D. Yacobaccio, 231–41.

Hinojosa, L. F. 2005. "Cambios Climáticos y Vegetacionales Inferidos a Partir de Paleofloras Cenozoicas del sur de Sudamérica." *Revista Geológica de Chile* 32:95–11.

Hinojosa, L. F., and C. Villagrán. 1997. "Historia de los Bosques del sur de Sudamérica, I: Antecedentes Paleobotánicos, Geológicos y Climáticos del Terciario del Cono sur de América." *Revista Chilena de Historia Natural* 70:225–39.

Horovitz, I., M. R. Sánchez-Villagra, T. Martin, and O. A. Aguilera. 2006. "The Fossil Record of *Phoberomys pattersoni* Mones 1980 (Mammalia, Rodentia) from Urumaco (Late Miocene, Venezuela), with an Analysis of its Phylogenetic Relationships." *Journal of Systematic Palaeontology* 4:293–306.

Johnson, C. N. 2002. "Determinants of Loss of Mammal Species During the Late Quaternary 'Megafauna' Extinctions: Life History and Ecology, but not Body Size." *Proceedings of the Royal Society of London, B, Biological Sciences* 269:2221–27.

———. 2009. "Ecological Consequences of Late Quaternary Extinctions of Megafauna." *Proceedings of the Royal Society of London, B, Biological Sciences* 276:2509–19.

Johnson, S., and R. Madden. 1997. "Uruguaytheriine Astrapotheres of Tropical South America." In *Vertebrate Paleontology in the Neotropics: The Miocene Fauna of La Venta, Colombia*, edited by R. Kay, R. Madden, R. Cifelli, and J. Flynn, 355–82. Washington, DC: Smithsonian Institution Press.

Kramarz, A. G., and M. Bond. 2008. "Revision of *Parastrapotherium* (Mammalia, Astrapotheria) and other Deseadan Astrapotheres of Patagonia." *Ameghiniana* 45:537–51.

Lessa, E. P., and R. A. Fariña. 1996. "Reassessment of Extinction Patterns Among the Late Pleistocene Mammals of South America." *Palaeontology* 39:651–62.

Marshall, L. G., and R. L. Cifelli. 1989. "Analysis of Changing Diversity Patterns in Cenozoic Land Mammal Age Faunas, South America." *Palaeovertebrata* 19:169–210.

Marshall, L. G., and C. de Muizon. 1988. "The Dawn of the Age of Mammals in South America." *National Geographic Research* 4:23–55.

Martin, P. S., and D. W. Steadman. 1999. "Prehistoric Extinctions on Islands and Continents." In *Extinctions in Near Time: Causes, Contexts and Consequences*, edited by R. D. E. MacPhee, 17–56. New York: Kluwer/Plenum.

McKenna, M. 1975. "Toward a Phylogenetic Classification of the Mammalia." In *Phylogeny of the Primates: A Multidisciplinary Approach*, edited by P. Luckett and F. Szalay, 21–46. New York: Plenum.

McNab, B. K. 1985. "Energetics, Population Biology, and Distribution of Xenarthrans, Living and Extinct." In *Evolution and Ecology of Armadillos, Sloths and Vermilinguas*, edited by G. G. Montgomery, 219–32. Washington, DC: Smithsonian Institution Press.

Menéndez, C. 1971. "Floras Terciarias de la Argentina." *Ameghiniana* 8:357–70.

Novacek, M. J. 1992. "Mammalian Phylogeny: Shaking the Tree." *Nature* 356:121–5.

Novacek, M. J., and A. R. Wyss. 1986. "Higher-Level Relationships of the Recent Eutherian Orders: Morphological Evidence." *Cladistics* 2:257–87.

Ortiz-Jaureguizar, E., and G. A. Cladera. 2006. "Paleoenvironmental Evolution of Southern South America During the Cenozoic." *Journal of Arid Environments* 66:498–532.

Pascual, R. 2006. "Evolution and Geography: The Biogeographic History of South American Land Mammals." *Annals of the Missouri Botanical Garden* 93:209–30.

Pascual, R., and E. Ortiz-Jaureguizar. 2007. "The Gondwanan and South American Episodes: Two Major and Unrelated Moments in the History of the South American Mammals." *Journal of Mammalian Evolution* 14:75–137.

Pascual, R., M. G. Vucetich, G. J. Scillato-Yané, and M. Bond. 1985. "Main Pathways of Mammalian Diversification in South America." In *The Great American Biotic Interchange*, edited by F. Stehli and S. D. Webb, 219–47. New York: Plenum.

Poux, C., P. Chevret, D. Huchon, W. W. de Jong, and E. J. P. Douzery. 2006. "Arrival and Diversification of Caviomorph Rodents and Platyrrhine Primates in South America." *Systematic Biology* 55:228–44.

Prasad, A. B., M. W. Allard, NISC Comparative Sequencing Program, and E. D. Green. 2008. "Confirming the Phylogeny of Mammals by Use of Large Comparative Sequence Data Sets." *Molecular Biology and Evolution* 25:1795–1808.

Prevosti, F. J., and S. F. Vizcaíno. 2006. "Paleoecology of the Large Carnivore Guild from the Late Pleistocene of Argentina." *Acta Palaeontologica Polonica* 51:407–22.

Ramos, M., M. Lanza, F. Bognann, and V. Helfer. 2008. "Implicancias Arqueológicas Respecto del Ganado Introducido y el Tráfico de los Cimarrones." *Tefros* 6:1–24.

Reguero, M. A. 1999. "El Problema de las Relaciones Sistemáticas y Filogenéticas de los Typotheria y Hegetotheria (Mammalia, Notoungulata): Análisis de los Taxones de Patagonia de la Edad-Mamífero Deseadense (Oligoceno)." Unpublished PhD diss., Universidad Nacional de Buenos Aires, Facultad de Ciencias Exactas y Naturales.

Rinderknecht, A., and R. E. Blanco. 2008. "The Largest Fossil Rodent." *Proceedings of the Royal Society of London, B, Biological Sciences* 275:923–8.

Romero, E. J. 1978. "Paleoecología y Paleofitogeografía de las Tafofloras del Cenofítico de Argentina y Áreas Vecinas." *Ameghiniana* 15:209–27.

———. 1986. "Paleogene Phytogeography and Climatology of South America." *Annals of Missouri Botanical Garden* 73:449–61.

Rose, K. D., and R. J. Emry. 1993. "Relationships of Xenarthra, Pholidota, and Fossil Edentates: The Morphological Evidence." In *Mammal Phylogeny: Placentals*, edited by F. Szalay, M. C. McKenna, and M. J. Novacek, 81–102. New York: Springer-Verlag.

Rose, K. D., R. J. Emry, T. J. Gaudin, and G. Storch. 2005. "Xenarthra and Pholidota." In *The Rise of Placental Mammals: Origins and Relationships of the Major Extant Clades*, edited by K. D. Rose and J. D. Archibald, 106–26. Baltimore: Johns Hopkins University Press.

Sánchez-Villagra, M. R., O. Aguilera, and I. Horovitz. 2003. "The Anatomy of the World's Largest Extinct Rodent." *Science* 301:1708–9.

Shockey, B. J. and Anaya, F. 2004. "*Pyrotherium macfaddeni*, sp. nov. (late Oligocene, Bolivia) and the Pedal Morphology of Pyrotheres." *Journal of Vertebrate Paleontology* 24:481–88.

Simpson, G. G. 1950. "History of the Fauna of Latin America." *American Scientist* 38:361–89.

———. 1980. *Splendid Isolation: The Curious History of South American Mammals*. New Haven: Yale University Press.

Steele, J., and G. Politis. 2009. "AMS 14C Dating of Early Human Occupation of Southern South America." *Journal of Archaeological Science* 36:419–29.

Tauber, A. A. 1997. "Bioestratigtrafía de la Formación Santa Cruz (Mioceno Inferior) en el Extremo Sudeste de la Patagonia." *Ameghiniana* 34:413–26.

Tonni, E. P., and A. L. Cione. 1997. "Did the Argentinean Pampean Ecosystem Exist in the Pleistocene?" *Current Research in the Pleistocene* 14:131–3.

Troncoso, A., and E. J. Romero. 1998. "Evolución de las Comunidades Florísticas en el Extremo sur de Sudamérica Durante el Cenofítico." In *Proceedings of the Congreso Latinoamericano de Botánica*, No. 6, Monographs in Systematic Botany, edited by R. Fortunato and N. Bacigalupo, 149–72. St Louis: Missouri Botanical Garden.

Villagrán, C., and L. F. Hinojosa. 1997. "Historia de los Bosques del sur de Sudamérica, II: Análisis Fitogeográfico." *Revista Chilena de Historia Natural* 70:241–67.

Vizcaíno, S. F. 2009. "The Teeth of the 'Toothless': Novelties and Key Innovations in the Evolution of Xenarthrans (Mammalia, Xenarthra)." *Paleobiology* 35:343–66.

Vizcaíno, S. F., M. S. Bargo, and G. H. Cassini. 2006. "Dental Occlusal Surface Area in Relation to Body Mass, Food Habits and Other Biological Features in Fossil Xenarthrans." *Ameghiniana* 43:11–26.

Vizcaíno, S. F., M. S. Bargo, and R. A. Fariña. 2008. "Form, Function and Paleobiology in Xenarthrans." In *The Biology of the Xenarthra*, edited by S. F. Vizcaíno and W. J. Loughry, 86–99. Gainesville: University Press of Florida.

Vizcaíno, S. F., R. A. Fariña, M. S. Bargo, and G. De Iuliis. 2004. "Functional and Phylogenetical Assessment of the Masticatory Adaptations in Cingulata (Mammalia, Xenarthra)." *Ameghiniana* 41:651–64.

Vizcaíno, S. F., R. A. Fariña, M. A. Zárate, M. S. Bargo, and P. Schultz. 2004. "Palaeoecological Implications of the Mid-Pliocene Faunal Turnover in the Pampean Region (Argentina)." *Palaeogeography, Palaeoclimatology, Palaeoecology* 213:101–13.

Vizcaíno, S. F., N. Milne, and M. S. Bargo. 2003. "Limb Reconstruction of *Eutatus seguini* (Mammalia: Dasypodidae): Paleobiological Implications." *Ameghiniana* 40:89–101.

Vizcaíno, S. F., M. A. Reguero, F. J. Goin, C. P. Tambussi and J. I. Noriega. 1998. "Community Structure of Eocene Terrestrial Vertebrates from Antarctica." *Paleógeno de América del Sur y de la Península Antártica. Asociación Paleontológica Argentina. Publicación Especial* 5, 30 (12):179–85.

Webb, S. D. 1991. "Ecogeography and the Great American Interchange." *Paleobiology* 17:266–80.

Wible, J. R., G. W. Rougier, M. J. Novacek, and R. J. Asher. 2007. "Cretaceous Eutherians and Laurasian Origin for Placental Mammals Near the K/T Boundary." *Nature* 447:1003–6.

Zachos, J., M. Pagani, L. Sloan, E. Thomas, and K. Billups. 2001. "Trends, Rhythms, and Aberrations in Global Climate 65 Ma to Present." *Science* 292:686–93.

Evolution of the South American Carnivores (Mammalia, Carnivora)
A Paleontological Perspective

Francisco J. Prevosti and Leopoldo H. Soibelzon

Abstract

Although the history of placental carnivores (order Carnivora) in South America is relatively short, they are a successful and diverse group. Carnivores, like many other taxa, entered South America from North America during an event called the "Great American Biotic Interchange" (GABI). Most families, genera, and species are recorded since the Early Pleistocene (~1.8 Ma), but the oldest records are represented by procyonids found in Late Miocene levels (6–7 Ma), followed by mustelids and canids in the Late Pliocene (~2.5 Ma). The available evidence suggests that the immigration of placental carnivores to South America is not related to the extinction of the native carnivores (Sparassodonta, Metatheria). We review the fossil record of South American carnivores based on the latest taxonomic, phylogenetic, and biostratigraphic studies to investigate their patterns of origin and diversification. During the Miocene-Pliocene, the diversity of carnivores is lower than in the Pleistocene and most species were small and omnivorous, but in the Early Pleistocene, diversity increased, reaching levels somewhat lower than in the present. Size and diet disparity were also augmented in the Early Pleistocene with the presence of hypercarnivore, omnivore, mesocarnivore, and piscivore species of various sizes (ranging from 1–1000 kg). The lacks of records or low diversity observed in several ages (e.g., Barrancalobian, Bonaerian, and Platan) are mostly related to taphonomic or analytical biases. The taphonomic bias against tropical areas is a key problem in the South American record, because almost all records before the Late Pleistocene come from the southern part of the continent. The available information suggests that Recent and fossil carnivores invaded South America from Central America in several independent events, but that speciation within South America also produced many species and several genera.

6.1 Introduction

South America possesses a rich fauna of eutherian carnivorans (Mammalia: Carnivora), comprising ~47 species, with a large diversity of canids, felids, and procyonids (Wozencraft 2005). Most of these species are endemic to South America, while others have a Neotropical distribution (e.g., *Eira barbara*) and a few are actually Pan-American (e.g., *Puma concolor*). This diversity is amazing, especially taking into account that the order Carnivora is a relatively recent group in South America. Like many other taxa of Holarctic origin, most of the carnivores arrived in South America through the Panamanian Isthmus as part of the biogeographic event called the Great American Biotic Interchange (GABI) (Marshall et al. 1982). However, this biogeographic event started before the disappearance of the marine barrier that separated South America and North America, around the Pliocene–Pleistocene boundary (ca. 4–2.5 Ma; see Woodburne, Cione, and Tonni 2006), as indicated by the fossil record of Procyonidae in South America. Thus, carnivores are recorded in South America from late Miocene to Recent times by representatives of the families Felidae, Canidae, Ursidae, Mustelidae, and Procyonidae (Soibelzon and Prevosti 2007).

Previous authors (Cione and Tonni 1995; Prevosti, Gasparini, and Bond 2006; Soibelzon and Prevosti 2007; Woodburne, Cione, and Tonni 2006) have suggested that the immigration of carnivorans to South America occurred in a "step-like" pattern. The first record corresponds to the Huayquerian (late Miocene) procyonids; then at the end of the Pliocene (Vorohueian), the canids (Caninae) and mustelids (Mustelinae) appeared. After the Plio-Pleistocene boundary (Ensenadan), the carnivore guild in South America peaked in diversity, when Ursidae, Felidae, Mephitinae, Lutrinae, and large canids were first recorded (fig. 6.1).

Before the arrival of carnivorans, the mammalian carnivore guild in South America was composed of metatherians. The last record of a large metatherian carnivore corresponds to *Parahyaenodon argentinus* and *Thylacosmilus atrox* in the Early and "middle" Pliocene, respectively (Forasiepi, Martinelli, and Goin 2007). The subsequent appearance of large eutherian carnivores (Canidae, Felidae, and Ursidae) has given rise to several interpretations about the ecological roles of eutherian carnivores relative to those of metatherians and the causes of extinction of sparassodont carnivores (e.g., "competitive displacement," "replacement," "enrichment;" see Simpson 1950, 1980; also Forasiepi, Martinelli, and Goin 2007; Marshall 1982; Marshall et. al. 1982; Patterson and Pascual 1972; Reig 1981; Webb 1985). Recent works have not supported competitive displacement, instead supporting a replacement scenario (Forasiepi, Martinelli, and Goin 2007; Prevosti et al. 2009).

The aim of this contribution is to review and update the fossil record of terrestrial Carnivora in South America to infer the order's origination and evolution on this continent. We base this review on a quantification of the diversity of the group through time, its first and latest taxonomic records, and ecological

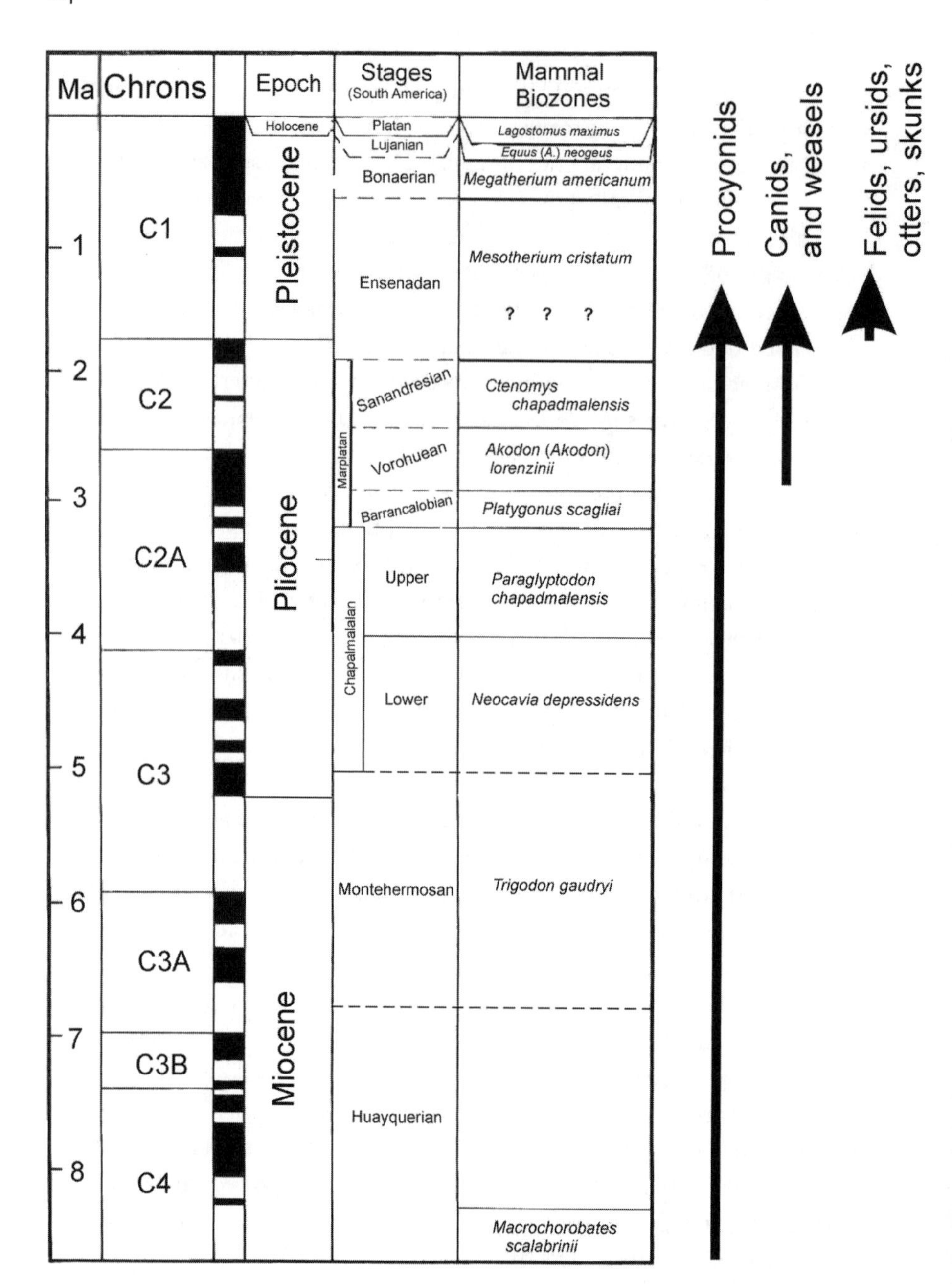

Figure 6.1 Paleomagnetic, biostratigraphic, and chronostratigraphic chart of South America, with the first occurrences of carnivores of North American origin during the Late Miocene-Pleistocene. Modified from Prevosti, Gasparini, and Bond (2006).

characteristics (diet and body mass) taken from the latest taxonomic and biostratigraphic studies. We also discuss the quality of the group's fossil record. Our review suggests that the diversity of fossil and Recent carnivores in South America is a consequence of several independent immigrations from North America (also within subfamilies and genera) and the local diversification of these immigrants. Two important extinction events are identified, one at the beginning of the Middle Pleistocene (Ensenadan), and the other at the end of the Pleistocene (Lujanian), when the largest mammals or mega-mammals disappeared. The South American carnivoran fossil record has several biases, with little or no representation in the Barrancalobian (Late Pliocene), Bonaerian (Middle-Late Pleistocene), and Holocene. During most of the time that these carnivores occurred in South America, their fossil record has been restricted to the Southern Cone, and only in the Late Pleistocene-Holocene has it included most of the continent.

6.2 Methods

We used the chronostratigraphy and biostratigraphy of Cione and Tonni (2005; modified by Woodburne, Cione, and Tonni 2006; fig. 6.1). The information on fossil taxa included in the analyses was taken from the literature (e.g., Berman 1994; Prevosti 2006a, 2006b; Prevosti and Rincón 2007; Seymour 1999; E. Soibelzon et al. 2008; L. Soibelzon 2002, 2004; L. Soibelzon and Rincón 2007; L. Soibelzon, Tonni, and Bond 2005) or from data associated with museum specimens. Institutions utilized were: CEHA–Centro de Estudios del Hombre Austral, Chile; GALY–Grupo de Arqueología del Liceo de Young, República Oriental del Uruguay; GP–Instituto de Geociencias, Universidade de Sao Paulo, Brazil; IGC–Instituto de Geociencias, Universidade Federal de Minas Gerais, Brazil; MACN–Vertebrate Paleontology, Museo Argentino de Ciencias Naturales "Bernardino Rivadavia," Argentina; MACN-zool–Mastozoology, Museo Argentino de Ciencias Naturales "Bernardino Rivadavia," Argentina; MARC–Museo y Archivo Regional Castelli, Argentina; MBLUZ–Museo de Biología de la Universidad del Zulia, Venezuela; MHJ–Museo Histórico de Junín, Argentina; MHNLP–Museo de Historia Natural de La Paz, Bolivia; MLP–Vertebrate Paleontology, Museo de La Plata, Argentina; MLP-M–Mastozoología, Museo de La Plata, Argentina; MMMP–Museo Municipal de Mar de Plata "Lorenzo Scaglia," Argentina; MMPH–Museo Municipal "Punta Hermengo," Argentina; MNHNP-PAM–Museum National de Histoire Naturelle, Pampean Collection, France; MPD–Museo Paleontológico de Daireux, Argentina; MPS–Museo

Paleontológico de San Pedro, Argentina; MPV–Museo Paleontológico de Valencia, Spain. NHM–Natural History Museum, London; PIMUZ–Paläontologisches Institut und Museum der Universität Zürich, Switzerland; UZM–Zoological Museum, University of Copenhagen, Denmark; and VF–Museo Royo y Gómez, Universidad Central de Venezuela, Venezuela.

We calculated diversity as the number of species per age, and also first and last species records. Several authors have suggested that raw diversity quantified in this way may be affected by different biases (Foote 2000; Palombo et al. 2008). To detect possible biases in the South America carnivore fossil record, we determined the number of "range-through taxon" (a taxon absent in an age or strata, but present in the overlying and underlying ones), the rank correlation (Spearman R_s) between diversity and other variables, and the number of sites per age and the age's temporal span (Foote 2000; Fortelius et al. 1996; Maas et al. 1995; Palombo et al. 2008).

Our taxonomy mainly follows Berman (1994), Berta (1989), Kraglievich (1930), Prevosti (2006a, 2006b), Seymour (1999), L. Soibelzon (2004), and other published information (e.g., Berta and Marshall 1978; Bond 1986; Mones and Rinderknecht 2004; Pomi and Prevosti 2005; Prevosti and Ferrero 2008; Prevosti and Pomi 2007; Prevosti and Rincón 2007; Prevosti, Zurita, and Carlini 2005; L. Soibelzon and Rincón 2007). Some groups, like foxes and procyonids, have only old or partial revisions, and some dubious taxa (e.g., *Dusicyon peruanus*) were excluded from this review. We also omitted taxa recognized as new but still unpublished (e.g., Berman 1987, 1989, 1994; Zetti 1972).

To characterize the ecological diversity of carnivore faunas, we used body mass and diet (cf. Van Valkenburgh 1988, 1991, 2007; Van Valkenburgh and Hertel 1998). We chose broad dietary classes because they were easy to apply to any taxon with the available information, but we recognize that these classes are segments of continuous variation and not discrete categories. Nevertheless, these categories offer adequate descriptions of the general dietary habits of carnivores and have been used in several papers (e.g., Berta 1989; Van Valkenburgh 1988, 1991, 2007; Van Valkenburgh and Hertel 1998). Carnivores were classified as: (1) hypercarnivorous, species that feed mostly on other vertebrates (mammals principally); (2) mesocarnivorous, species with diets mostly composed of vertebrates but incorporating some consistent amount of insects, fruits, or other nonvertebrate items; (3) omnivorous, species that incorporate a large proportion of nonvertebrate items like insects or vegetables; or (4) piscivorous, species with a diet composed mainly of fishes. The diet and the body mass data of extinct species were taken from the available literature (Berman 1994; Berta 1989; Figueirido and Soibelzon 2010; Prevosti 2006a; Prevosti and Vizcaíno 2006; L. Soibelzon and Tartarini 2009; Van Valkenburgh 1991; Van Valkenburgh and Hertel 1998; Van Valkenburgh and Koepfli 1993). For those species that lacked published data but have close living relatives of similar size and morphology, information on their relatives was used (e.g., *Lycalopex griseus* for the extinct *L. cultridens*). In cases where there was more than one recent representative, data were averaged. Body mass of *Cyonasua* and *Chapalmalania* was estimated following the equation for all Carnivora based on m1 mesio-distal length, as proposed by Van Valkenburgh (1990).

In order to distinguish immigration events, we used the criterion that if the taxon (or its closest relative) has an older record in Central or North America, it is counted as an immigrant. For example, the presence of *Smilodon* in South America during the Ensenadan is counted as an immigration event because *Smilodon* has older records and a sister taxon in North America. Conversely, the first occurrence of a taxon in South America was taken as an "in situ" event. To define "South American taxon," we used biogeographic and fossil evidence: these are taxa that are (and were) restricted to South America, and their oldest fossil record occurs on this continent. For example, the first record of *Smilodon populator* is counted as an in situ speciation. Phylogenetic analyses (Bardeleben, Moore, and Wayne 2005; Johnson et al. 2006; Koepfli et al. 2007, 2008; Krause et al. 2008; Prevosti 2006a) were used to classify these taxa. The poverty of the Central American fossil record may generate some false in situ speciation events, and thus our classification should be treated as a hypothesis to be tested with new fossil finds in the future. The distribution, body mass, and dietary class of Recent species was taken from Nowak (2005), Sillero Zubiri et al. (2004), Silva and Downing (1995), Sunquist and Sunquist (2002), and Wozencraft (2005). The number of sites per age was compiled from Cione and Tonni (1999), Cione, Tonni, and Soibelzon (2009), Croft (2009), Flynn and Swisher (1995), and Marshall et al. (1984). Age spans were taken from Cione and Tonni (2005) and Woodburne, Cione, and Tonni (2006), corresponding in most cases to the maximum temporal interval of each. The only exception was the Ensenadan, because most of the faunas of this age and the species included in this analysis were present in the upper part of the Ensenadan (between 1–0.5 Ma; see E. Soibelzon et al. 2008, 2009). If time averaging introduces a bias that generates higher diversity (e.g., Palombo et al. 2008), it is important to constrain such bias where possible.

6.3 Results

The diversity curve (fig. 6.2A; table 6.1) shows that eutherian carnivores were few during the Late Miocene-Pliocene (Huayquerian-Sanandresian) interval, with four or fewer species. There is no record of eutherian carnivores during the Barrancalobian (Late Pliocene). Throughout the Early-Middle Pleistocene (Ensenadan), the diversity increased steeply to nearly 20 species, and then rose to more than 35 species in the Late Pleistocene (Lujanian). Between the Lujanian and Ensenadan the number of species dropped to 7. Another drop to just over 20 species was observed from the Lujanian to the Holocene (Platan), reaching a figure similar to that of the Ensenadan. Finally, the living fauna (composed of 46 species) surpasses the fossils known from any age.

The first and last records follow more or less the same pattern as the diversity curve (fig. 6.2B). There were few first records between Late Miocene and Pliocene (Huayquerian to Marplatan), but these increased to almost 20 in the Early-Middle Pleistocene (Ensenadan) and nearly 30 in the Late Pleistocene (Lujanian). During the Bonaerian (Middle-Late Pleistocene) and Platan (Holocene), the records fell to 4 and 5, respectively. The present fauna includes 15 first records (i.e., new species not represented by fossils). Last records (extinctions) are roughly equal to first records during Late Miocene-Pliocene; during the Pleistocene the two curves are similar in shape, but there are fewer last records than new ones. It is notable that there is only one last record in the Holocene and another in the modern fauna (recent extinction). The same pattern is evident when first and last records were calculated as a percentage of diversity (fig. 6.2C). During Late Miocene-Pliocene, first and last records were roughly similar for each age, reaching 100% during Huayquerian, Montehermosan, and Chapadmalalan (only first records). During the Pleistocene-Recent, first records exceeded last records, but only reached 100% in the Ensenadan. The percentage of last records is highest in the Ensenadan.

The first immigration event was recorded in the Late Miocene. In the Late Pliocene (Vorohuean), two new immigration events are recorded, but it is in the Early to Middle Pleistocene (Ensenadan) and Late Pleistocene (Lujanian) that the immigration events increased abruptly to 8 (fig. 6.2D). No immigrations were recorded in the Middle Pleistocene (Bonaerian). After the Late Pleistocene (Lujanian), immigrations decreased to 2 in the Holocene, and 2 in recent times. The in situ originations follow the same pattern as the immigration events, but the numbers are higher for the Pleistocene and Recent, and lower for the Holocene. The percentage of immigration events and in situ speciation events

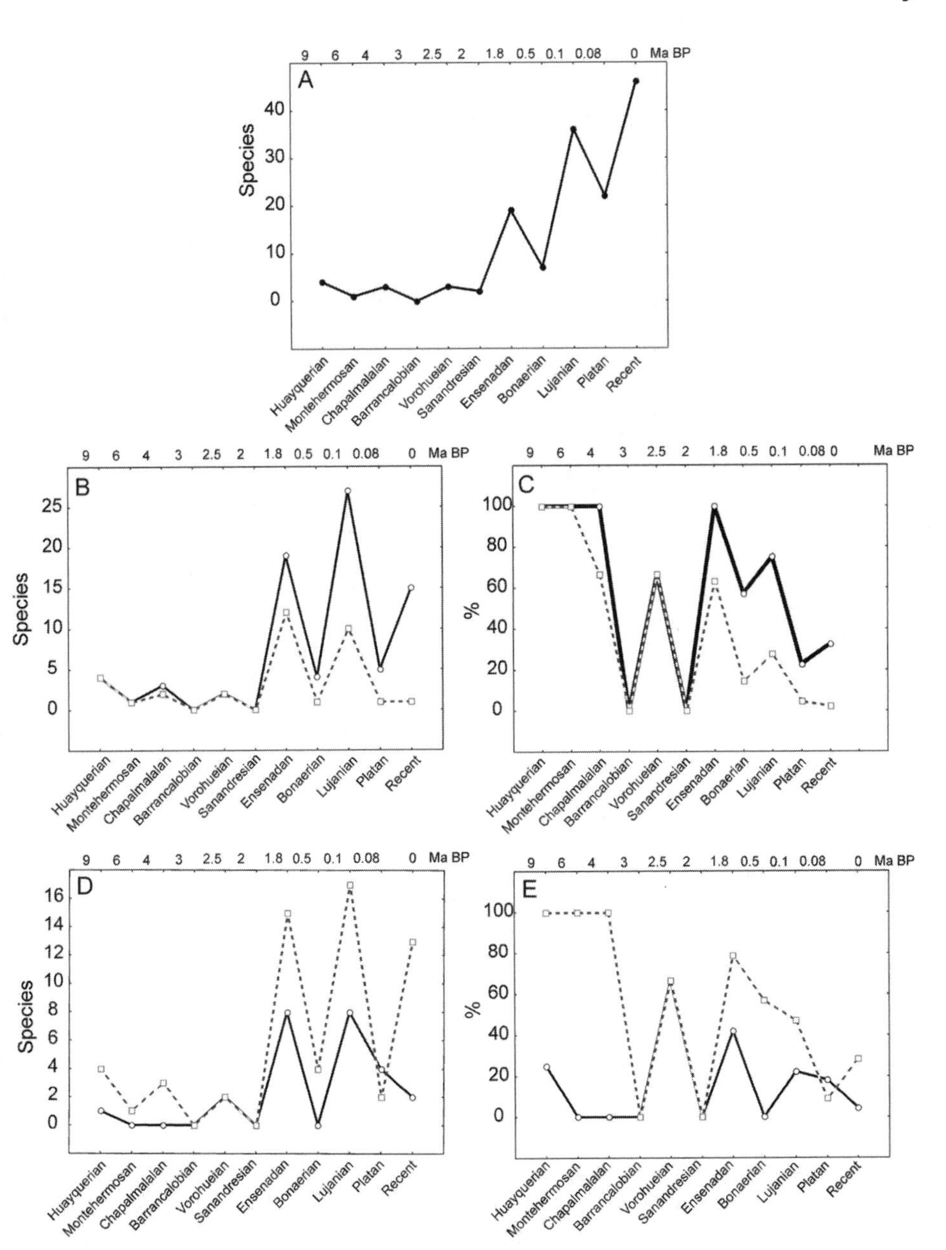

Figure 6.2 Diversity, first and last records, immigration, and in situ speciation of South American carnivores. (A): diversity; (B) and (C): absolute number and percentage of first (solid line) and last (broken line) appearances, respectively; (D) and (E): absolute number and percentage of immigration (solid line) and in situ speciation (broken line) events, respectively.

Table 6.1. Diversity, first and last records, immigration and in situ speciation events, diet, and "Lazarus taxa" of South America for each age.

Ages	Diversity	FR	LR	Mig	Spe	Laz	Hyp	Mes	Omn	Pisc	Body Mass			Number of sites	Time span (Ma)
											Median	SD	Range		
Huayquerian	4	4	4	1	4	0	0	0	4	0	6.38	0.34	5.69–6.38	12	3.80
Montehermosan	1	1	1	0	1	0	0	0	1	0	6.78			1	1.50
Chapadmalalan	3	3	2	0	3	0	0	0	3	0	22.50	9.12	6.70–22.50	9	1.70
Barrancalobian	0	0	0	0	0	1	0	0	0	0				1	0.40
Vorrohuean	3	2	2	2	2	0	1	1	1	0	3.33	11.43	2.12–22.50	2	0.50
Sanandresian	2	0	0	0	0	1	1	1	0	0	3.16	1.47	2.12–4.20	3	0.80
Ensenadan	19	19	12	8	15	0	12	3	3	1	10.33	210.22	0.23–900.00	13	(1.30) 0.50
Bonaerian	7	4	1	0	4	5	2	1	4	0	290.00	248.35	2.36–600.00	8	0.35
Lujanian	36	27	10	8	17	0	17	4	13	2	6.50	139.03	0.23–600.00	108	0.12
Platan	22	5	1	4	2	10	8	4	9	1	5.35	38.88	0.12–175.00	135	0.08
Recent	46	15	1	2	13	0	18	6	18	4	4.12	27.81	0.12–175.00		

Note: FR, first record; LR, last record; Mig, immigration events; Spe, in situ speciation events; Laz, Lazarus taxon; Hyp, hypercarnivores; Mes, mesocarnivores; Omn, omnivores; Pisc, piscivores; SD, standard deviation. Time span is in million years (MY). In the case of Ensenadan, there are two values; the duration of its later part, and between brackets its full span.

relative to diversity (fig. 6.2E) show the same pattern. In situ speciation is the only one to reach 100%, which occurred from the Huayquerian through the Chapadmalalan. During pre-Pleistocene ages both variables have similar values but in situ speciation is higher in the Pleistocene-Recent.

Although omnivorous species are the only eutherian carnivores during the Late Miocene-Middle Pliocene span (Huayquerian-Barrancalobian), their diversity was noticeably low (5 species; fig. 6.3A). This changed in the Late

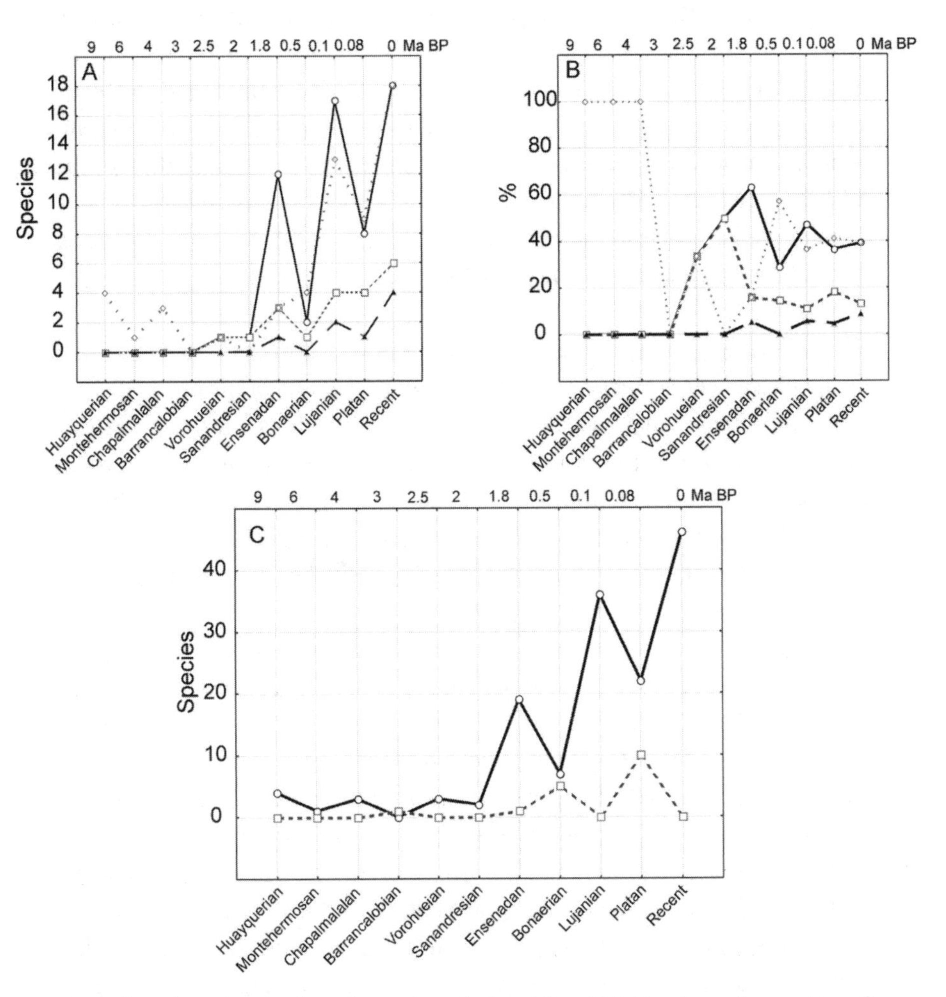

Figure 6.3 Diet and range-through taxa through time. (A) and (B): Number and percentage of hypercarnivorous (solid line and empty circles), omnivorous (dotted line and diamonds), mesocarnivorous (broken line and squares), and piscivorous species (broken lines and triangles), respectively; (C): diversity (solid line) and range-through taxa (broken line).

Pleistocene (Lujanian), when 13 omnivorous species were recorded. During the Holocene this number fell to 8, a drop that also affects other diet categories, but to different degrees. Presently, there are 14 omnivorous species in South America. The first hypercarnivorous eutherian carnivores occurred in the Late Pliocene (Vorohuean-Sanandresian), when a single species was recorded, but in the Ensenadan and Lujanian, 12 and 17 species were recorded, respectively. During the Bonaerian, the number of hypercanivores was limited to two. Currently, the number of hypercarnivores inhabiting South America is similar to that of the Lujanian. The first mesocarnivorous species were recorded in the Late Pliocene (Vorohuean-Sanandresian). Later, at the beginning of the Pleistocene, mesocarnivore diversity started to increase (3 species in the Ensenadan, 4 in the Bonaerian, 13 in the Lujanian), followed by a noticeable drop in the Holocene (8 species). At present, there are 18 species of mesocarnivores in South America. Piscivorous species are restricted to the Pleistocene-Holocene, but only in the Late Pleistocene (Lujanian) was there more than 1 species (and then only 2).

Interestingly, the percentage of diet types in each age (fig. 6.3B) shows that in the Pleistocene-Recent interval, hypercarnivorous and omnivorous taxa are better represented than mesocarnivorous and piscivorous ones. Omnivorous species dominated in pre-Pleistocene ages, at least until the Vorohuean, when hypercarnivorous and mesocarnivorous species are first recorded. Another remarkable thing is that during the Pleistocene (except in the Bonaerian), hypercarnivores were more diverse than omnivores, but in the Holocene-Recent they are less or equally represented.

Eutherian carnivores were small animals (body mass ~2–7 kg) during the Late Miocene-Pliocene, except for the giant procyonid *Chapalmalania* (~25 kg; fig. 6.4; table 6.1). The range of carnivoran body masses at each age during this interval was narrow, probably due to the low diversity and taxonomic uniformity seen prior to the Early Pleistocene. In the Early to Middle Pleistocene, this range increased dramatically, from ~0.20 kg to a maximum of ~1000 kg, although most species were in the 4–50 kg range (median = 10.3 kg). Post-Ensenadan faunas presented the same pattern, but the median and maximum values were lower (especially for Holocene to Recent faunas). The Bonaerian was exceptional because of the higher median values recorded (290 kg), even if the maximum value was lower (600 kg). However, this is an artifact of the relatively few species known from this age (7), coupled with the fact that 4 are large (3 bears and *Smilodon populator*).

There is one range-through taxon recorded from the Barrancalobian,

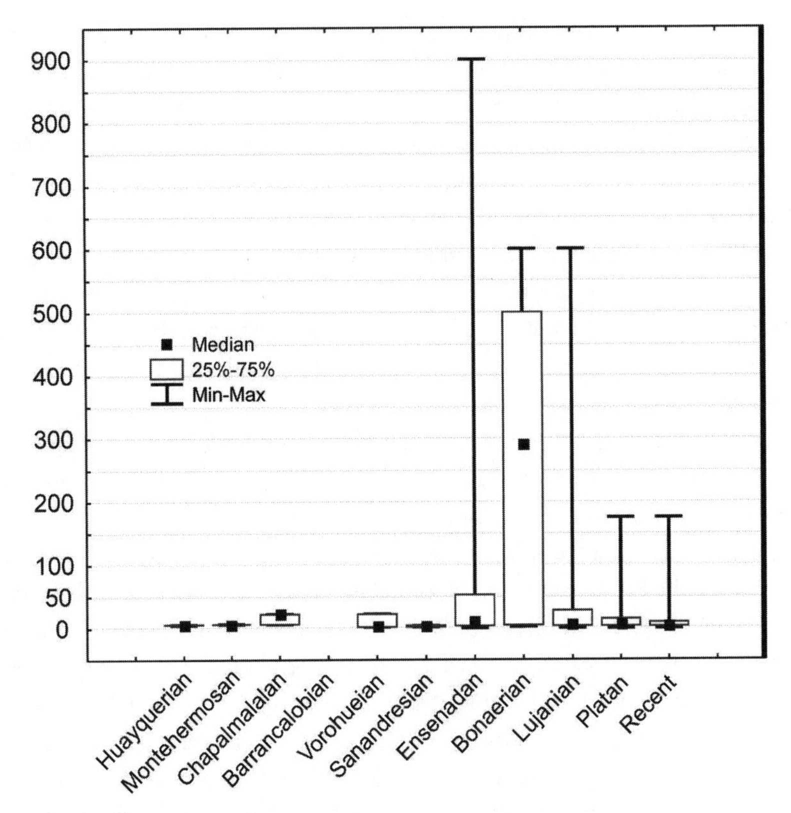

Figure 6.4 Carnivore body mass through time. First and third quartiles, 25% and 75%, respectively; Min-Max, minimum and maximum, respectively.

another from the Sanandresian, 5 from the Bonaerian, and 10 from the Holocene (fig. 6.3C). These peaks may be due to a lack of data or declines recorded in earlier variables (e.g., diversity, see earlier). Diversity and number of localities per age showed a strong and highly significant positive relationship ($r_s = 0.92$, $p < 0.001$). Age span presented an inverse but nonsignificant relationship with diversity, whether the temporal interval of the Ensenadan is constrained to its later part (i.e., 0.5 Ma, see table 6.1; $r_s = -0.52$, n.s.) or its full span is used (1.5 Ma; $r_s = -0.43$, n.s.).

6.4 Discussion

The results presented here regarding the diversity of carnivorans through time (fig. 6.2A) coincide with those published previously (e.g., Cione and Tonni

1995; Prevosti, Gasparini, and Bond 2006; L. Soibelzon and Prevosti 2007; Woodburne, Cione, and Tonni 2006). The first South American occurrence of the group was in Late Miocene, when it was represented by procyonids, followed by mustelids and canids in the Late Pliocene (Vorohuean), and then by several families (Felidae, Ursidae, Mephitidae), subfamilies (Lutrinae), and genera in the Early-Middle Pleistocene (Ensenadan). The low pre-Pleistocene diversity of carnivores contrasts with the higher numbers recorded in Pleistocene-Recent faunas. At first sight, diversity appears to increase gradually during the interval Pleistocene to Recent, but there were two drops: the first in the Bonaerian and the second in the Holocene. As discussed later, we believe that these drops are artifactual, due to biases in the fossil record.

Prior to the Pleistocene, the first and last records follow the diversity line in magnitude and pattern; this is because most species are restricted to one age (see Berman 1989, 1994; Cione and Tonni 1995; figs. 6.2B, 6.2C). The last records of the Pleistocene-Holocene implicate two major extinction events: one in the Middle Pleistocene (end of the Ensenadan) and the other at the end of the Lujanian. The first corresponds to a sudden change in the composition of carnivore faunas between the Ensenadan and Bonaerian. Several large hypercarnivorous canids (e.g., *Theriodictis*, *Protocyon scagliarum*, "*Canis*" *gezi*), foxes (e.g., *Dusicyon ensenadensis*), the giant short-faced bear (*Arctotherium angustidens*), some mustelids (e.g., *Lyncodon bosei*, *Galictis henningi*, *Stipanicicia*), and the last member of the procyonid *Cyonasua* group (*C. merani*) became extinct at this boundary (Prevosti 2006a, 2006b; L. Soibelzon, Tonni, and Bond 2005). However, several short-faced bears (*Arctotherium tarijense*, *A. bonariense*, *A. vetustum*, *A. wingei*), mustelids (e.g., *G. cuja*, *L. patagonicus*, *Eira*, *Pteronura*), foxes and wolf-like canids (e.g., *Canis dirus*, *D. avus*), procyonids (e.g., *Nasua*, *Procyon*), and mephitids (e.g., *Conepatus semistriatus*, *Conepatus humboldtii*) appeared after the Ensenadan-Bonaerian boundary. The extinction recorded at the end of the Lujanian corresponds to the Pleistocene-Holocene mass extinction that affected most American mammals, especially the larger ones (see Borrero 2009; Cione, Tonni, and Soibelzon 2009).

The immigration curve (fig. 6.2D, 6.2E) not only shows the previously recognized events of the Late Miocene, Late Pliocene, and Early-Middle Pleistocene, but also two others, in the Late Pleistocene and Holocene. The peak recorded in the Ensenadan corresponds to a massive immigration of North American mammals, which included several herbivorous lineages (see Cione and Tonni 1995; Woodburne, Cione, and Tonni 2006). The peak recorded in the Lujanian is a lesser-known immigration event, but for carnivores it was comparable in size to the previous one. Saber-toothed cats (*S. fatalis*), dire wolves (*C. dirus*), the North American grey fox (*Urocyon cinereoargenteus*), procyonids (*Nasua*, *Procyon*), and mustelids (*Eira*, *Pteronura*) were part or products of this immigration (see Kurtén and Werdelin 1990; Prevosti 2006a; Prevosti and Ferrero 2008; Prevosti and Rincón 2007). The first records of *Tremarctos*, *Mustela*, *Potos*, and *Canis familiaris* in the Holocene of South America may represent post-Pleistocene immigrations (L. Soibelzon, Tonni, and Bond 2005).

As can be seen in figs. 6.2D and 6.2E, in situ speciation explains much more diversity than immigration, but it is clear that both contributed to the past and present diversity of carnivores in South America. These carnivore faunas are a consequence of several immigration events from Central and North America, plus later speciation in situ (see Soibelzon and Prevosti 2007). Successive immigration and speciation events caused an enrichment of the South American carnivore fauna through time (figs. 6.2D, 6.2E). This pattern was not gradual (fig. 6.2A) and can be divided into two phases, one of low diversity before the Pleistocene, and another of high diversity in the Pleistocene-Recent.

There are several taxa that may represent "reverse" cases (i.e., that migrated from South America to Central and North America). For example *Herpailurus*, *Eira barbara*, *Procyon cancrivorus*, and *Speothos* all have older records and a wider Recent geographic distribution in South America, and may thus be interpreted as South American invaders into Central and North America. Webb (2004, 2006) postulated a Late Pleistocene invasion of Central America by South American intertropical mammals that might include these species. However, some of these taxa (e.g., *Herpailurus*) are widely distributed and also occur in temperate habitats, something that could point to a different time of immigration.

Diet and body mass also show a two-phase pattern (figs. 6.3A, 6.3B, 6.4). Pre-Pleistocene carnivores were mostly omnivorous, while Pleistocene-Recent faunas included taxa of all dietary types. The main difference between Pleistocene and Holocene-Recent faunas is that the former proportionally encompassed more hypercarnivores than the latter. This could be a result of the Pleistocene-Holocene mass extinction, where mainly carnivores larger than 30 kg disappeared, most of which were hypercarnivores. During pre-Pleistocene ages, most species were small (<10 kg), except for the Late Pliocene (Chapalmalalan-Vorohueian) *Chapadmalania*, which had a body mass of more than 20 kg. Pleistocene-Recent faunas show a wider range of body sizes, from 0.5 to more than 100 kg. Pleistocene species had higher maximum body masses than Holocene-Recent ones (600–900 versus 175 kg), a difference explained by the Pleistocene-Holocene extinction that removed most of the large carnivores (e.g., Cione, Tonni, and Soibelzon 2009; Van Valkenburgh and Hertel 1998). The greater variety of size and diets during Pleistocene-Recent indicates more complex paleosynecological relationships within the terrestrial carnivore guild, and between them and their prey (see Prevosti and Vizcaíno 2006; E. Soibelzon et al. 2009).

There is a clear bias in the fossil record against tropical regions (Behrensmeyer, Kidwell, and Gastaldo 2000; Marshall et al. 1982) that affects the South American record. Most pre-Lujanian sites are from the southern part of South America, especially Argentina, while Lujanian-Platan sites are more evenly distributed. Thus, provinciality can be expected to increase the diversity (and other variables) of the Lujanian-Recent faunas (figs. 6.2, 6.3). In fact, if we exclude all other faunas except those of Chile, Uruguay, and Argentina, the diversity, first records, and number of hypercarnivorous and omnivorous species of the Lujanian are similar to the Ensenadan ones, but last records, immigrations, and in situ speciation are lower. Something similar happened with the Recent fauna. As can be seen in figures 6.2 and 6.3, range-through taxon indicates that the absence of records during the Barrancalobian and the low diversity of the Bonaerian (Middle-Late Pleistocene) and Platan (Holocene) are artifacts due to poor-preservation conditions or the paucity of sites from these ages. This is supported by the strong positive relationship observed between documented diversity and number of sites per age. Bonaerian sites were restricted to a few localities in Buenos Aires province, which explains the low diversity of carnivores in this age. Some of the high numbers of last records of the Ensenadan and high numbers of first records, immigration, and in situ speciation in the Lujanian may be artifacts of the small sample from Bonaerian fossil sites. The same argument could be used for the low diversity (and other variables) of the Platan records: most first records, immigration, and in situ speciation events recorded for the Recent fauna are probably by-products of the preservation bias of the Platan. Another problem is the age of several cave deposits in Brazil that are typically referred to the Lujanian (also in this chapter), but that were recently dated to between 10 and 30 ka (Bonaerian-Lujanian; Auler et al. 2006). Excluding these sites produced results similar to those obtained when sites from the northern part of South America were excluded (see earlier).

One interesting result is the nonsignificant and inverse relationship between diversity and time span of each age in contrast to the expected bias (e.g., Palombo et al. 2008). If diversity is standardized by age span, the curve obtained presents a pattern similar to the one observed in figure 6.2A, but Lujanian and Platan show high values because they represent short spans (table 6.1). Using the complete span of the Ensenadan produces a lower ratio than the Bonaerian

(14.6 vs. 18.9, respectively), which contradicts the raw diversity numbers. But if the time span of the Ensenadan is constrained to the actual chronological data, it has a higher ratio than the Bonaerian (38.0 vs. 18.9). This suggests that the length of each span did not affect the pattern observed, at least to the limits of the available data.

This review is the first to explore the history of South American carnivores through quantification of the information in the fossil record. Systematic revisions, fieldwork, revision of biostratigraphic-chronostratigraphic schemes, and chronologic assignation of fossil sites should help to improve this reconstruction, and may provide a more complete picture of the history of carnivorans in South America. A deeper analysis of this data set will help to test the effect of the detected biases on its interpretation.

Acknowledgments

We thank Bruce Patterson and Leonora Costa for inviting us to participate in the Symposium "Historical Biogeography of Neotropical Mammals: The Setting" at the International Mammalogist Congress 2009. Reviewers Lars Werdelin and Richard Tedford, Lucas Pomi, Ulyses Pardiñas, Sergio Vizcaíno, and Eduardo Tonni offered discussions that helped to improve this contribution. The CONICET provided financial support. This is a contribution to the grants PIP 112 200801 01054 (CONICET) and PICT2007-00428 (Agencia-FONCYT).

Literature Cited

Auler, A. S., L. B. Piló, P. L. Smart, X. Wang, D. Hoffmann, D. A. Richards, R. L. Edwards et al. 2006. "U-Series Dating and Taphonomy of Quaternary Vertebrates from Brazilian Caves." *Palaeogeography, Palaeoclimatology, Palaeoecology* 240:508–22.

Bardeleben, C., R. L. Moore, and R. K. Wayne. 2005. "A Molecular Phylogeny of the Canidae Based on Six Nuclear Loci." *Molecular Phylogenetics and Evolution* 37:815–31.

Behrensmeyer, A. K., S. M. Kidwell, and R. A. Gastaldo. 2000. "Taphonomy and Paleobiology." In *Deep Time, Paleobiology's Perspective*, edited by D. H. Erwin and S. L. Wing, 103–47. Washington, DC: The Paleontological Society.

Berman, W. D. 1987. "Una Nueva Especie de Procyonidae (Mammalia, Carnivora) del Terciario Superior de la Provincia de Jujuy (Argentina). Consideraciones Sobre la Distribución Geográfica de *Cyonasua* Durante el Mioceno Tardío." *Boletín Informativo de la Asociación Paleontológica Argentina* 16:8–9.

———. 1989. "Notas Sobre la Sistemática y Paleobiogeografia del Grupo *Cyonasua* (Carnivora, Procyonidae). *VI Jornadas Argentinas de Paleontología Vertebrados* Actas:77–9.

———. 1994. "Los Carnívoros Continentales (Mammalia, Carnivora) del Cenozoico en la Provincia de Buenos Aires." Unpublished PhD diss., Facultad de Ciencias Naturales y Museo, Universidad Nacional de La Plata, La Plata.

Berta, A. 1989. “Quaternary Evolution and Biogeography of the Large South American Canidae (Mammalia: Carnivora).” *University of California Publications Geological Sciences* 132:1–149.

Berta, A., and L. G. Marshall. 1978. “South American Carnivora.” In *Fossilium Catalogus, I: Animalia*, edited by F. Westphal, 1–48. The Hague: Dr. W. Junk.

Bond, M. 1986. “Los Carnívoros Terrestres Fósiles de Argentina: Resumen de su Historia.” *Actas del IV Congreso Argentino de Paleontología y Bioestratigrafía* 2:167–71.

Borrero, L. A. 2009. “The Elusive Evidence: The Archeological Record of the South American Extinct Megafauna.” In *American Megafaunal Extinctions at the End of the Pleistocene*, edited by G. Haynes, 145–68. Dordrecht: Springer Science + Business Media.

Cione, A. L., and E. P. Tonni. 1995. “Chronostratigraphy and ‘Land Mammal-Ages’: The Uquian Problem.” *Journal of Paleontology* 69:135–59.

———, eds. 1999. “Quaternary Vertebrate Palaeontology in South America.” *Quaternary of South America and Antarctic Peninsula* 12:1–310.

———. 2005. “Bioestratigrafía Basada en Mamíferos del Cenozoico Superior de la Provincia de Buenos Aires, Argentina.” In *Geología y Recursos Minerales de la Provincia de Buenos Aires*, Relatorio 11, edited by R. E. de Barrio, R. O. Etcheverry, M. F. Caballé, and E. Llambías, 183–200. La Plata: 16° Congreso Geológico Argentino.

Cione, A. L., E. P. Tonni, and L. H. Soibelzon. 2009. “Did Humans Cause the Late Pleistocene-Early Holocene Mammalian Extinctions in South America in a Context of Shrinking Open Areas?” In *American Megafaunal Extinctions at the End of the Pleistocene*, edited by G. Haynes, 125–44. Dordrecht: Springer Science + Business Media.

Croft, D. A. 2009. “South American Mammal Diversities and Distributions During the Miocene.” *10th International Mammalogical Congress, Mendoza* Abstracts:310–11.

Flynn, J. J., and C. C. Swisher, III. 1995. “Cenozoic South American Land Mammal Ages: Correlation to Global Geochronologies.” In *Geochronology, Time Scales, and Global Stratigraphic Correlation*, edited by W. A. Berggren, D. V. Kent, M.-P. Aubry, and J. Hardenbol, 317–33. Tulsa: Society for Sedimentary Geology (SEPM Special Publication 54).

Figueirido, B., and L. H. Soibelzon. 2010. “Inferring Paleoecology in Extinct Tremarctine Bears (Carnivora, Ursidae) Via Geometric Morphometrics.” *Lethaia* 43:209–22.

Foote, M. 2000. “Origination and Extinction Components of Taxonomic Diversity: General Problems.” *Paleobiology* 26 (Suppl.):74–102.

Forasiepi, A. M., A. G. Martinelli, and F. J. Goin. 2007. “Revisión Taxonómica de *Parahyaenodon Argentinus* Ameghino y sus Implicancias en el Conocimiento de los Grandes Mamíferos Carnívoros del Mio-Plioceno de América de Sur.” *Ameghiniana* 44:143–59.

Fortelius, M., L. Werdelin, P. Andrews, R. L. Bernor, A. Gentry, L. Humphrey, H.-W. Mittmann, and S. Viranta. 1996. “Provinciality, Diversity, Turnover, and Paleoecology in Land Mammal Faunas of the Later Miocene of Western Eurasia.” In *The Evolution of Western Eurasian Neogene Mammal Faunas*, edited by R. L. Bernor Fahlbusch and H.-W Mittmann, 414–48. New York: Columbia University Press.

Johnson, W. E., E. Eizirik, J. Pecon-Slattery, W. J. Murphy, A. Antunes, E. Teeling, and S. J. O’Brien. 2006. “The Late Miocene Radiation of Modern Felidae: A Genetic Assessment.” *Science* 311:73–7.

Koepfli, K.-P., K. A. Deere, G. J. Slater, C. Begg, K. Begg, L. Grassman, M. Lucherini et al. 2008. "Multigene Phylogeny of the Mustelidae: Resolving Relationships, Tempo and Biogeographic History of a Mammalian Adaptive Radiation." *BMC Biology* 6:10.

Koepfli, K.-P., M. E. Gompper, E. Eizirik, C.-C. Ho, L. Linden, J. E. Maldonado, and R. K. Wayne. 2007. "Phylogeny of the Procyonidae (Mammalia: Carnivora): Molecules, Morphology and the Great American Interchange." *Molecular Phylogenetic and Evolution* 43:1076–95.

Kraglievich, L. 1930. "Craneometría y Clasificación de los Cánidos Sudamericanos, Especialmente los Argentinos Actuales y Fósiles." *Physis* 10:35–73.

Krause, J., T. Unger, A. Noçon, A.-S. Malaspinas, S. Kolokotronis, M. Stiller, L. H. Soibelzon et al. 2008. "Mitochondrial Genomes Reveal an Explosive Radiation of Extinct and Extant Bears Near the Miocene-Pliocene Boundary." *BMC Evolutionary Biology* 8:220.

Kurtén, B., and L. Werdelin. 1990. "Relationships Between North and South American *Smilodon*." *Journal of Vertebrate Paleontology* 10:158–69.

Maas, M. C., M. R. L. Anthony, P. D. Gingerich, G. F. Gunnell, and D. W. Krause. 1995. "Mammalian Generic Diversity and Turnover in the Late Paleocene and Early Eocene of the Bighorn and Crazy Mountains Basins, Wyoming and Montana (USA)." *Palaeogeography, Palaeoclimatology, Palaeoecology* 115:181–207.

Marshall, L. G. 1982. "Calibration of the Age of Mammals in South America." *Geobios, Memoire Special* 6:427–37.

Marshall, L. G., A. Berta, R. Hoffstetter, R. Pascual, O. A. Reig, M. Bombin, and A. Mones. 1984. "Mammals and Stratigraphy: Geochronology of the Continental Mammal-Bearing Quaternary of South America." *Paleovertebrata, Montpellier, Mémoires Extracts* 1984:1–76.

Marshall, L. G., S. D. Webb, J. J. Sepkoski, and D. M. Raup. 1982. "Mammalian Evolution and the Great American Interchange." *Science* 215:1351–7.

Mones, A., and A. Rinderknecht. 2004. "The First South American Homotheriini (Mammalia: Carnivora: Felidae)." *Comunicaciones Paleontológicas del Museo Nacional de Historia Natural y Antropología* 35:201–12.

Nowak, R. M. 2005. *Walker's Carnivores of the World*. 6th ed. Baltimore: Johns Hopkins University Press.

Palombo, M. R., M. T. Alberdi, B. Azanza, C. Giovinazzo, J. L. Prado, and R. Sardella. 2008. "How Did Environmental Disturbances Affect Carnivoran Diversity? A Case Study of the Plio-Pleistocene Carnivora of the North-Western Mediterranean." *Evolutionary Ecology* 23:569–89.

Patterson, B., and R. Pascual. 1972. "The Fossil Mammal Fauna of South America." In *Evolution, Mammals, and Southern Continents*, edited by A. Keast, F. C. Erk, and B. Glass, 247–309. Albany: State University of New York Press.

Pomi, L. H., and F. J. Prevosti. 2005. "Sobre el Status Sistemático de *Felis longifrons* Burmeister, 1866 (Carnivora: Felidae)." *Ameghiniana* 42:489–94.

Prevosti, F. J. 2006a. "Grandes Cánidos (Carnivora, Canidae) del Cuaternario de la Republica Argentina: Sistemática, Filogenia, Bioestratigrafíay Paleoecología." Unpublished PhD diss., Universidad Nacional de La Plata, La Plata.

———. 2006b. "New Materials of Pleistocene Cats (Carnivora, Felidae) from Southern South America, with Comments on Biogeography and the Fossil Record." *Geobios* 39:679–94.

Prevosti, F. J., and B. S. Ferrero. 2008. "A Pleistocene Giant River Otter from Argentina: Remarks on the Fossil Record and Phylogenetic Analysis." *Journal of Vertebrate Paleontology* 28:1171–81.

Prevosti, F. J., A. M. Forasiepi, L. H. Soibelzon, and N. Zimicz. 2009. "Sparassodonta vs. Carnivora: Ecological Relationships Between Carnivorous Mammals in South America." *10th International Mammalogical Congress, Mendoza* Abstracts:61–2.

Prevosti, F. J., G. M. Gasparini, and M. Bond. 2006. "Systematic Position of a Specimen Previously Assigned to Carnivora from the Pliocene of Argentina and its Implications for the Great American Biotic Interchange." *Neues Jahrbuch für Geologie und Paläontologie, Monatshefte und Abhandlungen* 242:133–44.

Prevosti, F. J., and L. H. Pomi. 2007. "*Smilodontidion riggii* (Carnivora, Felidae, Machairodontinae): Revisión Sistemática del Supuesto Félido Chapadmalalense." *Revista del Museo Argentino de Ciencias Naturales*, n.s., 9:67–77.

Prevosti, F. J., and A. D. Rincón. 2007. "Fossil Canid Assemblage from the Late Pleistocene of Northern South America: The Canids of the Inciarte Tar Pit (Zulia, Venezuela), Fossil Record and Biogeography." *Journal of Paleontology* 81:1053–65.

Prevosti, F. J., and S. Vizcaíno. 2006. "The Carnivore Guild of the Late Pleistocene of Argentina: Paleoecology and Carnivore Richness." *Acta Paleontologica Polonica* 51: 407–22.

Prevosti, F. J., A. E. Zurita, and A. A. Carlini. 2005. "Biostratigraphy, Systematics and Palaeoecology of the Species of *Protocyon* Giebel, 1855 (Carnivora, Canidae) in South America." *Journal of South American Earth Science* 20:5–12.

Reig, O. 1981. "Teoría del Origen y Desarrollo de la Fauna de Mamíferos de América del Sur." *Monographie Naturae* 1:1–162.

Seymour, K. L. 1999. "Taxonomy, Morphology, Paleontology and Phylogeny of the South American Small Cats (Mammalia: Felidae)." Unpublished PhD diss., University of Toronto, Toronto.

Sillero Zubiri, C., M. Hoffmann, and D. W. MacDonald. 2004. *Canids: Foxes, Wolves, Jackals and Dogs. Status Survey and Conservation Action Plan*. Gland: IUCN Species Programme.

Silva, M., and J. A. Downing. 1995. *CRC Handbook of Mammalian Body Masses*. Boca Raton: CRC Press Inc.

Simpson, G. G. 1950. "History of the Fauna of Latin America." *American Scientist* 38:361–89.

———. 1980. *Splendid Isolation: The Curious History of South American Mammals*. New Haven: Yale University Press.

Soibelzon, E., G. M. Gasparini, A. E. Zurita, and L. H. Soibelzon. 2008. "Las 'Toscas del Río de La Plata' (Buenos Aires, Argentina): Una Actualización Paleofaunística." *Revista del Museo Argentino de Ciencias Naturales "Bernardino Rivadavia"* 19:291–308.

Soibelzon, E., F. J. Prevosti, J. C. Bidegain, Y. Rico, D. H. Verzi, and E. P Tonni. 2009. "Correlation of Cenozoic Sequences of Southeastern Buenos Aires Province: Biostratigraphy and Magnetostratigraphy." *Quaternary International* 210:51–6.

Soibelzon, L. H. 2002. "Los Ursidae (Carnivora, Fissipedia) Fósiles de la República Argentina. Aspectos Sistemáticos y Paleoecológicos." Unpublished PhD diss., Universidad Nacional de La Plata, La Plata.

———. 2004. "Revisión Sistemática de los Tremarctinae (Carnivora, Ursidae) Fósiles de América del Sur." *Revista del Museo Argentino de Ciencias Naturales* 6:105–31.

Soibelzon, L. H., L. M. Pomi, E. P. Tonni, S. Rodriguez, and A. Dondas. 2009. "First Report of a Short-Faced Bears' Den (*Arctotherium Angustidens*). Palaeobiological and Palaeoecological Implications." *Alcheringa* 33:211–22.

Soibelzon, L. H., and F. J. Prevosti. 2007. "Los Carnívoros (Carnivora, Mammalia) Terrestres del Cuaternario de América del Sur." In *Geomorfología Litoral i Quaternari. Homenatge a Joan Cuerda Barceló*, edited by G. X. Pons and D. Vicens, 49–68. Palma de Mallorca: Monografia de la Societat d'Història Natural.

Soibelzon, L. H., and A. Rincón. 2007. "The Fossil Record of the Short-Faced Bears (Ursidae, Tremarctinae) from Venezuela: Systematic, Biogeographic, and Paleoecological Implications." *Neues Jahrbuch für Geologie und Paläontologie, Monatshefte und Abhandlungen* 245:287–98.

Soibelzon L. H., and V. B. Tartarini. 2009. "Estimación de la Masa Corporal de las Especies de Osos Fósiles y Actuales (Ursidae, Tremarctinae) de América del Sur." *Revista Museo Argentino de Ciencias Naturales "Bernardino Rivadavia"* 11:243–54.

Soibelzon, L. H., E. P. Tonni, and M. Bond. 2005. "The Fossil Record of the South American Short-Faced Bears (Ursidae, Tremarctinae)." *Journal of South American Earth Sciences* 20:105–13.

Sunquist, M., and F. Sunquist. 2002. *Wild Cats of the World*. Chicago: University of Chicago Press.

Van Valkenburgh, B. 1988. "Trophic Diversity Within Past and Present Guilds of Large Predatory Mammals." *Paleobiology* 11:406–28.

———. 1990. "Skeletal and Dental Predictors of Body Mass in Carnivores." In *Body Size in Mammalian Paleobiology: Estimation and Biological Implication*, edited by J. Damuth and B. J. Macfadden, 181–205. Cambridge: Cambridge University Press.

———. 1991. "Iterative Evolution of Hypercarnivory in Canids (Mammalia: Carnivore): Evolutionary Interactions Among Sympatric Predators." *Paleobiology* 17:340–62.

———. 2007. "Déjà Vu: The Evolution of Feeding Morphologies in the Carnivora." *Integrative and Comparative Biology* 47:147–63.

Van Valkenburgh, B., and F. Hertel. 1998. "The Decline of North American Predators During the late Pleistocene." In *Quaternary Paleozoology in the Northern Hemisphere*, edited by J. J. Saunders, B. W. Styles, and G. F. Baryshnikov, 357–74. Springfield: Illinois State Museum, Scientific Papers, 27.

Van Valkenburgh, B., and K. P. Koepfli. 1993. "Cranial and Dental Adaptation to Predation in Canids." *Symposia of the Zoological Society of London* 65:15–37.

Webb, S. D. 1985. "Late Cenozoic Mammal Dispersals Between the Americas." In *The Great American Biotic Interchange*, edited by F. G. Stehli and S. D. Webb, 357–84. New York: Plenum Press.

———. 2004. "El Gran Intercambio Americano de Fauna." In *Paseo Pantera: Una Historia de la Naturaleza y Cultura de Centroamérica*, edited by A. G. Coates, 107–36. Washington, DC: Smithsonian Books.

———. 2006. "The Great American Biotic Interchange: Patterns and Processes." *Annals of the Missouri Botanical Garden* 93:245–57.

Woodburne, M., A. L. Cione, and E. P. Tonni. 2006. "Central American Provincialism and the Great American Biotic Interchange." In *Advances in Late Tertiary Vertebrate Paleontology in Mexico and the Great American Biotic Interchange*, edited by O. Carranza-Castañeda and E. H. Lindsay, 73–101. Mexico City: Universidad Nacional Autónoma de México, Publicación Especial del Instituto de Geología y Centro de Geociencias, 4.

Wozencraft, W. C. 2005. "Order Carnivora." In *Mammal Species of the World: A Taxonomic and Geographic Reference*, 3rd ed., edited by D. E. Wilson and D. M. Reeder, 532–628. Baltimore: Johns Hopkins University Press.

Zetti, J. 1972. "Los Mamíferos Fósiles de Edad Huayqueriense (Plioceno Medio) de la Región Pampeana." Unpublished PhD diss., Facultad de Ciencias Naturales y Museo, Universidad Nacional de La Plata, La Plata.

A Molecular View on the Evolutionary History and Biogeography of Neotropical Carnivores (Mammalia, Carnivora)

Eduardo Eizirik

Abstract

The mammalian order Carnivora exhibits great ecomorphological diversity and a rather rich fossil record documenting interesting evolutionary and biogeographic patterns. Recent analyses of large molecular data sets have revealed a detailed picture of the evolution of extant carnivoran lineages, starting from the phylogenetic positioning of the order relative to other eutherians, moving into the resolution of interfamilial relationships among carnivores, and on to the investigation of recent radiations of genera and species. In this chapter, I address carnivoran diversification in the Neotropics by reviewing recent studies that employed molecular data sets for phylogenetic reconstructions and molecular dating. Most of the examples cover interfamilial and intergeneric splits, and discuss the implications of molecular-derived divergence dates for the interpretation of biogeographic patterns. In some cases, the discussion extends to intrageneric diversification, contrasting recent endemic clades (especially in the Felidae and Canidae) that show rapid divergences postdating the Great American Biotic Interchange (GABI) with others that have a more ancient history in the region. This is particularly the case with the Procyonidae, whose molecular dating analyses point not only to a much older origin than previously estimated, but also to splits between sister species that predate the GABI, contrary to traditional views on their historical biogeography. Overall, carnivores appear to be a remarkable group for investigating the diversity of biogeographic patterns and processes that characterize the history of the Neotropics, as different clades have experienced disparate histories, with contrasting influences derived from the GABI.

7.1 Introduction

The order Carnivora comprises a diverse array of mammals adapted to a broad variety of habitats and feeding strategies, most of which involve predation on other animals (Macdonald 2001; Nowak 1999). The monophyly of this taxonomic group has been solidly established, as has its phylogenetic placement within the eutherian superorder Laurasiatheria (Murphy and Eizirik 2009; Springer et al. 2007). Carnivorans diverged from their sister group, the pangolins (order Pholidota), ~80 Ma (Murphy and Eizirik 2009), most likely in Laurasia-derived landmasses. During the Paleocene (~60 Ma) this order split into two different lineages, currently recognized as suborders Caniformia and Feliformia. In the Eocene, ~48 Ma, the Caniformia lineage split into two branches, currently classified as superfamilies Cynoidea (containing the family Canidae) and Arctoidea (containing eight families). The diversification of extant lineages of Feliformia also began by 40 Ma, giving rise to seven currently recognized families (Eizirik and Murphy 2009; Eizirik et al. 2010).

Of the 16 extant carnivoran families that are now recognized, only 8 are present in the Neotropics (Felidae, Canidae, Mustelidae, Mephitidae, Otariidae, Phocidae, Procyonidae, and Ursidae). This is a consequence of the inferred fact that carnivorans arose and underwent their early diversification in northern continents, and only later did they colonize southern landmasses such as Africa and South America. There is no evidence that additional carnivoran families were ever present in the Neotropics, indicating that no other lineages from this order invaded this region in the past (Hunt 1996).

Interestingly, almost all the carnivoran lineages that invaded South America belong to the Caniformia, with only one feliform family (the Felidae) represented in the Neotropics. This is likely due to the fact that the origin and diversification of Feliformia took place in the Old World, and all families but Felidae are still restricted to that hemisphere. The fossil record indicates that only three feliform families ever occurred in North America: Felidae, Hyaenidae, and the extinct Nimravidae. The nimravids were cat-like feliforms that occurred in North America and Eurasia from the late Eocene to the late Miocene, with very limited incursions into southern landmasses, and no records in South America (Hunt 1996; Van Valkenburgh 1999). Hyaenids appeared much later in the North American fossil record (from the Pliocene to the Pleistocene) and never invaded South America (Hunt 1996). In contrast, stem lineages of the Felidae were present in North America in the middle Miocene (Hunt 1996), and subsequent invasions from Asia supplied North America with additional representatives of this family (Johnson et al. 2006), which in turn formed the basis for its later invasion of the Neotropics (see later). It may thus be concluded that the absence of most feliform lineages from North America during the Neogene hampered their opportunity to invade South America, as this seems to have been the only source for carnivorans colonizing the Neotropics.

A very different pattern emerges from the caniform fossil record, which shows that much of the early diversification of this carnivoran clade involved North American lineages and considerable biogeographic exchange with Eurasia (e.g., Flynn and Wesley-Hunt 2005; Hunt 1996; Koepfli et al. 2008; Van Valkenburgh 1999). The Canidae were endemic to North America throughout their early history, only colonizing the Old World from the Miocene onward. Arctoid clades such as Ursidae, Amphicyonidae, Procyonidae, and Mustelidae also have a long history of occurrence and diverse representatives in the North American fossil record, so that several lineages could serve as progenitors to South American colonists. The amphicyonids (bear-dogs) never invaded the Neotropics, likely because they went extinct by the late Miocene (~7–9 Ma), prior to the formation of a land connection between South and North America (Hunt 1996). Conversely, all terrestrial arctoid families that were present in North America by the Pliocene contained lineages that invaded the Neotropics, supporting the notion that this time period was critical for carnivoran colonization of South America. The timing of invasions seems to have varied among arctoid families, a factor that may have played a role in the phylogenetic and biogeographic patterns exhibited by each of the resulting Neotropical lineages. Terrestrial arctoid families containing multiple living representatives in the Neotropics (Procyonidae, Mustelidae, and Mephitidae) will be treated separately in the next section, while pinnipeds and ursids are briefly discussed here.

Pinnipeds are marine carnivores belonging to three families (Odobenidae, Otariidae, and Phocidae) that jointly form a monophyletic group within Arctoidea (Eizirik and Murphy 2009). The families Otariidae (containing fur seals and sea lions) and Phocidae (containing true seals) are represented in the Neotropics by a few species each. Neotropical otariids include six species of fur seals (*Arctocephalus*), one southern sea lion (*Otaria*), and another sea lion (*Zalophus*). Molecular phylogenetic studies have recently addressed their relationships and divergence times (Arnason et al. 2006; Wynen et al. 2001; Yonezawa, Kohno, and Hasegawa 2009), yielding the notion that otariids arose in the north Pacific, later colonizing the western South American coast and subsequently diversifying in the southern oceans. The most recent molecular dating analysis suggested that an important portion of their diversification process occurred in Neotropical waters during the Pliocene and early Pleistocene (Yonezawa, Kohno, and Hasegawa 2009), which may have been influenced by coastal changes associated with the closure of the Panama seaway. If affirmed by additional analyses targeting the precise timing of this radiation, this diversification process in Otariidae may serve as an interesting marine counterpart to the historical biogeography of their terrestrial relatives in the Neotropics.

In contrast to the patterns inferred for the otariids, the diversification of Phocidae does not seem to have involved a substantial Neotropical component. This family is currently represented in the region by resident populations or vagrant individuals from species belonging to the genera *Hydrurga*, *Leptonychotes*, *Lobodon*, and *Mirounga*, all of which occur in Antarctic or subantarctic areas (Wozencraft 2005), with records extending to southern South America (a noteworthy exception was the recently extinct Caribbean monk seal, *Monachus tropicalis*, which occurred in the northern Neotropics). Recent molecular studies have supported the placement of *Hydrurga*, *Leptonychotes*, and *Lobodon* in a monophyletic group dating back to the late Miocene, whose closest relative is *Ommatophoca rossii*, an Antarctic species (Arnason et al. 2006; Fulton and Strobeck 2006). This group comprises the tribe Lobodontini, which was inferred to have shared a common ancestor with *Mirounga* ~13 Ma (Arnason et al. 2006).

The most basal extant lineage of Arctoidea corresponds to the family Ursidae, which diverged from other living arctoids ~42 Ma (Eizirik et al. 2010). The Ursidae is presently represented in the Neotropics solely by the spectacled bear (*Tremarctos ornatus*), an Andean species that is the only living representative of subfamily Tremarctinae. Tremarctines include various fossil taxa endemic to the New World, including the short-faced bears that radiated in both North and South America (Soibelzon, Tonni, and Bond 2005). A recent molecular study has employed complete mitochondrial genome data to estimate the phylogeny and divergence dates for living ursids as well as two fossil bears (Krause et al. 2008). One of them was the extinct North American short-faced bear (*Arctodus simus*), which diverged from *T. ornatus* ~5.7 Ma (with a minimum bound of the 95% credibility interval of 4.3 Ma). This finding supports the view that the divergence of *T. ornatus* from other tremarctine bears took place prior to the formation of the Panamanian land bridge, even though their invasion of South America only occurred after this event. This is consistent with the fossil record, which indicates that the initial radiation of this clade likely occurred in North America, prior to invasion of South America, even though some autochthonous speciation events may have occurred in the latter continent (Krause et al. 2008; Soibelzon, Tonni, and Bond 2005).

As is the case with many other mammalian groups, a key issue in understanding the evolution of Neotropical carnivores is to assess the influence of the Great American Biotic Interchange (GABI) on their diversification (Marshall et al. 1982; see Prevosti and Soibelzon, chapter 6, this volume). Molecular phylogenies and associated inferences of divergence dates can illuminate the tempo and mode of lineage diversification and allow them to be compared with biogeographic patterns and historical phenomena thought to have influenced regional biotas. This is the case with the GABI, as it is interesting to determine the timing of each lineage's invasion of South America, and to place it within the context of this broader process of faunal exchange. How many carnivoran lineages independently invaded South America? How many autochthonous radiations occurred in the Neotropics after the GABI? Were there shared patterns among these radiations? Were there carnivoran lineages that invaded South America substantially before the GABI? If so, when and how did these invasions take place? These are some of the outstanding questions that may be tackled with the use of reliably dated molecular phylogenies. The next section reviews the current knowledge on the diversification of Neotropical lineages within each of the more diverse terrestrial carnivoran families. The focus is on reviewing results based on molecular data, including phylogenetic analyses and molecular dating, although some comments on fossil evidence will also be provided.

7.2 Intra-Familial Patterns of Diversification in the Neotropics

7.2.1 PROCYONIDAE

Procyonids seem to have been the first members of the Carnivora to invade South America. Their colonization of this continent prior to other carnivoran families has long been recognized, given the allocation of late Miocene fossils such as *Chapalmalania* and *Cyonasua* to this arctoid clade (Baskin 2004; Hunt 1996; Simpson 1980). These fossils support the inference that this family colonized South America prior to the GABI, possibly via island-hopping at the same time as other mammals (e.g., ground sloths) crossed from South to North America. Given their present and fossil distribution, it might be hypothesized that much of the evolutionary diversification of extant procyonids occurred in the Neotropics, possibly in South America, after invasion of this continent in the late Miocene.

Recent molecular studies offer a surprising perspective on this issue, revealing that the age of crown procyonids is much older than previously expected. A data set focused on the evolution of this carnivoran family (Koepfli et al. 2007) indicated that its most basal extant lineage is represented by the kinkajou (*Potos flavus*), which diverged from other procyonids ~20 Ma (fig. 7.1). A subsequent study employing a larger data set also yielded the same inference, supporting a basal position for *Potos* and a divergence time from other procyonids deep in the Early Miocene (Eizirik et al. 2010). Remarkably, both studies indicated that all

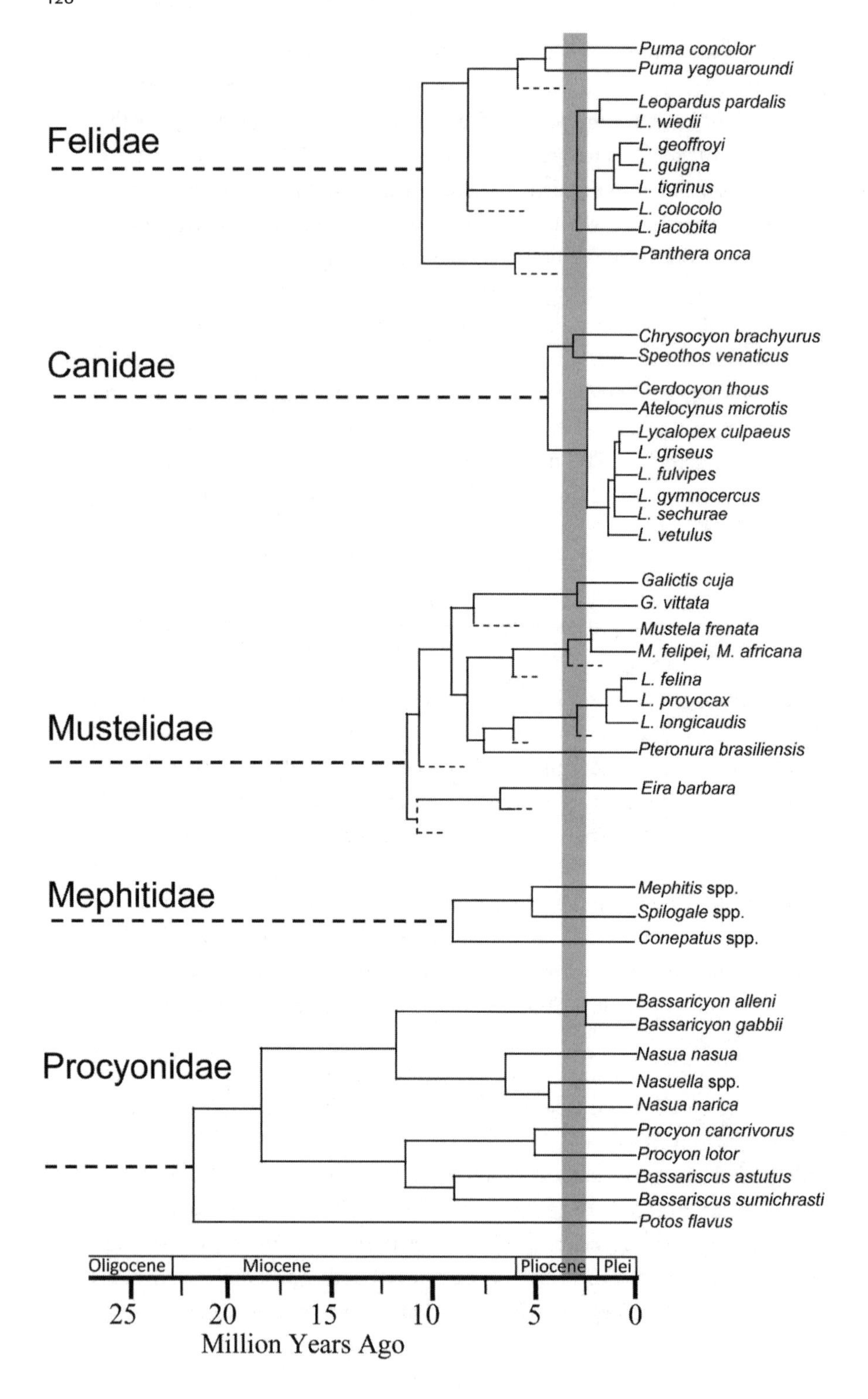
Felidae
Puma concolor
Puma yagouaroundi
Leopardus pardalis
L. wiedii
L. geoffroyi
L. guigna
L. tigrinus
L. colocolo
L. jacobita
Panthera onca
Canidae
Chrysocyon brachyurus
Speothos venaticus
Cerdocyon thous
Atelocynus microtis
Lycalopex culpaeus
L. griseus
L. fulvipes
L. gymnocercus
L. sechurae
L. vetulus
Mustelidae
Galictis cuja
G. vittata
Mustela frenata
M. felipei, M. africana
L. felina
L. provocax
L. longicaudis
Pteronura brasiliensis
Eira barbara
Mephitidae
Mephitis spp.
Spilogale spp.
Conepatus spp.
Procyonidae
Bassaricyon alleni
Bassaricyon gabbii
Nasua nasua
Nasuella spp.
Nasua narica
Procyon cancrivorus
Procyon lotor
Bassariscus astutus
Bassariscus sumichrasti
Potos flavus
Oligocene
Miocene
Pliocene
Plei
25
20
15
10
5
0
Million Years Ago

major genus-level lineages in the Procyonidae appeared in the Miocene, clearly predating the GABI. Furthermore, several divergence events within genera were also inferred to have occurred before the GABI (Koepfli et al. 2007), pushing back in time much of this family's inferred diversification (fig. 7.1).

The only sister species pair analyzed whose divergence seems to have post-dated the GABI was *Bassaricyon alleni* and *B. gabbii*, which are inferred to have split 2.5 Ma (Koepfli et al. 2007). This was possibly connected to an episode of inter-American colonization, whose polarity is currently difficult to assess (see later). However, species delimitation within the genus *Bassaricyon* is still poorly established (Wozencraft 2005; K. Helgen, pers. comm.), which is needed to better characterize current geographic ranges, as well to precisely assess relationships and divergence times among these lineages.

Divergence dates between species in the genera *Procyon* and *Nasua* were estimated by Koepfli et al. (2007) to be 5 Ma and 7 Ma, respectively (fig. 7.1), which contrasts with expectations based on their current geographic ranges. In both genera there is a clear and parallel biogeographic split between sister species: *P. cancrivorus* and *N. nasua* are essentially restricted to South America, whereas *P. lotor* and *N. narica* mostly occur in North and Central America. Therefore, it could be hypothesized that the underlying speciation events were influenced by the GABI, or occurred subsequent to colonization of South America. However,

---

Figure 7.1 (*facing page*) Contrasting patterns of diversification among Neotropical carnivores inferred from molecular data sets. Five families of terrestrial carnivores (Felidae, Canidae, Mustelidae, Mephitidae, and Procyonidae) are included, all containing multiple Neotropical species. Each phylogeny is a timetree (chronogram) depicting the phylogenetic relationships among Neotropical species of the family, as well as the date inferred for each divergence event (i.e., branch lengths are proportional to time; timescale shown at the bottom). Only lineages containing Neotropical representatives are explicitly indicated; dashed lines indicate one or more additional lineages that do not occur in the Neotropics. For each family, the tree was redrawn as a synthesis of current phylogenetic and divergence date information, and often based on more than one source paper, as follows: Felidae (Johnson et al. 2006); Canidae (Lindblad-Toh 2005; Perini, Russo, and Schrago 2010); Mustelidae (Harding and Smith 2009; Koepfli et al. 2008); Mephitidae (Eizirik et al. 2010); Procyonidae (Helgen et al. 2009; Koepfli et al. 2007). In two cases, the divergence date was inferred or adjusted here relative to the original reference. First, in Mustelidae, the chronogram is based on dates reported by Koepfli et al. (2008), with the insertion of *Mustela felipei* and *M. africana* at the position reported by Harding and Smith (2009); their divergence relative to *M. frenata* was recalculated here as an approximate function of the previous node's age in Koepfli et al. (2008), to normalize the estimates with respect to the latter study. Second, in Procyonidae, the position of the genus *Nasuella* as a lineage nested within *Nasua* follows Helgen et al. (2009), with the divergence date relative to *N. narica* approximated based on the sequence divergence reported in that study, calibrated by that of the previous node. The gray bar indicates the time span (2.5–5 Ma) commonly associated with the Great American Biotic Interchange (GABI).

estimated divergence dates suggest that both events occurred prior to the GABI, followed by episodes of inter-American (or trans-Andean) dispersal. Again, the polarity of such events is difficult to assess, as it is possible that these pre-GABI splits occurred in North and Central America before a full connection with South America was established. Nevertheless, the reverse is also plausible, given that procyonids have a fossil record in South America prior to the GABI.

An interesting observation reported by Koepfli et al. (2007) pertains to the inference that two procyonid divergences (*Nasua–Bassaricyon* and *Procyon–Bassariscus*) occurred almost simultaneously in the mid-Miocene, ~11–12 Ma. This suggests that some historical event, possibly connected to a concordant biogeographic barrier, led to a parallel split in both lineages (Koepfli et al. 2007). Although the exact nature and location of such an event is presently difficult to determine, this observation creates an interesting hypothesis that might help decipher the history of procyonid diversification. Even though later analyses by Eizirik et al. (2010) did not corroborate such a close coincidence between these two nodes, the credibility intervals of both studies overlapped substantially, suggesting that this issue should be the focus of further scrutiny employing molecular dating approaches with expanded data sets.

A recent study has added representatives of the genus *Nasuella* to the molecular phylogeny of procyonids (Helgen et al. 2009). Interestingly, *Nasuella* seems to be nested within the genus *Nasua*, as a sister to *N. narica* (fig. 7.1), challenging the monophyly of the latter and underscoring the need for a taxonomic revision. Although Helgen et al. (2009) did not report detailed dating analyses, their estimates of molecular divergences allowed an indirect calculation of the age of this split, which appears to have occurred before the GABI (fig. 7.1). Given current geographic distributions, it may be hypothesized that this lineage began its diversification in South America in the late Miocene, with *N. nasua* and *Nasuella* spp. (see Helgen et al. 2009 for evidence of two species in this genus) remaining in this continent, and *N. narica* subsequently colonizing Central America. Additional molecular studies focusing on this group should help test this biogeographic hypothesis, as well as to more precisely date these divergences.

Even though the phylogenetic relationships and divergence dates established for extant procyonid lineages by recent molecular studies seem consistent, their biogeographic history remains uncertain in several respects. Given the paucity of fossils from most of these lineages, one is compelled to hypothesize patterns of diversification based on current geographic ranges, evolutionary relationships, and divergence dates, bringing molecular data to the center of this investigation. Current results indicate that most of the diversification that led to extant procyonid lineages occurred in the Miocene, beginning ~20 Ma. Could it be possible that most of these splits occurred in South America, implying a much older arrival in this continent than currently known? Or did they all occur in North America (leaving very little fossil evidence behind), followed by massive colonization of multiple lineages into South America during the GABI? Several of these Miocene splits might have occurred in southern North America and Central America, during the formation of the putative island chain that preceded the final closure of the Panama Isthmus.

7.2.2 MUSTELIDAE

Mustelids comprise a speciose carnivoran lineage, including terrestrial, arboreal, and semiaquatic forms that occupy diverse habitats across most of the globe. Eleven species occur in the Neotropics, including an endemic radiation of otters (genus *Lontra*), a divergent lineage also belonging to subfamily Lutrinae (the giant river otter *Pteronura brasiliensis*), the semiarboreal tayra (*Eira barbara*—a species related to the Holarctic martens), along with smaller, weasel-like groups included in the genera *Galictis*, *Lyncodon*, and *Mustela*. Each of these groups likely invaded South America independently, possibly during or after the GABI, and will be briefly discussed later.

Neotropical otters of the genus *Lontra* seem to be an example of autochthonous post-GABI diversification. The three Neotropical species belonging to this genus (*L. longicaudis*, *L. felina*, and *L. provocax*) constitute a monophyletic group, whose basal diversification occurred <2.5 Ma (Koepfli et al. 2008). Their sister group is the North American otter *L. canadensis*, from which their ancestor diverged ~2.8 Ma (Koepfli et al. 2008) during the GABI (see fig. 7.1). It thus seems likely that a single ancestor for this group invaded South America during the GABI, and subsequently radiated into the three extant species belonging to this clade.

A different picture emerges from the assessment of the relationships of the giant river otter (*Pteronura brasiliensis*). This species currently represents the most basal lineage in the subfamily Lutrinae, having diverged from other otters ~7.4 Ma (Koepfli et al. 2008). Given that its current range is restricted to South America, it could be hypothesized that this lineage was an early carnivoran colonist of this continent (possibly contemporaneous with Miocene procyonid immigrants), diverging from other otter lineages as it became isolated in this region. However, fossil evidence and biogeographic reconstructions seem to favor an alternative hypothesis that postulates the ancestry of this lineage in

North America or Eurasia (see Koepfli et al. 2008 and references therein). If this hypothesis is corroborated, it indicates that *Pteronura* invaded South America more recently (during or after the GABI) where it established at least one successful species, but left no extant representatives (and little fossil evidence) in its region of origin.

Another mustelid group that became established as a typical component of Neotropical carnivoran communities is the genus *Galictis*, comprising two species of grison (*G. cuja* and *G. vittata*) that are endemic to this biogeographic region (Wozencraft 2005). Molecular data indicate that grisons diverged from other mustelids ~8 Ma, and that the two extant species are the product of a split that occurred ~2.8 Ma (Koepfli et al. 2008). These dates would suggest a scenario in which an ancestral North American species of *Galictis* invaded South America during the GABI and soon afterward diverged into the two extant lineages (fig. 7.1). However, an interesting complement to investigations of this lineage will be adding the Patagonian genus *Lyncodon* to the molecular phylogeny of mustelids. Preliminary analyses indicate that the single extant species (occurring in Argentina and Chile, still very poorly known) constitutes the sister group to *Galictis* (Wolsan and Sato 2009), suggesting that this lineage as a whole may have diversified in South America. Additional work on the phylogeny, dating, and phylogeography of this group will allow the testing of this hypothesis, as well as the assessment of whether a single ancestor for these species could have invaded South America during the GABI.

Another poorly known group of Neotropical mustelids is the genus *Mustela*. Two species are endemic to South America (*M. africana* and *M. felipei*), while a third (*M. frenata*) occurs there as well as in Central and North America (Wozencraft 2005). Molecular data are now available for these taxa (Harding and Smith 2009; Koepfli et al. 2008), allowing an assessment of their evolutionary history. Jointly, these two studies suggest that these species comprise a monophyletic group whose basal divergence occurred ~2.4 Ma (see fig. 7.1 and comments there with respect to dating adjustments). They in turn are sister to the American mink (*Neovison*), from which they diverged ~3.2 Ma. If corroborated by additional analyses, these dates would be consistent with a single colonization of South America during the GABI, prior to diversification of the Neotropical species.

Finally, the tayra (*Eira*) represents the basal-most living member of the marten lineage, having diverged from other species ~6.8 Ma (Koepfli et al. 2008). It almost certainly derives from an independent colonization of the Neotropics by a Nearctic ancestor. Although this lineage plausibly evolved in North

America in the late Miocene and only invaded the Neotropics during the GABI, it is tempting to speculate that it may have been an early colonist of South America, as its divergence date from a northern sister group is similar to that estimated for *Pteronura*, and also fits the appearance of procyonids in the South American fossil record.

In summary, on the basis of current data, it may be inferred that at least five separate invasions of South America have led to the current Neotropical mustelid fauna: (1) the common ancestor for genera *Galictis* and *Lyncodon*, (2) a representative of *Mustela*, (3) a representative of *Lontra*, (4) a *Pteronura*, and (5) an *Eira*. With the possible exceptions of *Pteronura* and *Eira* (as discussed earlier), current molecular data indicate independent entries of these lineages during or after the GABI, followed by autochthonous diversification in the genera *Galictis*, *Mustela*, and *Lontra*.

7.2.3 MEPHITIDAE

Traditional carnivoran taxonomy has usually placed skunks within the family Mustelidae. However, in the 1980s and 1990s, growing molecular evidence led to the recognition that they comprise a separate lineage within Arctoidea, not immediately related to the mustelids (Dragoo and Honeycutt 1997; Flynn et al. 2000; Wayne et al. 1989). Recent molecular studies have consolidated this view and support the placement of skunks in their own family, Mephitidae (Eizirik and Murphy 2009; Eizirik et al. 2010; Wozencraft 2005).

The Mephitidae is subdivided into two major branches, whose divergence took place ~20 Ma (Eizirik et al. 2010). One branch is now represented by the stink badgers (*Mydaus* spp.) and is endemic to Southeast Asia. All other extant lineages (comprising the genera *Conepatus*, *Mephitis*, and *Spilogale*) are restricted to the New World, and shared a common ancestor ~9 Ma (Eizirik et al. 2010). Of these, the basal-most lineage is *Conepatus*, which occurs in North, Central, and South America. The other two genera (*Mephitis* and *Spilogale*) occur in North and Central America, and diverged from each other ~5 Ma. Fossil evidence suggests that *Conepatus* was already present in South America ~2 Ma (Hunt 1996), and the molecular dates are consistent with a scenario of North American diversification followed by invasion of South America by this genus during the GABI. Species-level molecular phylogenies are not yet available for this group, precluding an in-depth assessment of their diversification patterns in the Neotropics. As these data become available, it will be interesting to reconstruct the tempo of within-genus divergences in skunks, which may aid understanding their biogeographic history in the Neotropics.

7.2.4 CANIDAE

Canids are an important component of Neotropical carnivoran communities (Eisenberg and Redford 1999; MacDonald 2001). South America currently harbors a very diverse assemblage of this family, with 11 different species, 10 of which are endemic (Wozencraft 2005). The additional species is the gray fox (*Urocyon cinereoargenteus*), which occurs in North and Central America, extending its range into northern South America. The endemic Neotropical canids are usually placed in genera *Atelocynus*, *Cerdocyon*, *Chrysocyon*, *Lycalopex*, and *Speothos*, whose phylogenetic placement within the Canidae has been extensively investigated using molecular tools (e.g., Wayne et al. 1997). Although initial analyses led to conflicting results, recent molecular studies have indicated that the endemic Neotropical canids comprise a monophyletic group (Lindblad-Toh et al. 2005; Perini, Russo, and Schrago 2010). The age of this group seems slightly older than the GABI (Perini, Russo, and Schrago 2010), suggesting that two different canid lineages from North America may have invaded the Neotropics during this faunal exchange (fig. 7.1). Additional analyses are required to further test whether a single autochthonous origin for this Neotropical clade can be refuted. Other than the group's basal-most node, all remaining divergences in this clade seem to have occurred during or after the GABI (fig. 7.1), highlighting the impact that the faunal exchange process may have had on the diversification of this lineage by providing opportunities for radiation after the invasion of South America.

A remarkable feature of Neotropical canid diversity is the endemic radiation of the genus *Lycalopex*, which currently comprises 6 fox species whose range is restricted to South America (Wozencraft 2005). Recent molecular dating analyses (Perini, Russo, and Schrago 2010) indicate that this group underwent a very rapid and recent diversification process, likely within the last 1–1.5 Ma. This has led to considerable difficulty in resolving the phylogenetic relationships among these species, a problem still not fully settled (fig. 7.1). Most recent analyses place *L. vetulus* as the most basal lineage in this clade (Lindblad-Toh et al. 2005), which suggests an initial split in the genus isolating this species in central Brazil (where it currently occurs) from the progenitor of the remaining lineages, possibly in Argentina. Given the speed of this diversification process, it is likely that complete phylogenetic resolution of all nodes will require very large data sets, as well as the application of analytical methods that accommodate issues such as incomplete lineage sorting. Still, the full resolution of this phylogeny should prove very useful to investigate the biogeographic patterns and processes underlying this radiation.

7.2.5 FELIDAE

The Felidae is represented in the Neotropics by 10 recognized species, which are placed in 3 different phylogenetic lineages (currently assigned to genera *Leopardus*, *Panthera*, and *Puma*, respectively; Wozencraft 2005; Johnson et al. 2006). Our current understanding is that each of these lineages represents a separate invasion of the Neotropics, stemming from different Nearctic ancestors that colonized South America during or after the GABI. Each is treated separately later.

The most speciose genus of Neotropical felids is *Leopardus*, a clade that is endemic to this biogeographic region and likely results from an autochthonous radiation postdating the GABI (fig. 7.1). Seven species are currently recognized (or nine if we consider the subdivision of *L. colocolo* proposed by Garcia-Perea [1994] and provisionally followed by Wozencraft [2005]). All of these species occur either exclusively in South America or extend their ranges into Central America and southern North America (Mexico and southernmost United States). Molecular dating analyses indicate that this lineage diverged from other felids ~8 Ma, possibly in North America, but only radiated into the currently extant species ~2.9 Ma (Johnson et al. 2006). This radiation most likely took place in South America, which is where most of the present diversity of the genus is observed (e.g., four species completely confined to this continent, two of which are restricted to its southern portion). It is thus likely that a single felid species present in North America in the Pliocene invaded South America during the GABI, giving rise to this radiation of small to medium-sized wild cats that currently occur in most Neotropical biomes.

The biogeographic pattern of this evolutionary radiation has not yet been fully characterized, even though most phylogenetic branches relating these species have been robustly established. There seems to be a phylogenetic division into two main subgroups, one comprising the ocelot (*L. pardalis*) and margay (*L. wiedii*), which occupy mostly tropical habitats from Argentina and southern Brazil to southern North America, and the other including the remaining species. Given their broad and mostly concordant ranges, it is difficult to define the precise geographic origin of ocelots and margays, although phylogeographic evidence suggests that northern South America (i.e., north of the Amazon River) harbors considerable intraspecific diversity at least in ocelots, with some indication of south-to-north colonization of Central America (Eizirik et al. 1998). It can thus be postulated that this subgroup of the genus evolved in forested areas of northern South America, from where it expanded both south and north. Additional phylogeographic analyses are required to further investigate this hypothesis. The second subgroup includes *L. colocolo*, *L. tigrinus*, *L. geoffroyi*, and *L. guigna*, all of which are restricted to South America, with the exception of *L. tigrinus*, which ranges into Panama and Costa Rica. Interestingly, molecular data have revealed that *L. tigrinus* samples from Costa Rica are evolutionarily very distinct from those collected in Brazil (Johnson et al. 1999; Trigo et al. 2008), questioning the monophyly of this species, and posing additional challenges to the biogeographic reconstruction of this radiation. Nevertheless, given the fact that *L. geoffroyi* and *L. guigna* are restricted to the "Southern Cone" of South America (Wozencraft 2005) and that *L. colocolo* ranges mostly in the Andes and open habitats of Argentina, Brazil, and Uruguay, it is likely that this group has evolved in this southerly region of the continent. Finally, the lone unstable branch in this phylogeny is the Andean mountain cat (*L. jacobita*), which has been variably placed in different molecular phylogenetic studies (Johnson et al. 1998, 2006), precluding a meaningful reconstruction of its biogeography.

Another felid lineage occurring in the Neotropics is represented by the sister species puma (*Puma concolor*) and jaguarundi (*P. yagouaroundi*). The most recent molecular analysis of felid relationships generated an estimate of 4.2 Ma for the age of their divergence, with a credibility interval ranging from 3.2–6 Ma (Johnson et al. 2006). The point estimate reported in that study would thus indicate that these felids diverged from each other prior to the GABI, suggesting that this split occurred in North America followed by independent colonization of South America by the descendent species. However, the credibility interval approached the time span of the GABI, allowing for an alternative hypothesis in which a single lineage might have colonized South America, where the subsequent split could have occurred.

The genus *Panthera* is currently represented in the Neotropics by a single species, the jaguar (*P. onca*). This large felid is closely related to the lion (*P. leo*), tiger (*P. tigris*), and leopard (*P. pardus*), all presently confined to the Old World. The joint interpretation of molecular and paleontological evidence indicates that this group radiated in Eurasia, most likely starting from Southeast Asia, from where it expanded its range into most other continents (Johnson et al. 2006). Jaguar ancestors seem to have invaded North America via Beringia ~1.5 Ma (Christiansen and Harris 2009), leading to the colonization of South America well after the GABI. It has been hypothesized that the North American fossil pantherines ascribed to *P. atrox* may have been more closely related to the jaguar than to the lion, leading to the inference of a single invasion of the New World (south of Beringia) by this genus (Christiansen and Harris 2009). However, a recent molecular study that included DNA samples from ancient specimens of this group has provided phylogeographic evidence against this hypothesis, and supported the view that *P. atrox* was in fact more closely related to modern lions (*P. leo*) and particularly to the extinct Eurasian and Beringian lions ascribed to the taxon *P. spelaea* (or *P. leo spelaea*) than to jaguars (Barnett et al. 2009). These data indicate that at least two different lineages of *Panthera* invaded the New World in the Pleistocene, one of which constituted the ancestors of jaguars that went on to colonize the Neotropics.

7.3 Comparative Analysis of Biogeographic Patterns

Contrasting patterns of phylogenetic structure are observed among the different carnivoran families that currently occur in the Neotropical region (fig. 7.1). On the one extreme is Procyonidae, with deep divergences among most genera, all predating the GABI. In particular, the base of this group lies deep in the Early Miocene, challenging the hypothesis that this group could represent an autochthonous radiation in South America after a late Miocene invasion of that continent. Two hypotheses may be raised to explain this pattern: (1) procyonid ancestors invaded South America much earlier than currently accepted, at the beginning of the Miocene, and then underwent a gradual diversification in situ; or (2) the basal diversification of procyonids occurred in North or Central America, giving rise to all extant lineages, which subsequently invaded South America multiple times, leaving in many cases no evidence of their Nearctic origin.

On the other extreme, Canidae and Felidae provide examples of rapid and recent radiations that seem to result from invasions during the GABI and subsequent endemic diversification in South America (fig. 7.1). The most dramatic cases are those of *Lycalopex* in the Canidae and *Leopardus* in the Felidae, each comprising six to nine extant species. Such parallel radiations provide an interesting case study for the investigation of rapid diversification processes following the invasion of a new continent. Future comparative studies targeting a detailed assessment of the rates and patterns of diversification in both genera may provide interesting insights into the processes influencing speciation in these groups following an invasion during the GABI.

Intermediate scenarios may be gleaned from the assessment of the remaining carnivoran lineages that are part of the Neotropical biota. Multiple episodes of invasion may be inferred for the Mustelidae, and to a lesser degree also to the Felidae, Canidae, and Mephitidae. In several cases the age of Neotropical clades is consistent with an invasion and diversification associated with the GABI (see fig. 7.1), highlighting the role of this faunal exchange process in the formation of present-day carnivoran diversity. Such comparisons of divergence dates across carnivoran clades are only now becoming possible, with the accumulation of molecular data for each lineage coupled with integrated analyses of fossil evidence. As more lineages are sampled for multiple molecular markers, more precise estimations of divergence dates should be achieved, allowing for detailed comparisons across families and direct testing of at least some biogeographic hypotheses.

7.4 Challenges and Prospects for Future Studies

Recent advances derived from molecular studies have contributed significantly to our understanding of carnivoran biogeography in the Neotropics. Still, there are many aspects of their evolution in the region that remain unexplored or poorly understood, and most would benefit from additional molecular studies. These topics may be categorized as follows: (1) phylogenetic relationships, (2) divergence dating, (3) species delimitation, (4) assessment of current geographic ranges, (5) assessment of spatial and temporal ranges of extinct species, and (6) intraspecific phylogeography.

Although the phylogenetic relationships among most of these taxa have now been robustly estimated using molecular and morphological data, some nodes in the respective family-level trees remain unresolved (fig. 7.1), and should be the target of future studies aiming to refine and consolidate them. To achieve consistent resolution, future molecular studies will likely include multiple individuals per species, a large set of genomic markers, and sophisticated analyses using diverse phylogenetic approaches. These expanded data sets should also allow for more precise and accurate estimates of divergence times among carnivoran lineages, employing a variety of dating approaches whose performance is constantly improving.

An important point is that species-level delimitation is still not completely settled for several carnivoran lineages (e.g., *Bassaricyon*, *Conepatus*, *Galictis*, and *Nasua*), therefore requiring additional attention with the combined use of morphological, molecular, and ecological approaches. Such delimitation is critical not only to define all the lineages that must be included in future phylogenetic and dating studies, but also to improve the understanding of current geographic ranges of Neotropical carnivores. This refined information would be very important to propose and assess biogeographic hypotheses related to the diversification of these taxa in the Neotropics. It would, of course, be even better to be able to base hypotheses not only on current but also on historical geographic ranges of each lineage. Therefore, another area that should receive increased attention is the carnivoran fossil record in the Neotropics. Paleontological studies may allow the discovery and characterization of extinct species, as well as a direct assessment of the evolutionary history of specific lineages. They are also critical to provide reliable fossil calibrations for any molecular dating analysis, and are thus an important component of advances in this field as well. I hope the near future will see improvements in both molecular and fossil data with the integration between these fields allowing more detailed and accurate assessments of carnivoran evolutionary history.

Finally, studies addressing the intraspecific phylogeography of Neotropical carnivorans may also shed light onto the history of this group in the region. Comparative analyses of the distribution of genetic diversity in natural populations may allow the reconstruction of historical processes of expansion and colonization, in some cases perhaps helping to polarize episodes of inter-American dispersal associated with the GABI. Moreover, such geographic patterns of genetic diversity may illuminate the tempo, mode, and location of post-GABI endemic radiations, providing interesting insights onto the processes that shaped the present Neotropical carnivoran communities.

Acknowledgments

I would like to thank Warren Johnson, Klaus-Peter Koepfli, Stephen O'Brien, Robert Wayne, William Murphy, and Kris Helgen for relevant discussions on carnivoran phylogeny and phylogeography. I also thank my students Tatiane Trigo, Ligia Tchaicka, Manoel Rodrigues, Cristine Trinca, Renata Bornholdt, Eunice Matte, Mirian Tsuchiya-Jerep, Henrique Figueiró, Marina Favarini, Anelisie Santos, and Carla Pires for their efforts to investigate and characterize phylogenetic and phylogeographic patterns of Neotropical carnivores, which should hopefully help shed more light onto the evolutionary history of this group in the near future. Finally, I thank Bruce Patterson and Leonora Costa for their encouragement and for comments on a previous version of this chapter, as well as one anonymous reviewer who also contributed valuable suggestions to improve the text.

Literature Cited

Arnason, U., A. Gullberg, A. Janke, M. Kullberg, N. Lehman, E. A. Petrov, and R. Vainola. 2006. "Pinniped Phylogeny and a New Hypothesis for Their Origin and Dispersal." *Molecular Phylogenetics and Evolution* 41:345–54.

Barnett, R., B. Shapiro, I. Barnes, S. Y. W. Ho, J. Burger, N. Yamaguchi , T. F. G. Higham et al. 2009. "Phylogeography of Lions (*Panthera leo* ssp.) Reveals Three Distinct Taxa and a Late Pleistocene Reduction in Genetic Diversity." *Molecular Ecology* 18:1668–77.

Baskin, J. A. 2004. "*Bassariscus* and *Probassariscus* (Mammalia, Carnivora, Procyonidae) from the Early Barstovian (Middle Miocene)." *Journal of Vertebrate Paleontology* 24:709–20.

Christiansen, P., and J. M. Harris. 2009. "Craniomandibular Morphology and Phylogenetic Affinities of *Panthera atrox*: Implications for the Evolution and Paleobiology of the Lion Lineage." *Journal of Vertebrate Paleontology* 29:934–45.

Dragoo, J. W., and R. L. Honeycutt. 1997. "Systematics of Mustelid-Like Carnivores." *Journal of Mammalogy* 78:426–43.

Eisenberg, J. F., and K. H. Redford. 1999. *Mammals of the Neotropics, Vol. 3, The Central Neotropics: Ecuador, Peru, Bolivia, Brazil*. Chicago: University of Chicago Press.

Eizirik, E., S. L. Bonatto, W. E. Johnson, P. G. Crawshaw Jr., J. Vie, D. M. Brousset, S. J. O'Brien, and F. M. Salzano. 1998. "Phylogeographic Patterns and Evolution of the Mitochondrial DNA Control Region in Two Neotropical Cats (Mammalia, Felidae)." *Journal of Molecular Evolution* 47:613–24.

Eizirik, E., and W. J. Murphy. 2009. "Carnivores (Carnivora)." In *The Timetree of Life*, edited by S. B. Hedges and S. Kumar, 504–7. New York: Oxford University Press.

Eizirik, E., W. J. Murphy, K.-P. Koepfli, W. E. Johnson, J. W. Dragoo, R. K. Wayne, and S. J. O'Brien. 2010. "Pattern and Timing of Diversification of the Mammalian Order Carnivora Inferred from Multiple Nuclear Gene Sequences." *Molecular Phylogenetics and Evolution* 56:49–63.

Flynn, J. J., M. A. Nedbal, J. W. Dragoo, and R. L. Honeycutt. 2000. "Whence the Red Panda?" *Molecular Phylogenetics and Evolution* 17:190–99.

Flynn, J. J., and G. D. Wesley-Hunt. 2005. "Carnivora." In *The Rise of Placental Mammals: Origins and Relationships of the Major Extant Clades*, edited by K. D. Rose and J. D. Archibald, 175–98. Baltimore: Johns Hopkins University Press.

Fulton, T. L., and C. Strobeck. 2006. "Molecular Phylogeny of the Arctoidea (Carnivora): Effect of Missing Data on Supertree and Supermatrix Analyses of Multiple Gene Data Sets." *Molecular Phylogenetics and Evolution* 41:165–81.

García-Perea, R. 1994. "The Pampas Cat Group (Genus *Lynchailurus* Severtzov, 1858) (Carnivora: Felidae): A Systematic and Biogeographic Review." *American Museum Novitates* 3096:1–35.

Harding, L. E., and F. A. Smith. 2009. "*Mustela* or *Vison*? Evidence for the Taxonomic Status of the American Mink and a Distinct Biogeographic Radiation of American Weasels." *Molecular Phylogenetics and Evolution* 52:632–42.

Helgen, K. M., R. Kays, L. E. Helgen, M. T. N. Tsuchiya-Jerep, C. M. Pinto, K.-P. Koepfli, E. Eizirik, and J. E. Maldonado. 2009. "Taxonomic Boundaries and Geographic Distributions Revealed by an Integrative Systematic Overview of the Mountain Coatis, *Nasuella* (Carnivora: Procyonidae)." *Small Carnivore Conservation* 41:65–74.

Hunt, R. M., Jr. 1996. "Biogeography of the Order Carnivora." In *Carnivore Behavior, Ecology and Evolution*, 2nd ed., edited by J. L. Gittleman, 485–541. Ithaca, NY: Cornell University Press.

Johnson, W. E., M. Culver, A. Iriarte, E. Eizirik, K. Seymour, and S. J. O'Brien. 1998. "Tracking the Evolution of the Elusive Andean Mountain Cat with Mitochondrial DNA Sequences." *Journal of Heredity* 89:227–32.

Johnson, W. E., E. Eizirik, J. Pecon-Slattery, W. J. Murphy, A. Antunes, E. Teeling, and S. J. O'Brien. 2006. "The Late Miocene Radiation of the Modern Felidae: A Genetic Assessment." *Science* 311:73–77.

Johnson, W. E., J. Pecon-Slattery, E. Eizirik, J. Kim, M. Menotti-Raymond, C. Bonacic, R. Cambre et al. 1999. "Disparate Phylogeographic Patterns of Molecular Genetic Variation in Four Closely Related South American Small Cat Species." *Molecular Ecology* 8:S79–S94.

Koepfli, K.-P., K. A. Deere, G. J. Slater, C. Begg, K. Begg, L. Grassman, M. Lucherini et al. 2008. "Multigene Phylogeny of the Mustelidae: Resolving Relationships, Tempo and Biogeographic History of a Mammalian Adaptive Radiation." *BMC Biology* 6:10.

Koepfli, K.-P., M. E. Gompper, E. Eizirik, C.-C. Ho, L. Linden, J. E. Maldonado, and R. K. Wayne. 2007. "Phylogeny of the Procyonidae (Mammalia: Carnivora): Molecules, Morphology and the Great American Interchange." *Molecular Phylogenetics and Evolution* 43:1076–95.

Krause, J., T. Unger, A. Noçon, A. S. Malaspinas, S. O. Kolokotronis, M. Stiller, L. Soibelzon et al. 2008. "Mitochondrial Genomes Reveal an Explosive Radiation of Extinct and Extant Bears Near the Miocene-Pliocene Boundary." *BMC Evolutionary Biology* 8:220.

Lindblad-Toh, K., C. M. Wade, T. S. Mikkelsen, E. K. Karlsson, D. B. Jaffe, M. Kamal, M. Clamp et al. 2005. "Genome Sequence, Comparative Analysis and Haplotype Structure of the Domestic Dog." *Nature* 438:803–19.

Macdonald, D. W. 2001. *The Encyclopedia of Mammals*, 2nd ed. New York: Barnes and Noble Books.

Marshall, L. G., S. D. Webb, J. J. Sepkoski Jr., and D. M. Raup. 1982. "Mammalian Evolution and the Great American Interchange." *Science* 215:1351–7.

Murphy, W. J., and E. Eizirik. 2009. "Placental Mammals (Eutheria)." In *The Timetree of Life*, edited by S. B. Hedges and S. Kumar, 471–74. New York: Oxford University Press.

Nowak, R. M. 1999. *Walker's Mammals of the World*, 6th ed. Baltimore: Johns Hopkins University Press.

Perini, F. A., C. A. M. Russo, and C. G. Schrago. 2010. "The Evolution of South American Endemic Canids: A History of Rapid Diversification and Morphological Parallelism." *Journal of Evolutionary Biology* 23:311–22.

Simpson, G. G. 1980. *Splendid Isolation: The Curious History of South American Mammals*. New Haven: Yale University Press.

Soibelzon, L. H., E. P. Tonni, and M. Bond. 2005. "The Fossil Record of South American Short-Faced Bears (Ursidae, Tremarctinae)." *Journal of South American Earth Sciences* 20:105–13.

Springer, M. S., W. J. Murphy, E. Eizirik, O. Madsen, M. Scally, C. J. Douady, E. Teeling et al. 2007. "A Molecular Classification for the Living Orders of Placental Mammals and the Phylogenetic Placement of Primates." In *Primate Origins: Adaptations and Evolution*, edited by M. J. Ravosa and M. Dagosto, 1–28. New York: Springer-Verlag.

Trigo, T. C., T. R. O. Freitas, G. Kunzler, L. Cardoso, J. C. R. Silva, W. E. Johnson, S. J. O'Brien et al. 2008. "Inter-Species Hybridization Among Neotropical Cats of the Genus *Leopardus*, and Evidence for an Introgressive Hybrid Zone Between *L. geoffroyi* and *L. tigrinus* in Southern Brazil." *Molecular Ecology* 17:4317–33.

Van Valkenburgh, B. 1999. "Major Patterns in the History of Carnivorous Mammals." *Annual Review of Earth and Planetary Sciences* 27:463–93.

Wayne, R. K., R. E. Benveniste, D. N. Janczewski, and S. J. O'Brien. 1989. "Molecular and Biochemical Evolution of the Carnivora." In *Carnivore Behavior, Ecology, and Evolution*, 2nd ed., edited by J. L. Gittleman, 465–494. Ithaca: Comstock Publishing Associates.

Wayne, R. K., E. Geffen, D. J. Girman, K.-P. Koepfli, L. M. Lau, and C. R. Marshall. 1997. "Molecular Systematics of the Canidae." *Systematic Biology* 46:622–53.

Wolsan, M., and J. Sato. 2009. "Multilocus DNA Phylogeny of Mustelidae and the Ancestry of South American Species." *10th International Mammalogical Congress, Mendoza, Argentina* Abstracts:59.

Wozencraft, W. C. 2005. "Carnivora." In *Mammal Species of the World: A Taxonomic and Geographic Reference*, 3rd ed., edited by D. E. Wilson and D. M. Reeder, 532–628. Baltimore: Johns Hopkins University Press.

Wynen, L. P., S. D. Goldsworthy, S. J. Insley, M. Adams, J. W. Bickham, J. Francis, J. P. Gallo et al. 2001. "Phylogenetic Relationships Within the Family Otariidae (Carnivora)." *Molecular Phylogenetics and Evolution* 21:270–84.

Yonezawa, T., N. Kohno, and M. Hasegawa. 2009. "The Monophyletic Origin of Sea Lions and Fur Seals (Carnivora; Otariidae) in the Southern Hemisphere." *Gene* 441: 89–99.

Part 2
Regional Patterns

Hierarchical Organization of Neotropical Mammal Diversity and its Historical Basis

Sergio Solari, Paúl M. Velazco, and Bruce D. Patterson

The Neotropics are home to roughly 1550 living species of mammals, 30% of all extant species (Wilson and Reeder 2005), including groups found nowhere else on Earth. Endemics include rodent, primate, and bat families, as well as two orders of marsupials and one of edentates. This richness and uniqueness can be attributed to the amazing diversity of biomes in the Neotropics, including tropical rain forests, highland grasslands, deserts, savannas, and scrublands, many of them influenced to some degree by the Andes Mountains. In addition, the South American portion of the Neotropics was isolated during the late Mesozoic and most of the Cenozoic, interrupted by a sequence of continental connections that permitted faunal interchanges with Africa, Antarctica and Australia, and North America at different times. As discussed in the first part of this volume, its endemic families and orders appeared in the Paleogene, constituting many of the taxa that Simpson (1980) allocated to Strata 1 and 2.

Biogeographic classifications should reflect how history, climate, and geology have shaped diversification and distributions. General classifications may summarize how major geological and climatic events affected entire biotas, but different taxa present different diversification patterns: Müller's (1973) fauna-wide classification differs from Cracraft's (1985) for birds and those of Hershkovitz (1958, 1969) for mammals. In addition, improved taxonomic and distributional understanding should lead to refined definition and composition of biogeographic units. Continued species discovery, geographic range revision, and expanded phylogenetic analyses have eclipsed the efforts of Simpson (1980) and Mares (1992) to provide a comprehensive treatment of Neotropical mammals. With improved methods to understand the diversification of lineages and their historic scenarios, we have additional tools to recover diversification patterns. In particular, genetic data permit fine-scaled biogeographic analyses, but only a few mammal groups are known well enough to offer a comprehensive pattern of diversity and distribution in the Neotropics.

Here, we summarize the distributions of living mammals at a general, gross scale, using a recent synopsis of mammal taxonomy (Wilson and Reeder 2005) and previously identified biogeographic units (Cabrera and Willink 1980; Morrone 2001) to identify general patterns reflected in the distributions of supraspecific groups. Analyses of diversity, endemism, and origin of these biotas are addressed in eight subsequent chapters in this volume. Knowledge of species and their ranges is still too inadequate to generate a meaningful hierarchical classification of Neotropical regions and subregions directly from the distributions themselves. Contributors to this volume have used their knowledge and experience to analyze each regional fauna to provide the clearest overview of the biogeographic history of its members. For our analysis, we approach this complex topic with overall approximations of both taxonomic and geographic diversity.

8.1 Historical Overview on Provincialism and Regions of Endemism

The Neotropical zoogeographic region was first recognized and defined by the works of Sclater (1858) and Wallace (1876). Wallace (1876) identified four subregions: Mexican (including all Middle American territories), Antillean (for the West Indian Archipelago), Brazilian (for Amazonian forests), and Chilean (for Patagonia and related nonforested habitats). They and other authors noted the similarity of the northern half of South America to southern Mexico and Central America, calling it the Middle American province of the Brazilian Subregion (see Hershkovitz 1958). Simpson (1956) considered the West Indies more closely related to the Neotropics, with components coming from both Central and South America (Mexican and Brazilian subregions, respectively), but excluded the Greater Antilles from this region.

Hershkovitz (1958; fig. 8.1) reclassified the subregions as follows: (1) Brazilian, including Middle American, Colombian, Guianan, and Amazonia provinces, among others; (2) Patagonian; and (3) West Indian. No provinces were recognized within the Patagonian or West Indian subregions. A persistent problem with this classification is the absence of natural barriers bounding these elements, especially between the Brazilian and Patagonian regions (see Cabrera and Willink 1980). The distinction of the two regions appears to be based on a climatic and vegetation gradient from very humid, forested habitats (in the Brazilian subregion) to semiarid and arid open ones (Patagonian), but there are exceptions (e.g., Caatinga and Valdivian Forest, respectively). Although Hershkovitz (1969) did not subdivide the subregions, he discussed eight faunal centers, referring to three of them (Guianan, Paraná-Paraguay Valley, and the Altiplano) as districts, but offered no definition or justification for this classification.

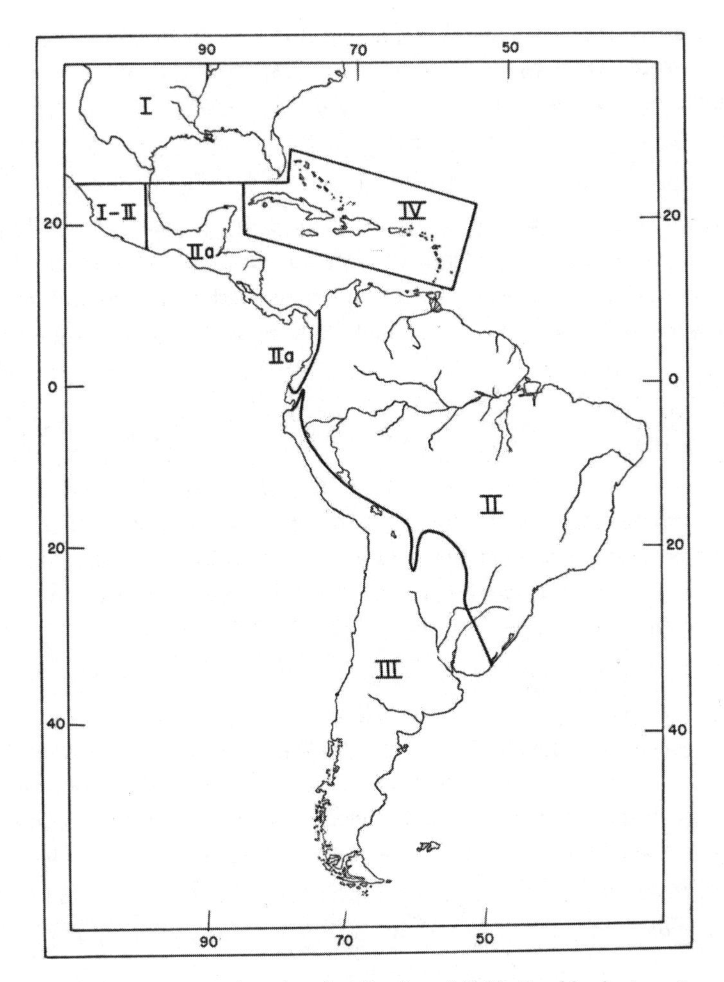

Figure 8.1 Neotropical Region classification of Philip Hershkovitz (1958), based on the distributions of Recent mammals. I, southern part of the Nearctic Subregion of the Holarctic Region; I–II, Neotropical-Holarctic Transition Zone; II, Brazilian Subregion; IIa, Middle American Province of Brazilian Subregion; III, Patagonian Subregion; IV, West Indian Subregion (reprinted with permission from *Fieldiana: Zoology*, 36[6]).

Mares (1992) compared mammalian diversity among South America's major macrohabitats, which were loosely based on biogeographic regions. He recognized: (1) Amazonian lowlands, (2) western montane forests, (3) Atlantic rain forests, (4) upland semideciduous forests, (5) southern mesophytic forests, and (6) drylands. However, the last one comprised a heterogeneous assemblage of very distinctive biomes (e.g., Cerrado, pampas, deserts), rendering it useless for historical biogeography.

Faunal-wide approaches generate very different results than these mammalian perspectives. In his review of South American zoogeography, Fittkau (1969) discussed Hershkovitz's (1958) views and recognized tropical Guianan-Brazilian and temperate Andean-Patagonian subregions. Udvardy (1975) proposed not only a new hierarchical system with biogeographic regions (e.g., Neotropical Realm) containing biotic provinces (e.g., Pacific Desert Biogeographic Province), but also identified 37 provinces within the continental Neotropics (plus nine for oceanic islands and one for Lake Titicaca). Cabrera and Willink (1980) also presented a detailed subdivision, including five dominions within the Neotropics: (1) Caribbean, from central Mexico through Central America to the northern coast of South America (La Guajira) and the Antilles; (2) Amazonian, the largest one and subdivided in nine provinces; (3) Guianan; (4) Chaco, including several noncontinuous dry to semiarid habitats; and (5) Andean-Patagonian. Most of the mammalian taxa chosen to characterize these units came from Hershkovitz (1969). Finally, Morrone (2001) identified 3 regions within Latin America and the Caribbean: Nearctic, Neotropical, and Andean. Only the latter 2 correspond to the Neotropical Region of previous authors—each of these regions was then subdivided into 4 subregions, and a total of 65 provinces were recognized within them.

Knowledge of regional mammal faunas in South America has been shaped more by accessibility than by scientific goals (Hershkovitz 1987). The sampling localities chosen by the first explorers were biased, most being near large cities (e.g., Rio de Janeiro, Maldonado, Bogotá, Cayenne), roads, or large rivers (also used for transportation). Only later were inland basins and remote montane regions sampled. In addition, the definition of diversity patterns has been modulated by changing systematic theory, especially species concepts. During the Modern Synthesis, widespread use of the polytypic species concept and geographic "races" to describe morphological variation has limited the resolution of geographic patterns among Neotropical mammals. Many of these geographic races have subsequently become recognized as full species, but many other groups are in need of revision (Patterson 1994). By the end of the twentieth century, only a handful of tropical rain forest localities in the Neotropics had been fully inventoried for their mammal species (see Voss and Emmons 1996; Solari et al. 2006; Lim, chapter 11, this volume).

Because this strong bias makes some geographic regions and taxonomic groups better known, analyzing specific patterns of distribution in a truly unbiased fashion would require an elaborate analysis, which is beyond the goals of this simple review. Instead, we simply compiled endemic distributions following the taxonomy used by contributors to Wilson and Reeder (2005), updated as necessary. As a result, our basic data set includes 1421 species of land mammals in 337 genera; of these, 946 species were identified as endemic and allocated among 11 subregions based on their distributions (fig. 8.2). Our choice and delimitation of subregions followed those proposed for the Neo-

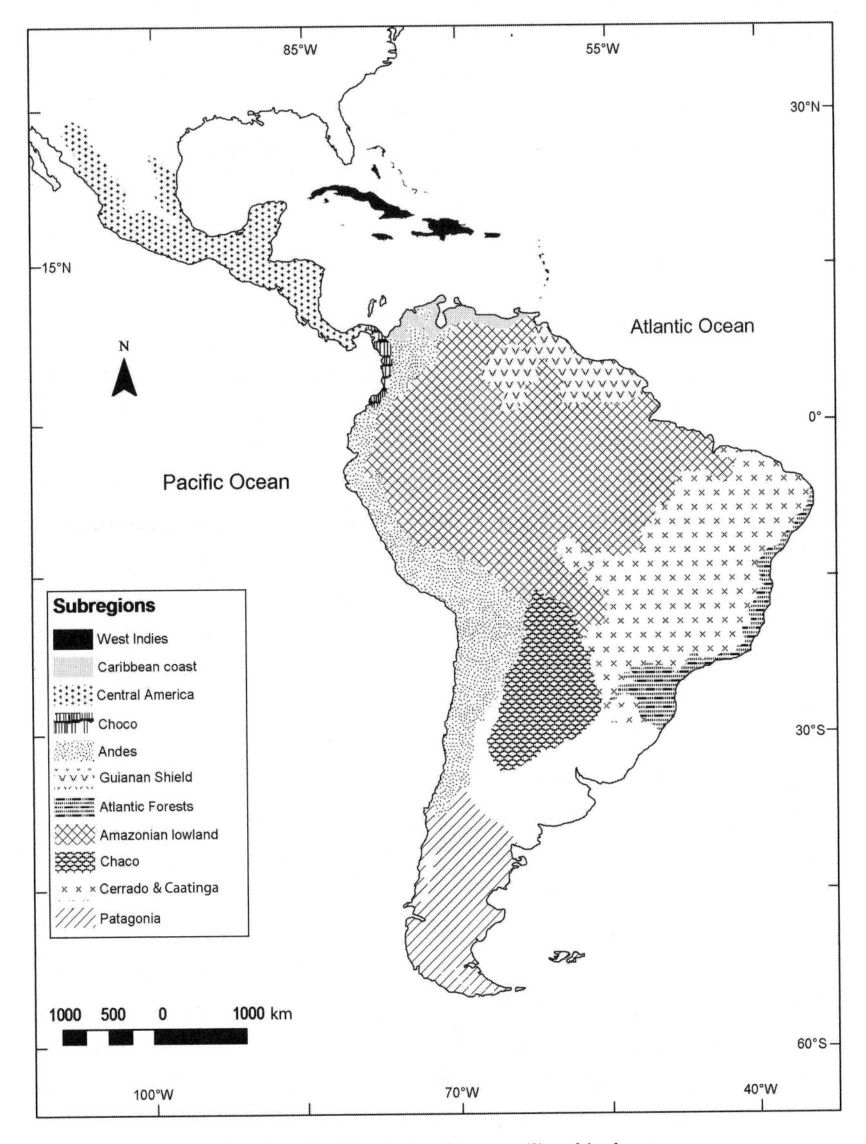

Figure 8.2 Neotropical Region classification used in compiling this chapter.

tropical region (e.g., Cabrera and Willink 1980), avoiding smaller subdivisions (see Morrone 2001) that did not appear useful for mammals. For this analysis, the Andes constitutes a single biogeographic subregion that includes the western and eastern slopes, as well as the highland habitats. The 10 others we examined include: (1) the Chocó; (2) the Caribbean coast of northern South America; (3) the Amazonian lowlands; (4) the Guiana Shield; (5) the Cerrado and Caatinga complex; (6) the Chaco; (7) Patagonia, including the Valdivian forests of southern Chile; (8) the Atlantic forests of southeastern Brazil; (9) the West Indies; and (10) Central America. We did not include the Pampas or the Monte Espinoso in our analysis or on the map (these appear as white areas), as no chapters were devoted to them; however, Carmingnotto et al. (chapter 14, this volume) discuss the relationships of these southern open formations to the Cerrado and Caatinga.

8.2 Patterns of Endemism and Shared Diversity

The complex geography of the Neotropics makes mammal faunas not only rich, but also heterogeneous. The different biogeographic units in the Neotropics harbor unique faunas, but part of the fauna is shared between units at different levels. Including aquatic and marine mammals, the Neotropics are home to at least 1550 mammal species, grouped in 15 orders (Gardner 2008; Wilson and Reeder 2005). Three of these 15 orders are endemic: Microbiotheria (1 genus), Paucituberculata (3 genera), and Pilosa (5 genera), and 2 others (Didelphimorphia and Cingulata) each have only a single living Nearctic member. The most speciose orders in the Neotropics are the Rodentia (15 families, 154 genera, and 711 species) and Chiroptera (9 families, 89 genera, and 317 species); together, rodents and bats account for ~75% of all species of mammals in the Neotropics.

The region with the greatest number of endemics is the Andes (29 genera), followed by the West Indies, Atlantic Forest, and Patagonia, with 16, 14, and 12 genera, respectively. Amazonia and Central America each support 10 endemic genera, whereas the Cerrado and Caatinga have 7, the Chaco has 6, and the Chocó and the Guianas each harbor a lone endemic genus. Endemism at the species level is parallel, with the Andes, Amazonia, Central America, and Atlantic Forest having the highest numbers of endemic species, with 269, 144, 108, and 105 endemic species, respectively. At the specific level, the Caribbean coast and the Chocó exhibit the fewest endemics, with 15 and 11 endemic species, respectively. Although these numbers reflect the age, area, and latitude of the regions, the diversity of these regions is profoundly affected by their geometry and interconnections as well.

8.3 Subregions of Endemism

In view of its tiny land area, the West Indies supports an impressive number of endemic genera and species. Sixteen endemic genera are present: 10 bats (*Ardops*, *Ariteus*, *Brachyphylla*, *Erophylla*, *Monophyllus*, *Phyllonycteris*, *Phyllops*, *Stenoderma*, *Chilonatalus*, and *Nyctiellus*), 5 rodents (*Capromys*, *Geocapromys*, *Mesocapromys*, *Mysateles*, and *Plagiodontia*), and 1 soricomorph (*Solenodon*). Many others are recently extinct (e.g., the rodents *Boromys* and *Brotomys* and the soricomorph *Nesophontes*). The West Indies present a higher diversity of endemic species than the Cerrado and the Guiana Shield with 52. These species are distributed in 3 orders: 36 bats in 19 genera, 14 rodents in the 5 endemic genera mentioned earlier, and 2 surviving soricomorphs in the endemic genus *Solenodon*. Although the total numbers of endemic species and genera in the West Indies are modest in Neotropical terms, the proportions of the fauna that are endemic are very high and often highly endangered. In chapter 9 (this volume), Dávalos and Turvey review phylogenetic, population genetic, and radiocarbon dating studies to explore the geological and biological drivers of colonization, speciation, and extinction in this area.

Central America may be one of the youngest subregions of the Neotropical realm, founded by the formation of the Panamanian land bridge (see Almendra and Rogers, chapter 10, this volume). Nevertheless a large number of endemic genera are present, including 3 monotypic bats (*Bauerus*, *Hylonycteris*, and *Ectophylla*) and 7 rodents (*Isthmomys*, *Nyctomys*, *Otonyctomys*, *Ototylomys*, *Rheomys*, *Scotinomys*, and *Syntheosciurus*). At the specific level, Central America supports 69 endemic rodents (in 27 genera), 14 bats (in 11 genera), 11 shrews (including 10 *Cryptotis* and 1 *Sorex*), 4 primates (2 *Alouatta*, *Ateles geoffroyi*, and *Saimiri oerstedii*), 4 carnivores (2 *Bassaricyon*, *Bassariscus sumichrasti*, and *Procyon pygmaeus*), 3 lagomorphs (2 *Sylvilagus* and *Lepus flavigularis*), 2 marsupials (*Marmosa mexicana* and *Marmosops invictus*), and 1 sloth (*Bradypus pygmaeus*). A number of other lineages derived from Neotropical stocks now range from Central America into the Nearctic Region or Subregion (the marsupial *Tlacuatzin* and 2 bats, *Choeronycteris* and *Musonycteris*), and a large number of Nearctic groups also extend into this transition zone. In chapter 10 (this volume), Almendra and Rogers review the geological and chronological settings for this impressive mixing and diversification.

Although it supports some of the world's richest biotas (e.g., Lim and Engstrom 2005), the Guiana Shield supports only modest numbers of endemics. Only a single genus is endemic, the monotypic rodent *Podoxymys*, while 32 endemic species are distributed in 5 orders: 5 opossums (*Didelphis*, *Monodelphis*, *Marmosa*, and *Marmosops*), 4 primates (*Saguinus*, *Chiropotes*, *Alouatta*, and *Ateles*), 5 bats (*Saccopteryx*, *Lophostoma*, *Platyrrhinus*, *Molossus*, and *Lasiurus*), a carnivore (*Bassaricyon*), and 17 rodents (*Sciurus*, *Neacomys*, *Neusticomys*, *Oecomys*, *Podoxymys*, *Rhipidomys*, *Sphiggurus*, *Dasyprocta*, *Echimys*, *Isothrix*, and *Proechimys*). Focusing analysis on the Guianas proper, Lim (chapter 11, this volume) considers how the Guianan plateau (>500 m elevation) has influenced diversification of mammals in this region, both as an area of endemism and as a geographic barrier.

Amazonian Forest is the largest subregion in the Neotropics. It supports 10 endemic genera: 3 rodents (*Amphinectomys*, *Lonchothrix*, and *Toromys*), 2 monkeys (*Callimico* and *Cacajao*), 2 marsupials (*Caluromysiops* and *Glironia*), 2 bats (*Scleronycteris* and *Neonycteris*), and 1 carnivore (*Atelocynus*). The diversity of endemic species in Amazonia exceeds Central America and Atlantic Forest but is less than the Andes. The majority of endemic species (70) are primates, distributed among 13 genera, with 12 in *Saguinus*, 13 in *Callithrix*, and 21 in *Callicebus*. Rodents rank second with 48 species, distributed in 25 genera (including 4 *Makalata*, 5 *Sciurus*, and 9 *Proechimys*). There are 13 endemic species of opossums, in 6 genera, and 11 species of bats, in 8 genera. Finally, there are only 2 endemic species of carnivores, *A. microtis* and *Mustela africana*. Bonvicino and Weksler (chapter 12, this volume) evaluate the fit of numerous hypotheses for the diversification of Amazonian mammals using phylogenetic, geological, and paleoclimate data.

The Atlantic Forest of Brazil and adjacent Paraguay and Argentina harbors a unique biota (see appendix 13.1). Of the 14 endemic mammal genera, 12 are rodents (*Abrawayaomys*, *Blarinomys*, *Brucepattersonius*, *Delomys*, *Euryzygomatomys*, *Juliomys*, *Lundomys*, *Phaenomys*, *Thaptomys*, *Callistomys*, *Kannabateomys*, and *Chaetomys*) and 2 are primates (*Brachyteles* and *Leontopithecus*). At the species level, the Atlantic Forest supports 68 endemic rodents (including 6 *Akodon*, 7 *Brucepattersonius*, 11 *Phyllomys*, and 10 *Trinomys*), 18 endemic primates (including 4 *Callithrix*, 5 *Callicebus*, and 4 *Leontopithecus*), 11 species of opossums (in 7 genera), 6 species of bats (in 6 genera), 1 sloth (*Bradypus torquatus*), and 1 deer (*Mazama bororo*), ranking fourth in endemic species in the Neotropics. Costa and Leite (chapter 13, this volume) examine the genetic structure of Atlantic Forest mammals and show that this biogeographic region is subdivided into two distinct historical components.

The Cerrado and Caatinga complex includes the most extensive woodland and savanna in the Neotropics. At the generic level, its endemics are known to include 1 bat (*Xeronycteris*) and 6 rodents (*Juscelinomys*, *Microakodontomys*, *Thalpomys*, *Wiedomys*, *Kerodon*, and *Carterodon*). The 45 endemic species are distributed in 6 orders: 4 opossums (*Gracilinanus*, *Monodelphis*, and *Thylamys*), 1 armadillo (*Tolypeutes*), 2 monkeys (*Callithrix* and *Cebus*), 4 bats (*Glyphonycteris*, *Lonchophylla*, *Micronycteris*, and *Xeronycteris*), a fox (*Lycalopex*), and 33 rodents (in 20 genera). Carmignotto et al. (chapter 14, this volume) show that the mammalian faunas of the Carrado and Caatinga are composed of endemic taxa derived from both rain forest and open-country lineages. Although they share species with both the Amazon and Atlantic forests, their oldest and most distinctive members are open-country lineages.

The richest and most diverse subregion in the Neotropics (and possibly on Earth), the Andes harbor at least 29 endemic genera and 269 endemic species. Twenty genera (*Aepeomys*, *Andinomys*, *Anotomys*, *Auliscomys*, *Chibchanomys*, *Chilomys*, *Chinchillula*, *Eremoryzomys*, *Galenomys*, *Lenoxus*, *Mindomys*, *Neotomys*, *Octodontomys*, *Oreoryzomys*, *Pattonomys*, *Punomys*, *Santamartamys*, *Thomasomys*, *Cuscomys*, and *Olallamys*) and 192 species are rodents. There are 3 endemic ungulate genera (*Lama*, *Vicugna*, and *Hippocamelus*), with 4 species of Camelidae and 5 of Cervidae. Also endemic are 2 genera of shrew opossums, the monotypic *Lestoros* and 4 species of *Caenolestes*. The only endemic carnivore genus is the monotypic *Tremarctos*; the other 4 endemic species belong to the widespread genera *Leopardus*, *Mustela*, and *Nasuella*. In addition, monotypic *Oreonax* represents the lone endemic genus of monkey. Two bat genera are endemic (*Platalina* and *Amorphochilus*), whereas 26 endemic bat species are known among *Balantiopteryx*, *Saccopteryx*, *Anoura*, *Lonchophylla*, *Mimon*, *Sturnira* (with 7 species), *Vampyressa*, *Molossops*, *Mormopterus*, *Eptesicus*, *Histiotus*, and *Myotis*. Endemism at the species level is also marked among the opossums (15 species in *Didelphis*, *Gracilinanus*, *Marmosa*, *Marmosops*, *Monodelphis*, and *Thylamys*), armadillos (2 species in *Chaetophractus* and *Dasypus*), shrews (7 species of *Cryptotis*), and tapirs (*Tapirus pinchaque*). In chapter 15 (this volume), we explore the bases and timing of mammalian diversification in the Andes, as well as the broader role of this cordillera as a distributional barrier to lowland mammals.

The Patagonian mammal fauna includes a number of endemic genera and species. The order Microbiotheria, with 1 extant species *Dromiciops gliroides*, is endemic to this region, and 11 other genera are also endemic to Patagonia: *Lestodelphys* (Didelphimorphia), *Rhyncholestes* (Paucituberculata), *Chlamyphorus*,

Zaedyus (Cingulata), *Lyncodon* (Carnivora), *Deltamys*, *Geoxus*, *Loxodontomys*, *Notiomys*, *Octomys*, and *Pearsonomys* (Rodentia). The 65 endemic species represent 7 orders: 1 opossum (*Lestodelphys*), 1 shrew opossum (*Rhyncholestes*), 2 armadillos (*Chlamyphorus* and *Zaedyus*), 2 bats (*Histiotus* and *Myotis*), 3 carnivores (*Lontra*, *Lyncodon*, and *Conepatus*), 2 ungulates (*Hippocamelus* and *Pudu*), and 52 rodents representing 22 genera (*Ctenomys* being the most diverse with 17 species). Lessa et al. (chapter 16, this volume) examine the genetic structure of Patagonian rodents for clues on how Pleistocene glaciation influenced the diversification of Patagonian lineages.

8.4 Gaps in our Current State of Knowledge

Although a full analysis of patterns of diversity and endemism is beyond the focus of this introductory chapter, our preliminary assessment of updated data for the distribution of Neotropical mammal species clearly supports the conventional view that the Andes and Amazonia are the richest regions in the Neotropics (Antonelli et al. 2010; Patterson and Velazco 2008). However, the Atlantic Forest and Middle America (including the Chocó) show striking diversity and their endemism is higher than that of other equivalent regions, as the reader will see in the ensuing chapters. These current patterns are consistent with the unique histories each region has had since South America reached its equatorial position and reconnected with North America through the Isthmus of Panama 3–6 Ma.

Systematic revisions including phylogeographic analyses are being completed or are in progress for a number of groups, but additional taxa and parts of their geographic range need to be covered. Broader taxonomic and geographic sampling will provide more robust tests for the patterns described in this chapter and volume. Some of these analyses have already changed our views, for example, by tying a Guianan endemic to an Andean radiation (Lim et al. 2010) or an Andean endemic to an Amazon-Orinoco-Guianan radiation (Patterson and Velazco 2008). The chapters in this volume should foster and expedite this kind of research.

Acknowledgments

We would like to thank Janet Voight for permission to reprint fig. 8.1 and Wilmar Munera, of Universidad de Antioquia, Medellin, Colombia, for his help in constructing fig. 8.2.

Literature Cited

Antonelli, A., A. Quijada-Mascareñas, A. J. Crawford, J. M. Bates, P. M. Velazco, and W. Wüster. 2010. "Molecular Studies and Phylogeography of Amazonian Tetrapods and Their Relation to Geological and Climatic Models." In *Amazonia, Landscape and Species Evolution: A Look Into the Past*, edited by C. Hoorn and F. P. Wesselingh, 386–404. San Francisco: John Wiley and Sons.

Cabrera, A. L., and A. Willink. 1980. *Biogeografía de América Latina*. O.E.A. Serie Monográfica 13, Serie de Biología. Washington, DC: Organización de Estados Americanos.

Cracraft, J. 1985. "Historical Biogeography and Patterns of Differentiation Within the South American Avifauna: Areas of Endemism." In *Neotropical Ornithology*, edited by P. A. Buckley, M. S. Foster, E. S. Morton, R. S. Ridgely, and F. G. Buckley, 49–84. Washington, DC: American Ornithologists Union, *Ornithological Monographs* 36.

Fittkau, E. J. 1969. "The Fauna of South America." In *Biogeography and Ecology in South America*, Vol. 2, edited by E. J. Fittkau, J. Illies, H. Klinge, G. H. Schwabe, and H. Sioli, 624–58. The Hague: Dr. W. Junk.

Gardner, A. L., ed. 2008. *Mammals of South America, Vol. 1: Marsupials, Xenarthrans, Shrews and Bats*. Chicago: University of Chicago Press.

Hershkovitz, P. 1958. "A Geographical Classification of Neotropical Mammals." *Fieldiana: Zoology* 36:581–646.

———. 1969. "The Recent Mammals of the Neotropical Region." *Quarterly Review of Biology* 44:1–70.

———. 1987. "A History of the Recent Mammalogy of the Neotropical Region from 1492 to 1850." In *Studies in Neotropical Mammalogy: Essays in Honor of Philip Hershkovitz*, edited by B. D. Patterson and R. M. Timm, 11–98. Chicago: Field Museum of Natural History. *Fieldiana: Zoology*, n.s., 39.

Lim, B. K., and M. D. Engstrom. 2005. "Mammals of Iwokrama Forest." *Proceedings of the Academy of Natural Sciences of Philadelphia* 154:71–108.

Lim, B. K., M. D. Engstrom, J. C. Patton, and J. W. Bickham. 2010. "Molecular Phylogenetics of Reig's Short-Tailed Opossum (Monodelphis Reigi) and its Distributional Range Extension Into Guyana." *Mammalian Biology* 75:287–93.

Mares, M. A. 1992. "Neotropical Mammals and the Myth of Amazonian Diversity." *Science* 255:976–79.

Morrone, J. J. 2001. *Biogeografía de América Latina y el Caribe*. M & T Manuales y Tesis SEA, vol. 3. Zaragoza: CYTED, ORCYT-UNESCO, and SEA.

Müller, P. 1973. *The Dispersal Centres of Terrestrial Vertebrates in the Neotropical Realm: A Study in the Evolution of the Neotropical Biota and its Native Landscapes*. The Hague: Dr. W. Junk.

Patterson, B. D. 1994. "Accumulating Knowledge on the Dimensions of Biodiversity: Systematic Perspectives on Neotropical Mammals." *Biodiversity Letters* 2:79–86.

Patterson, B. D., and P. M. Velazco. 2008. "Phylogeny of the Rodent Genus *Isothrix* (Hystricognathi, Echimyidae) and its Diversification in Amazonia and the Eastern Andes." *Journal of Mammalian Evolution* 15:181–201.

Sclater, P. L. 1858. "On the General Geographic Distribution of the Members of the Class Aves." *Journal of the Linnean Society: Zoology* 2:130–45.

Simpson, G. G. 1956. "Zoogeography of West Indian Land Mammals." *American Museum Novitates* 1759:1–28.

———. 1980. *Splendid Isolation: The Curious History of South American Mammals*. New Haven: Yale University Press.

Solari, S., V. Pacheco, L. Luna, P. M. Velazco, and B. D. Patterson. 2006. "Mammals of the Manu Biosphere Reserve." In *Mammals and Birds of the Manu Biosphere Reserve, Peru*, edited by B. D. Patterson, D. F. Stotz, and S. Solari, 13–22. Chicago: Field Museum of Natural History. *Fieldiana: Zoology*, n.s., 110.

Udvardy, M. D. F. 1975. "A Classification of the Biogeographical Provinces of the World." *IUCN Occasional Papers* 18:1–49.

Voss, R. S., and L. H. Emmons. 1996. "Mammalian Diversity in Neotropical Lowland Rainforests: A Preliminary Assessment." *Bulletin of the American Museum of Natural History* 230:1–115.

Wallace, A. R. 1876. *The Geographical Distribution of Animals*, 2 vols. London: McMillan & Co.

Wilson, D. E., and D. M. Reeder, eds. 2005. *Mammal Species of the World: A Taxonomic and Geographic Reference*, 3rd ed. Baltimore: Johns Hopkins University Press.

West Indian Mammals
The Old, the New, and the Recently Extinct

Liliana M. Dávalos and Samuel T. Turvey

Abstract

The West Indian mammal fauna has played a key role in the development of biogeographic ideas for over a century, but a synthesis explaining regional patterns of mammal diversity and distribution in a historical framework has not emerged. We review recent phylogenetic, population genetic, and radiocarbon dating studies of West Indian mammals and explore the biological and historical drivers of colonization, speciation, and extinction in this region of endemism. We also present the first complete list of all its extant and extinct mammals. The mammalian biota is older than was earlier presumed, with many ancient endemic lineages, even among highly vagile organisms such as bats. Land bridges, Cenozoic eustatic sea-level changes, and Pleistocene glacial cycles have been proposed to explain the colonization of the islands, but phylogenetic divergence analyses often conflict with the timing of these events and favor alternative biogeographic histories. The loss of West Indian biodiversity is incompletely understood, but new radiometric chronologies indicate that anthropogenic impacts rather than glacial-interglacial environmental changes are responsible for most Quaternary extinction and extirpation events involving land mammals. However, many outstanding questions of historical biogeography remain unresolved, including appropriate methods for interpreting phylogenies and divergence estimates in a biogeographic context, and whether to use vicariance or dispersal as the null hypothesis when investigating regional patterns of colonization, speciation, and extinction in comparative analyses. We propose synthetic approaches drawing from phylogenetics, population genetics, paleogeography, paleontology, and even archaeology to resolve persisting questions in Caribbean biogeography.

9.1 Introduction

The mammals of the West Indies have been crucial to the development of biogeography from its very inception (Wallace 1876). The endemicity of the Antillean biota, for example, led Wallace to propose land interconnections and subsequent subsidence that isolated the islands first from South America and later from Central America (Wallace 1876). This was the first cogent—if incorrect—biogeographic hypothesis for the region. Even this early biogeographic work highlighted two critical aspects of the Caribbean mammalian fauna: its poverty in comparison to continental areas of equal size, and its sharp divergence from the nearby continental fauna. By proposing an ancient interconnection, severed in the Miocene, his geological hypothesis accounted for the endemicity of the mammals, while the subsidence of a large proportion of the land mass was proposed to explain the small number of surviving lineages (Wallace 1876).

Although Caribbean mammals, particularly bats, were key to developing the equilibrium theory of island biogeography (e.g., Koopman 1958; MacArthur and Wilson 1967), the importance of endemicity and geological changes in regional biogeographic studies declined with growing interest in mechanistic explanations of island diversity (MacArthur and Wilson 1963). The goal of explaining the origin and diversity of West Indian mammals in a historical framework, however, was not completely forgotten, especially among systematists (e.g., Williams 1952; Williams and Koopman 1951). Initial systematic zoogeography reviews (e.g., Baker and Genoways 1978; Koopman 1989) gave way to increasingly formal biogeographic analyses (e.g., Griffiths and Klingener 1988; Woods 1989; Woods, Ottenwalder, and Oliver 2001), culminating in the explicit use of phylogenies to infer biogeographic history (e.g., Dávalos 2004b, 2005, 2006, 2007; Roca et al. 2004). Whether informed by phylogeny or not, these studies have shared a historical perspective and focus on the origin and diversification of multiple mammalian lineages.

Aside from questions on the origins and colonization routes of Caribbean mammals, another main focus of biogeographic research has been quantifying and explaining extinction. As Quaternary fossil findings accumulated (e.g., Anthony 1918; Koopman and Williams 1951; MacPhee and Iturralde-Vinent 1995a, 1995b; MacPhee, White, and Woods 2000; MacPhee, Iturralde-Vinent, and Gaffney 2003; Miller 1918, 1922, 1929a, 1929b; Williams and Koopman 1951), new competing hypotheses on the drivers of regional Pleistocene and Holocene extinctions were proposed (MacPhee and Marx 1997; Pregill and Olson 1981; Steadman et al. 2005). An extensive literature has sought to explain the extinction and extirpation of numerous terrestrial mammal species (MacPhee 2008; MacPhee, Ford, and McFarlane 1989; Morgan 2001; Morgan and Woods 1986; Turvey, Grady, and Rye 2006, Turvey et al. 2007). At present, three main questions remain on the region's historical biogeography: (1) What is the geographic origin of the endemic mammal fauna?; (2) How did mammals, especially nonvolant ones, reach the Antilles?; and (3) What drove most of the nonvolant mammal fauna to extinction? In this chapter, we synthesize recent evidence from molecular phylogenetics, population genetics, paleontology, zoology, and archaeology to address these questions. Our goals are to present the status of historical biogeography of Caribbean mammals and to point to new methodological and analytical approaches that will resolve persistent gaps in understanding Antillean historical biogeography.

9.1.1 GEOGRAPHIC AND TAXONOMIC SCOPE

In this chapter, "West Indies," "Antilles," and "Caribbean" refer to the islands of the Caribbean Sea that have an insular biota (Koopman 1989; Morgan 2001; Morgan and Woods 1986), including San Andrés, (Old) Providence, and Swan Island (see fig. 9.1). Phylogenetic studies and analyses of fossil remains have overlapped most frequently in the Greater Antilles—Cuba, Jamaica, Hispaniola, and Puerto Rico—so we devote particular attention to these islands. The mammal faunas of Trinidad, Tobago, Margarita, Aruba, Bonaire, and Curaçao are not discussed here because these islands are characterized by a South American biota (e.g., Hooijer 1959, 1966, 1967; Trejo-Torres and Ackerman 2001; Vázquez-Miranda, Navarro-Sigüenza, and Morrone 2007; Voss and Weksler 2009). We have included data on extinct rice rat species from Grenada and the Grenadines because the terrestrial mammal fauna of these islands, though poorly known, is apparently endemic (Turvey et al. 2010). The bats of those islands, however, are not discussed, as most of these insular populations maintain gene flow with South American populations and are better thought of as being at the northern margin of their distributions (Genoways, Phillips, and Baker 1998; Koopman 1989; Presley and Willig 2008). A total of 55 extant and 12 regionally or globally extinct bats have been recorded in the West Indian Holocene, representing about 45 independent lineages. Only 16% (16 species) of an estimated 99 Quaternary nonvolant mammals survive to this day. Fossil and subfossil remains have been described for many lineages, both volant and not, but the bat lineages are better covered than other mammals in molecular phylogenetic and population genetic analyses, largely because of ongoing problems with extracting sequence data from degraded Caribbean Quaternary subfossil and zooarchaeological mammal material. We review both endemic and widespread species, briefly summarize the diversity of all native West Indian mammals (table 9.1), and provide a complete mammal species list in appendix 9.1.

Table 9.1 Orders and families of native Holocene (or putatively Holocene) West Indian mammals. Taxonomy follows contributors to Wilson and Reeder (2005) and Turvey (2009), with additional data from White and MacPhee (2001) and Rega et al. (2002) for sloths, MacPhee and Flemming (2003), Borroto-Páez et al. (2005), and Turvey et al. (2006) for rodents, and Dávalos (2006), Larsen et al. (2007), Morgan (2001), and Tejedor (2006) for bats.

Order	Family	Common name	Genera		Species		
			Endemic	Total	Endemic	Extinct	Total
Pilosa	Megalonychidae	Sloths	6	6	15	15	15
Soricomorpha	Solenodontidae	Solenodons	1	1	4	2	4
	Nesophontidae	Island shrews	1	1	8	8	8
Rodentia	Echimyidae	Spiny rats	4	4	6	6	6
	Capromyidae	Hutias	9	9	39	25	39
	Heptaxodontidae	Giant hutias	4	4	4	4	4
	Cricetidae	Rice rats	2	≥4	≤18	≤18	≤18
	incertae sedis	"Giant hutias"	2	2	2	2	2
Primates	Pitheciidae	Antillean monkeys	3	3	3	3	3
Chiroptera	Natalidae	Funnel-eared bats	2	3	8	0	8
	Noctilionidae	Fishing bat	0	1	0	0	1
	Mormoopidae	Ghost-faced and mustached bats	0	2	9	4	12
	Phyllostomidae	New World leaf-nosed bats	9	16	23	7	30
	Vespertilionidae	Various	0	5	8	1	12
	Molossidae	Free-tailed bats	0	5	1	0	8
Total			43	66	148	95	170

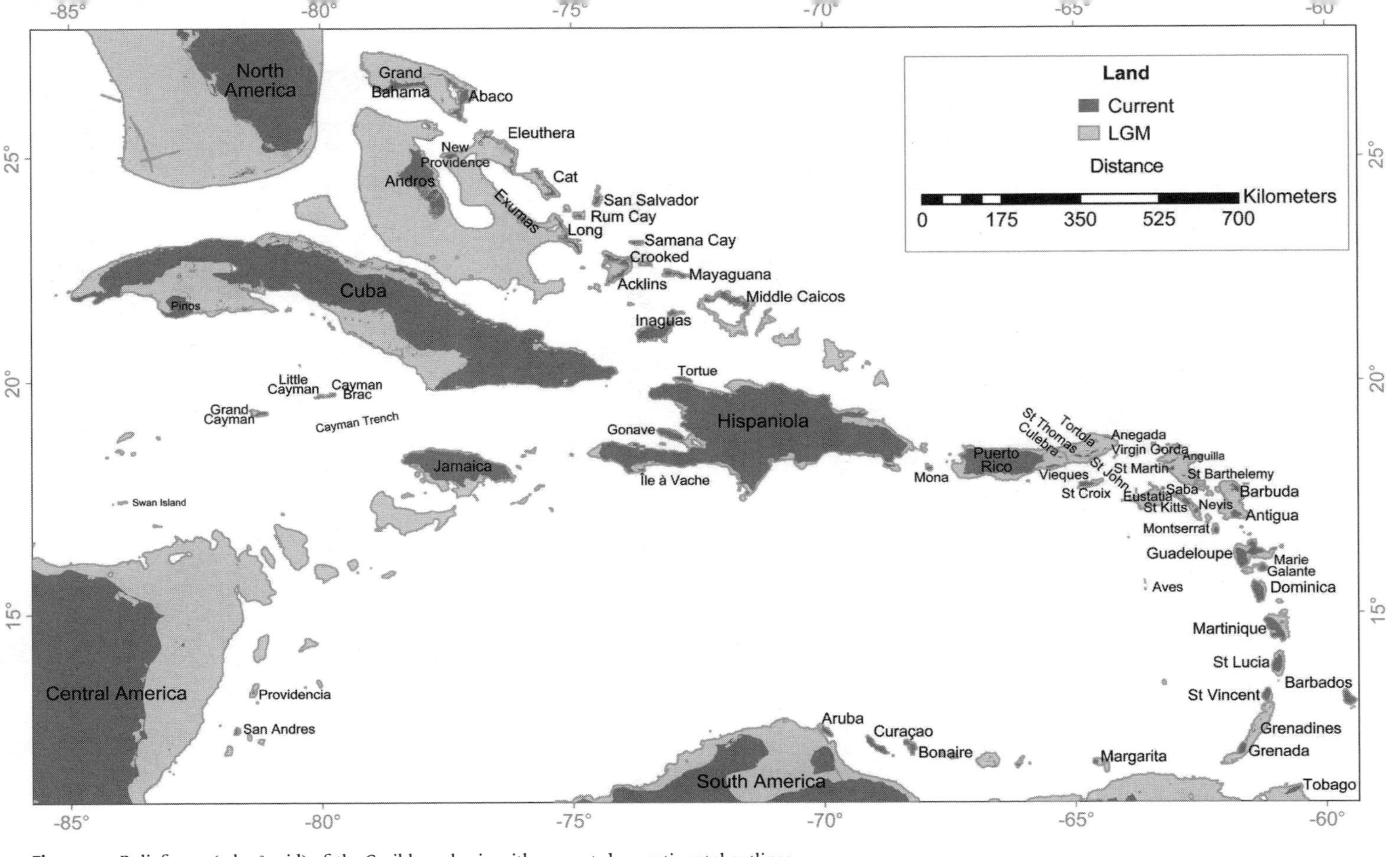

Figure 9.1 Relief map (1 km² grid) of the Caribbean basin with present-day continental outlines.

9.2 The Old: Origin and Phylogeny of West Indian Mammals

Early proponents of West Indian vicariance cited the vulnerability of mammals to refute "flotsam or jetsam dispersal" and argue instead for land interconnections, both between islands and between islands and continents (Allen 1911; Barbour 1916). In contrast, proponents of dispersal as the main mechanism responsible for biotic assembly pointed to the low diversity and peculiar composition of the Antillean biota compared to continental islands, such as Trinidad, or island-sized continental regions (Darlington 1938; Matthew 1918). By the time of Simpson's (1956) review of the West Indian mammal fauna, dispersal explanations held sway, but vicariant mechanisms were current enough to merit a thorough rebuttal. Despite advances in Antillean geology, paleontology, and mammalogy, the arguments for vicariance or dispersal relied on similar evidence, which remained virtually unchanged over the first half of the twentieth century: (1) classification or, at best, evolutionary systematics; (2) static continents (Simpson 1943); and (3) estimates of probability of dispersal across water gaps and, sometimes, their relationship to hurricanes, ocean currents, and drainage basins.

The development of phylogenetics (Edwards and Cavalli-Sforza 1964) and the establishment of plate tectonics as the mechanism underlying continental drift (Hess 1962) helped revive vicariance in the Antilles but did not close the debate on the prevalent mechanism of biotic assembly. Formal studies of Caribbean biogeography started with the first biogeographic methods using a form of phylogenetic information, such as generalized tracks (Rosen 1975). That analysis compiled the distribution of dozens of monophyletic or presumed monophyletic groups (mainly vertebrates) to identify patterns of overlap across independent groups of close relatives. The patterns were then interpreted to support a multistep vicariant explanation for the origin of the insular biota (Rosen 1975, 1985). In particular, a proto-Antillean archipelago bridging North and South America was postulated in the Cretaceous, followed by separation of the three landmasses by the Oligocene, concluding with the consolidation of Central America and closure of the Isthmus of Panama in the late Cenozoic. Almost immediately, Rosen's vicariant model was criticized for its outdated geological framework, very ancient dates for the majority of lineages, and inability to explain the absence of major continental groups on the islands (Pregill 1981). Despite its gaps, the proto-Antillean archipelago hypothesis was the basis for the first phylogeny-based biogeography of Antillean insectivores (MacFadden 1980). In fact, because of the complexity of Caribbean plate tectonics, Rosen's geological framework was, at the time, considered plausible and in the mainstream of biogeographic explanations (Hedges 1982; MacFadden 1981).

The more recent West Indian dispersal versus vicariance debate traces back to the early 1990s, when molecular clocks were first applied to date the colonization of multiple amphibian and reptile lineages (Hedges et al. 1992). The absence of phylogenies in that initial salvo was quickly identified as a key methodological problem, requiring reanalysis and reinterpretation using cladistic biogeography methods (Page and Lydeard 1994). The land bridge hypothesis of MacPhee and Iturralde-Vinent (1995b) emerged as the vicariant alternative to the molecular clock-based dispersal model. Articulated more fully elsewhere (Iturralde-Vinent and MacPhee 1999), the Greater Antillean and Aves Ridge hypothesis—or GAARlandia—drew on both geological and biological lines of evidence to postulate a temporary land bridge connecting the Greater Antillean Ridge and northwestern South America through the Aves Ridge. The GAARlandia hypothesis proposed a two-step mechanism to explain the patterns of diversity and distribution of land mammals in the West Indies. Initially, the land bridge enabled dispersal from the mainland without crossing ocean barriers, and the eventual disappearance of the bridge then led to vicariant speciation and subsequent independent evolution of the Antillean lineages.

As at the beginning of the twentieth century, the dispersal counterhypothesis invoked prevailing ocean currents and river drainages to explain repeated dispersal across ocean barriers, again criticizing the new land bridge hypothesis for its inability to explain the diversity and distribution of the Caribbean fauna (Hedges 1996). If a land bridge existed, why wasn't the fauna a random subsample of the continental fauna? This question overlooks the possibility of ecological filtering leading to the dispersal and establishment of some lineages but not others; for example, primates but not marsupials. Another tenet of the contemporary dispersal hypothesis is that the entire pre-Tertiary Caribbean biota went extinct because of the dust clouds, tsunamis, and earthquakes that would have followed the asteroid impact at nearby Chicxulub (Yucatán) 65 Ma (Alvarez et al. 1980; Grajales Nishimura et al. 2000) and the subsidence of the West Indies in the Eocene (Graham 2003).

Although the resurrected dispersal and vicariance hypotheses initially lacked phylogenies, multiple morphology-based phylogenies for Antillean mammals soon became available (Horovitz and MacPhee 1999; MacPhee, Iturralde-Vinent, and Gaffney 2003; White and MacPhee 2001; Woods, Borroto, and Kilpatrick 2001), and the first targeted molecular phylogenies soon followed (Dávalos

2005, 2006, 2007; Roca et al. 2004). A reconciled-tree approach applied to Caribbean mammal phylogenies identified a few instances of congruence with the GAARlandia hypothesis, but also pointed to contradictory nodes across several trees, and difficulties reconciling dated molecular phylogenies with the timing of the land bridge (Dávalos 2004b). In particular, the colonization of most nonvolant West Indian lineages was dated to the middle Miocene, but not to the late Eocene/early Oligocene boundary as required by GAARlandia (Dávalos 2004b; Iturralde-Vinent 2006). At the same time, a third alternative, the interconnection of North America and South America through the proto-Antilles in the Cretaceous, was revived by Mesozoic-age molecular divergence estimates for the soricomorph *Solenodon* and the xantusiid lizard *Cricosaura* (Roca et al. 2004). Congruence with GAARlandia has further eroded, as the timing of mammalian colonization for most remaining lineages has been dated to either before or after the proposed land bridge (Dávalos 2010; table 9.2). Finally, the timing of divergence between insular and continental bat lineages has been traced to periods of low sea level, contributing to a fourth biogeographic model of facilitated dispersal (Dávalos 2010).

Rather than revisit the implications of the hypothetical land bridge (Dávalos 2004b), eustatic sea level changes (Dávalos 2010), or the Cenozoic fossil record in the Caribbean (MacPhee 2005; Turvey 2009), we focus instead on the avenues to resolve outstanding biogeographic questions. Current biogeography studies face two practical challenges: how best to use phylogenies to inform biogeography, and how to incorporate fossil calibrations in phylogeny and then interpret the results of divergence analyses. Beyond these methodological concerns, however, lies the question of which conceptual framework is appropriate in historical biogeography.

First, we concur with earlier analyses that reaffirm the central role of phylogeny in biogeographic research (Page and Lydeard 1994). There seems to be little debate on this point, as even the strongest recent proponents of dispersal currently rely on phylogenies to show that divergences between West Indian endemics and their closest extant mainland relatives are not clustered, but rather interspersed through time (Heinicke, Duellman, and Hedges 2007). Although phylogenies have become available in recent years for bats (Dávalos 2005, 2006, 2007) and *Solenodon* (Roca et al. 2004), phylogenies for rodents and vespertilionid and molossid bats are still lacking. How to use phylogenies to inform biogeography remains an open question. Mapping areas as characters (Dávalos 2007, 2010; Roca et al. 2004), dispersal-vicariance analyses (Dávalos 2005, 2006), and phylogeny-reconciliation approaches (Dávalos 2004b) have all

Table 9.2 Geographic origin of Antillean mammals (endemic species or higher-level taxa), and estimated age of divergence from mainland taxa.

Antillean lineage	Closest mainland relatives	Inferred geographic origin	Molecular divergence (Ma)	Fossil divergence (Ma)	Sources
Pilosa					
Choloepodinae (*Acratocnus* and *Neocnus*)	*Choloepus*	South America		$\geq$33–34?	MacPhee and Iturralde-Vinent 1995b; White and MacPhee 2001
Megalocninae (*Megalocnus* and *Parocnus*)	*Bradypus*	South America		$\geq$33–34?	MacPhee and Iturralde-Vinent 1995b; White and MacPhee 2001
Soricomorpha					
Solenodon	Eulipotyphlan insectivores (Talpidae + (Erinaceidae + Soricidae)	Proto-Antilles plus North America	76 (72–81)		Roca et al. 2004
Nesophontes	Soricidae?	Unknown			Asher 1999, 2005; Asher, Emry, and McKenna 2005
Rodentia					
Capromyidae (+ Heptaxodontidae?)	*Clyomys* + *Euryzygomatomys* (*Myocastor?*)	South America	18 (11–27)	$\geq$17.5–18.5	Galewski et al. 2005; MacPhee, Iturralde-Vinent, and Gaffney 2003; Woods. Borroto, and Kilpatrick 2001
Heteropsomyinae (*Boromys*, *Brotomys* and *Heteropsomys*)	Mainland echimyid rodents/ capromyids	South America/ Greater Antilles			Wilson and Reeder 2005; Woods, Borroto, and Kilpatrick 2001
"*Megalomys*" *audreyae*	Unknown	South America?			Turvey et al. 2010

(continued)

Table 9.2 *(continued)*

Antillean lineage	Closest mainland relatives	Inferred geographic origin	Molecular divergence (Ma)	Fossil divergence (Ma)	Sources
Megalomys desmarestii and *luciae*	*Sigmodontomys aphrastus*	South America			Turvey et al. 2010
Oligoryzomys victus	Other *Oligoryzomys* spp.	South America			Turvey et al. 2010
Oryzomys antillarum	*Oryzomys couesi*	Mesoamerica			Morgan 1993
Pennatomys nivalis	Oryzomyini "Clade D" (*Aegialomys*, *Amphinectomys*, *Melanomys*, *Nectomys* + *Sigmodontomys*)	South America			Turvey et al. 2010
Primates					
Xenotrichini	*Callicebus*	South America		≥17.5–18.5	MacPhee and Horovitz 2004
Chiroptera					
Nyctiellus, *Chilonatalus*, *Natalus*	Mainland Natalidae	Equivocal: Eurasia/North America	50 (45–56)		Teeling et al. 2005
Mormoops blainvillei	*Mormoops megalophylla*	Equivocal: Mesoamerica/northern South America/West Indies	15 (11–24)		Dávalos 2006, 2010
Mormoops magna	*Mormoops megalophylla*	Unknown			Silva-Taboada 1974
Pteronotus parnellii sensu lato (Antillean spp.)	*Pteronotus parnellii sensu lato* (mainland spp.)	Equivocal: Mesoamerica/northern South America	5 (3–8)		Dávalos 2006, 2010

Table 9.2 (*continued*)

Antillean lineage	Closest mainland relatives	Inferred geographic origin	Molecular divergence (Ma)	Fossil divergence (Ma)	Sources
Pteronotus pristinus	*Pteronotus parnellii*	Unknown			Simmons and Conway 2001
Pteronotus quadridens and *macleayii*	*Pteronotus davyi* and *P. gymnonotus*	Equivocal: Mesoamerica/northern South America/Cuba/Jamaica	14 (9–21)		Dávalos 2006, 2010
Desmodus puntajudensis	*Desmodus stocki* and *D. archaeodaptes*	North America			Suárez 2005
Macrotus waterhousii sensu lato (Antillean spp.)	*Macrotus waterhousii sensu lato* (mainland sp.)	Mexico			Fleming, Murray, and Carstens 2010
Palynophil (*Erophylla*, *Phyllonycteris*, *Brachyphylla*)	*Glossophaga*	Equivocal: Mesoamerica/northern South America/West Indies	17 (12–26)		Dávalos 2010
Monophyllus	*Glossophaga*	Equivocal: Mesoamerica/northern South America/West Indies	14 (10–22)		Dávalos 2010
Sturnira thomasi	*Sturnira luisi* sp. complex	Northern South America			Villalobos and Valerio 2002; C. Iudica, pers. comm.
Chiroderma improvisum	*C. villosum*	Northern South America			Baker et al. 1994
Artibeus anthonyi	Other *Artibeus*	Unknown			based on systematics—see Simmons 2005

(*continued*)

Table 9.2 *(continued)*

Antillean lineage	Closest mainland relatives	Inferred geographic origin	Molecular divergence (Ma)	Fossil divergence (Ma)	Sources
Stenodermatina: *Ardops, Ariteus, Cubanycteris, Phyllops, Stenoderma*	*Artibeus*	Equivocal: Mesoamerica/northern South America/West Indies	10 (7–16)		Dávalos 2007, 2010
Myotis dominicensis, and *martiniquensis*	*M. atacamensis, M. yumanensis* and *M. velifer*	Neotropical	4 (3–5)		Stadelmann et al. 2007
Eptesicus guadeloupensis	*E. fuscus* sp. complex	Unknown			Jones et al. 2002
Lasiurus degelidus	*L. seminolus*	North America			Baker et al. 1988
Lasiurus insularis	*L. ega* and *L. intermedius*	North America (including Mexico)			Morales and Bickham 1995
Lasiurus minor	*L. borealis, L. blossevillii*, and *L. seminolus*	North America			based on systematics—see Simmons 2005
Lasiurus pfeifferi	*L. seminolus*	North America			Morales and Bickham 1995
Nycticeius cubanus	*Nycticeius humeralis*	North America			inferred from systematics—see Simmons 2005
Mormopterus minutus	*M. phrudus* and *M. kalinowskii*	South America			Jones et al. 2002

been used recently, but these methods used neither branch lengths (and hence dates—Ree and Smith 2008) nor accounted for environmental change in areas over time (Yesson and Culham 2006). The use of character-mapping methods that do account for branch lengths seems a logical next step in phylogeny-based biogeography, because there is greater probability of change over long branches than over short branches (e.g., Brumfield and Edwards 2007; McGuire et al. 2007). Molecular dating estimates place the earliest divergence of Caribbean clades as far back as the Mesozoic (Roca et al. 2004), so the net effect of applying these methods might be to increase uncertainty on the geographic origin, mechanism of range expansion, and subsequent diversification of mammals. Greater uncertainty might be more consistent with the dynamic geological history and complex biotic history of the region than a single all-encompassing dispersal or vicariance model.

Second, a similar reassessment of molecular divergence dates is in order, as their uncertainty is often large. Rather than confirm a particular date, molecular divergence dates can only exclude periods of time, such as when GAARlandia would have existed. There are many dating methods, and most rely on hard boundaries set by the date of last occurrence as inferred from the fossil record (Rutschmann 2006). There is also uncertainty associated with the stratigraphy of the fossils used in these calibrations, a fact not always considered in molecular analyses. A minimal requirement for reporting molecular divergence dates should be a sensitivity analysis of the dates to the fossils available (Dávalos 2010), the use of soft boundaries to account for stratigraphic uncertainty (Yang and Rannala 2006), and the use of a frequency distribution rather than a point estimate of the age of each fossil calibration (Drummond et al. 2006). Divergence dates varied up to 10% depending on parameters modeling the age of the root of a clade in a sensitivity analysis of molecular dates for Caribbean mammals (Dávalos 2010). If the biogeographic hypothesis being tested requires precise divergence times, sensitivity analyses could reveal greater variation, thereby increasing uncertainty in biogeographic inference.

Third, what is the appropriate null for testing historical biogeography? At the core of the vicariance versus dispersal debate is the struggle to define the null model of biogeography. This choice is crucial. Individual phylogenies can simultaneously be congruent with vicariance and dispersal hypotheses because both processes can give rise to indistinguishable patterns. Congruence in biogeographic patterns between phylogenies is taken to indicate vicariance, another way of saying that vicariance should be the null hypothesis. This makes intuitive sense, but it ignores the possibility of congruent dispersal. Conversely, dispersal as null hypothesis is difficult to test because it requires somehow testing the predictions of isolated events across different lineages.

One quantitative approach to dispersal, the equilibrium theory of island biogeography (MacArthur and Wilson 1963), has been largely overlooked in discussions of West Indian mammal biogeography despite the early role of Caribbean continental-shelf islands in developing the theory (Koopman 1958). One early island biogeography analysis calibrated species-area curves with extant Caribbean bats (Griffiths and Klingener 1988), without estimating colonization or extinction rates from the data. Another application included both extinct and extant mammals to calibrate species-area curves, and calculated extinction and immigration rates based on then-available phylogenies and rough estimates of colonization times (Morgan and Woods 1986).

Island biogeographic analyses of Caribbean mammals have been limited, perhaps because paleontologists, phylogenetic systematists, and ecologists have questioned the central tenets of island biogeography, particularly as it applies over evolutionary history (Brown and Lomolino 2000; Heaney 2000, 2007; Olson and James 1982; Steadman 1995). The equilibrium theory of island biogeography has been criticized because: (1) both extant and extinct island faunas reveal many instances of nonequilibrium; (2) the theory reduces individual islands to their isolation and area; and (3) by reducing individual species into interchangeable units, the theory ignores the differences in speciation, extinction, and colonization rates arising from a species' characteristics. But it is precisely the expectation of equilibrium and factoring out of phylogeny that makes island biogeography appropriate as a null model for quantitative tests of dispersal. If periods of high dispersal are driven by lowered sea levels, then immigration rates should not be uniform but instead show nonequilibrium dynamics with peaks around glacial periods. Changes in sea level should also dictate the size of islands, producing higher rates of extinction during periods of high sea levels. This approach has been used to analyze community assembly in Lesser Antillean birds (Ricklefs and Bermingham 2001), identifying a rise in colonization rate or a mass extinction event before the last glaciation.

Analyzing dispersal in the framework of island biogeography would complement, not replace, vicariance analyses. Vicariance models are appropriate null hypotheses when paleoclimate or paleogeographic reconstructions indicate continuous habitats at certain periods (e.g., between Grand Terre and Basse Terre in Guadeloupe, or between the Exumas in the Bahamas). Choosing a dispersal or vicariance hypothesis as a null model will depend on geological data. When geological data indicate a plausible mechanism of vicariance, this can be the null model. In the absence of such data, and based on island biogeographic theory, dispersal becomes the null model, and quantitative tests based on dated phylogenies become possible.

9.3 The New: Population Genetics of West Indian Mammals

Historical biogeography has traditionally focused on the study of endemic taxa and their origins (Dávalos 2004b; Kluge 1989; Page and Lydeard 1994); the overlapping ranges of endemics help to outline areas of endemism (Platnick 1991). With the increasing use of molecular markers in biogeography, it has become possible to analyze patterns of population expansion, stasis, or contraction, as well as origins of island populations (Lessa, Cook, and Patton 2003; Russell et al. 2007; Russell, Goodman, and Cox 2008; Russell et al. 2008b). In the Antilles, only two population-genetic studies, for *Artibeus jamaicensis* and *Macrotus waterhousii*, have encompassed both continental and island populations, with different results in each case (Fleming, Murray, and Carstens 2010; Larsen et al. 2007). The population structure of the widespread frugivorous phyllostomid *Artibeus jamaicensis* has been studied using mitochondrial restriction sites and RFLP (restriction fragment length polymorphisms) mapping (Phillips et al. 1989; Phillips et al. 1991; Pumo et al. 1988), mitochondrial sequences (Carstens et al. 2004; Fleming, Murray, and Carstens 2010; Pumo et al. 1988; Pumo et al. 1996), and, more recently, amplified fragment-length polymorphisms (AFLPs; Larsen, Marchán-Rivadeneira, and Baker 2010). Traditionally, thirteen subspecies of *Artibeus jamaicensis* have been recognized, three—*parvipes*, *jamaicensis*, and *schwartzi*—confined to the West Indies (Koopman 1994; Simmons 2005). Large amounts of genetic variation were detected in the earliest genetic analyses of *Artibeus jamaicensis* (e.g., Phillips et al. 1989; Pumo et al. 1988), with highly divergent haplotypes found coexisting in the Lesser Antilles (Carstens et al. 2004; Pumo et al. 1996).

Multiple hypotheses have been proposed to account for these highly divergent haplotypes: (1) hybridization between differentiated subspecies of a single species (Jones 1989); (2) relictual diversity from an ancient invasion that was subsequently swamped by new arrivals (Phillips et al. 1989; Pumo et al. 1996); and (3) a ring species arriving from the confluence in the Lesser Antilles of an eastward Mesoamerican invasion and a northward South American invasion (Carstens et al. 2004). An analysis of mitochondrial sequences offered an alternative interpretation: species-level recognition for three reciprocally-monophyletic clades, two of them sympatric in St. Kitts, Nevis, Montserrat, St. Lucia, and Barbados, and all three present in St. Vincent (Larsen et al. 2007). The proposed species limits did not follow the traditionally recognized subspecies, but they accommodate the clades obtained in analyses of the mitochondrial cytochrome-*b* gene. This last interpretation implied that populations of *Artibeus jamaicensis*, *A. planirostris*, and *A. schwartzi* were cryptic in the Lesser Antilles and have converged on a similar phenotype, despite divergent phylogenetic and geographic origins (Larsen et al. 2007). A subsequent study of nuclear markers revealed that patterns of genomic variation in *A. schwartzi* are consistent with this population originating through hybridization between *A. jamaicensis* and *A. planirostris* (Larsen, Marchán-Rivadeneira, and Baker 2010). Hybridization would have been possible by imperfect reproductive isolation between the two parent species. The biogeographic origin of the parent species would indicate an eastward invasion of *A. jamaicensis* from Mesoamerica, with *planirostris* originating in South America. Targeted studies are needed to determine if these populations have recently grown, as expected from recent colonization leading to hybridization.

While *Artibeus jamaicensis* appears to maintain gene flow with mainland populations on Mesoamerica, Antillean populations of the insectivorous phyllostomid *Macrotus waterhousii* seem completely isolated from the mainland counterparts (Fleming, Murray, and Carstens 2010). Although *Macrotus waterhousii* has been thought to comprise one widespread population from Mexico through the Bahamas and Greater Antilles, rapidly evolving mitochondrial sequences represent at least four reciprocally monophyletic groups, each corresponding to islands or island banks. These populations are effectively isolated, without shared haplotypes. Morphology-based systematics would suggest a very recent colonization from Mexico (Griffiths and Klingener 1988; Koopman 1989), but the molecular data showed no evidence of recent population expansion and dozens of fixed differences with respect to the mainland population, indicating a more ancient colonization date than previously thought (Fleming, Murray, and Carstens 2010). As with *Artibeus*, additional markers and more research on the ecology and morphology of *Macrotus* are needed to understand how these highly divergent allopatric populations maintain their nearly identical morphology.

Population genetic analyses are available for only three other phyllostomid endemics: flower-visiting *Brachyphylla* (Carstens et al. 2004; Dávalos 2004a) and *Erophylla* (Fleming and Murray 2009; Fleming, Murray, and Carstens 2010) and the frugivorous *Ardops* (Carstens et al. 2004). Mitochondrial sequences of *Brachyphylla* revealed the reciprocal monophyly of *Brachyphylla nana* populations on Cuba and Grand Cayman versus Hispaniola and Middle Caicos (Dávalos 2004a). Conversely, detailed analyses of the population genetic structure of *Brachyphylla cavernarum* showed no evidence of interisland monophyly and instead were consistent with incomplete lineage sorting following recent expansion into the Lesser Antilles from Puerto Rico (Carstens et al. 2004). A mirror image of this pattern of stasis in one part of the range and expansion in another is shown by *Erophylla*, whose western populations (Bahamas, Cuba, Caymans, and Jamaica) have expanded recently, in contrast with the stable populations of Hispaniola and Puerto Rico (Fleming, Murray, and Carstens 2010). Unlike *Brachyphylla*, *Erophylla* populations have not attained reciprocal monophyly, indicating much more recent isolation and only incipient speciation. In contrast with these two species, which maintain gene flow across most shallow water barriers, the Lesser Antillean *Ardops* is relatively well-differentiated on individual islands. Coalescent analyses could not reject island monophyly across the northern Antilles (Carstens et al. 2004).

None of these endemic genera—*Brachyphylla*, *Erophylla*, and *Ardops*—have a sister genus on the mainland, and *Brachphylla cavernarum* and *Ardops* originated west of their current range. This can be inferred for *Brachyphylla* from its basal relatives in Hispaniola and Cuba (Dávalos 2004a), and for *Ardops* from its common ancestry with the Jamaican *Ariteus* (Dávalos 2007). In contrast, in *Erophylla*, it is the western populations that are recent, likely as a result of sea-level changes that made the banks of the Bahamas and Cuba much larger than at present (Fleming, Murray, and Carstens 2010). *Brachyphylla* and *Erophylla* share a common ancestor with the (mostly) Mesoamerican *Glossophaga* and *Leptonycteris*, indicating an origin in that Neotropical subregion. *Ardops* could either be part of an Antillean endemic radiation or the descendent of an ancient mainland colonizer (table 9.1; Dávalos 2007, 2010).

Population genetic approaches are needed in Caribbean mammal biogeography to close two gaps in higher-level analyses: to delimit species and identify species complexes and to expand biogeographic understanding of widespread species. One example of the first gap is the recent revision of *Artibeus jamaicensis*. Although superficially similar, the island populations hitherto called *Artibeus jamaicensis* have complex evolutionary histories and rightfully should be called a species complex. In-depth examination of Antillean *Natalus* has revealed isolated populations on Cuba, Jamaica, and Hispaniola (Dávalos 2005; Tejedor, Tavares, and Silva-Taboada 2005). The last time these ancient lineages exchanged genes was 1.3 million years ago (Dávalos 2010). Other populations, including *Lasiurus* (Morales and Bickham 1995), *Eptesicus*, and *Tadarida*, remain to be studied and might also reveal much greater diversity than currently recognized. Although related to the gap in species delimitation, the biogeography of widespread species is indispensable to understand community assembly, current structure, and to prioritize areas for conservation (Gannon et al. 2005; Rodríguez Durán and Kunz 2001). By identifying routes of colonization and providing estimates of the age of colonization, population genetic studies can help resolve whether recent colonizers have outcompeted and replaced endemic lineages (e.g., *Artibeus* replacing *Brachyphylla* in Jamaica; Koopman and Williams 1951), if the absence of a species on an island indicates extinction or failure to colonize (e.g., *Natalus* in Puerto Rico; Tejedor 2006), or whether colonization patterns are similar among ecologically distinct genera (Fleming, Murray, and Carstens 2010).

9.4 The Recently Extinct: Caribbean Mammal Species Losses

In addition to its importance in developing key ideas in biogeography, the Antillean biota has also been used to identify fundamental ecological processes of faunal turnover and extinction (Ricklefs 1970; Ricklefs and Bermingham 2001; Ricklefs and Cox 1972), and the region's historical land mammal fauna has been the subject of considerable investigation into mammalian extinction dynamics. The Tertiary terrestrial fossil record of the insular Caribbean is still highly incomplete, posing a major obstacle to understanding ancient patterns of colonization and biogeography across the region (MacPhee, Iturralde-Vinent, and Gaffney 2003; Portell, Donovan, and Domning 2001). Conversely, from the mid-nineteenth century onwards (Castro 1864; Leidy 1868), investigation of Quaternary deposits on numerous Caribbean islands has revealed increasingly diverse assemblages of recently extinct mammal species, containing both megafaunal and pygmy arboreal sloths, an endemic Caribbean clade of primates, and extensive insular radiations of rodents and insectivores, as well as numerous bats (see appendix 9.1). Most Quaternary fossils from the Caribbean have been reported from cave deposits, but additional material is also known from asphalt seeps and sinkholes (Iturralde-Vinent et al. 2000; Steadman et al. 2007).

It is still difficult to generate an accurate estimate of the diversity of the prehuman Caribbean mammal fauna. Extinct Late Quaternary mammal species continue to be discovered from all of the major Caribbean islands (e.g., MacPhee and Flemming 2003; Mancina and Garcia-Rivera 2005; Rega et al. 2002; Suárez and Díaz-Franco 2003; Turvey, Grady, and Rye 2006), and large numbers of additional species, notably Lesser Antillean oryzomyine rice rats, remain undescribed (see appendix 9.1), often because they have been studied only by zooarchaeologists (Newsom and Wing 2004; Pregill, Steadman, and Watters 1994; Turvey 2009; Turvey et al. 2010). There are also major unresolved problems with available taxonomies for extinct Caribbean mammals, notably for Cuban and Hispaniolan capromyid rodents, and many supposed species are likely to represent synonyms (Díaz-Franco 2001; Rímoli 1976). However, despite these taxonomic uncertainties, over 100 species or island populations of volant and nonvolant land mammals can be interpreted as having become extinct during the Late Quaternary (Morgan 2001; Turvey 2009). Nevertheless, it is clear that the Caribbean mammal fauna has experienced the highest level of recent species loss of any mammal fauna in the world, both for the period following 1500 AD and across the entire Holocene (MacPhee 2008; MacPhee and Flemming 1999; Morgan 2001; Turvey 2009). For example, the Lesser Antillean Windward and Leeward Islands alone have lost approximately 20 island populations of oryzomyine rice rats, many of which were probably distinct species; these rice rat extinctions are equivalent in magnitude to the much better known historical-era loss of marsupials and rodents in Australia (Johnson 2006; MacPhee and Flemming 1999; Turvey 2009; Turvey et al. 2010), but they comprise only part of the much greater series of land mammal extinctions so far documented across Caribbean islands.

Investigation of the Late Quaternary Caribbean mammal mass extinction event may provide novel insights into the putatively human-caused, Late Pleistocene megafaunal extinctions in North America (see Barnosky et al. 2004; Martin 1984; Martin and Steadman 1999), and a wider base for developing appropriate conservation management plans for surviving Caribbean hutias and insectivores, almost all of which are threatened with extinction (International Union for Conservation of Nature [IUCN] 2008). As with megafaunal extinction on the mainland, two major competing hypotheses have been proposed to account for Quaternary Caribbean mammal extinctions. Pregill and Olson (1981) noted that many now-extinct terrestrial vertebrates (particularly reptiles and birds) present in Late Quaternary deposits in the West Indies were characteristic of xeric habitats (arid savanna, grassland, and scrub forest) and obligate xerophiles that are still extant had wider distributions in the Recent fossil record. Nonanthropogenic environmental change at the Pleistocene-Holocene boundary at the end of the last glaciation, notably a large-scale shift to more mesic forested habitats, may have been a major driver of faunal extinction in the region. This hypothesis was adopted to explain West Indian bat extinctions by Morgan (2001), who demonstrated that most regional bat population or species losses affected obligate or facultative cave-dwelling species—these extinctions may therefore have been driven by changes in cave microclimates or the inundation of large cave systems by rising sea levels or erosional collapse during the Pleistocene-Holocene climatic transition. In contrast, other authors have considered that most or all of the region's mammal extinctions occurred later in time, and were instead driven by mid-late Holocene anthropogenic actions such as overhunting, habitat destruction, and introduction of exotic predators, competitors, and diseases following the arrival of humans in the Caribbean around 6000 BP (Burney, Burney, and MacPhee 1994; MacPhee 2008; Morgan and Woods 1986; Wilson 2007).

The question of the timing and causation of Caribbean mammal extinctions is truly interdisciplinary, with potential contributions from paleontology, zoology, and archaeology. However, analyses based on approaches such as population genetics have not yet been able to provide useful insights into this question, in part because of the continuing difficulty of extracting DNA from Caribbean Quaternary specimens, as well as the challenge of obtaining sufficient genetic samples from extant but threatened and cryptic land mammal species. Distinguishing between the two extinction hypotheses requires establishing "last-occurrence" dates for extinct species based on historical, radiometric, or constrained stratigraphic data; meaningful extinction chronologies are lacking for most of the region's extinct mammal fauna (MacPhee 2008; MacPhee, Ford, and McFarlane 1989). Some last-occurrence dates are available for a handful of mammal species that persisted into the nineteenth or twentieth centuries (e.g., *Megalomys* rice rats and the hutia, *Geocapromys thoracatus*; Allen 1942; Clough 1976), but there are few records from earlier centuries and they seldom identify particular species with any accuracy (MacPhee and Fleagle 1991; Miller 1929a). Dubious twentieth-century reports of several now-extinct mammal species have generated additional confusion (MacPhee et al. 1999; Miller 1930; Raffaele 1979; Woods, Ottenwalder, and Oliver 1985). Collagen degradation under moist, humid subtropical conditions has hindered radiometric dating of even young subfossil material (Turvey et al. 2007), and direct radiometric last-occurrence dates are published for 11 extinct Caribbean insular mammal species. A wider series of terminus post quem dates have been generated with reasonable confidence from the apparent stratigraphic co-occurrence of extinct species with introduced mammals (particularly *Rattus rattus*) in superficial cave sediments, although this approach may also be problematic and open to alternative interpretations (MacPhee and Flemming 1999; Woods and Ottenwalder 1992). Information on key variables such as pre-Columbian human population densities and prehistoric levels of habitat conversion is also highly speculative (e.g., Watts 1987), and evidence on past human exploitation of most native Caribbean mammal species, especially large-bodied mammals, is typically lacking.

Despite these obstacles, stratigraphic studies and applied dating efforts since the 1980s (e.g., Steadman, Pregill, and Olson 1984) have provided direct or indirect evidence that most of the region's Late Quaternary mammal species persisted into the Holocene. These studies have therefore disproved the environmental change hypothesis of Pregill and Olson (1981) as a general explanation for Caribbean mammal extinctions, and have led to the development of a two-stage human-driven extinction model for nonvolant Caribbean land mammals. Although there is evidence for relatively intensive pre-Columbian Amerindian exploitation of some rodents, notably in the Lesser Antilles (Newsom and Wing 2004), many or most of the extinct small- and medium-sized rodent and insectivore species (nesophontid island-shrews, heteropsomyine echimyids, hutias, rice rats), and the Jamaican monkey *Xenothrix mcgregori* are now thought to have survived until around the time of European arrival. Few of these appear to have survived much beyond first European contact (Flemming and MacPhee 1999; MacPhee and Flemming 1999; MacPhee et al. 1999; McFarlane et al. 2000; Turvey et al. 2007; Turvey 2009).

It is probable that the extinction of most of the smaller nonvolant land mammal fauna was driven by interactions with *Rattus rattus*, which reached the Caribbean by the early 1500s, although the subsequent deliberate introduction of the mongoose *Herpestes javanicus* and massive forest clearing for sugarcane and other crops were also key drivers in extinctions of some native small mammals. Although further data are required to clarify the ecological mechanism(s) by which exotic *Rattus* species cause extinctions, rats have been implicated in the disappearance of small mammals and many other taxa on island systems across the world through competition, predation, disease transmission, and habitat modification (Drake and Hunt 2009; Harris 2009; Harris, Gregory, and Macdonald 2006; Harris and Macdonald 2007; Towns, Atkinson, and Daugherty 2006; Wyatt et al. 2008).

Recent direct radiometric studies have also demonstrated the protracted survival of Caribbean large-bodied mammals (MacPhee, Iturralde-Vinent, and Vazquez 2007; Steadman et al. 2005; Turvey et al. 2007), with at least some megalonychid sloths (*Megalocnus rodens*, *Neocnus comes*) and heptaxodontid rodents (*Elasmodontomys obliquus*) apparently persisting for millennia beyond first human arrival in the Greater Antilles. These taxa apparently became extinct through attrition, possibly driven by low-level exploitation before Columbus or by indirect factors such as progressive habitat modification—"sitzkrieg"-style events (sensu Diamond 1989)—rather than through a rapid "blitzkrieg"-style overkill following Amerindian colonization.

This two-stage pattern of extinction—protracted survival after human arrival, but eventually leading to extinction—may reflect either different levels of human exploitation of large-bodied and small-bodied mammal taxa or the intrinsically higher vulnerability of larger-bodied species to human impacts due to size-dependent scaling of ecological and life-history traits (Cardillo et al. 2005). The delayed extinction of even the large-bodied Caribbean land mammals contrasts markedly with rapid extinctions of other large-bodied insular taxa overexploited by early hunters (e.g., New Zealand moas; Holdaway and Jacomb 2000), and instead resembles the protracted late Holocene declines of large-bodied mammals on Madagascar (Burney et al. 2004). Several extinct Caribbean bats are known to have persisted into the Holocene (Jiménez Vázquez, Condis, and García 2005; Steadman, Pregill, and Olson 1984), and there is little or no evidence that humans ever consumed bats in the Caribbean (Mickleburgh, Waylen, and Racey 2009). It should be noted that many Caribbean bats (e.g., *Natalus*) are severely threatened by invasive mammals such as feral cats (Tejedor et al. 2005) and by loss of foraging habitat through deforestation (Gannon et al. 2005). These anthropogenic factors may also have contributed to past bat extirpations and extinctions in the region. Large congregations of cave-roosting bats may enhance their vulnerability to introduced predators such as rats, cats, and mongooses. This risk becomes particularly acute as natural or anthropogenic change confines populations to single caves (e.g., *Natalus* in Cuba or Jamaica; Tejedor, Silva-Taboada, and Rodriguez Hernandez 2004; Tejedor, Tavares, and Silva-Taboada 2005).

Although most Caribbean land mammal extinctions have been caused by prehistoric and historic-era human impacts during an interval of relatively modest environmental change, there is some evidence to suggest that other regional extinction events may have occurred as a result of environmental change before humans reached the islands. Uranium-series disequilibrium dates support a nonanthropogenic Late Pleistocene extinction for the giant hutia *Amblyrhiza inundata*, probably caused by inundation of the Anguilla Bank at the end of the last glaciation (McFarlane, MacPhee, and Ford 1998). Pre-Holocene extinctions have also been postulated for other species, including the Jamaican giant rodent *Clidomys osborni*, the Puerto Rican rodent *Puertori-*

comys corozalus, and the Cuban monkey *Paralouatta varonai*, on the basis of the heavy fossilization of all known specimens and their absence from well-studied Late Quaternary deposits (MacPhee and Meldrum 2006; MacPhee, Ford, and McFarlane 1989; Morgan and Wilkins 2003; Williams and Koopman 1951). However, the apparent absence of these species may also reflect incomplete palaeontological sampling rather than early extinction. Although none of the Greater Antillean islands experienced catastrophic late Quaternary fluctuations in exposed areas from eustatic changes in sea level, severe climatic events in the Antilles have been postulated for the last interglacial period (McFarlane and Lundberg 2004)—glaciers may even have formed at higher elevations on the largest islands (Schubert and Medina 1982)—with unexplored consequences for the regional fauna. Understanding of Caribbean mammal extinctions is thus still incomplete and in need of further investigation and testing; we expect that the temporal framework for extinctions we have sketched will have to be revised as new data emerge.

9.5 Prospects for the Historical Biogeography of West Indian Mammals

There are three main gaps in our understanding of Antillean historical biogeography: (1) resolving species limits and phylogenetic relationships for several endemic lineages (e.g., hutias, *Lasiurus*); (2) undertaking revisionary morphological and population genetics analyses of widespread lineages (e.g., *Mormoops megalophylla* and *Eptesicus fuscus*); and (3) combining phylogenetics, population genetics, and ecological biogeography with paleontology, zoology, and archaeology to detail the history of both colonization and extinction across Caribbean communities. Despite recent progress in resolving relationships among the majority of the endemic West Indian lineages (e.g., Dávalos 2007; Horovitz and MacPhee 1999; White and MacPhee 2001), the largest radiation of nonvolant mammals—the hutias—remains only partly resolved (Woods, Borroto, and Kilpatrick 2001). Revisionary work (e.g., Turvey, Grady, and Rye 2006) has demonstrated the need to examine species limits of both extant and fossil material. This also holds for lineages with fewer species, such as *Lasiurus*, that have never been included simultaneously in a phylogenetic analyses and whose phylogenetic relationships are still poorly understood (Simmons 2005). Quantifying rates of colonization, speciation, and extinction over time requires as many phylogenies and instances of colonization as possible, and a phylogeny of hutias would add the largest number of nonvolant species in the region. The phylogenies of these taxa would be much more than single data points in historical biogeography—they would illuminate the mechanisms of dispersal from the mainland, interisland colonization, and further clarify drivers of extinction risk.

A second remaining gap in Caribbean mammal biogeography involves widespread species and species complexes whose population genetic structure and geographic origin remain largely unexplored. Both morphological measurements and the few mitochondrial sequences available suggest that insular populations of *Pteronotus parnellii* constitute distinct species (Dávalos 2006; Lewis-Oritt, Porter, and Baker 2001). *Mormoops megalophylla* requires similar revisionary work to determine if the fossils found throughout North America, the West Indies, and northern South America are conspecific (Czaplewski and Cartelle 1998; Silva-Taboada 1974). Although population structure in large, vagile, wide-ranging species such as *Tadarida brasiliensis* or *Eptesicus fuscus* should be detectable only at very broad geographic scales, the distances and depths separating West Indian and mainland populations might still prove to be significant barriers to gene flow (Russell and McCracken 2006; Russell, Medellin, and McCracken 2005).

Closing the third gap will require extending population genetic studies to continental populations, and using ecological modeling, radiometric dating of fossils and subfossils, and even sequencing genetic samples from extinct populations. Despite several recent studies tackling the population genetics and phylogeography of West Indian bats (e.g., Carstens et al. 2004; Fleming and Murray 2009; Fleming, Murray, and Carstens 2010; Larsen et al. 2007; Larsen, Marchán-Rivadeneira, and Baker 2010), there has been no research investigating both population genetic structure and timing of colonization in widespread bats (e.g., Russell, Goodman, and Cox 2008). By combining timing of divergence or colonization from population genetics analyses with modeling of the climatic niche of the populations in question, new insights on the relationship between climate change, colonization, and persistence are possible (e.g., Carnaval and Moritz 2008; Carnaval et al. 2009). Dating fossil remains (McFarlane and Lundberg 2004), and even the use of ancient DNA from subfossils (Shapiro et al. 2004), would refine the timescale of decline and extinction of populations. These synthetic analyses would test climate change as the primary driver of extinction in the West Indies, particularly among bats (Morgan 2001). By directly evaluating the availability of suitable habitat for different species under alternative climate conditions, such studies would help clarify the relative roles of overexploitation and habitat change in past Caribbean mammal extinctions (MacPhee, Iturralde-Vinent, and Vazquez 2007; Turvey et al. 2007).

Both traditional systematic approaches (e.g., to resolve species limits using morphological data, and syntheses of more novel methods) along with combining historical and ecological biogeography with radiometric dating of fossil remains, will be necessary to resolve the outstanding questions on the geographic origins and drivers of extinction in this biota. After reviewing the century-long history of the vicariance versus dispersal debate, we conclude that quantitative approaches, including equilibrium models hitherto absent from most historical analyses, can better serve as null hypotheses than single-mechanism hypotheses such as dispersal or vicariance.

Literature Cited

Allen, G. M. 1911. "Mammals of the West Indies." *Bulletin of the Museum of Comparative Zoology* 54:175–263.

———.1942. *Extinct and Vanishing Mammals of the Western Hemisphere with the Marine Species of All the Oceans*. Washington, DC: American Committee for International Wildlife Protection, *Special Publication*, 11.

Alvarez, L. W., W. Alvarez, F. Asaro, and H. V. Michel. 1980. "Extraterrestrial Cause for the Cretaceous Tertiary Extinction." *Science* 208:1095–108.

Anthony, H. E. 1918. "The Indigenous Land Mammals of Porto Rico, Living and Extinct." *Memoirs of the American Museum of Natural History* 2:331–435.

Asher, R. J. 1999. "A Morphological Basis for Assessing the Phylogeny of the 'Tenrecoidea' (Mammalia, Lipotyphla)." *Cladistics* 15:231–52.

———. 2005. "Insectivoran-Grade Placentals." In *The Rise of Placental Mammals: Origins and Relationships of the Major Extant Clades*, edited by K. D. Rose and J. D. Archibald, 50–70. Baltimore: Johns Hopkins University Press.

Asher, R. J., R. J. Emry, and M. C. McKenna. 2005. "New Material of *Centetodon* (Mammalia, Lipotyphla) and the Importance of (Missing) DNA Sequences in Systematic Paleontology." *Journal of Vertebrate Paleontology* 25:911–23.

Baker, R. J., and H. H. Genoways. 1978. "Zoogeography of Antillean Bats." *Special Publication, Academy of Natural Sciences of Philadelphia* 13:53–97.

Baker, R. J., J. C. Patton, H. H. Genoways, and J. W. Bickham. 1988. "Genic Studies of *Lasiurus* (Chiroptera: Vespertilionidae)." *Occasional Papers, Museum of Texas Tech University* 117:1–15.

Baker, R. J., V. A. Taddei, J. L. Hudgeons, and R. A. Van Den Bussche. 1994. "Systematic Relationships within *Chiroderma* (Chiroptera: Phyllostomidae) Based on Cytochrome *b* Sequence Variation." *Journal of Mammalogy* 75:321–27.

Barbour, T. 1916. "Some Remarks upon Matthew's 'Climate and Evolution.'" *Annals of the New York Academy of Sciences* 27:1–15.

Barnosky, A. D., P. L. Koch, R. S. Feranec, S. L. Wing, and A. B. Shabel. 2004. "Assessing the Causes of Late Pleistocene Extinctions on the Continents." *Science* 306:70–75.

Borroto-Páez, R., C. A. Woods, and C. W. Kilpatrick. 2005. "Sistemática de las Jutías de las Antillas (Rodentia, Capromyidae)." In *Proceedings of the International Symposium Insular Vertebrate Evolution: the Palaeontological Approach*, edited by J. A. Alcover and P. Bover, 33–50. Palma de Mallorca: Monografies de la Societat d'Història Natural de les Balears, 12.

Brown, J. H., and M. V. Lomolino. 2000. "Concluding Remarks: Historical Perspective and the Future of Island Biogeography Theory." *Global Ecology and Biogeography* 9:87–92.

Brumfield, R. T., and S. V. Edwards. 2007. "Evolution into and out of the Andes: A Bayesian Analysis of Historical Diversification in *Thamnophilus* antshrikes." *Evolution* 61:346–67.

Burney, D. A., L. P. Burney, L. R. Godfrey, W. L. Jungers, S. M. Goodman, H. T. Wright, and A. J. T. Jull. 2004. "A Chronology for Late Prehistoric Madagascar." *Journal of Human Evolution* 47:25–63.

Burney, D. A., L. P. Burney, and R. D. E. MacPhee. 1994. "Holocene Charcoal Stratigraphy from Laguna Tortuguero, Puerto Rico, and the Timing of Human Arrival on the Island." *Journal of Archaeological Science* 21:273–81.

Cardillo, M., G. M. Mace, K. E. Jones, J. Bielby, O. R. P. Bininda-Emonds, W. Sechrest, C. D. L. Orme, et al. 2005. "Multiple Causes of High Extinction Risk in Large Mammal Species." *Science* 309:1239–41.

Carnaval, A. C., M. J. Hickerson, C. F. B. Haddad, M. T. Rodrigues, and C. Moritz. 2009. "Stability Predicts Genetic Diversity in the Brazilian Atlantic Forest Hotspot." *Science* 323:785–89.

Carnaval, A. C., and C. Moritz. 2008. "Historical Climate Modelling Predicts Patterns of Current Biodiversity in the Brazilian Atlantic Forest." *Journal of Biogeography* 35:1187–201.

Carstens, B. C., J. Sullivan, L. M. Dávalos, P. A. Larsen, and S. C. Pedersen. 2004. "Exploring Population Genetic Structure in Three Species of Lesser Antillean Bats." *Molecular Ecology* 13:2557–66.

Castro, M. F. de 1864. De la Existencia de Grandes Mamíferos Fósiles en la Isla de Cuba. *Anales de la Academia de Ciencias Médicas, Físicas y Naturales de la Habana* 1:17–21.

Clough, G. C. 1976. "Current Status of Two Endangered Caribbean Rodents." *Biological Conservation* 10:43–47.

Czaplewski, N. J., and C. Cartelle. 1998. "Pleistocene Bats from Cave Deposits in Bahia, Brazil." *Journal of Mammalogy* 79:784–803.

Darlington, P. J., Jr. 1938. "The Origin of the Fauna of the Greater Antilles, with Discussion of Dispersal of Animals over Water and through the Air." *Quarterly Review of Biology* 13:274–300.

Dávalos, L. M. 2004a. "Historical Biogeography of the Antilles: Earth History and Phylogenetics of Endemic Chiropteran Taxa." Unpublished PhD diss., Columbia University, New York.

———.2004b. "Phylogeny and Biogeography of Caribbean Mammals." *Biological Journal of the Linnean Society* 81:373–94.

———.2005. "Molecular Phylogeny of Funnel-Eared Bats (Chiroptera: Natalidae), with Notes on Biogeography and Conservation." *Molecular Phylogenetics and Evolution* 37:91–103.

———. 2006."The Geography of Diversification in the Mormoopids (Chiroptera: Mormoopidae)." *Biological Journal of the Linnean Society* 88:101–18.

———. 2007."Short-Faced Bats (Phyllostomidae: Stenodermatina): A Caribbean Radiation of Strict Frugivores." *Journal of Biogeography* 34:364–75.

———. 2010."Earth History and the Evolution of Caribbean Bats." In *Island Bats: Ecology, Evolution, and Conservation*, edited by T. H. Fleming and P. A. Racey, 96–115. Chicago: University of Chicago Press.

Diamond, J. M. 1989. "Quaternary Megafaunal Extinctions: Variations on a Theme by Paganini." *Journal of Archaeological Science* 16:167–75.

Díaz-Franco, S. 2001. "Situación taxonómica de *Geocapromys megas* (Rodentia: Capromyidae)." *Caribbean Journal of Science* 37:72–80.

Drake, D. R., and T. L. Hunt. 2009. "Invasive Rodents on Islands: Integrating Historical and Contemporary Ecology." *Biological Invasions* 11:1483–87.

Drummond, A. J., S. Y. W. Ho, M. J. Phillips, and A. Rambaut. 2006. "Relaxed Phylogenetics and Dating with Confidence." *PLoS Biology* 4:e88.

Edwards, A. W. F., and L. L. Cavalli-Sforza. 1964. "Reconstruction of Evolutionary Trees." In *Phenetic and Phylogenetic Classification*, edited by V. H. Heywood and J. McNeill, 67–76. London: Systematics Association.

Fleming, T. H., and K. L. Murray. 2009. "Population and Genetic Consequences of Hurricanes for Three Species of West Indian Phyllostomid Bats." *Biotropica* 41:250–56.

Fleming, T. H., K. L. Murray, and B. C. Carstens. 2010. "Phylogeography and Genetic Structure of Three Evolutionary Lineages of West Indian Phyllostomid Bats." In *Island Bats: Evolution, Ecology, and Conservation*, edited by T. H. Fleming and P. A. Racey, 116–50. Chicago: University of Chicago Press.

Flemming, C., and R. D. E. MacPhee. 1999. "Redetermination of Holotype of *Isolobodon portoricensis* (Rodentia, Capromyidae), with Notes on Recent Mammalian Extinctions in Puerto Rico." *American Museum Novitates* 3278:1–11.

Galewski, T., J.-F. Mauffrey, Y. L. R. Leite, J. L. Patton, and E. J. P. Douzery. 2005. "Ecomorphological Diversification Among South American Spiny Rats (Rodentia; Echimyidae): A Phylogenetic and Chronological Approach." *Molecular Phylogenetics and Evolution* 34:601–15.

Gannon, M. R., A. Kurta, A. Rodriguez Duran, and M. R. Willig. 2005. *Bats of Puerto Rico: An Island Focus and a Caribbean Perspective*. Lubbock: Texas Tech University Press.

Genoways, H. H., C. J. Phillips, and R. J. Baker. 1998. "Bats of the Antillean Island of Grenada: A New Zoogeographic Perspective." *Occasional Papers, Museum of Texas Tech University* 177:1–28.

Graham, A. 2003. "Geohistory Models and Cenozoic Paleoenvironments of the Caribbean Region." *Systematic Botany* 28:378–86.

Grajales Nishimura, J. M., E. Cedillo Pardo, C. Rosales Dominguez, D. J. Moran Zenteno, W. Alvarez, P. Claeys, J. Ruiz Morales et al. 2000. "Chicxulub Impact: The Origin of Reservoir and Seal Facies in the Southeastern Mexico Oil Fields." *Geology* 28:307–10.

Griffiths, T. A., and D. Klingener. 1988. "On the Distribution of Greater Antillean Bats." *Biotropica* 20:240–51.

Harris, D. B. 2009. "Review of Negative Effects of Introduced Rodents on Small Mammals on Islands." *Biological Invasions* 11:1611–30.

Harris, D. B., S. D. Gregory, and D. W. Macdonald. 2006. "Space Invaders? A Search for Patterns Underlying the Coexistence of Alien Black Rats and Galápagos Rice Rats." *Oecologia* 149:276–88.

Harris, D. B., and D. W. Macdonald. 2007. "Interference Competition between Introduced Black Rats and Endemic Galápagos Rice Rats." *Ecology* 88:2330–44.

Heaney, L. R. 2000. "Dynamic Disequilibrium: A Long-Term, Large-Scale Perspective on the Equilibrium Model of Island Biogeography." *Global Ecology and Biogeography* 9:59–74.

———. 2007. "Is a New Paradigm Emerging for Oceanic Island Biogeography?" *Journal of Biogeography* 34:753–57.

Hedges, S. B. 1982. "Caribbean Biogeography: Implications of Recent Plate Tectonic Studies." *Systematic Zoology* 31:518–22.

———. 1996. "Historical Biogeography of West Indian Vertebrates." *Annual Review of Ecology and Systematics* 27:163–96.

Hedges, S. B., C. A. Hass, and L. R. Maxson. 1992. "Caribbean Biogeography: Molecular Evidence for Dispersal in West Indian Terrestrial Vertebrates." *Proceedings of the National Academy of Sciences, USA* 89:1909–13.

Heinicke, M. P., W. E. Duellman, and S. B. Hedges. 2007. "Major Caribbean and Central American Frog Faunas Originated by Ancient Oceanic Dispersal." *Proceedings of the National Academy of Sciences, USA* 104:10092–97.

Hess, H. H. 1962. "History of Ocean Basins." In *Petrologic Studies: A Volume in Honor of A. F. Buddington*, edited by A. E. J. Engel, H. L. James, and B. F. Leonard, 599–620. New York: Geological Society of America.

Holdaway, R. N., and C. Jacomb. 2000. "Rapid Extinction of the Moas (Aves: Dinornithiformes): Model, Test, and Implications." *Science* 287:2250–54.

Hooijer, D. A. 1959. "Fossil Rodents from Curaçao and Bonaire." *Studies on the Fauna of Curaçao and Other Caribbean Islands* 9:1–27.

———. 1966. "Fossil Mammals of the Netherlands Antilles." *Archives Néerlandaises de Zoologie* 16:531–32.

———. 1967. "Pleistocene Vertebrates of the Netherlands Antilles." In *Pleistocene Extinctions: The Search for a Cause*, edited by P. S. Martin and H. E. Wright Jr., 399–406. New Haven: Yale University Press.

Horovitz, I., and R. D. E. MacPhee. 1999. "The Quaternary Cuban Platyrrhine *Paralouatta varonai* and the Origin of Antillean Monkeys." *Journal of Human Evolution* 36:33–68.

International Union for Conservation of Nature. (IUCN). 2008. "2008 IUCN Red List of Threatened Species." Available at http://www.iucnredlist.org/.

Iturralde-Vinent, M. A. 2006. "Meso-Cenozoic Caribbean Paleogeography: Implications for the Historical Biogeography of the Region." *International Geology Review* 48:791–827.

Iturralde-Vinent, M. A., and R. D. E. MacPhee. 1999. "Paleogeography of the Caribbean Region: Implications for Cenozoic Biogeography." *Bulletin of the American Museum of Natural History* 238:1–95.

Iturralde-Vinent, M. A., R. D. E. MacPhee, S. Díaz-Franco, R. Rojas-Consuegra, W. Suárez, and A. Lomba. 2000. "Las Breas de San Felipe, a Quaternary Fossiliferous Asphalt Seep Near Martí (Matanzas Province, Cuba)." *Caribbean Journal of Science* 36:300–23.

Jiménez Vázquez, O., M. M. Condis, and C. E. García. 2005. "Vertebrados Post-glaciales en un Residuario Fósil de *Tyto alba scopoli* (Aves: Tytonidae) en el Occidente de Cuba." *Revista Mexicana de Mastozoología* 9:85–112.

Johnson, C. 2006. "Australia's Mammal Extinctions: A 50,000 Year History." Cambridge: Cambridge University Press.

Jones, J. K., Jr. 1989. "Distribution and Systematics of Bats in the Lesser Antilles." In *Biogeography of the West Indies: Past, Present, and Future*, edited by C. A. Woods, 645–60. Gainesville: Sandhill Crane Press.

Jones, K. E., A. Purvis, A. MacLarnon, O. R. P. Bininda-Emonds, and N. B. Simmons. 2002. "A Phylogenetic Supertree of the Bats (Mammalia: Chiroptera)." *Biological Reviews* 77:223–59.

Kluge, A. G. 1989. "A Concern for Evidence and a Phylogenetic Hypothesis of Relationships Among *Epicrates* (Boidae, Serpentes)." *Systematic Zoology* 38:7–25.

Koopman, K. F. 1958. "Land Bridges and Ecology in Bat Distribution on Islands off the Northern Coast of South America." *Evolution* 12:429–39.

———. 1989. "A Review and Analysis of the Bats of the West Indies." In *Biogeography of the West Indies: Past, Present, and Future*, edited by C. A. Woods, 635–44. Gainesville: Sandhill Crane Press.

———. 1994. "Chiroptera: Systematics." *Handbuch der Zoologie* 8:1–217.

Koopman, K. F., and E. E. Williams. 1951. "Fossil Chiroptera Collected by H. E. Anthony in Jamaica, 1919–1920." *American Museum Novitates* 1519:1–29.

Larsen, P. A., S. R. Hoofer, M. C. Bozeman, S. C. Pedersen, H. H. Genoways, C. J. Phillips, D. E. Pumo et al. 2007. "Phylogenetics and Phylogeography of the *Artibeus jamaicensis* Complex Based on Cytochrome-b DNA Sequences." *Journal of Mammalogy* 88:712–27.

Larsen, P. A., M. R. Marchán-Rivadeneira, and R. J. Baker. 2010. "Natural Hybridization Generates Mammalian Lineage with Species Characteristics." *Proceedings of the National Academy of Sciences, USA* 107:11447–52.

Leidy, J. 1868. "Notice of Some Vertebrate Remains from the West Indian Islands." *Proceedings of the Academy of Natural Sciences of Philadelphia* 20:178–80.

Lessa, E. P., J. A. Cook, and J. L. Patton. 2003. "Genetic Footprints of Demographic Expansion in North America, But Not Amazonia, During the Late Quaternary." *Proceedings of the National Academy of Sciences, USA* 100:10331–34.

Lewis-Oritt, N., C. A. Porter, and R. J. Baker. 2001. "Molecular Systematics of the Family Mormoopidae (Chiroptera) Based on Cytochrome *b* and Recombination Activating Gene 2 Sequences." *Molecular Phylogenetics and Evolution* 20:426–36.

MacArthur, R. H., and E. O. Wilson. 1963. "An Equilibrium Theory of Insular Zoogeography." *Evolution* 17:373–87.

———. 1967. *The Theory of Island Biogeography. Monographs in Population Biology*, 1. Princeton: Princeton University Press.

MacFadden, B. J. 1980. "Rafting Mammals or Drifting Islands?: Biogeography of the Greater Antillean Insectivores *Nesophontes* and *Solenodon*." *Journal of Biogeography* 7:11–22.

———. 1981. "Comments on Pregill's Appraisal of Historical Biogeography of Caribbean Vertebrates: Vicariance, Dispersal, or Both?" *Systematic Zoology* 30:370–72.

MacPhee, R. D. E. 2005. "'First' Appearances in the Cenozoic Land-Mammal Record of the Greater Antilles: Significance and Comparison with South American and Antarctic Records." *Journal of Biogeography* 32:551–64.

———. 2008. "*Insulae Infortunatae*: Establishing a Chronology for Late Quaternary Mammal Extinctions in the West Indies." In *American Megafaunal Extinctions at the End of the Pleistocene*, edited by G. Haynes, 169–93. Dordrecht: Springer Science + Business Media B.V.

MacPhee, R. D. E., and J. G. Fleagle. 1991. "Postcranial Remains of *Xenothrix mcgregori* (Primates, Xenotrichidae) and Other Late Quaternary Mammals from Long Mile Cave, Jamaica." *Bulletin of the American Museum of Natural History* 206:287–321.

MacPhee, R. D. E., and C. Flemming. 1999. "*Requiem Æternam*: The Last Five Hundred Years of Mammalian Species Extinctions." In *Extinctions in Near Time: Causes, Contexts, and Consequences*, edited by R. D. E. MacPhee, 333–71. New York: Kluwer Academic/Plenum.

———. "A Possible Heptaxodontine and Other Caviidan Rodents from the Quaternary of Jamaica." *American Museum Novitates* 3422:1–42.

MacPhee, R. D. E., C. Flemming, D. P. Domning, R. W. Portell, and B. Beatty. 1999. "Eocene ?Primate Petrosal from Jamaica: Morphology and Biogeographical Implications." *Journal of Vertebrate Paleontology* 19:61.

MacPhee, R. D. E., D. C. Ford, and D. A. McFarlane. 1989. "Pre-Wisconsinan Mammals from Jamaica and Models of Late Quaternary Extinction in the Greater Antilles." *Quaternary Research* 31:94–106.

MacPhee, R. D. E., and I. Horovitz. 2004. "New Craniodental Remains of the Quaternary Jamaican Monkey *Xenothrix mcgregori* (Xenotrichini, Callicebinae, Pitheciidae), with a Reconsideration of the *Aotus* Hypothesis." *American Museum Novitates* 3434:1–51.

MacPhee, R. D. E., and M.-A. Iturralde-Vinent. 1995a. "Earliest Monkey from Greater Antilles." *Journal of Human Evolution* 28:197–200.

———. 1995b. "Origin of the Greater Antillean Land Mammal Fauna, 1: New Tertiary Fossils from Cuba and Puerto Rico." *American Museum Novitates* 3141:1–30.

MacPhee, R. D. E., M. A. Iturralde-Vinent, and E. S. Gaffney. 2003. "Domo de Zaza, an Early Miocene Vertebrate Locality in South-Central Cuba, with Notes on the Tectonic Evolution of Puerto Rico and the Mona Passage." *American Museum Novitates* 3394:1–42.

MacPhee, R. D. E., M. A. Iturralde-Vinent, and O. J. Vazquez. 2007. "Prehistoric Sloth Extinctions in Cuba: Implications of a New 'Last' Appearance Date." *Caribbean Journal of Science* 43:94–98.

MacPhee, R. D. E., and P. A. Marx. 1997. "The 40,000-Year Plague: Humans, Hyperdisease, and First-Contact Extinctions." In *Natural Change and Human Impact in Madagascar*, edited by S. M. Goodman and B. D. Patterson, 169–217. Washington, DC: Smithsonian Institution Press.

MacPhee, R. D. E., and J. Meldrum. 2006. "Postcranial Remains of the Extinct Monkeys of the Greater Antilles, with Evidence for Semiterrestriality in *Paralouatta*." *American Museum Novitates* 3516:1–65.

MacPhee, R. D. E., J.-L. White, and C.-A. Woods. 2000. "New Megalonychid Sloths (Phyllophaga, Xenarthra) from the Quaternary of Hispaniola." *American Museum Novitates* 3303:1–32.

Mancina, C. A., and L. Garcia-Rivera. 2005. "New Genus and Species of Fossil Bat (Chiroptera: Phyllostomidae) from Cuba." *Caribbean Journal of Science* 41:22–27.

Martin, P. S. 1984. "Prehistoric Overkill: The Global Model." In *Quaternary Extinctions: A Prehistoric Revolution*, edited by P. S. Martin and R. G. Klein, 354–404. Tucson: University of Arizona Press.

Martin, P. S., and D. W. Steadman. 1999. "Prehistoric Extinctions on Islands and Continents." In *Extinctions in Near Time: Causes, Contexts, and Consequences*, edited by R. D. E. MacPhee, 17–55. New York: Kluwer Academic/Plenum.

Matthew, W. D. 1918. "Affinites and Origin of the Antillean Mammals." *Bulletin of the Geological Society of America* 29:657–66.

McFarlane, D. A., and J. Lundberg. 2004. "*Reliquiae Diluviane Alter*: Last Interglacial Flood Deposits in Caves of the West Indies." In *Studies of Cave Sediments*, edited by I. D. Sasowsky and J. Mylroie, 313–22. New York: Kluwer/Plenum Press.

McFarlane, D. A., R. D. E. MacPhee, and D. C. Ford. 1998. "Body Size Variability and a Sangamonian Extinction Model for *Amblyrhiza*, a West Indian Megafaunal Rodent." *Quaternary Research* 50:80–89.

McFarlane, D. A., A. Vale, K. Christenson, J. Lundberg, G. Atilles, and S. E. Lauritzen. 2000. "New Specimens of Late Quaternary Extinct Mammals from Caves in Sanchez Ramirez Province, Dominican Republic." *Caribbean Journal of Science* 36:163–66.

McGuire, J. A., C. C. Witt, D. L. Altshuler, and J. V. Remsen Jr. 2007. "Phylogenetic Systematics and Biogeography of Hummingbirds: Bayesian and Maximum Likelihood Analyses of Partitioned Data and Selection of an Appropriate Partitioning Strategy." *Systematic Biology* 56:837–56.

Mickleburgh, S., K. Waylen, and P. Racey. 2009. "Bats as Bushmeat: A Global Review." *Oryx* 43:217–34.

Miller, G. S. 1918. "Three New Bats from Haiti and Santo Domingo." *Proceedings of the Biological Society of Washington* 31:39–40.

———. 1922. "Remains of Mammals from Caves in the Republic of Haiti." *Smithsonian Miscellaneous Contributions* 74:1–8.

———. 1929a. "A Second Collection of Mammals from Caves Near St. Michel, Haiti." *Smithsonian Miscellaneous Contributions* 81:1–30.

———. 1929b. "Mammals Eaten by Indians, Owls, and Spaniards in the Coast Region of the Dominican Republic." *Smithsonian Miscellaneous Contributions* 82:1–16.

———. 1930. “Three Small Collections of Mammals from Hispaniola.” *Smithsonian Miscellaneous Contributions* 82:1–15.

Morales, J. C., and J. W. Bickham. 1995. “Molecular Systematics of the Genus *Lasiurus* (Chiroptera: Vespertilionidae) Based on Restriction-Site Maps of the Mitochondrial Ribosomal Genes.” *Journal of Mammalogy* 76:730–49.

Morgan, G. S. 1993. “Quaternary Land Vertebrates of Jamaica.” In *Biostratigraphy of Jamaica*, edited by R. M. Wright and E. Robinson, 417–42. Boulder: Geological Society of America. *Memoir* 182.

———. 2001. “Patterns of Extinction in West Indian Bats.” In *Biogeography of the West Indies: Patterns and Perspectives*, edited by C. A. Woods and F. E. Sergile, 369–407. Boca Raton: CRC Press.

Morgan, G. S., and L. Wilkins. 2003. “The Extinct Rodent *Clidomys* (Heptaxodontidae) from a Late Quaternary Cave Deposit in Jamaica.” *Caribbean Journal of Science* 39:34–41.

Morgan, G. S., and C. A. Woods. 1986. “Extinction and the Zoogeography of West Indian Land Mammals.” In *Island Biogeography of Mammals. Biological Journal of the Linnean Society*, 28, edited by L. R. Heaney and B. D. Patterson, 167–203. New York: Academic Press.

Newsom, L. A., and E. S. Wing. 2004. *On Land and Sea: Native American Uses of Biological Resources in the West Indies*. Tuscaloosa: University of Alabama Press.

Olson, S. L., and H. F. James. 1982. “Fossil Birds from the Hawaiian Islands: Evidence for Wholesale Extinction by Man Before Western Contact.” *Science* 217:633–35.

Page, R. D. M., and C. Lydeard. 1994. “Towards a Cladistic Biogeography of the Caribbean.” *Cladistics* 10:21–41.

Phillips, C. J., D. E. Pumo, H. H. Genoways, and P. E. Ray. 1989. “Caribbean Island Zoogeography: A New Approach Using Mitochondrial DNA to Study Neotropical Bats.” In *Biogeography of the West Indies: Past, Present, and Future*, edited by C. A. Woods, 661–84. Gainesville: Sandhill Crane Press.

Phillips, C., J. D. E. Pumo, H. H. Genoways, P. E. Ray, and C. Briskey. 1991. “Mitochondrial DNA Evolution and Phylogeography in Two Neotropical Fruit Bats, *Artibeus jamaicensis* and *Artibeus lituratus*.” In *Latin American Mammalogy: History, Biodiversity, and Conservation*, edited by M. Mares and D. Schmidly, 97–123. Norman: University of Oklahoma Press.

Platnick, N. I. 1991. “On Areas of Endemism.” *Australian Systematic Botany* 4:ix–xii.

Portell, R. W., S. K. Donovan, and D. P. Domning. 2001. “Early Tertiary Vertebrate Fossils from Seven Rivers, Parish of St. James, Jamaica, and their Biogeographical Implications.” In *Biogeography of the West Indies*, edited by C. A. Woods and F. E. Sergile, 191–200. Boca Raton: CRC Press.

Pregill, G. K. 1981. “An Appraisal of the Vicariance Hypothesis of Caribbean Biogeography and its Application to West Indian Terrestrial Vertebrates.” *Systematic Zoology* 30:147–55.

Pregill, G. K., and S. L. Olson. 1981. “Zoogeography of West Indian Vertebrates in Relation to Pleistocene Climatic Cycles.” *Annual Review of Ecology and Systematics* 12:75–98.

Pregill, G. K., D. W. Steadman, and D. R. Watters. 1994. “Late Quaternary Vertebrate

Faunas of the Lesser Antilles: Historical Components of Caribbean Biogeography." *Bulletin of Carnegie Museum of Natural History* 30:1–51.

Presley, S. J., and M. R. Willig. 2008. "Composition and Structure of Caribbean Bat (Chiroptera) Assemblages: Effects of Inter-Island Distance, Area, Elevation and Hurricane-Induced Disturbance." *Global Ecology and Biogeography* 17:747–57.

Pumo, D. E., E. Z. Goldin, B. Elliot, C. Phillips, and H. H. Genoways. 1988. "Mitochondrial DNA Polymorphism in Three Antillean Island Populations of the Fruit Bat, *Artibeus jamaicensis*." *Molecular Biology and Evolution* 5:79–89.

Pumo, D. E., K. Iksoo, J. Remsen, C. J. Phillips, and H. H. Genoways. 1996. "Molecular Systematics of the Fruit Bat, *Artibeus jamaicensis*: Origin of an Unusual Island Population." *Journal of Mammalogy* 77:491–503.

Raffaele, H. A. 1979. "The Status of Some Endangered Species in Puerto Rico with Particular Emphasis on *Isolobodon* (Rodentia)." In *Memorias del Tercer Simposio*, 100–104. San Juan: Departamento de Recursos Naturales.

Ree, R. H., and S. A. Smith. 2008. "Maximum-Likelihood Inference of Geographic Range Evolution by Dispersal, Local Extinction, and Cladogenesis." *Systematic Biology* 57:4–14.

Rega, E., D. A. McFarlane, J. Lundberg, and K. Christenson. 2002. "A New Megalonychid Sloth from the Late Wisconsinan of the Dominican Republic." *Caribbean Journal of Science* 38:11–19.

Ricklefs, R. E. 1970. "Stage of Taxon Cycle and Distribution of Birds on Jamaica, Greater Antilles." *Evolution* 24:475–77.

Ricklefs, R. E., and E. Bermingham. 2001. "Nonequilibrium Diversity Dynamics of the Lesser Antillean Avifauna." *Science* 294:1522–24.

Ricklefs, R. E., and G. C. Cox. 1972. "Taxon Cycles in the West Indian Avifauna." *American Naturalist* 106:195–219.

Rímoli, R. O. 1976. "Roedores Fosiles de la Hispaniola." *Universidad Central del Este, Serie Cientifica III* 95:1–93.

Roca, A. L., G. Kahila Bar-Gal, E. Eizirik, K. M. Helgen, R. Maria, M. S. Springer, S. J. O'Brien et al. 2004. "Mesozoic Origin for West Indian Insectivores." *Nature* 429:649–51.

Rodríguez Durán, A., and T. H. Kunz. 2001. "Biogeography of West Indian Bats: An Ecological Perspective." In *Biogeography of the West Indies: Patterns and Perspectives*, edited by C. A. Woods and F. E. Sergile, 355–68. Boca Raton: CRC Press.

Rosen, D. E. 1975. "A Vicariance Model of Caribbean Biogeography." *Systematic Zoology* 24:431–64.

———. 1985. "Geological Hierarchies and Biogeographic Congruence in the Caribbean." *Annals of the Missouri Botanical Garden* 72:636–59.

Russell, A., and G. F. McCracken. 2006. "Population Genetic Structure of Very Large Populations: The Brazilian Free-Tailed Bat, *Tadarida brasiliensis*." In *Functional and Evolutionary Ecology of Bats*, edited by A. Zubaid, G. F. McCracken, and T. H. Kunz, 227–47. New York: Oxford University Press.

Russell, A. L., R. A. Medellin, and G. F. McCracken. 2005. "Genetic Variation and Migration in the Mexican Free-Tailed Bat (*Tadarida brasiliensis mexicana*)." *Molecular Ecology* 14:2207–22.

Russell, A. L., J. Ranivo, E. P. Palkovacs, S. M. Goodman, and A. D. Yoder. 2007. "Working at the Interface of Phylogenetics and Population Genetics: A Biogeographical Analysis of *Triaenops* spp. (Chiroptera: Hipposideridae)." *Molecular Ecology* 16:839–51.

Russell, A. L., S. M. Goodman, and M. P. Cox. 2008. "Coalescent Analyses Support Multiple Mainland-to-Island Dispersals in the Evolution of Malagasy *Triaenops* bats (Chiroptera: Hipposideridae)." *Journal of Biogeography* 35:995–1003.

Russell, A. L., S. M. Goodman, I. Fiorentino, and A. D. Yoder. 2008. "Population Genetic Analysis of *Myzopoda* (Chiroptera: Myzopodidae) in Madagascar." *Journal of Mammalogy* 89:209–21.

Rutschmann, F. 2006. "Molecular Dating of Phylogenetic Trees: A Brief Review of Current Methods That Estimate Divergence Times." *Diversity and Distributions* 12:35–48.

Schubert, C., and E. Medina. 1982. "Evidence of Quaternary Glaciation in the Dominican Republic: Some Implications for Caribbean Paleoclimatology." *Palaeogeography, Palaeoclimatology, Palaeoecology* 39:281–94.

Shapiro, B., A. J. Drummond, A. Rambaut, M. C. Wilson, P. E. Matheus, A. V. Sher, O. G. Pybus et al. 2004. "Rise and Fall of the Beringian Steppe Bison." *Science* 306:1561–65.

Silva-Taboada, G. 1974. "Fossil Chiroptera from Cave Deposits in Central Cuba, with Description of Two New Species (Genera *Pteronotus* and *Mormoops*) and the First West Indian Record of *Mormoops megalophylla*." *Acta Zoologica Cracoviensia* 19:33–73.

Simmons, N. B. 2005. "Order Chiroptera." In *Mammal Species of the World: A Taxonomic and Geographic Reference*, 3rd ed., edited by D. E. Wilson and D. M. Reeder, 313–529. Baltimore: Johns Hopkins University Press.

Simmons, N. B., and T. M. Conway. 2001. "Phylogenetic Relationships of Mormoopid Bats (Chiroptera: Mormoopidae) Based on Morphological Data." *Bulletin of the American Museum of Natural History* 258:1–97.

Simpson, G. G. 1943. "Mammals and the Nature of Continents." *American Journal of Science* 241:1–31.

———. 1956. "Zoogeography of West Indian Land Mammals." *American Museum Novitates* 1759:1–28.

Stadelmann, B., L. K. Lin, T. H. Kunz, and M. Ruedi. 2007. "Molecular Phylogeny of New World *Myotis* (Chiroptera, Vespertilionidae) Inferred from Mitochondrial and Nuclear DNA Genes." *Molecular Phylogenetics and Evolution* 43:32–48.

Steadman, D. W. 1995. "Prehistoric Extinctions of Pacific Island Birds: Biodiversity Meets Zooarchaeology." *Science* 267:1123–31.

Steadman, D. W., R. Franz, G. S. Morgan, N. A. Albury, B. Kakuk, K. Broad, S. E. Franz et al. 2007. "Exceptionally Well Preserved Late Quaternary Plant and Vertebrate Fossils from a Blue Hole on Abaco, the Bahamas." *Proceedings of the National Academy of Sciences, USA* 104:19897–902.

Steadman, D. W., P. S. Martin, R. D. E. MacPhee, A. J. T. Jull, H. G. McDonald, C. A. Woods, M. Iturralde-Vinent et al. 2005. "Asynchronous Extinction of Late Quaternary Sloths on Continents and Islands." *Proceedings of the National Academy of Sciences, USA* 102:11763–68.

Steadman, D. W., G. K. Pregill, and S. L. Olson. 1984. "Fossil Vertebrates from Antigua, Lesser Antilles: Evidence for Late Holocene Human-Caused Extinctions in the West Indies." *Proceedings of the National Academy of Sciences, USA* 81:4448–51.

Suárez, W. 2005. "Taxonomic Status of the Cuban Vampire Bat (Chiroptera: Phyllostomidae: Desmodontinae: *Desmodus*)." *Caribbean Journal of Science* 41:761–67.

Suárez, W., and S. Díaz-Franco. 2003. "A New Fossil Bat (Chiroptera: Phyllostomidae) from a Quaternary Cave Deposit in Cuba." *Caribbean Journal of Science* 39:371–77.

Teeling, E. C., M. S. Springer, O. Madsen, P. Bates, S. J. O'Brien, and W. J. Murphy. 2005. "A Molecular Phylogeny for Bats Illuminates Biogeography and the Fossil Record." *Science* 307:580–84.

Tejedor, A. 2006. "The Type Locality of *Natalus stramineus* (Chiroptera: Natalidae): Implications for the Taxonomy and Biogeography of the Genus *Natalus*." *Acta Chiropterologica* 8:361–80.

———. 2011. "Systematics of Funnel-Eared Bats (Chiroptera: Natalidae)." *Bulletin of the American Museum of Natural History* 353:1–40.

Tejedor, A., G. Silva-Taboada, and D. Rodriguez Hernandez. 2004. "Discovery of Extant *Natalus major* (Chiroptera: Natalidae) in Cuba." *Mammalian Biology* 69:153–62.

Tejedor, A., V. d. C. Tavares, and G. Silva-Taboada. 2005. "Taxonomic Revision of Greater Antillean Bats of the Genus *Natalus*." *American Museum Novitates* 3493:1–22.

Towns, D., I. Atkinson, and C. Daugherty. 2006. "Have the Harmful Effects of Introduced Rats on Islands Been Exaggerated?" *Biological Invasions* 8:863–91.

Trejo-Torres, J. C., and J. D. Ackerman. 2001. "Biogeography of the Antilles Based on a Parsimony Analysis of Orchid Distributions." *Journal of Biogeography* 28:775–94.

Turvey, S. T. 2009. "Holocene Mammal Extinctions." In *Holocene Extinctions*, edited by S. T. Turvey, 41–61. Oxford: Oxford University Press.

Turvey, S. T., F. V. Grady, and P. Rye. 2006. "A New Genus and Species of 'Giant Hutia' (*Tainotherium valei*) from the Quaternary of Puerto Rico: An Extinct Arboreal Quadruped?" *Journal of Zoology* 270:585–94.

Turvey, S., J. Oliver, Y. Narganes Storde, and P. Rye. 2007. "Late Holocene Extinction of Puerto Rican Native Land Mammals." *Biology Letters* 3:193–96.

Turvey, S. T., M. Weksler, E. L. Morris, and M. Nokkert. 2010. "Taxonomy, Phylogeny, and Diversity of the Extinct Lesser Antillean Rice Rats (Sigmodontinae: Oryzomyini), with Description of a New Genus and Species." *Zoological Journal of the Linnean Society* 160:748–72.

Vázquez-Miranda, H., A. G. Navarro-Sigüenza, and J. J. Morrone. 2007. "Biogeographical Patterns of the Avifaunas of the Caribbean Basin Islands: A Parsimony Perspective." *Cladistics* 23:180–200.

Villalobos, F., and A. A. Valerio. 2002. "The Phylogenetic Relationships of the Bat Genus *Sturnira* Gray, 1842 (Chiroptera: Phyllostomidae)." *Mammalian Biology* 67:268–75.

Voss, R. S., and M. Weksler. 2009. "On the Taxonomic Status of *Oryzomys curasoae* Mc-

Farlane and Debrot, 2001, with Remarks on the Phylogenetic Relationships of *O. gorgasi* Hershkovitz, 1971." *Caribbean Journal of Science* 45:1–7.

Wallace, A. R. 1876. *The Geographical Distribution of Animals: With a Study of the Relations of Living and Extinct Faunas as Elucidating the Past Changes of the Earth's Surface*. New York: Harper & Brothers.

Watts, D. 1987. *The West Indies: Patterns of Development, Culture and Environmental Change Since 1492*. Cambridge: Cambridge University Press.

White, J. L., and R. D. E. MacPhee. 2001. "The Sloths of the West Indies: A Systematic and Phylogenetic Overview." In*Biogeography of the West Indies*, edited by C. A. Woods and F. E. Sergile, 201–35. Boca Raton: CRC Press.

Williams, E. E. 1952. "Additional Notes on Fossil and Subfossil Bats from Jamaica." *Journal of Mammalogy* 33:171–79.

Williams, E. E., and K. F. Koopman. 1951. "A New Fossil Rodent from Puerto Rico." *American Museum Novitates* 1515:1–9.

Wilson, D. E., and D. M. Reeder, eds. 2005. *Mammal Species of the World: A Taxonomic and Geographic Reference*, 3rd ed. Baltimore: Johns Hopkins University Press.

Wilson, S. M. 2007. *The Archaeology of the Caribbean*. Cambridge: Cambridge University Press.

Woods, C. A. 1989. "The Biogeography of West Indian Rodents." In *Biogeography of the West Indies: Past, Present, and Future*, edited by C. A. Woods, 741–98. Gainesville: Sandhill Crane Press.

Woods, C. A., R. Borroto, and C. Kilpatrick. 2001. "Insular Patterns and Radiations of West Indian Rodents." In *Biogeography of the West Indies*, edited by C. A. Woods and F. E. Sergile, 335–53. Boca Raton: CRC Press.

Woods, C. A., J. A. Ottenwalder, and W. L. R. Oliver. 1985. "Lost Mammals of the Greater Antilles: The Summarised Findings of a Ten Weeks Field Survey in the Dominican Republic, Haiti and Puerto Rico." *Dodo* 22:23–42.

Woods, C. E., and J. A. Ottenwalder. 1992. *The Natural History of Southern Haiti*. Gainesville: Florida Museum of Natural History.

Wyatt, K. B., P. F. Campos, M. T. P. Gilbert, S.-O. Kolokotronis, W. H. Hynes, R. DeSalle, P. Daszak et al. 2008. "Historical Mammal Extinction on Christmas Island (Indian Ocean) Correlates with Introduced Infectious Disease." *PLoS ONE* 3:e3602.

Yang, Z., and B. Rannala. 2006. "Bayesian Estimation of Species Divergence Times under a Molecular Clock Using Multiple Fossil Calibrations with Soft Bounds." *Molecular Biology and Evolution* 23:212–26.

Yesson, C., and A. Culham. 2006. "Phyloclimatic Modeling: Combining Phylogenetics and Bioclimatic Modeling." *Systematic Biology* 55:785–802.

Appendix 9.1

Table A9.1 Complete taxonomic list, status, and distribution of extinct and extant West Indian mammals. Taxonomy follows contributors to Wilson and Reeder (2005) and Turvey (2009), with additional data from White and MacPhee (2001) and Rega et al. (2002) for sloths, MacPhee and Flemming (2003), Borroto-Páez, Woods, and Kilpatrick (2005), and Turvey et al. (2006, 2010) for rodents, and Dávalos (2006), Larsen et al. (2007), Morgan (2001), and Tejedor (2011) for bats. Named subspecies of *Pteronotus parnellii*, *Chilonatalus micropus*, and *Brachyphylla nana* are treated as distinct based on morphological and molecular data in Dávalos (2004a, 2006), Morgan (2001), and Tejedor (2011). Five bat species found in the Antilles only on Grenada (*Anoura geoffroyi*, *Artibeus glaucus*, *Micronycteris megalotis*, and *Peropteryx macrotis*) are not included in this list. An online version of this table is available at http://sites.google.com/site/lmdavalos/appendix9_1.csv.

ORDER, FAMILY, and *Genus*	SPECIES	ENDEMIC	EXTINCT	DISTRIBUTION († indicates extirpated from an island)
PILOSA				
MEGALONYCHIDAE				
Acratocnus	*odontrigonus*	+	†	Puerto Rico
	antillensis	+	†	Cuba
	ye	+	†	Hispaniola
	simorhynchus	+	†	Hispaniola
Galerocnus	*jaimezi*	+	†	Cuba
Megalocnus	*rodens*	+	†	Cuba
	zile	+	†	Hispaniola, Île de la Tortue
Neocnus	*glirifomis*	+	†	Cuba
	major	+	†	Cuba
	comes	+	†	Hispaniola
	dousman	+	†	Hispaniola
	toupiti	+	†	Hispaniola
Paramiocnus	*riveroi*	+	†	Cuba
Parocnus	*serus*	+	†	Hispaniola, Île de la Tortue, Île de la Gonave
	browni	+	†	Cuba

(continues)

Table A9.1 *(continued)*

ORDER, FAMILY, and *Genus*	SPECIES	ENDEMIC	EXTINCT	DISTRIBUTION († indicates extirpated from an island)
SORICOMORPHA				
NESOPHONTIDAE		+	†	
Nesophontes	*edithae*	+	†	Puerto Rico, Vieques, St. John, St. Thomas
	hypomicrus	+	†	Hispaniola, Île de la Gonave
	major	+	†	Cuba
	micrus	+	†	Cuba, Isle of Pines
	paramicrus	+	†	Hispaniola
	zamicrus	+	†	Hispaniola, Île de la Gonave
	sp. nov. A	+	†	Cayman Brac
	sp. nov. B	+	†	Grand Cayman
SOLENODONTIDAE		+		
Solenodon	*arredondoi*	+	†	Cuba
	cubanus	+		Cuba
	marcanoi	+	†	Hispaniola
	paradoxus	+		Hispaniola, Île de la Gonave
RODENTIA				
CAPROMYIDAE		+		
Capromys	*antiquus*	+	†	Cuba
	arredondoi	+	†	Cuba
	latus	+	†	Cuba
	pappus	+	†	Cuba
	pilorides	+		Cuba, Isle of Pines, other Cuban offshore islands
	robustus	+	†	Cuba
	sp. nov.	+		Cayo Ballentino del Medio (Camaguey, Cuba)

ORDER, FAMILY, and *Genus*	SPECIES	ENDEMIC	EXTINCT	DISTRIBUTION († indicates extirpated from an island)
Geocapromys	*brownii*	+		Jamaica
	columbianus	+	†	Cuba
	ingrahami	+		Acklins, Crooked Island, Middle Caicos, Andros, Cat Island, Eleuthera Island, Great Exuma, Little Exuma, Long Island, New Providence, Ragged Island, Great Abaco, East Plana Cay, San Salvador, Samana Cay
	pleistocenicus	+	†	Cuba
	thoracatus	+	†	Little Swan Island
	sp. nov. A	+	†	Cayman Brac
	sp. nov. B	+	†	Grand Cayman
Hexolobodon	*phenax*	+	†	Hispaniola, Île de la Gonave
gen. nov.? (aff. *Hexolobodon*)	sp. nov.	+	†	Hispaniola
Isolobodon	*montanus*	+	†	Hispaniola, Île de la Gonave
	portoricensis	+	†	Hispaniola, Île de la Gonave, Île de la Tortue, Mona, Guana, Jost van Dyke, Puerto Rico, St. John, St. Thomas, Tortola, Vieques, St. Croix
Mesocapromys	*angelcabrerai*	+		Cayos de Ana Maria (Cuba)
	auritus	+		Cayo Fragoso (Archipielago de Samana, Cuba)
	barbouri	+	†	Cuba
	beatrizae	+	†	Cuba
	delicatus	+	†	Cuba
	gracilis	+	†	Cuba
	kraglievichi	+	†	Cuba
	melanurus	+		Cuba
	minimus	+	†	Cuba
	nanus	+		Cuba, Isle of Pines
	sanfelipensis	+		San Felipe Cays (Cuba)
	silvai	+	†	Cuba

(continues)

Table A9.1 *(continued)*

ORDER, FAMILY, and *Genus*	SPECIES	ENDEMIC	EXTINCT	DISTRIBUTION († indicates extirpated from an island)
Mysateles	*garridoi*	+		Cayo Maya, Cayo Largo, Cayo de la Piedra (Cuba)
	gundlachi	+		Isle of Pines
	jaumei	+	†	Cuba
	meridionalis	+		Isle of Pines (Cuba)
	prehensilis	+		Cuba
Plagiodontia	*aedium*	+		Hispaniola, Île de la Gonave
	araeum	+	†	Hispaniola, Île de la Gonave
	ipnaeum	+	†	Hispaniola
Rhizoplagiodontia	*lemkei*	+	†	Hispaniola
Cricetidae				
gen. nov.?	sp. nov.?	+	†	Anguilla, St. Martin, Tintamarre
gen. nov.?	sp. nov.?	+	†	Barbados
gen. nov.?	sp. nov.?	+	†	Carriacou
gen. nov.?	sp. nov.?	+	†	Grenada
gen. nov.?	sp. nov.?	+	†	Grenada
gen. nov.?	sp. nov.?	+	†	Guadeloupe
gen. nov.?	sp. nov.?	+	†	La Desirade
gen. nov.?	sp. nov.?	+	†	Marie Galante
gen. nov.?	sp. nov.?	+	†	Montserrat
gen. nov.?	sp. nov.?	+	†	Montserrat
gen. nov.?	sp. nov.?	+	†	Saba
gen. nov.? ("*Ekbletomys*")	sp. nov.?	+	†	Antigua, Barbuda
Megalomys	*audreyae*	+	†	Barbuda
	desmarestii	+	†	Martinique
	luciae	+	†	St. Lucia

ORDER, FAMILY, and *Genus*	SPECIES	ENDEMIC	EXTINCT	DISTRIBUTION († indicates extirpated from an island)
Oligoryzomys	*victus*	+	†	St. Vincent
Oryzomys	*antillarum*	+	†	Jamaica
Pennatomys	*nivalis*	+	†	Nevis, St. Eustatius, St. Kitts
ECHIMYIDAE				
Boromys	*offella*	+	†	Cuba
	torrei	+	†	Cuba, Isle of Pines
Brotomys	*contractus*	+	†	Hispaniola
	voratus	+	†	Hispaniola, Île de la Gonave
Heteropsomys	*insulans*	+	†	Puerto Rico, Vieques
Puertoricomys	*corozalus*	+	†	Puerto Rico
HEPTAXODONTIDAE		+	†	
Clidomys	*osborni*	+	†	Jamaica
Elasmodontomys	*obliquus*	+	†	Puerto Rico
Quemisia	*gravis*	+	†	Hispaniola
Xaymaca incertae sedis	*fulvopulvis*	+	†	Jamaica
gen. nov.	sp. nov.	+	†	Jamaica
Tainotherium	*valei*	+	†	Puerto Rico
PRIMATES				
PITHECIIDAE				
Antillothrix	*bernensis*	+	†	Hispaniola
Paralouatta	*varonai*	+	†	Cuba
Xenothrix	*mcgregori*	+	†	Jamaica

(*continues*)

Table A9.1 *(continued)*

ORDER, FAMILY, and *Genus*	SPECIES	ENDEMIC	EXTINCT	DISTRIBUTION († indicates extirpated from an island)
CHIROPTERA				
MOLOSSIDAE				
Eumops	*auripendulus*			Jamaica
	glaucinus			Cuba, Jamaica
	perotis			Cuba
Molossus	*molossus*			Cayman Brac, Cuba, Isle of Pines, Grand Cayman, Hispaniola, Ïle de la Gonave, Jamaica, Culebra, Guana, Puerto Rico, St. John, St. Thomas, Tortola, Vieques, Virgin Gorda, St. Croix, Anguilla, St. Barthelemy, St. Eustatius, St. Martin, Antigua, Barbuda, Barbados, Dominica, Carriacou, Grenada, Union, Guadeloupe, La Desirade, Marie Galante, Martinique, Montserrat, Saba, St. Lucia, St. Vincent, Nevis, St. Kitts
Mormopterus	*minutus*	+		Cuba
Nyctinomops	*laticaudatus*			Cuba
	macrotis			Cuba, Jamaica, Hispaniola
Tadarida	*brasiliensis*			Acklins, Crooked Island, Fortune Island, Middle Caicos†, Eleuthera Island, Great Exuma, Little Exuma, Long Island, New Providence†, Great Abaco, Little Abaco, Cuba, Isle of Pines, Grand Cayman, Hispaniola, Jamaica, Puerto Rico, St. John, Anguilla, St. Barthelemy, St. Eustatius, St. Martin, Antigua, Barbuda, Dominica, Guadeloupe, La Desirade, Martinique, Montserrat, Saba, St. Lucia, St. Vincent, Nevis, St. Kitts
MORMOOPIDAE				
Mormoops	*blainvillei*	+		Little Exuma†, New Providence†, Great Abaco†, Cuba, Hispaniola, Ïle de la Gonave†, Jamaica, Mona, Puerto Rico, Anguilla†, Antigua†, Barbuda†

ORDER, FAMILY, and *Genus*	SPECIES	ENDEMIC	EXTINCT	DISTRIBUTION († indicates extirpated from an island)
	magna	+	†	Cuba
	megalophylla		†	Andros, Great Abaco, Cuba, Hispaniola, Jamaica
Pteronotus	*davyi*			Dominica, Grenada, Marie Galante, Martinique
	macleayii	+		New Providence†, Cuba, Isle of Pines, Jamaica
	parnellii parnellii	+		New Providence†, Great Abaco†, Cuba, Isle of Pines†, Grand Cayman†, Jamaica
	parnellii portoricensis	+		Mona, Puerto Rico, Antigua†
	parnellii pusillus	+		Hispaniola, Île de la Gonave†
	parnellii rubiginosus			St. Vincent
	pristinus	+	†	Cuba
	quadridens	+		Andros†, New Providence†, Great Abaco†, Cuba, Hispaniola, Jamaica, Puerto Rico
	sp. nov.	+	†	Hispaniola
NATALIDAE				
Chilonatalus	*tumidifrons*	+		Andros, Cat Island†, Great Exuma†, New Providence†, Great Abaco, San Salvador
	micropus micropus	+		Hispaniola, Jamaica, Providencia
	micropus macer	+		Cuba, Isle of Pines, Grand Cayman†
Natalus	*stramineus*	+		Anguilla, St. Martin, Antigua, Barbuda, Dominica, Guadeloupe, Marie Galante, Martinique, Montserrat, Saba, Nevis
	jamaicensis	+		Jamaica
	major	+		Hispaniola, Middle Caicos†
	primus	+		Andros†, New Providence†, Great Abaco†, Cuba, Isle of Pines†, Grand Cayman†
Nyctiellus	*lepidus*	+		Andros†, Cat Island, Eleuthera Island, Great Exuma†, Little Exuma, Long Island, Cuba, Isle of Pines

(*continues*)

Table A9.1 (*continued*)

ORDER, FAMILY, and *Genus*	SPECIES	ENDEMIC	EXTINCT	DISTRIBUTION († indicates extirpated from an island)
NOCTILIONIDAE				
Noctilio	*leporinus*			Great Inagua, Cuba, Isle of Pines, Hispaniola, Jamaica, Mona, Culebra, Puerto Rico, St. John, St. Thomas, Vieques, St. Croix, St. Martin, Antigua, Barbuda, Barbados, Dominica, Carriacou, Grenada, Guadeloupe, Marie Galante, Martinique, Montserrat, St. Lucia, St. Vincent, Nevis, St. Kitts
PHYLLOSTOMIDAE				
Ardops	*nichollsi*	+		St. Eustatius, St. Martin, Dominica, Guadeloupe, Marie Galante, Martinique, Montserrat, Saba, St. Lucia, St. Vincent, Nevis, St. Kitts
Ariteus	*flavescens*	+		Jamaica
Artibeus	*jamaicensis*			Providenciales, Great Inagua, Little Inagua, Mayaguana, Cayman Brac, Little Cayman, Cuba, Isle of Pines, Grand Cayman, Hispaniola, Île de la Gonave, Jamaica, Anegada, Culebra, Guana, Puerto Rico, St. John, St. Thomas, Tortola, Vieques, Virgin Gorda, St. Croix, Anguilla, St. Barthelemy, St. Eustatius, St. Martin, Antigua, Barbuda, Barbados, Dominica, Bequia, Carriacou, Grenada, Mustique, Union, Guadeloupe, La Desirade, Marie Galante, Martinique, Montserrat, Saba, St. Lucia, St. Vincent, Nevis, St. Kitts, Providencia, San Andres
	lituratus			St. Vincent
	anthonyi	+	†	Cuba
	schwartzi	+		Barbados, Montserrat, St. Lucia, St. Vincent, Nevis, St. Kitts
	planirostris			St. Vincent
Brachyphylla	*cavernarum*	+		Guana, Puerto Rico, St. John, St. Thomas, St. Croix, Anguilla, St. Barthelemy, St. Eustatius, St. Martin, Antigua, Barbuda, Barbados, Dominica, Guadeloupe, La Desirade, Marie Galante, Martinique, Montserrat, Saba, St. Lucia, St. Vincent, Nevis, St. Kitts
	nana nana	+		Andros†, New Providence†, Cayman Brac†, Cuba, Isle of Pines,
	nana pumila	+		Middle Caicos, Cayman Brac†, Hispaniola, Jamaica†

ORDER, FAMILY, and Genus	SPECIES	ENDEMIC	EXTINCT	DISTRIBUTION († indicates extirpated from an island)
Chiroderma	*improvisum*	+		Guadeloupe, Montserrat
Cubanycteris	*silvai*	+	†	Cuba
Desmodus	*puntajudensis*	+	†	Cuba
Erophylla	*bombifrons*	+		Hispaniola, Puerto Rico
	sezekorni	+		Acklins, Crooked Island, East Caicos, Middle Caicos, North Caicos, Providenciales, Andros, Cat Island, Eleuthera Island, Great Exuma, Little Exuma, Long Island, New Providence, Great Inagua, Grand Bahama, Great Abaco, Mayaguana, East Plana Cay, San Salvador, Cayman Brac, Cuba, Isle of Pines, Grand Cayman, Jamaica
Glossophaga	*longirostris*			Carriacou, Grenada, Union, St. Vincent
	soricina			Jamaica
Macrotus	*waterhousii*	+		Acklins, Crooked Island, East Caicos, Middle Caicos†, North Caicos, Providenciales, Andros, Cat Island, Darby, Eleuthera Island, Great Exuma, Little Exuma, Long Island, New Providence, Great Inagua, Great Abaco, East Plana Cay, San Salvador, Cayman Brac, Little Cayman, Cuba, Isle of Pines, Grand Cayman, Hispaniola, Ile de la Gonave†, Jamaica, Navassa, Puerto Rico†, Anguilla†
Monophyllus	*plethodon*	+		Puerto Rico†, Anguilla, St. Barthelemy, St. Martin, Antigua, Barbuda, Barbados, Dominica, Guadeloupe, Martinique, Montserrat, Saba, St. Lucia, St. Vincent, Nevis, St. Kitts
	redmani	+		Acklins, Crooked Island, Middle Caicos, North Caicos, Providenciales, Andros†, New Providence†, Great Abaco†, Cayman Brac†, Cuba, Isle of Pines, Grand Cayman†, Hispaniola, Ïle de la Gonave†, Jamaica, Puerto Rico
Phyllonycteris	*aphylla*	+		Jamaica
	poeyi	+		New Providence†, Great Abaco†, Cayman Brac†, Cuba, Isle of Pines, Hispaniola
	major	+	†	Puerto Rico

(continues)

Table A9.1 *(continued)*

ORDER, FAMILY, and *Genus*	SPECIES	ENDEMIC	EXTINCT	DISTRIBUTION († indicates extirpated from an island)
Phyllops	*falcatus*	+		Cayman Brac, Cuba, Isle of Pines†, Grand Cayman, Hispaniola
	silvai	+	†	Cuba
	vetus	+	†	Cuba
Stenoderma	*rufum*	+		Puerto Rico, St. John, St. Thomas, Vieques, St. Croix
Sturnira	*lilium*			Dominica, Grenada, Martinique, St. Lucia, St. Vincent
	thomasi	+		Guadeloupe, Montserrat
Tonatia	*saurophila*		†	Jamaica
VESPERTILIONIDAE				
Antrozous	*pallidus*			Cuba
Eptesicus	*fuscus*			Grand Bahama, Great Abaco, San Salvador, Cayman Brac, Cuba, Isle of Pines, Grand Cayman, Hispaniola, Jamaica, Puerto Rico, Dominica
	guadeloupensis	+		Guadeloupe
Lasiurus	*degelidus*	+		Jamaica
	insularis	+		Cuba
	minor	+		Providenciales, Andros, Cat Island, Long Island, New Providence, Great Inagua, Grand Bahama, Mayaguana, Hispaniola, Puerto Rico
	pfeifferi	+		Cuba
	intermedius			Cuba, Isle of Pines, Hispaniola†
Myotis	*dominicensis*	+		Dominica, Guadeloupe
	martiniquensis	+		Martinique, Barbados
	cf. *austroriparius*	+	†	Abaco
Nycticeius	*cubanus*	+		Cuba

Biogeography of Central American Mammals
Patterns and Processes

Ana Laura Almendra and Duke S. Rogers

Abstract

Relative to its land area, Central America contains a disproportionate amount of biodiversity owing to its complex topography and geologic history and its position between the Nearctic and Neotropical realms. Although our understanding of Central American faunas is far from complete, and biogeographic and phylogeographic patterns for mammals are not well-articulated, some conclusions are emerging from analyses of molecular data. For example, the actual biodiversity for mammals, particularly for rodents, is likely much higher than currently documented. The historical events and geographic features that have shaped Central America seem to have affected mammals and other groups in similar fashion. These include dispersal events both prior and subsequent to the permanent land bridge between North and South America; the northern Andean orogeny, in situ divergence both between and within the northern and southern Central American mountain ranges, as well as between Atlantic and Pacific lowlands separated by these highland areas; and the barriers represented by three areas (Isthmus of Tehuantepec, Nicaraguan Depression, and Central Panama) that were submerged for various times in the past. In general, mid- and high-elevation faunas are relatively diverse and contain higher levels of endemism than lowland faunas, although radiations have occurred among both lowland and montane taxa. Rodents exhibit more genetic structure than do bats, ungulates, and primates over comparable geographic sampling. In many cases, estimated levels of molecular divergence correspond to events that occurred in the early Pleistocene or late Pliocene. Unfortunately, continued and rapid change in land-use practices throughout Central America may preclude a complete and accurate reconstruction of the region's historical biogeography.

No phenomenon in the whole realm of nature forced itself earlier upon the notice of man than certain facts of geographic distribution. —C. H. MERRIAM (1892, 3)

10.1 Introduction

Central America extends more than 1500 km from southeastern Mexico to eastern Panama and encompasses habitats ranging from low elevation savanna, semiarid scrub, and humid tropical forests to montane habitats that exceed 4000 m in elevation (Savage 1982) and with no fewer than 15 recognized physiographic provinces (Marshall 2007). This region represents a bridge, both literally and figuratively, between the Nearctic and Neotropical biogeographic realms (Halffter 1987; Marshall and Liebherr 2000; Morrone 2006; Webb 2006). As a result of its location, complex physiography, and climatic fluctuations, Central America is one of the most biologically diverse regions in the world (Marshall and Liebherr 2000; Patterson 2001; Webb 2006). The region supports more than 6% of the world's mammalian diversity (Ceballos 2007; Ceballos, Arroyo-Cabrales, and Medellín 2002; Reid 2009), but represents only 0.4% of its total land surface (Marshall 2007). Furthermore, at least 100 mammal species (~ 30% of the total) are endemic to Central America (Jenkins and Giri 2008; Reid 2009). It has long been known that the number of taxa with South American origins decreases northwards, with the reverse being true for North American species (Merriam 1892). Despite earlier contributions to our understanding of Central American biogeography by Rosen (1978), Halffter (1987), and Savage (1982)—and by Marshall et al. (1982) and Webb (1991, 2006) for mammals in particular—many details regarding the historical processes involved in shaping mammalian distributions and diversification are lacking. Most molecular studies of mammals within the region have focused on estimating genealogical relationships rather than explicitly examining biogeographic hypotheses; they largely preclude rigorous evaluation of alternate scenarios leading to diversification. However, molecular phylogeographic studies generally agree with patterns recovered for other organisms (discussed later). Certainly, molecular studies of mammals demonstrate that the biodiversity within Central America is currently underestimated. This finding is in agreement with Carleton, Sánchez, and Vidales (2002), who predicted that the mammalian diversity in the Central American highlands is likely 30–40% greater than currently understood. Undoubtedly, levels of endemism also are greater than is currently appreciated, as are amounts of evolutionary change (i.e., branch lengths in a phylogenetic tree) expected from biodiversity estimates alone (Sechrest et al. 2002).

In the following, we summarize the geologic history and major events leading to faunal diversification in Central America based primarily on fossil and distributional data for mammals. We then describe phylogeographic patterns of mammalian taxa, restricting our discussion to studies relying on sequence data and incorporating results from other animal and plant groups as appropriate. Finally, we summarize these findings and comment on new directions and methodologies that may be applied to the new and still-developing field of molecular biogeography.

10.2 Geologic Processes and the Great American Biotic Interchange

The paleogeology of Central America is complex, and our understanding is far from complete. In the north, it involved subduction of the Cocos plate beneath the western margin of the Caribbean plate, together with the Motagua-Polochic fault system that transects central Guatemala roughly east to west and demarcates the boundary between the North American (Maya Block) and Caribbean (Chortis Block) plates (Marshall 2007). Tectonic activity has resulted in the formation of the Central Massif, which includes the Sierra Madre de Chiapas and the Maya highlands (Briggs 1994; Halffter 1987; Marshall 2007). However, there is disagreement over whether these events took place during the Paleocene, 65–55 Ma (Moran-Zenteno 1994; Raven and Axelrod 1974), or the Miocene, 5.7–2.2 Ma (Escalante et al. 2007). Southern Central America consists of the Chorotega Block (Costa Rica and western Panama) and the Chocó Block (eastern Panama), which are situated within an area of tectonic interactions among the Caribbean, Cocos, Nazca, and South American plates (Marshall 2007). In this region lies the Cordillera de Talamanca, an area of uplift that stretches across Costa Rica and Panama. This area may have begun its deformation in the late Cretaceous (65 Ma) and continued through most of the Cenozoic as part of the volcanic arc that extended along the Pacific coast from Guatemala to Panama (Coates and Obando 1996). However, major uplifts occurred as a result of the underthrusting of the Cocos Ridge and subduction of the Nazca Plates (Silver et al. 1990), events that occurred during the end of the Miocene and beginning of the Pliocene, 5.7–3 Ma (Abratis and Wörner 2001; Coates and Obando 1996; Gräfe et al. 2002).

The terrestrial and freshwater biotas of South America were isolated from those of North America throughout the Cretaceous until the Pliocene. This ended with the formation of the Panamanian land bridge in the middle Pliocene 3.5–3.1 Ma at the junction of the Pacific and Caribbean plates (Coates and Obando 1996; Coates et al. 2004). These geological events represented an intercontinental corridor for terrestrial dispersal (Briggs 1994; Vermeij 1991; Wallace 1876; Webb 2006), which initiated the episode known as the Great American Biotic Interchange or GABI (Stehli and Webb 1985). Correspondingly, the final closure of the isthmus had oceanographic and climatic effects and coincided with the beginning of the major northern hemisphere glaciations (Cronin and Dowsett 1996). Miller, Fairbanks, and Mountain (1987) estimated a 4°C decline of global temperatures from the mid-Pliocene to the Pleistocene. This resulted in an increase in cool-to-cold-climate plant pollens found in Central America and northern South American palynofloras (Graham 1999). Analysis of pollen and spore flora from central Panama ~3 Ma document the presence of grass species indicative of tropical dry forest or savanna habitats (Graham 1991), and fossil leaves and paleosols from the mid-Miocene in Central Panama indicate a cooler and drier climate than currently exists in the region (Retallack and Kirby 2007). According to Graham (1989), upland habitats in the isthmian region were in place ~2.5 Ma. It is therefore likely that considerable topographic and corresponding habitat diversity existed during, or even prior to, the formation of a permanent land bridge connecting the two continents and may have facilitated dispersal by both tropical savanna and tropical forest-adapted taxa.

The process of connecting North and South America is hypothesized to have begun in the early Oligocene (30 Ma), with volcanic islands along the present position of Central America. Thus, there could have been a connection between western Panama and North America as early as 10–12 Ma (Raven and Axelrod 1974). Whether this corridor was a discontinuous, island arc through present-day lower Central America (Coates and Obando 1996) or a peninsula connecting North America with portions of present-day western Panama (Whitmore and Stewart 1965) remains contentious. However, recent studies support a peninsular configuration extending to central Panama. Kirby and MacFadden (2005) compared mammalian tooth size (as a surrogate for body mass) between counterparts of six fossil species in Panama and Texas. They argued that if central Panama was at one time an island, fossil mammals should exhibit the "island rule" (Van Valen 1973), in which mammals >1 kg in body mass become smaller, whereas those <1 kg evolve to a larger size (Damuth 1993; Lomolino 1985). Results of their study showed no difference in tooth size between taxa found in Texas and Panama, failing to support the archipelago model. Kirby, Jones, and MacFadden (2008) presented lithostratigraphic, biostratigraphic, and chemostratigraphic evidence for a Central American peninsula that existed as early as 19 Ma. Whichever model is correct, it must account for dispersal of taxa prior to the closure of the Panamanian straits. The reason for this is that fossil evidence documents movement of mammals as long ago as the middle Miocene (~9 Ma). For example, *Thinobadistes*, an extinct mylodontid sloth, and

Pliometanastes, a megalonychid sloth, arrived in North America ~9 Ma from South America. Conversely, *Cyonasua*, a large procyonid from North America, appears in the fauna of northwestern Argentina in the middle to late Miocene (9–7 Ma; Marshall 1988; Webb 1991, 2006). Based on fossil evidence, some investigators hypothesize that members of the rodent family Cricetidae entered South America prior to the formation of the Panamanian bridge (Marshall 1979; Woodburne and Swisher 1995). Others suggest that cricetids likely diversified in southern Central America and that once the Panamanian land bridge was completed in the Pliocene (3–2.7 Ma), one or several lineages of cricetids were among the first mammalian groups to enter South America (Pardiñas, D'Elia, and Ortiz 2002; Steppan, Adkins, and Anderson 2004). Recently, Verzi and Montalvo (2008) described late Miocene fossils belonging to the rodent subfamily Sigmodontinae and the carnivore family Mustelidae from the Cerro Azul Formation in Caleufú, Argentina (5.8–5.7 Ma). However, Prevosti and Pardiñas (2009) argued that the age of this site is not well-established and provided evidence that the purported carnivore is in fact a didelphimorph marsupial.

After the formation of the Panamanian land bridge, representatives of the family Tayassuidae migrated into South America and from the Pliocene until the middle Pleistocene (4–1.5 Ma; see Gibbard, Head, and Walker 2010), a large number of taxa colonized South America including members of the Camelidae, Canidae, Cervidae, Equidae, Felidae, Gomphotheriidae (mastodonts), Heteromyidae, Tapiridae, and Ursidae (Marshall et al. 1982; Webb 2006), whereas Pascual (2006) suggested that representatives of the family Heteromyidae entered South America during the Holocene. Pleistocene glaciations are assumed to have modified the biotic patterns (Betancourt, Van Devender, and Martin 1990; Horn 1990), and during the Holocene, four additional families of small mammals (Geomyidae, Leporidae, Sciuridae, and Soricidae) are hypothesized to have entered South America (Marshall et al. 1982).

During the late Pliocene and Pleistocene (4.7–1.8 Ma), the South American "legions" (sensu Marshall et al. 1982)—including members of the families Dasypodidae, Erethizontidae, Glyptodontidae, and Hydrochoeridae—migrated to North America, followed by the families Didelphidae and Megatheriidae in the middle to late Pleistocene (1.8–0.3 Ma). During the Holocene, the families Atelidae, Bradypodidae, Callitrichidae, Cebidae, Choleopodidae, Dasyproctidae, Echimyidae, and Myrmecophagidae are also hypothesized to have migrated from South America northward (Marshall et al. 1982), although molecular evidence suggests earlier entries. However, relatively few families became established north of Central America. Lone members of the families Dasypodidae and Erethizontidae and two species each in the families Atelidae and Didelphidae occur west and north of the Isthmus of Tehuantepec in Mexico. Moreover, three other South American families (Megatheriidae, Glyptodontidae, and Toxodontidae) dispersed into Central America but became extinct during the Pleistocene (Arroyo-Cabrales et al. 2009; Webb 2006).

It is evident that the process of intercontinental colonization is more complex than explained by the GABI alone. Many mammals diversified in northern South America as a result of the Andean orogeny, with east and west lineages (cis- and trans-disjunctions, respectively, following Haffer 1967) that began ~12 Ma (Albert, Lovejoy, and Crampton 2006; Patterson, Solari, and Velazco, chapter 15, this volume), but this cordillera had reached only half its modern elevation by 10 Ma (Gregory-Wodzicki 2000). Therefore, considerable habitat diversity must have existed in northwestern South America since the late Miocene. Ford (2006) proposed three post-land bridge dispersal events from northwestern Colombia into Central America for primates. Similarly, Santos et al. (2009) demonstrated repeated colonization of Central America prior to the formation of the Panamanian isthmus by amphibian lineages that originated in the Amazonian and Chocó regions of South America. This biogeographic hypothesis of recurrent colonization coincides with geological periods of isolation and potential connections between North America and Central America both before and after the formation of the Panamanian land bridge (Coates and Obando 1996). These events are assumed to have led to rapidly changing distributions of the new immigrants and were accompanied by subsequent diversification of the lineages that colonized Central America (Marshall and Liebherr 2000; Webb 2006). Partly as a result of dispersal and subsequent in situ diversification, one-third of the mammalian fauna in Central America is comprised of endemic species. Accordingly, the region is viewed as a hotspot of species richness and endemism (Ceballos and Ehrlich 2006; Jenkins and Giri 2008; Mittermeier et al. 2004; Myers et al. 2000; Sechrest et al. 2002).

10.3 Mexico and Central America

The Isthmus of Tehuantepec represents a biogeographic demarcation between Central America and areas to the north and west in Mexico. This pattern is well-supported for a variety of taxa (see Weir et al. 2008 for a recent summary). A series of papers have examined relationships among montane taxa distributed across the isthmus; these have typically concluded that climatic changes (Toledo 1982), coupled perhaps with the most recent marine incursion (Beard,

Sangree, and Smith 1982; Maldonado-Koerdell 1964), are the vicariant events likely responsible for divergences.

To date, all examples of molecular-based, species-level divergence among mammals associated with the Isthmus of Tehuantepec involve montane rodent taxa that originated in North America. Sullivan, Markert, and Kilpatrick (1997) determined that samples of *Peromyscus aztecus* south and east of the Isthmus of Tehuantepec represented a distinct species-level clade from samples of *P. aztecus* from Mexican Sierra Madre Oriental and Oaxacan Highlands west of the isthmus. Sullivan, Arellano, and Rogers (2000), Hardy et al. (forthcoming), and Arellano, González-Cozátl, and Rogers (2005) likewise recovered what they considered species-level divergence among allopatric samples of *Reithrodontomys sumichrasti* and *R. microdon*, respectively, distributed on either side of the isthmus. Ordóñez-Garza et al. (2010) hypothesized that the Isthmus of Tehuantepec formed a vicariant barrier separating two clades within the *Peromyscus mexicanus* species group and both Rogers et al. (2007) and León-Paniagua et al. (2007) uncovered comparable evidence for species of *Habromys*. Edwards and Bradley (2002) also found what they considered a species-level split for allopatric populations of *Neotoma* on either side of the isthmus based on cytochrome-*b* (the same relationship was not recovered with a nuclear marker in a follow-up study by Longhofer and Bradley 2006). Sullivan, Arellano, and Rogers (2000) determined that the deepest node separating two codistributed taxa (*P. aztecus* and *R. sumichrasti*) from their congeners corresponded geographically to the Isthmus of Tehuantepec. According to Sullivan, Arellano, and Rogers (2000), the amount of sequence divergence was consistent with isolation by an early Pleistocene, trans-isthmus marine barrier (Barber and Klicka 2010). Estimates for the timing of separation between samples of *R. sumichrasti* on either side of the isthmus (3.4–1.8 Ma) are consistent with this hypothesis (Hardy et al., forthcoming). Thus far, the only exception to this general pattern is evidenced by the genus *Glaucomys*. In this instance, samples spanning the Isthmus of Tehuantepec do not approach species-level divergence based on cytochrome-*b* (Kerhoulas and Arbogast 2010) or the mitochondrial control region (Ceballos et al. 2010), prompting Kerhoulas and Arbogast (2010) to propose that occupation of the Sierra Madre de Chiapas by flying squirrels was a relatively recent event.

This scenario also is consistent with the phylogeographic structure recovered for toads (Mulcahy, Morrill, and Mendelsen 2006), suggesting that the isthmus may represent a barrier to taxa that occur in both lowland and montane habitats. However, based on molecular data, Rogers and González (2010) recovered a clade within Desmarest's spiny pocket mouse known from two disjunct localities in Oaxaca and Chiapas, Mexico. Both sites lie within a transition zone between Cloud Forest and Tropical Evergreen Forest (Leopold, 1950) and span the Isthmus of Tehuantepec with only minimal cytochrome-*b* divergence, suggesting that this habitat type may have been continuous across the Isthmus in the recent past. Molecular analyses of rodent species whose origins likely are southern Central America indicate that the Isthmus of Tehuantepec does not represent a barrier. Although sampling was limited, both *Oligoryzomys fulvescens* sensu stricto (Rogers et al. 2009) and *Oryzomys couesi* (Hanson et al. 2010) occur in lowland habitats throughout Central America and northward along the eastern and western coasts of Mexico. Rogers et al. (2009) hypothesized a relatively recent northward dispersal of *Oligoryzomys fulvescens* from Central America to northern Mexico. Lack of genetic divergence across the isthmus for both *Oligoryzomys fulvescens* and *Oryzomys couesi* is best explained by dispersal once the isthmian marine barrier no longer existed (Beard, Sangree, and Smith 1982).

10.4 Northwestern South America and Central America

Portions of Central America and the Pacific coastal region of South America extending to southern Ecuador were first recognized by Hershkovitz (1958) as a separate zoogeographic region. A series of molecular phylogenetic studies have recovered what are considered codistributed species-level clades that link Central America with western (trans-Andean) South America. Examples include the marsupial *Marmosa isthmica* (Gutiérrez, Jansa, and Voss 2010; Rossi, Voss, and Lunde 2010), the monkeys *Alouatta palliata* (Cortés-Ortiz et al. 2003) and *Ateles geoffroyi* (Collins and Dubach 2000a, 2000b), the bats *Artibeus jamaicensis* (Larsen et al. 2007), *Carollia brevicauda*, and *C. castanea* (Hoffman and Baker 2003), *Dermanura rava* (Solari et al. 2009), *Glossophaga soricina* (Hoffman and Baker 2001), *Uroderma bilobatum* (Hoffman, Owen, and Baker 2003), *Vampyressa thyone* (Hoofer and Baker 2006), and the rodents *Orthogeomys dariensis* (Sudman and Hafner 1992) and *Reithrodontomys mexicanus* (Arellano, González-Cozátl, and Rogers 2005).

Other studies have recovered a sister-group relationship between species distributed in western Colombia and Ecuador and those found in Central America. Examples include *Artibeus fraterculus* and *A. hirsutus* (Larsen et al. 2007; Redondo et al. 2008), *Dermanura rosenbergi* and *D. watsoni* (Solari et al. 2009), as well as the *Heteromys anomalus* and *H. desmarestianus* species groups (Rogers and González 2010). This pattern is replicated in other groups such as amphibians (Vallinoto et al. 2010), birds (Cracraft and Prum 1988; Marks, Hackett, and Capparella 2002; Ribas et al. 2005), and snakes (Zamudio and Green 1997), and likely represents relatively more ancient vicariant events that were contemporaneous with the Andean orogeny (Velazco and Patterson 2008). Hanson and Bradley (2008) examined phylogenetic relationships among samples of *Melanomys caliginosus* from Costa Rica, Ecuador, Nicaragua, Panama, and Venezuela. They recovered an Ecuadorian clade that was basal to all other samples, including two *Sigmodontomys* (one sample each from Panama and Ecuador), which formed a sister group to samples of *M. caliginosus* from Panama and Venezuela. If species identifications are correct, then this genealogical pattern renders *Melanomys* paraphyletic and is suggestive of a cis- and trans-Andean split followed by differentiation among samples west of the Andes in Central America and northern South America. However, limited geographic sampling and lack of estimates for the timing of these events preclude additional interpretation.

10.5 Amazon and Central America

Several alternative explanations have been described to account for the pattern of genetic diversity documented among taxa distributed in Amazonia and Central America. The first involves a western Amazon diversification followed by entry into Central America. This pattern has been recovered for the opossums *Marmosa* (Gutiérrez, Jansa, and Voss 2010), *Micoureus* (Patton and Costa 2003), *Philander* (Patton and da Silva 1997); the mice *Oligoryzomys* (Miranda et al. 2008; Rogers et al. 2009); and the monkeys *Ateles* (Collins and Dubach 2000b, 2001), *Alouatta* (Cortés-Ortiz et al. 2003), and *Saimiri* (Lavergne et al. 2010). Precursors of the three primate lineages (*Alouatta*, *Ateles*, and *Saimiri*) are thought to have migrated from South America shortly after the completion of the Panamanian land bridge (Collins and Dubach 2000a, 2000b; Ford 2006; Lavergne et al. 2010). These genera are unrelated to fossil primates found in the Greater Antilles, which comprise a monophyletic group whose closest mainland relative is the South American genus *Callicebus* (Horovitz and MacPhee 1999; MacPhee and Horovitz 2004).

The molecular phylogeny for the genus *Didelphis* by Patton and Costa (2003) can best be explained by two dispersal events into Central America—the first by a precursor to the modern *D. virginiana*, followed by the more recent entry of *D. marsupialis*. Based on more limited sampling, the phylogeographic pattern recovered for *Carollia perspicillata* indicates a South American origin and relatively recent range expansion into Central America (Hoffman and Baker 2003). Baker et al. (1994) showed that *Chiroderma salvini* is sister to the remaining species in the genus, and hypothesized that this could be explained by its isolation in Central America from the common ancestor to the remaining *Chiroderma* species. This latter pattern is similar to that identified by Hanson and Bradley (2008) for *Melanomys caliginosus*, in which Central American samples from Nicaragua and Costa Rica formed a sister group with those from Panama and Venezuela. Based on the phylogenetic reconstruction of relationships among members of the genus *Myotis*, Stadelmann et al. (2007) proposed a complex scenario that involved "early *Myotis* lineages" colonizing South America ~10–7 Ma and subsequently dispersing northward across the Isthmus of Panama.

The phylogenetic relationships among samples of the widely distributed *Desmodus rotundus* recovered by Martins et al. (2009) indicated that Central American populations are not most closely related to vampire bats from northern South America or even the remainder of Amazonia. Instead, Central American samples form a sister group with vampire bats from the Brazilian Pantanal, prompting these authors to hypothesize gene flow along the Andes cordillera. This particular pattern has not been replicated in other taxa. Other molecular data for mammalian taxa support the notion that movement from South America northward occurred primarily after the formation of the Panamanian land bridge. The relatively recent and rapid northward expansion out of South America is a pattern shared by parrots (Eberhard and Bermingham 2004), freshwater fish (Bermingham and Martin 1998; Perdices et al. 2002; Reeves and Bermingham 2006), and caimans (Venegas-Anaya et al. 2008).

Relatively deep molecular divergences have been recovered among species in North or Central American lineages that entered South America. Arellano, González-Cozátl, and Rogers (2005) examined phylogeographic relationships among samples of *Reithrodontomys mexicanus* sensu stricto and recovered the sample from Colombia as basal relative to individuals from Central and Middle America. Gongora et al. (2006) recovered two clades of South American peccaries, which currently are allocated to separate genera. Perini, Russo, and Schrago (2010) proposed that the initial diversification of South American canids (4 Ma) predates the Panamanian land bridge and proposed that these two closely related lineages entered South America. In turn, these lineages gave rise to an extant fauna that includes five endemic genera of canids (see also Eizirik, chapter 7, this volume). Rogers and González (2010) confirmed monophyly of the *Heteromys anomalus* species group, a basal clade within the genus that is distributed primarily in northern South America. Sequence divergence between the *H. anomalus* group and other basal clades in the genus support a single Pliocene entry into South America and thus might predate completion of the Panamanian land bridge, as suggested by Engel et al. (1998) for sigmodontine rodents. Regardless of the timing of these events, it is clear that entry into South America by many mammalian groups occurred earlier than was assumed previously.

10.6 Divergence within Central America

Some species endemic to Central America exhibit considerable geographic structure in DNA sequence data. The phylogeographic patterns recovered indicate in situ isolation of lowland taxa by the Sierra Madre de Chiapas and Central Massif and the Cordillera de Talamanca (fig. 10.1), separation of montane taxa that occur in both of these upland areas (as well as isolation within the complex Cordillera de Talamanca), and separation of lowland taxa by a marine incursion in southern Nicaragua. For example, Cortés-Ortiz et al. (2003) determined that occupation of Mesoamerica by *Alouatta* coincides with formation of the Panamanian land bridge and the 3 Ma split between the two Mesoamerica *Alouatta* clades. Lower sea levels ~2.5 Ma (Haq, Hardenbol, and Vail 1987) likely facilitated range expansions of these taxa. Baumgarten and Williamson (2007) contended that ancestral populations of *Alouatta* were isolated by a cooling period that separated the Yucatan Peninsula (*A. pigra*) from lowlands to the south (*A. palliata*); this vicariant event was driven by the Sierra Madre de Chiapas and the Maya highlands (fig. 10.1), and also may be responsible for isolating one of two clades of *Marmosa mexicana* recovered by Gutiérrez, Jansa, and Voss (2010). In addition, Baumgarten and Williamson (2007) argued that sea level increases ~2 Ma (Bermingham and Martin 1998; Perdices et al. 2002) may have reinforced separation of *A. pigra* and *A. palliata* (Ford 2006). Rogers and Vance (2005) documented a deep split among samples of *Liomys salvini* from the dry forests along the Pacific versant of Chiapas, Mexico, compared to samples from similar habitats in Honduras and Costa Rica. They also confirmed the sister group relationship between *L. salvini* and *L. adspersus*, the latter species known only from savanna habitat in central Panama. The relatively deep phylogenetic split between *L. salvini* and *L. adspersus* is consistent with the proposition that wet and dry forest habitats have existed in southern Central America since the late Miocene or early Pliocene (Crawford, Bermingham, and Polania 2007).

Differentiation among rodents in Central America has been extensive. Hardy et al. (forthcoming) recovered two well-differentiated clades of *R. sumichrasti* corresponding to the Central Massif and the Talamancan range in Costa Rica

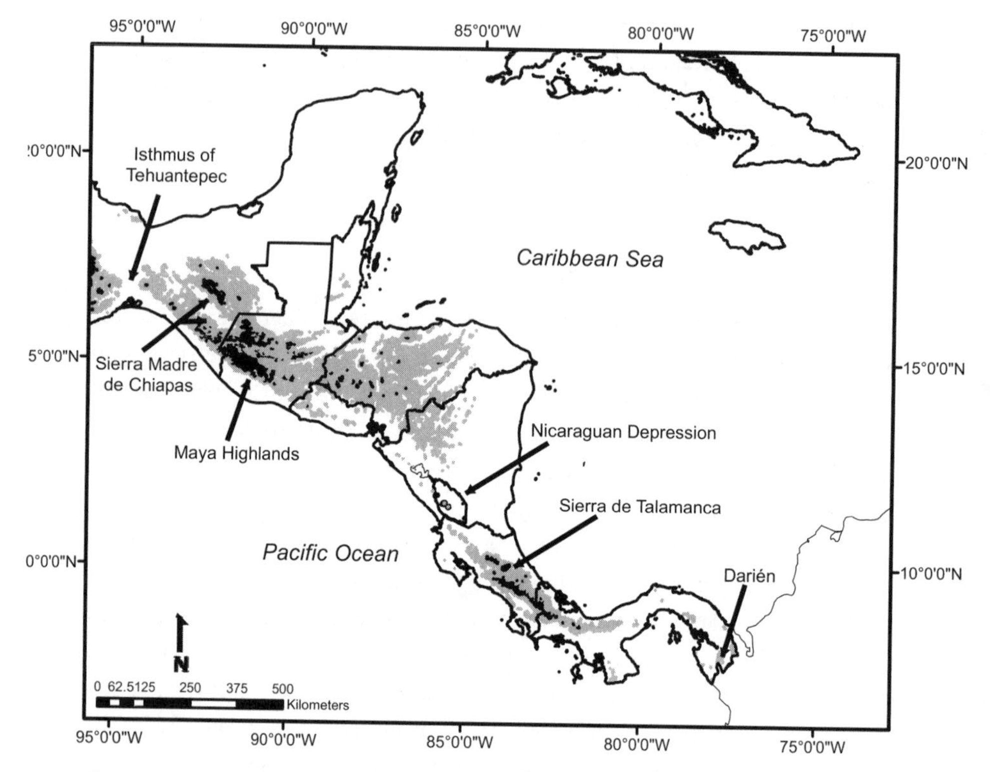

Figure 10.1 Map of Central America illustrating geologic features discussed in text. Elevation ranges are white (<800 m), gray (800–2000 m), and black (>2000 m). Stippled areas indicate lakes Managua and Nicaragua.

and Panama. Rogers et al. (2009) determined that Central American populations of *Oligoryzomys fulvescens* (sensu stricto) and the Costa Rican-southern Nicaraguan endemic *Oligoryzomys vegetus* were sister taxa. Cladogeneis in these groups may have resulted from a marine gap (Nicaraguan Depression) during the Miocene and most of the Pliocene (Coates and Obando 1996; Iturralde-Vincent 2006). The Nicaraguan Depression currently serves as a physiographic break for a variety of vertebrate, insect, and plant taxa (Castoe et al. 2009 and references therein).

Rogers and González (2010) evaluated phylogenetic relationships within the broadly distributed lowland rodent species *Heteromys desmarestianus*, which they recovered as paraphyletic. They documented three possible species within this taxon; one from the Atlantic drainage of Costa Rica, another from the Pacific slopes of Costa Rica and Panama, and a third from the Darién region of eastern Panama (fig. 10.1). Each of these candidate species corresponded to a different physiographic province as delimited by Marshall (2007). Well-supported genetic subdivisions also were recovered within *H. desmarestianus* sensu stricto, including separation of northern Central American samples from those in Costa Rica (Rogers and González 2010), a pattern reminiscent of that described by Hoffman and Baker (2003) for the bat *Carollia sowelli*.

Although limited in geographic sampling, other strictly molecular phylogenetic studies have uncovered a series of species-level rodent taxa in lower Central America. Arellano, González-Cozátl, and Rogers (2005) recognized *Reithrodontomys cherrii* (formerly a subspecies of *R. mexicanus*) as a deeply divergent clade from the Cordillera de Talamanca, Costa Rica, with affinities to the *R. tenuirostris* species group. The species-level status of *R. cherrii* was confirmed by Gardner and Carleton (2009), based on detailed examination of morphological evidence. Miller and Engstrom (2008) identified two undescribed species of *Reithrodontomys*, one from the Cerro de la Carpintera and another from Volcán Poas in Costa Rica. Rogers and Gonzalez (2010) confirmed the species status of *Heteromys nubicolens*, a species known only from the Cordillera de Tilarán and Cordillera de Guanacaste, Costa Rica, and whose sister taxon, *H. oresterus*, occurs to the south in the Cordillera de Talamanca (Anderson and Timm 2006). At least for rodents, it appears that vicariant events driven by climatic oscillations during the Pleistocene (or earlier) were sufficient to drive speciation. Panama's Darién region (fig. 10.1) likely was isolated from South America until ~13 to 7 Ma (Coates et al. 2004) and from Central America until the formation of the Panamanian land bridge. As such, the Darién is regarded as a separate physiographic province by Marshall (2007). A series of species-level splits have been identified within several rodent taxa distributed in eastern Panama, compared with populations in western Panama and Costa Rica. These include *Melanomys caliginosus* (Hanson and Bradley 2008) and *Heteromys desmarestianus* and *H. australis* (Rogers and González 2010).

A pattern similar to that found for *Alouatta pigra* and *A. palliata* also was recovered for mammal species whose distributions are not restricted to Central America. Bradley, Henson, and Durish (2008) determined that *Sigmodon toltecus* (generally distributed north of the Central American Highland Massif) and *S. hirsutus* (southern Central America and northern South America) were sister taxa. Hanson et al. (2010) evaluated genealogical relationships among samples of *Oryzomys couesi* from Mexico and Central America. Four species-level clades were identified, one each from the Atlantic and Pacific versants in northern Central America, a third from the Atlantic coast of Costa Rica, and a fourth from the Pacific coast of central Panama. Hoffman, Owen, and Baker (2003) documented isolation between two chromosomal races of *Uroderma bilobatum* from Central America: one generally distributed on the Pacific slopes of southern Mexico, Guatemala, and El Salvador, and the other found throughout the rest of Central America as well as western Colombia and Ecuador. Hoffman, Owen, and Baker (2003) estimated that their isolation occurred in the Pleistocene (0.9–0.2 Ma). This distribution pattern is identical to the sister-group relationship for *Carollia sowelli* and *C. subrufa* identified by Hoffman and Baker (2003), although *C. sowelli* is not known to occur in South America. In addition, Hoffman and Baker (2003) identified a well-supported phylogenetic split between populations of *C. sowelli* from northern Central America compared to those sampled from Costa Rica and Panama. Taken together, these genetic data indicate that multiple species-level clades exist within taxa that would not have been recovered based solely on morphological data.

10.7 Summary and Prospectus

The impacts of GABI are relatively well understood for mammals compared to other groups (Marshall et al. 1982; Vrba 1992), due in major part to their relatively abundant fossil record. Unfortunately, there are few molecular studies of mammals designed specifically to decipher phylogeographic patterns within Central America compared to other vertebrates (Patten and Smith-Patten 2008). The primary focus for most of these molecular studies has been phylogenetic reconstruction. As a result, detailed geographic sampling and estimates of divergence times among clades often are lacking. Despite these drawbacks, molecular studies have been useful in identifying some biogeographic (or phylogeographic) patterns within Central American mammals. In general, nonvolant small mammals exhibit greater genetic diversity over comparable geographic areas than do bats, primates, or artiodactyls. This pattern of relatively low levels of intraspecific divergence among larger mammals and bats is comparable to that of birds (Ditchfield and Burns 1998) and is not attributable to differences in rate of mitochondrial DNA evolution among mammalian groups (Ditchfield 2000). Rather, differences in vagility apparently explain this marked pattern among taxa. In general, levels of molecular differentiation between closely related taxa are consistent with biogeographic boundaries delimited by a variety of nonmolecular methods. These barriers include the Sierra Madre de Chiapas and Maya highlands from the Talamancan Range and the Nicaraguan Depression (Halffter 1987; Luna-Vega et al. 2001; Patten and Smith-Patten

2008; Rosen 1978), as well as eastern Panama (Coates 1997) and the Western Andean Cordillera (Patten and Smith-Patten 2008). For nonmammals, species-level biodiversity seems associated with tectonic and climatic events that predate the Pleistocene or even the Pliocene (Castoe et al. 2009). Whether or not this pattern holds for mammals should be tested rigorously, but preliminary results support species-level diversification occurring during the Pleistocene or late Pliocene, contra Savage (2002). Overall, results from a handful of molecular studies of Central American mammals have documented extensive in situ diversification that is driven by vicariant events. These findings are concordant with similar investigations for a variety of vertebrate (García-Moreno et al. 2006 and references therein) and plant taxa (Novick et al. 2003) and underscores the need for detailed sampling throughout Central America to fully appreciate its mammalian biodiversity.

The majority of molecular studies for mammals have used one or several mitochondrial markers. Although mitochondrial sequences offer advantages such as relatively rapid coalescence times and generally lack the problem of reticulation, these phylogenetic estimates represent gene trees rather than species trees, which can potentially be problematic (Degnan and Rosenberg 2006; Pamilo and Nei 1988). A subset of studies have used both mitochondrial and nuclear gene segments (Collins and Dubach 2000a, 2001; Cortés-Ortiz et al. 2003; Gongora et al. 2006; Hanson et al. 2010; Martins et al. 2009; Miller and Engstrom 2008; Redondo et al. 2008; Rogers and González 2010; Stadelmann et al. 2007; Velazco and Patterson 2008) to estimate phylogenetic relationships. However, the nuclear sequences employed typically yielded fewer phylogenetically informative characters (and less resolved genealogies) than gene trees obtained from mitochondrial sequence data; when concatenated with mitochondrial sequences in a combined evidence approach, the resulting trees tended to reflect clades based on mitochondrial sequences alone (Rogers and González 2010; Stadelmann et al. 2007).

Central America is experiencing some of the highest deforestation rates in the world. Recent studies estimate that only 20% of the original forested vegetation remains intact (Mittermeier et al. 2004). Moreover, only 12.6% of the land area is afforded some level of environmental protection, and only 3% is under protection that prevents alteration of native vegetation (Jenkins and Giri 2008). Unfortunately, the locations and sizes of these reserves apparently were selected without first obtaining data for species richness, biodiversity, distribution, or dispersal requirements of the mammals that were the focus of the conservation effort. As a result, mammals with small ranges (and most vulnerable to extinction) are found largely outside reserves. Given that species ranges within Central America generally are not well-known and often are fragmented, incorporating inferential methods that provide predictive information of geographic distribution such as ecological niche modeling (Graham et al. 2004; Peterson, Soberón, and Sánchez-Cordero 1999) are essential and can be especially useful in prioritizing conservation areas (Esselman and Allan 2010). Paleoclimate and future climate change models can be combined with molecular phylogeographic studies (Solomon et al. 2008) and coalescent simulations (Carstens and Richards 2007) to infer speciation events and address conservation issues. Incorporation of highly variable nuclear markers (Carstens and Knowles 2006; Shaffer and Thomson 2007; Thomson, Wang, and Johnson 2010) would result in more accurate estimates of a group's evolutionary history. This is particularly true for phylogeographic studies due to the relatively shallow genealogical patterns typically recovered. Likewise, recent developments in estimating species trees (Degnan and Rosenberg 2009; Heled and Drummond 2010; Liu et al. 2009), even with substantial incongruence among individual gene trees (Knowles 2009) and incomplete lineage sorting (Carstens and Knowles 2007b; McCormack, Huang, and Knowles 2009), promise to revolutionize our understanding of biological processes that have shaped evolutionary history (Knowles 2009) of mammals in the region. Sequence data from multiple, unlinked loci also should be employed to estimate divergence times more precisely using Bayesian MCMC (Markov chain Monte Carlo methods) or maximum likelihood analyses of molecular sequences (Carstens and Knowles 2007a; Drummond and Rambaut 2007; Lemmon and Lemmon 2008; Pyron 2010) and other statistical phylogeographic approaches. Relatively few molecular studies of Central American mammals were designed to test a priori biogeographic or phylogeographic hypotheses (e.g., Sullivan, Arellano, and Rogers 2000). Fortunately, new approaches in molecular phylogenetics (Johnson and Crandall 2009; Riddle et al. 2008) together with recent advances in methods and modeling techniques (Carstens, Stoute, and Reid 2009; Richards, Carstens, and Knowles 2007) have enabled investigators to develop biologically realistic phylogeographic hypotheses, even in the absence of a well-corroborated fossil record for the group under study. Finally, detailed phylogeographic studies should incorporate well-justified sampling designs together with an emphasis on analysis of ecological components (Buckley 2009). Studies such as those conducted by Robertson and Zamudio (2009) and Robertson, Duryea, and Zamudio (2009) should serve as templates for examining mammalian systems in this incredibly biodiverse region.

Acknowledgments

We thank the organizers of IMC-10, especially Bruce D. Patterson and Leonora Pires Costa, for the opportunity to present this summary. Nicole Lewis-Rogers and three anonymous reviewers read this chapter and provided useful comments.

Literature Cited

Abratis, M., and G. Wörner. 2001. "Ridge Collision, Slab-Window Formation, and the Flux of Pacific Asthenosphere into the Caribbean Realm." *Geology* 29:127–30.

Albert, J. S., N. R. Lovejoy, and W. G. R. Crampton. 2006. "Miocene Tectonism and the Separation of Cis- and Trans-Andean River Basins: Evidence from Neotropical Fishes." *Journal of South American Earth Sciences* 21:14–27.

Anderson, R. P., and R. M. Timm. 2006. "A New Montane Species of Spiny Pocket Mouse (Rodentia: Heteromyidae: *Heteromys*) from Northwestern Costa Rica." *American Museum Novitates* 3509:1–38.

Arellano, E., F. X. González-Cozátl, and D. S. Rogers. 2005. "Molecular Systematics of Middle American Harvest Mice *Reithrodontomys* (Muridae), Estimated from Mitochondrial Cytochrome *b* Gene Sequences." *Molecular Phylogenetics and Evolution* 37:529–40.

Arroyo-Cabrales, J., O. J. Polaco, E. Johnson, and I. Ferrusquía-Villafranca. 2009. "A Perspective on Mammal Biodiversity and Zoogeography in the Late Pleistocene of México." *Quaternary International* 212:187–97.

Baker, R. J., V. A. Taddei, J. L. Hudgeons, and R. A. Van Den Bussche. 1994. "Systematic Relationships Within *Chiroderma* (Chiroptera: Phyllostomidae) Based on Cytochrome *b* Sequence Variation." *Journal of Mammalogy* 75:321–27.

Barber, B. R., and J. Klicka. 2010. "Two Pulses of Diversification across the Isthmus of Tehuantepec in a Montane Mexican Bird Fauna." *Proceedings of the Royal Society of London, B, Biological Sciences* 277:2675–81.

Baumgarten, A., and G. B. Williamson. 2007. "The Distributions of Howling Monkeys (*Alouatta pigra* and *A. palliata*) in Southeastern Mexico and Central America." *Primates* 48:310–15.

Beard, J. H., J. B. Sangree, and L. A. Smith. 1982. "Quaternary Chronology, Paleoclimate, Depositional Sequences, and Eustatic Cycles." *American Association of Petroleum Geologists Bulletin* 66:158–69.

Bermingham, E., and A. P. Martin. 1998. "Comparative mtDNA Phylogeography of Neotropical Freshwater Fishes: Testing Shared History to Infer the Evolutionary Landscape of Lower Central America." *Molecular Ecology* 7:499–517.

Betancourt, J. L., T. R. Van Devender, and P. S. Martin. 1990. *Packrat Middens: The Last 40,000 Years of Biotic Change*. Tucson: University of Arizona Press.

Bradley, R. D., D. D. Henson, and N. D. Durish. 2008. "Re-evaluation of the Geographic Distribution and Phylogeography of the *Sigmodon hispidus* Complex Based on Mitochondrial DNA Sequences." *Southwestern Naturalist* 53:301–10.

Briggs, J. C. 1994. "The Genesis of Central America: Biology versus Geophysics." *Global Ecology and Biogeography Letters* 4:169–72.

Buckley, D. 2009. "Toward an Organismal, Integrative, and Iterative Phylogeography." *BioEssays* 31:784–93.

Carleton, M. D., O. Sánchez, and G. U. Vidales. 2002. "A New Species of *Habromys* (Muriodea: Neotominae) from México, with Generic Review of Species Definitions and Remarks on Diversity Patterns among Mesoamerican Small Mammals Restricted to Humid Montane Forests." *Proceedings of the Biological Society of Washington* 115:488–533.

Carstens, B. C., and L. L. Knowles. 2006. "Variable Nuclear Markers for *Melanoplus oregonensis* Identified from Screening of a Genomic Library." *Molecular Ecology Notes* 6:683–85.

———. 2007a. "Estimating Species Phylogeny from Gene-Tree Probabilities Despite Incomplete Lineage Sorting: An Example from *Melanoplus* Grasshoppers." *Systematic Biology* 56:400–11.

———. 2007b. "Shifting Distributions and Speciation: Species Divergence During Rapid Climate Change." *Molecular Ecology* 16:619–27.

Carstens, B. C., and C. L. Richards. 2007. "Integrating Coalescent and Ecological Niche Modeling in Comparative Phylogeography." *Evolution* 61:1439–54.

Carstens, B. C., H. N. Stoute, and N. H. Reid. 2009. "An Information-Theoretical Approach to Phylogeography." *Molecular Ecology* 18:4270–82.

Castoe, T. A., J. M. Daza, E. N. Smith, M. M. Sasa, U. Kuch, J. A. Campbell, P. T. Chippindale, and C. L. Parkinson. 2009. "Comparative Phylogeography of Pitvipers Suggests a Consensus of Ancient Middle American Highland Biogeography." *Journal of Biogeography* 36:88–103.

Ceballos, G. 2007. "Conservation Priorities for Mammals in Megadiverse Mexico: The Efficiency of Reserve Networks." *Ecological Applications* 17:569–78.

Ceballos, G., J. Arroyo-Cabrales, and R. A. Medellín. 2002. "The Mammals of Mexico: Composition, Distribution, and Conservation Status." *Occasional Papers, Museum of Texas Tech University* 218:1–27.

Ceballos, G., and P. R. Ehrlich. 2006. "Global Mammal Distributions, Biodiversity Hotspots, and Conservation." *Proceedings of the National Academy of Sciences, USA* 103: 19374–79.

Ceballos, G., P. Manzano, F. M. Méndez-Harclerode, M. L. Haynie, D. H. Walker, and R. D. Bradley. 2010. "Geographic Distribution, Genetic Diversity, and Conservation Status of the Southern Flying Squirrel (*Glaucomys volans*) in México." *Occasional Papers, Museum of Texas Tech University* 299:1–15.

Coates, A. G. 1997. "The Forging of Central America." In *Central America: A Natural and Cultural History*, edited by A.G. Coates, 1–37. New Haven: Yale University Press.

Coates, A. G., L. S. Collins, M.-P. Aubry, and W. A. Berggren. 2004. "The Geology of the Darien, Panama, and the Late Miocene-Pliocene Collision of the Panama Arc with Northwestern South America." *Geological Society of America Bulletin* 116:1327–44.

Coates, A. G., and J. A. Obando. 1996. "The Geologic Evolution of the Central American Isthmus." In *Evolution and Environment in Tropical America*, edited by J. B. C. Jackson, A. F. Budd, and A. G. Coates, 21–56. Chicago: University of Chicago Press.

Collins, A. C., and J. M. Dubach. 2000a. "Biogeographic and Ecological Forces Responsible for Speciation in *Ateles*." *International Journal of Primatology* 21:421–44.

———. 2000b. "Phylogenetic Relationships Among Spider Monkeys (*Ateles*) Based on Mitochondrial DNA Variation." *International Journal of Primatology* 21:381–420.

———. 2001. "Nuclear DNA Variation in Spider Monkeys (*Ateles*)." *Molecular Phylogenetics and Evolution* 19:67–75.

Cortés-Ortiz, L., E. Bermingham, C. Rico, E. Rodríguez-Luna, I. Sampaio, and M. Ruiz-García. 2003. "Molecular Systematics and Biogeography of the Neotropical Monkey Genus, *Alouatta*." *Molecular Phylogenetics and Evolution* 26:64–81.

Cracraft, J., and R. O. Prum. 1988. "Patterns and Processes of Diversification: Speciation and Historical Congruence in Some Neotropical Birds." *Evolution* 42:603–20.

Crawford, A. J., E. Bermingham, and C. Polania. 2007. "The Role of Tropical Dry Forest as a Long-Term Barrier to Dispersal: A Comparative Phylogeographical Analysis of Dry Forest Tolerant and Intolerant Frogs." *Molecular Ecology* 16:4789–807.

Cronin, T. M., and H. J. Dowsett. 1996. "Biotic and Oceanographic Response to the Pliocene Closing of the Central American Isthmus." In *Evolution and Environment in Tropical America*, edited by J. B. C. Jackson, A. F. Budd, and A. G. Coates, 76–104. Chicago: University of Chicago Press.

Damuth, J. 1993. "Cope's Rule, the Island Rule and the Scaling of Mammalian Population Density." *Nature* 365:748–50.

Degnan, J. H., and N. A. Rosenberg. 2006. "Discordance of Species Trees with Their Most Likely Gene Trees." *PLOS Genetics* 2:762–68.

———. 2009. "Gene Tree Discordance, Phylogenetic Inference, and the Multispecies Coalescent." *Trends in Ecology and Evolution* 24:332–40.

Ditchfield, A. D. 2000. "The Comparative Phylogeography of Neotropical Mammals: Patterns of Intraspecific Mitochondrial DNA Variation Among Bats Contrasted with Nonvolant Small Mammals." *Molecular Ecology* 9:1307–18.

Ditchfield, A. D., and K. Burns. 1998. "DNA Sequences Reveal Phylogeographic Similarities of Neotropical Bats and Birds." *Journal of Comparative Biology* 3:165–70.

Drummond, A. J., and A. Rambaut. 2007. "BEAST: Bayesian Evolutionary Analysis of Sampling Trees." *BMC Evolutionary Biology* 7:214.

Eberhard, J. R., and E. Bermingham. 2004. "Phylogeny and Biogeography of the *Amazona ochrocephala* (Aves: Psittacidae) Complex." *Auk* 121:318–32.

Edwards, C. W., and R. D. Bradley. 2002. "Molecular Systematics and Historical Phylobiogeography of the *Neotoma mexicana* Species Group." *Journal of Mammalogy* 83: 20–30.

Engel, S. R., K. M. Hogan, J. F. Taylor, and S. K. Davis. 1998. "Molecular Systematics and Paleobiogeography of South American Sigmodontine Rodents." *Molecular Biology and Evolution* 15:35–49.

Escalante, T., G. Rodríguez, N. Cao, M. C. Ebach, and J. J. Morrone. 2007. "Cladistic Biogeographic Analysis Suggests an Early Caribbean Diversification in Mexico." *Naturwissenschaften* 94:561–65.

Esselman, P. C., and J. D. Allan. 2010. "Application of Species Distribution Models and

Conservation Planning Software to the Design of a Reserve Network for the Riverine Fishes of Northeastern Mesoamerica." *Freshwater Biology* 56:71–88.

Ford, S. M. 2006. "The Biogeographic History of Mesoamerican Primates." In *New Perspectives in the Study of Mesoamerican Primates: Distribution, Ecology, Behavior and Conservation*, edited by A. Estrada, P. A. Garber, M. S. M. Pavelka, and L. Luecke, 81–114. New York: Springer.

García-Moreno, J., A. G. Navarro-Sigüenza, A. T. Peterson, and L. A. Sánchez-González. 2006. "Genetic Variation Coincides with Geographic Structure in the Common Bush-Tanager (*Chlorospingus opthalmicus*) Complex from Mexico." *Molecular Phylogenetics and Evolution* 33:186–96.

Gardner, A. L., and M. D. Carleton. 2009. "A New Species of *Reithrodontomys*, Subgenus *Aporodon* (Cricetidae: Neotominae), from the Highlands of Costa Rica, with Comments on Costa Rican and Panamanian *Reithrodontomys*." In *Systematic Mammalogy: Contributions in Honor of Guy G. Musser*, edited by R. S. Voss and M. D. Carleton, 157–82. New York: *Bulletin of the American Museum of Natural History*, vol. 331.

Gibbard, P. L., M. L. Head, and M. J. C. Walker. 2010. "Formal Ratification of the Quaternary System/Period and the Pleistocene Series/Epoch with a Base at 2.58 Ma." *Journal of Quaternary Science* 25:96–102.

Gongora, J., S. Morales, J. E. Bernal, and C. Moran. 2006. "Phylogenetic Divisions Among Collared Peccaries (*Pecari tajacu*) Detected Using Mitochondrial and Nuclear Sequences." *Molecular Phylogenetics and Evolution* 41:1–11.

Gräfe, K., W. Frisch, I. M. Villa, and M. Meschede. 2002. "Geodynamic Evolution of Southern Costa Rica Related to Low-Angle Subduction of the Cocos Ridge: Constraints from Thermochronology." *Tectonophysics* 348:187–204.

Graham, A. 1989. "Late Tertiary Paleoaltitudes and Vegetational Zonation in Mexico and Central America." *Acta Botanica Neerlandica* 38:417–24.

———. 1991. "Studies in Neotropical Paleobotany. X. The Pliocene Communities of Panama—Composition, Numerical Representation, and Paleocommunity Paleoenvironmental Reconstruction." *Annals of the Missouri Botanical Garden* 78:465–75.

———. 1999. "The Tertiary History of the Northern Temperate Element in the Northern Latin America Biota." *American Journal of Botany* 86:32–38.

Graham, C. H., S. Ferrier, F. Huettman, C. Moritz, and A. T. Peterson. 2004. "New Developments in Museum-Based Informatics and Applications in Biodiversity Analysis." *Trends in Ecology and Evolution* 19:497–503.

Gregory-Wodzicki, K. M. 2000. "Uplift History of the Central and Northern Andes: A Review." *Geological Society of America Bulletin* 112:1091–105.

Gutiérrez, E. E., S. A. Jansa, and R. S. Voss. 2010. "Molecular Systematics of Mouse Opossums (Didelphidae: *Marmosa*): Assessing Species Limits Using Mitochondrial DNA Sequences, with Comments on Phylogenetic Relationships and Biogeography." *American Museum Novitates* 3692:1–22.

Haffer, J. 1967. "Speciation in Colombian Forest Birds West of the Andes." *American Museum Novitates* 2294:1–57.

Halffter, G. 1987. "Biogeography of the Montane Entomofauna of Mexico and Central America." *Annual Review of Entomology* 32:95–114.

Hanson, J. D., and R. D. Bradley. 2008. "Molecular Diversity within *Melanomys caliginosus* (Rodentia: Oryzomyini): Evidence for Multiple Species." *Occasional Papers, Museum of Texas Tech University* 275:1–11.

Hanson, J. D., J. L. Indorf, V. J. Swier, and R. D. Bradley. 2010. "Molecular Divergence within the *Oryzomys palustris* Complex: Evidence for Multiple Species." *Journal of Mammalogy* 91:336–47.

Haq, B. U., J. Hardenbol, and P. R. Vail. 1987. "Chronology of Fluctuating Sea Levels Since the Triassic." *Science* 235:1156–67.

Hardy, D. K., F. X. González-Cózatl, E. Arellano, and D. S. Rogers. Forthcoming. "Molecular Phylogeographic Structure and Phylogenetics of Sumichrast's Harvest Mouse (*Reithrodontomys sumichrasti*: Family Cricetidae) Based on Mitochondrial and Nuclear DNA Sequences." *Molecular Phylogenetics and Evolution*.

Heled, J., and A. J. Drummond. 2010. "Bayesian Inference of Species Trees from Multilocus Data." *Molecular Biology and Evolution* 27:570–80.

Hershkovitz, P. 1958. "A Geographic Classification of Neotropical Mammals." *Fieldiana, Zoology* 36:583–620.

Hoffman, F. G., and R. J. Baker. 2001. "Systematics of the Genus *Glossophaga* (Chiroptera: Phyllostomidae) and Phylogeography in *Glossophaga soricina* Based on the Cytochrome-*b* Gene." *Journal of Mammalogy* 82:1092–101.

———. 2003. "Comparative Phylogeography of Short-Tailed Bats (*Carollia*: Phyllostomidae)." *Molecular Ecology* 12:3403–14.

Hoffman, F. G., J. G. Owen, and R. J. Baker. 2003. "mtDNA Perspective of Chromosomal Diversification and Hybridization in Peters' Tent-Making Bat (*Uroderma bilobatum*: Phyllostomidae)." *Molecular Ecology* 12:2981–93.

Hoofer, S. R., and R. J. Baker. 2006. "Molecular Systematics of Vampyressine Bats (Phyllostomidae: Stenodermatinae) with Comparison of Direct and Indirect Surveys of Mitochondrial DNA Variation." *Molecular Phylogenetics and Evolution* 39:424–38.

Horn, S. P. 1990. "Timing of Deglaciation in the Cordillera de Talamanca, Costa Rica." *Climate Research* 1:81–83.

Horovitz, I., and R. D. E. MacPhee. 1999. "The Quaternary Cuban Platyrrhine *Paralouatta varonai* and the Origin of the Antillean Monkeys." *Journal of Human Evolution* 36:33–68.

Iturralde-Vincent, M. 2006. "Meso-Cenozoic Caribbean Paleogeography: Implications for the Historical Biogeography of the Region." *International Geology Review* 48:791–827.

Jenkins, C. N., and C. Giri. 2008. "Protection of Mammal Diversity in Central America." *Conservation Biology* 22:1037–44.

Johnson, J. B., and C. A. Crandall. 2009. "Expanding the Toolbox for Phylogeographic Analysis." *Molecular Ecology* 18:4137–39.

Kerhoulas, N. J., and B. S. Arbogast. 2010. "Molecular Systematics and Pleistocene Biogeography of Mesoamerican Flying Squirrels." *Journal of Mammalogy* 91:654–67.

Kirby, M. X., D. S. Jones, and B. J. MacFadden. 2008. "Lower Miocene Stratigraphy Along the Panama Canal and its Bearing on the Central American Peninsula." *PLoS One* 3:e2791.

Kirby, M. X., and B. J. MacFadden. 2005. "Was Southern Central America an Archipelago or a Peninsula in the Middle Miocene? A Test Using Land-Mammal Body Size." *Palaeogeography, Palaeoclimatology, Palaeoecology* 228:193–202.

Knowles, L. L. 2009. "Estimating Species Trees: Methods of Phylogenetic Analysis When There is Incongruence across Genes." *Systematic Biology* 58:463–67.

Larsen, P. A., S. R. Hoofer, M. C. Bozeman, S. C. Pedersen, H. H. Genoways, C. J. Phillips, D. E. Pumo, and R. J. Baker. 2007. "Phylogenetics and Phylogeography of the *Artibeus jamaicensis* Complex Based on Cytochrome-*b* DNA Sequences." *Journal of Mammalogy* 88:712–27.

Lavergne, A., M. Ruiz-García, F. Catzeflis, S. Lacote, H. Contamin, O. E. Mercereau-Puijalon, C. LaCoste, and B. De Thoisy. 2010. "Phylogeny and Phylogeography of Squirrel Monkeys (Genus *Saimiri*) Based on Cytochrome *b* Genetic Analysis." *American Journal of Primatology* 72:242–53.

Lemmon, A. R., and E. M. Lemmon. 2008. "A Likelihood Framework for Estimating Phylogeographic History on a Continuous Landscape." *Systematic Biology* 57:544–61.

León-Paniagua, L., A. G. Navarro-Sigüenza, B. E. Hernández-Baños, and J. C. Morales. 2007. "Diversification of the Arboreal Mice of the Genus *Habromys* (Rodentia: Cricetidae: Neotominae) in the Mesoamerican Highlands." *Molecular Phylogenetics and Evolution* 42:653–64.

Leopold, A. S. 1950. "Vegetation Zones of Mexico." *Ecology* 31:507–18.

Liu, L., L. Yu., L. Kubatko, D. K. Pearl, and S. C. Edwards. 2009. "Coalescent Methods for Estimating Phylogenetic Trees." *Molecular Phylogenetics and Evolution* 53:320–28.

Lomolino, M. V. 1985. "Body Size of Mammals on Islands: The Island Rule Reexamined." *American Naturalist* 125:310–16.

Longhofer, L. K., and R. D. Bradley. 2006. "Molecular Systematics of the Genus *Neotoma* Based on DNA Sequences from Intron 2 of the Alcohol Dehydrogenase Gene." *Journal of Mammalogy* 87:961–70.

Luna-Vega, I., J. J. Morrone, O. A. Ayala, and D. E. Organista. 2001. "Biogeographical Affinities Among Neotropical Cloud Forests." *Plant Systematics and Evolution* 228:229–39.

MacPhee, R. D. E., and I. Horovitz. 2004. "New Craniodental Remains of the Quaternary Jamaican Monkey *Xenothrix mcgregroi* (Xenotrichini, Callicebinae, Pitheciidae), with a Reconsideration of the *Aotus* Hypothesis." *American Museum Novitates* 3434:1–51.

Maldonado-Koerdell, M. 1964. "Geohistory and Paleogeography of Middle America." In *Handbook of Middle American Indians: Natural Environment and Early Cultures*, edited by R. Wauchope and R. C. West, 3–32. Austin: University of Texas Press.

Marks, B. D., S. J. Hackett, and A. P. Capparella. 2002. "Historical Relationships Among Neotropical Lowland Forest Areas of Endemism as Determined by Mitochondrial DNA Sequence Variation within the Wedge-Billed Woodcreeper (Aves: Dendrocolaptidae: *Glyphorynchus spirurus*)." *Molecular Phylogenetics and Evolution* 24:153–67.

Marshall, C. J., and J. K. Liebherr. 2000. "Cladistic Biogeography of the Mexican Transition Zone." *Journal of Biogeography* 27:203–16.

Marshall, J. S. 2007. "The Geomorphology and Physiographic Provinces of Central America." In *Central America: Geology, Resources, and Hazards*, edited by J. Bundschuh and G. Alvarado, 75–122. London: Taylor and Francis.

Marshall, L. G. 1979. "A Model for Paleobiogeography of South American Cricetine Rodents." *Paleobiology* 5:126–32.

———. 1988. "Land Mammals and the Great American Interchange." *American Scientist* 76:380–88.

Marshall, L. G., S. D. Webb, J. J. Sepkoski Jr., and D. M. Raup. 1982. "Mammalian Evolution and the Great American Interchange." *Science* 215:1351–57.

Martins, F. M., A. R. Templeton, A. C. O. Pavan, B. C. Kohlbach, and J. S. Morgante. 2009. "Phylogeography of the Common Vampire Bat (*Desmodus rotundus*): Marked Population Structure, Neotropical Pleistocene Vicariance and Incongruence Between Nuclear and mtDNA Markers." *BMC Evolutionary Biology* 9:294–307.

McCormack, J. E., H. Huang, and L. L. Knowles. 2009. "Maximum Likelihood Estimates of Species Trees: How Accuracy of Phylogenetic Inference Depends Upon the Divergence History and Sampling Design." *Systematic Biology* 58:501–08.

Merriam, C. H. 1892. "The Geographic Distribution of Life in North America with Special Reference to the Mammalia." *Proceedings of the Biological Society of Washington* 7:1–64.

Miller, J. R., and M. D. Engstrom. 2008. "The Relationships of Major Lineages within Peromyscine Rodents: A Molecular Phylogenetic Hypothesis and Systematic Reappraisal." *Journal of Mammalogy* 89:1279–95.

Miller, K. G., R. G. Fairbanks, and G. S. Mountain. 1987. "Tertiary Oxygen Isotope Synthesis, Sea Level History, and Continental Margin Erosion." *Paleoceanography* 2:1–19.

Miranda, G. B., L. F. B. Oliveira, J. Andrades-Miranda, A. Langguth, S. M. Callegari-Jacques, and M. S. Mattevi. 2008. "Phylogenetics and Phylogeographic Patterns in Sigmodontine Rodents of the Genus *Oligoryzomys*." *Journal of Heredity* 100:309–21.

Mittermeier, R. A., P. R. Gil, M. Hoffman, J. D. Pilgrim, T. M. Brooks, C. G. Mittermeier, J. Lamoreux, and G. A.B. da Fonseca. 2004. *Hotspots Revisited: Earth's Biologically Richest and Most Endangered Terrestrial Ecoregions*. Monterrey: CEMEX.

Moran-Zenteno, D. J. 1994. "Geology of the Mexican Republic." *American Association of Petroleum Geologists Studies in Geology* 39:1–160.

Morrone, J. J. 2006. "Biogeographic Areas and Transition Zones of Latin America and the Caribbean Islands Based on Panbiogeographic and Cladistic Analyses of the Entomofauna." *Annual Review of Entomology* 51:467–94.

Mulcahy, D. G., B. H. Morrill, and J. R. Mendelsen III. 2006. "Historical Biogeography of Lowland Species of Toads (*Bufo*) across the Trans-Mexican Neovolcanic Belt and the Isthmus of Tehuantepec." *Journal of Biogeography* 33:1889–904.

Myers, N., R. A. Mittermeier, C. G. Mittermeier, G. A. B. da Fonseca, and J. Kent. 2000. "Biodiversity Hotspots for Conservation Priorities." *Nature* 403:853–58.

Novick, R. R., C. W. Dick, M. R. Lemes, C. Navarro, A. Caccone, and E. Bermingham. 2003. "Genetic Structure of Mesoamerican Populations of Big-Leaf Mahogany (*Swietenia macrophylla*) Inferred from Microsatellite Analysis." *Molecular Ecology* 12:2875–83.

Ordóñez-Garza, N., J. O. Matson, R. E. Strauss, R. D. Bradley, and J. Salazar-Bravo. 2010.

"Patterns of Phenotypic and Genetic Variation in Three Species of Endemic Mesoamerican *Peromyscus* (Rodentia: Cricetidae)." *Journal of Mammalogy* 91:848–59.

Pamilo, P., and M. Nei. 1988. "Relationships Between Gene Trees and Species Trees." *Molecular Biology and Evolution* 5:568–83.

Pardiñas, U. F. J., G. D. D'Elia, and P. E. Ortiz. 2002. "Sigmodontinos Fósiles (Rodentia, Muroidea, Sigmodontinae) de América del Sur: Estado Actual de su Conocimiento y Prospectiva." *Mastozoología Neotropical* 9:209–52.

Pascual, R. 2006. "Evolution and Geography: The Biogeographic History of South American Land Mammals." *Annals of the Missouri Botanical Garden* 93:209–30.

Patten, M. A., and B. D. Smith-Patten. 2008. "Biogeographical Boundaries and Monmonier's Algorithm: A Case Study in the Northern Neotropics." *Journal of Biogeography* 35:407–16.

Patterson, B. D. 2001. "Fathoming Tropical Biodiversity: The Continuing Discovery of Neotropical Mammals." *Diversity and Distributions* 7:191–96.

Patton, J. L., and L. P. Costa. 2003. "Molecular Phylogeography and Species Limits in Rainforest Didelphid Marsupials of South America." In *Predators with Pouches: The Biology of Carnivorous Marsupials*, edited by M. Jones, C. Dickman, and M. Archer, 63–81. Sydney: CSIRO Publishing.

Patton, J. L., and M. N. F. da Silva. 1997. "Definition of Species of Pouched Four-Eyed Opossums (Didelphidae, *Philander*)." *Journal of Mammalogy* 78:90–102.

Perdices, A., E. Birmingham, A. Montilla, and I. Doadrio. 2002. "Evolutionary History of the Genus *Rhamdia* (Teleostei: Pimelodidae) in Central America." *Molecular Phylogenetics and Evolution* 25:172–89.

Perini, F. A., C. A. M. Russo, and C. G. Schrago. 2010. "The Evolution of South American Endemic Canids: A History of Rapid Diversification and Morphological Parallelism." *Journal of Evolutionary Biology* 23:311–22.

Peterson, A. T., J. Soberón, and V. Sánchez-Cordero. 1999. "Conservation of Ecological Niches in Evolutionary Time." *Science* 285:1265–67.

Prevosti, F. J., and U. F. J. Pardiñas. 2009. "Comment on 'The Oldest South American Cricetidae (Rodentia) and Mustelidae (Carnivora): Late Miocene Faunal Turnover in Central Argentina and the Great Biotic Interchange' by D. H. Verzi and C. I. Montalvo [*Palaeogeography, Palaeoclimatology, Palaeoecology*, 267 (2008) 284–291]." *Palaeogeography, Palaeoclimatology, Palaeoecology* 280:543–47.

Pyron, R. A. 2010. "A Likelihood Method for Assessing Molecular Divergence Time Estimates and the Placement of Fossil Calibrations." *Systematic Biology* 59:185–94.

Raven, P. H., and D. I. Axelrod. 1974. "Angiosperm Biogeography and Past Continental Movements." *Annals of the Missouri Botanical Garden* 61:539–673.

Redondo, R. A. F., L. P. S. Brina, R. F. Silva, A. D. Ditchfield, and F. R. Santos. 2008. "Molecular Systematics of the Genus *Artibeus* (Chiroptera: Phyllostomidae)." *Molecular Phylogenetics and Evolution* 49:44–58.

Reeves, R. G., and E. Bermingham. 2006. "Colonization, Population Expansion, and Lineage Turnover: Phylogeography of Mesoamerican Characiform Fish." *Biological Journal of the Linnean Society* 88:235–55.

Reid, F. A. 2009. *A Field Guide to the Mammals of Central America and Southeast Mexico*, 2nd ed. New York: Oxford University Press.

Retallack, G. J., and M. X. Kirby. 2007. "Middle Miocene Global Change and Paleogeography of Panama." *PALAIOS* 22:667–79.

Ribas, C. C., R. Gaban-Lima, C. Y. Miyaki, and J. Cracraft. 2005. "Historical Biogeography and Diversification within the Neotropical Parrot Genus *Pionopsitta* (Aves: Psittacidae)." *Journal of Biogeography* 32:1409–27.

Richards, C. L., B. C. Carstens, and L. L. Knowles. 2007. "Distribution Modeling and Statistical Phylogeography: An Integrative Framework for Generating and Testing Alternative Biogeographic Hypotheses." *Journal of Biogeography* 34:1833–45.

Riddle, B. R., M. N. Dawson, E. A. Hadly, D. J. Hafner, M. J. Hickerson, S. J. Mantooth, and A. D. Yoder. 2008. "The Role of Molecular Genetics in Sculpting the Future of Integrative Biogeography." *Progress in Physical Geography* 32:173–202.

Robertson, J. M., M. C. Duryea, and K. M. Zamudio. 2009. "Discordant Patterns of Evolutionary Differentiation in Two Neotropical Treefrogs." *Molecular Ecology* 18:1375–95.

Robertson, J. M., and K. R. Zamudio. 2009. "Genetic Diversification, Vicariance, and Selection in a Polytypic Frog." *Journal of Heredity* 100:715–31.

Rogers, D. S., E. A. Arenas, F. X. González-Cózatl, D. K. Hardy, J. D. Hanson, and N. Lewis-Rogers. 2009. "Molecular Phylogenetics of *Oligoryzomys fulvescens* Based on Cytochrome *b* Gene Sequences, with Comments on the Evolution of the Genus *Oligoryzomys*." In *60 Años de la Colección Nacional de Mamíferos del Instituto de Biología, UNAM: Aportaciones al Conocimiento y Conservación de los Mamíferos Mexicanos*, edited by F. A. Cervantes, J. Vargas-Cuenca, and Y. Hortelano-Moncada, 179–92. Mexico City: Universidad Autonoma de Mexico.

Rogers, D. S., C. C. Funk, J. R. Miller, and M. D. Engstrom. 2007. "Molecular Phylogenetic Relationships Among Crested-Tailed Mice (Genus *Habromys*)." *Journal of Mammalian Evolution* 14:37–55.

Rogers, D. S., and M. W. González. 2010. "Phylogenetic Relationships Among Spiny Pocket Mice (*Heteromys*) Inferred from Mitochondrial and Nuclear Sequence Data." *Journal of Mammalogy* 91:914–30.

Rogers, D. S., and V. L. Vance. 2005. "Phylogenetics of Spiny Pocket Mice (Genus *Liomys*): Analysis of Cytochrome *b* Based on Multiple Heuristic Approaches." *Journal of Mammalogy* 86:1085–94.

Rosen, D. E. 1978. "Vicariant Patterns and Historical Explanation in Biogeography." *Systematic Zoology* 27:159–88.

Rossi, R. V., R. S. Voss, and D. P. Lunde. 2010. "A Revision of the Didelphid Marsupial Genus *Marmosa*. Part 1. The Species in Tate's '*mexicana*' and '*mitis*' Sections and Other Closely Related Forms." *Bulletin of the American Museum of Natural History* 334:1–83.

Santos, J. C., L. A. Coloma, K. Summers, J. P. Caldwell, and R. Ree. 2009. "Amazonian Amphibian Diversity is Primarily Derived from Late Miocene Andean Lineages." *PLoS Biology* 7:0448–61.

Savage, J. M. 1982. "The Enigma of the Central American Herpeotfauna: Dispersals or Vicariance?" *Annals of the Missouri Botanical Garden* 69:464–547.

———. 2002. *The Amphibians and Reptiles of Costa Rica: A Herpetofauna Between Two Continents, Between Two Seas*. Chicago: University of Chicago Press.

Sechrest, W., T. M. Brooks, G. A. B. da Fonseca, W. R. Konstant, R. A. Mittermeier, A. Purvis, A. B. Rylands, and J. L. Gittleman. 2002. "Hotspots and the Conservation of Evolutionary History." *Proceedings of the National Academy of Sciences, USA* 99:2067–71.

Shaffer, H. B., and R. C. Thomson. 2007. "Delimiting Species in Recent Radiations." *Systematic Biology* 56:896–906.

Silver, E. A., D. L. Reed, J. E. Tagudin, and D. J. Heil. 1990. "Implications of the North and South Panama Thrust Belts for the Origin of the Panama Orocline." *Tectonics* 9:261–81.

Solari, S., S. R. Hoofer, P. A. Larsen, A. D. Brown, R. J. Bull, J. A. Guerrero, J. Ortega, J. P. Carrera, R. D. Bradley, and R. J. Baker. 2009. "Operational Criteria for Genetically Defined Species: Analysis of the Diversification of the Small Fruit-Eating Bats, *Dermanura* (Phyllostomidae: Sternodermatinae)." *Acta Chiropterologica* 11:279–88.

Solomon, S. E., M. Bacci Jr., J. Martins Jr., G. G. Vinha, and U. G. Mueller. 2008. "Paleodistributions and Comparative Molecular Phylogeography of Leafcutter Ants (*Atta* spp.) Provide New Insight into the Origins of Amazonian Diversity." *PLoS One* 3:e2738.

Stadelmann, B., L.-K. Lin, T. H. Kunz, and M. Ruedi. 2007. "Molecular Phylogeny of New World *Myotis* (Chiroptera, Vespertilionidae) Inferred from Mitochondrial and Nuclear DNA Genes." *Molecular Phylogenetics and Evolution* 43:32–48.

Stehli, F. G., and S. D. Webb, eds. 1985. *The Great American Biotic Interchange*. New York: Plenum Press.

Steppan, S. J., R. M. Adkins, and J. Anderson. 2004. "Phylogeny and Divergence-Date Estimates of Rapid Radiations in Muroid Rodents Based on Multiple Nuclear Genes." *Systematic Biology* 53:533–53.

Sudman, P. D., and M. S. Hafner. 1992. Phylogenetic Relationships Among Middle American Pocket Gophers (Genus *Orthogeomys*) Based on Mitochondrial DNA Sequences." *Molecular Phylogenetics and Evolution* 1:17–25.

Sullivan, J., E. Arellano, and D. S. Rogers. 2000. "Comparative Phylogeography of Mesoamerican Highland Rodents: Concerted Versus Independent Response to Past Climatic Fluctuations." *American Naturalist* 155:755–68.

Sullivan, J., J. A. Markert, and C. W. Kilpatrick. 1997. "Phylogeography and Molecular Systematics of the *Peromyscus aztecus* Species Group (Rodentia: Muridae) Inferred Using Parsimony and Likelihood." *Systematic Biology* 46:426–40.

Thomson, R. C., I. J. Wang, and J. R. Johnson. 2010. "Genome-Enabled Development of DNA Markers for Ecology, Evolution and Conservation." *Molecular Ecology* 19:2184–95.

Toledo, V. M. 1982. "Pleistocene Changes of Vegetation in Tropical Mexico." In *Biological Diversification in the Tropics*, edited by G. T. Prance, 93–111. New York: Columbia University Press.

Vallinoto, M., F. Sequeira, D. Sodré, J. A. R. Bernardi, I. Sampaio, and H. Schneider. 2010. "Phylogeny and Biogeography of the *Rhinella marina* Species Complex (Amphibia, Bufonidae) Revisited: Implications for Neotropical Diversification Hypotheses." *Zoologica Scripta* 39:128–40.

Van Valen, L. 1973. "Pattern and the Balance of Nature." *Evolutionary Theory* 1:31–49.

Velazco, P. M., and B. D. Patterson. 2008. "Phylogenetics and Biogeography of the Broad-Nosed Bats, Genus *Platyrrhinus* (Chiroptera: Phyllostomidae)." *Molecular Phylogenetics and Evolution* 49:749–59.

Venegas-Anaya, M., A. J. Crawford, A. H. E. Galván, O. I. Sanjur, L. D. Densmore III, and E. Bermingham. 2008. "Mitochondrial DNA Phylogeography of *Caiman crocodilus* in Mesoamerica and South America." *Journal of Experimental Biology* 309:1–14.

Vermeij, G. J. 1991. "When Biotas Meet: Understanding Biotic Interchange." *Science* 253:1099–104.

Verzi, D. H., and C. I. Montalvo. 2008. "The Oldest South American Cricetidae (Rodentia) and Mustelidae (Carnivora): Late Miocene Faunal Turnover in Central Argentina and the Great American Biotic Interchange." *Palaeogeography, Palaeoclimatology, Palaeoecology* 267:284–91.

Vrba, E. S. 1992. "Mammals as a Key to Evolutionary Theory." *Journal of Mammalogy* 73:1–28.

Wallace, A. R. 1876. *The Geographical Distribution of Animals: With a Study of the Relations of Living and Extinct Faunas as Elucidating the Past Changes of the Earth's Surface*. New York: Harper and Brothers.

Webb, S. D. 1991. "Ecogeography and the Great American Interchange." *Paleobiology* 17: 266–80.

———. 2006. "The Great American Biotic Interchange: Patterns and Processes." *Annals of the Missouri Botanical Garden* 93:245–57.

Weir, J. J., E. Bermingham, M. J. Miller, J. Klicka, and M. A. González. 2008. "Phylogeography of a Morphologically Diverse Neotropical Montane Species, the Common Bush-Tanager (*Chlorospingus ophthalmicus*)." *Molecular Phylogenetics and Evolution* 47:650–64.

Whitmore, F. C., Jr., and R. H. Stewart. 1965. "Miocene Mammals and Central American Seaways." *Science* 148:180–85.

Woodburne, M. O, and C. C. Swisher III. 1995. "Land Mammal High-Resolution Geochronology, Intercontinental Overland Dispersals, Sea Level, Climate, and Vicariance." In *Geochronology, Time-Scales, and Global Stratigraphic Correlations: Unified Temporal Framework for an Historical Geology*, edited by W. A. Berggren, D. V. Kent, M.-P. Aubry, and J. Hardenbol, 335–64. Tulsa: Society for Sedimentary Geology Special Publication 54.

Zamudio, K. R., and H. W. Greene. 1997. "Phylogeography of the Bushmaster (*Lachesis muta*: Viperidae): Implications for Neotropical Biogeography, Systematics, and Conservation." *Biological Journal of the Linnean Society* 62:421–42.

11 Biogeography of Mammals from the Guianas of South America

Burton K. Lim

Abstract

The Guianas of northern South America consist of French Guiana, Suriname, Guyana, and southeastern Venezuela (states of Delta Amacuro, Bolivar, and Amazonas). This region is bounded on the south by the mountains that separate the northward coastal drainage from the Amazon basin, and covers almost 1 million km.[2] It therefore excludes a portion of the Guiana Shield that is here considered as part of the Amazon basin.

There are 284 species representing 12 orders of terrestrial mammals currently documented from the Guianas. Over half of the species (147) are bats, with rodents accounting for about 20 percent and other orders each representing less than 10 percent of the mammalian species. More than one-third (105) of the species are found in all six political units (countries or states) and are considered widely distributed in the region. There are 16 species endemic to the Guianas, of which 9 are rodents, 5 are bats, and 2 are marsupials. Six of the endemic species occur only in upland areas >500 m in elevation, whereas 10 species occur only in lowland areas. Of the lowland endemics, 7 species are restricted to the western portion, 2 restricted to the eastern portion, and 1 is widely distributed. The Guiana plateau (>500 m elevation) is the most prominent biogeographic feature. It has influenced diversification of mammals in this region by fostering endemism and functioning as a geographic barrier. Within South America, the ancient Guiana Shield has acted as a stable core area for range expansions from both the Andes and the Amazon during periods of environmental change beginning in the Miocene.

11.1 Introduction

One of the world's few remaining pristine wilderness areas that still has contiguous tracts of intact rain forest and high levels of biodiversity is located in the Guianas of northern South America (Mittermeier et al. 1998). It is mostly identifiable by the countries lying wholly within this region and associated with the old English colonial name of Guiana, including present-day Guyana (British Guiana), Suriname (Dutch Guiana), and French Guiana (overseas territory of France), but there is also Spanish Guayana in southeastern Venezuela, in-

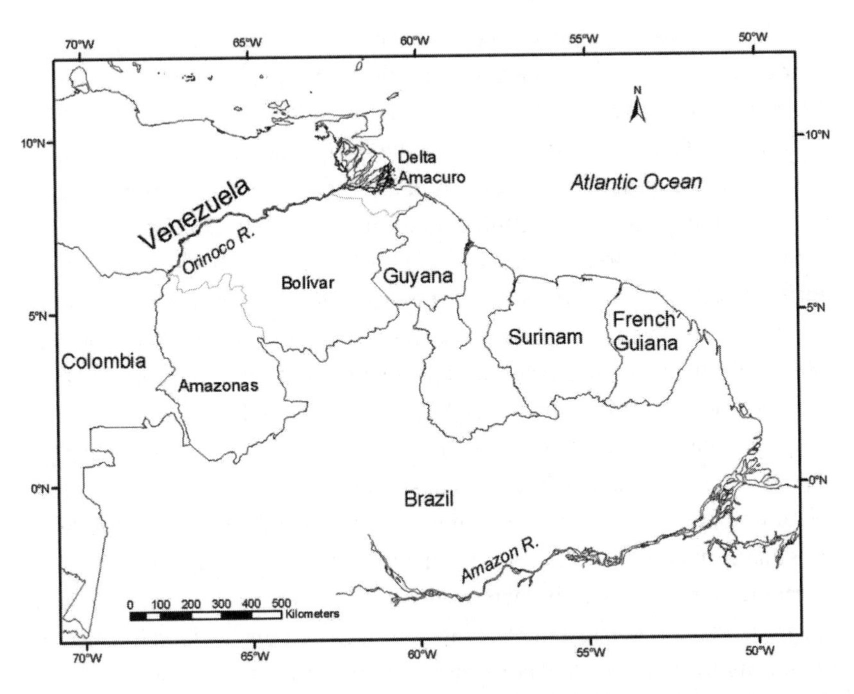

Figure 11.1 Map of the Guianas in northern South America, including southeastern Venezuela (states of Amazonas, Bolivar, and Delta Amacuro), Guyana, Suriname, and French Guiana.

cluding the states of Amazonas, Bolivar, and Delta Amacuro (fig. 11.1). The region is geographically delimited by the major north-draining rivers of the Orinoco, Essequibo, Courantyne, Maroni, and Oyapock. The Orinoco and Essequibo Rivers bisect Venezuela and Guyana, respectively. The Courantyne River forms the border between Guyana and Suriname, the Maroni River separates Suriname and French Guiana, and the Oyapock separates French Guiana and Brazil. The southern watershed is formed by the mountains of the Parima, Pakaraima, Acarai, and Tumucumac, from west to east, that separate the Guianas from Brazil and the Amazon drainage basin. The Guianas comprise nearly 1 million km², with the area split almost equally between the three states of Venezuelan Guayana and the three Guianan countries (Guyana, Suriname, and French Guiana). Broader definitions that encompass the Guianas are the zoogeographical Guiana subregion extending south into Brazil east of the Rio Negro and north of the Amazon River (Voss and Emmons 1996), and the even more inclusive geological Guiana Shield extending further west into Colombia to Serranía Chiribiquete (Gibbs and Barron 1993). For the purposes of this regional synopsis of the modern Neotropical mammal fauna, I am using the more restricted definition of the Guianas based on the northern coastal watershed to differentiate it from and to minimize overlap with Amazonia.

The Guianas are emerging as one of the last frontiers of undisturbed natural habitat in the tropics, primarily because they are relatively undeveloped, have a low human population, and minimal infrastructure. Comparatively little is known of this region because of the attention attracted by the larger neighboring Amazon basin to the south. Rain forest predominates in the Guianas, but there is also substantial savanna vegetation. The three largest blocks of grasslands in descending order of area are the Llanos of Venezuela; the contiguous Rupununi-Rio Branco-Gran Sabana of Guyana, Brazil, and Venezuela (respectively); and the Sipaliwini-Paru savanna on the Suriname-Brazil border. There is also savanna paralleling the coast of the Guianas near major areas of human habitation. The most notable topological feature in the region is the Guiana uplands, which is a plateau beginning at approximately 500 m in elevation characterized by highland flat-topped mountains (tepuis) beginning at approximately 1500 m and reaching 3000 m. There are lowland forested regions roughly southwest and northeast of the plateau, referred to as the Negro-Amazonian forest and the east Guiana forest by Tate (1939). These regions correspond to the western, eastern, and central phytogeographic provinces of Huber (2006), and I follow his elevational limits of lowlands <500 m, uplands between 500 and 1500 m, and highlands (or pantepui) >1500 m. My objectives here are to summarize our current knowledge of Guianan mammals from taxonomic, distributional, and biogeographic perspectives.

11.2 Taxonomic Composition

There are 284 species representing 12 orders of terrestrial mammals currently documented from the Guianas (appendix 11.1). Over half (147 species) are bats, with rodents accounting for about 20 percent (58 species), and other orders each representing less than 10 percent of the mammal species (23 species of opossums, 17 carnivores, 15 primates, 6 armadillos, 6 sloths and anteaters, 5 artiodactyls, 2 manatees, 2 rabbits, 2 dolphins, and 1 tapir). In terms of overall diversity, the Guianas account for approximately 17% of the 1657 species of Neotropical mammals (Wilson and Reeder 2005). However, some orders are better represented than others. Both species of manatees and freshwater dolphins are found in the Guianas, as are 6 of the 10 species of anteaters and sloths. However, of the more speciose orders, bats are the best represented, at 45% of the species from all 9 families found in the Neotropics. In contrast, only 7% of the Neotropical rodents occur in the Guianas, with 11 of 18 families absent. Other orders represented include one-third of the tapirs, 29% of the armadillos, 25% of the opossums, 22% of the carnivores, 20% of the artiodactyls, 12.5% of the rabbits, and 12% of the primates. Of the three remaining orders of Neotropical mammals, there are no species of shrews and no species of microbiotherian or paucituberculate marsupials found in the Guianas.

The most recent summary of the mammalian biodiversity from the Guianas reported 282 species (Lim, Engstrom, and Ochoa 2005), but did not fully incorporate the taxonomy of the latest edition of "Mammal Species of the World" (Wilson and Reeder 2005). Updates accounting for the currently recognized 284 species include taxonomic changes such as the descriptions of *Philander deltae* and *P. mondolfi* as new species distinct and allopatrically distributed from *P. opossum* (Lew, Pérez-Hernández, and Ventura 2006). *Philander deltae* was originally listed as an undescribed species endemic to Delta Amacuro by Lim, Engstrom, and Ochoa (2005b); however, its type locality is in the eastern portion of the neighboring state Monagas, and the species is more broadly distributed onto the alluvial plains of the Orinoco Delta and the Gulf of Paria to the west. *Philander mondolfii* is distributed from the Venezuelan Guayana to the Andean foothills of Colombia (Lew, Pérez-Hernández, and Ventura 2006). This revised taxonomy further supports the earlier suggestion that *P. opossum* is a widely distributed composite and that the nominal taxon should be restricted to the eastern Guianas (Guyana, Suriname, and French Guaina) and the eastern Amazon of Brazil (Patton and da Silva 1997).

Other taxonomic changes include the replacement of *Lonchorhina aurita* with *L. inusitata* in the Guiana Shield (Williams and Genoways 2008). Editing errors in Lim, Engstrom, and Ochoa (2005b) resulted in the omission of the widely distributed bat *Vampyrodes caraccioli* from all six political units, the erroneous inclusion of the bat *Vampyressa pusilla* in the Guianas, and the erroneous inclusion of *V. thyone* from Delta Amacuro and Suriname. The new species of disc-winged bat *Thyroptera* has been formally described as *T. devivoi* and currently known only from the savannas of the Rupununi in Guyana and the Cerrado in Brazil (Gregorin et al. 2006). Considered a subspecies of *Eumops bonariensis* by Eger (1977), *E. nanus* has been elevated to species status (Eger 2008). The once-widely distributed primate *Callicebus torquatus* is now restricted to south of the Rio Negro, and *C. lugens* is considered to replace it north of the Rio Negro (but with one record south), including Amazonas and Bolivar states in Venezuela (Casado, Bonvicino, and Seuánez 2007; van Roosmalen, van Roosmalen, and Mittermeier 2002). Similarly, the primate *Cacajao melanocephalus* has been separated into three allopatrically distributed species, including a new species *C. hosomi* north of the Rio Negro in Brazil and east of the Canal Cassiquiare in Amazonas, Venezuela. *Cacajao melanocephalus* is restricted to a small portion of Amazonas state of Venezuela west of the Canal Cassiquiare in Colombia and south of the Rio Negro in Brazil (Boubli et al. 2008). The previously widely distributed primate *Chiropotes satanus* was split into four allopatrically distributed species (Bonvicino et al. 2003). Two species, *C. satanus* and *C. utahicki*, are found south of the Lower Amazon; *C. israelita* occurs north of the Rio Negro and west of the Rio Branco in Brazil to the Orinoco River in Amazonas and Bolivar states of Venezuela; and *C. chiropotes* is found north of the Amazon River and east of the Rio Branco in Brazil to Guyana, Suriname, and French Guiana. The rodent species previously known as *Akodon urichi* was assigned to the genus *Necromys* following Smith and Patton (1999). The brush-tailed rat *Isothrix orinoci* is considered endemic to the upper Orinoco River drainage area of the Guiana Shield and is a distinct allopatrically distributed species from *I. bistriata* based on cranial morphometric differences (Patterson and Velazco 2008; Patton and Emmons 1985).

11.3 Geographic Limits

Since the summary by Lim, Engstrom, and Ochoa (2005), recent surveys have documented four additional species of bats (*Cyttarops alecto*, *Saccopteryx gymnura*, *Diaemus youngii*, and *Lasiurus egregius*) that are new to Suriname (Lim 2009). This increases the mammalian biodiversity in Suriname to 194 species. Similarly, a recent summary for French Guiana documents 189 species of mammals (F. Catzeflis, pers. comm.). For the other countries and states, there are 222 species of mammals reported in Guyana, 147 species in Delta Amacuro state, 241 species in Bolivar state, and 208 species in Amazonas state. In terms of area, Bolivar is the largest political unit in the Guianas (with 238,000 km^2) and Delta Amacuro is the smallest (with 40,000 km^2). There is a positive correlation ($r = 0.96$) between the area of a political subdivision and the number of species it supports (fig. 11.2). The Guianas are nearly 1 million km^2 in area and represent about 5% of the more than 20 million km^2 area of the Neotropical region, which is broadly defined as South America, Central America, Mexico, and the Caribbean.

In terms of distribution, over one-third (106 species) of the mammalian biodiversity is found in all six political units (countries or Venezuelan states)

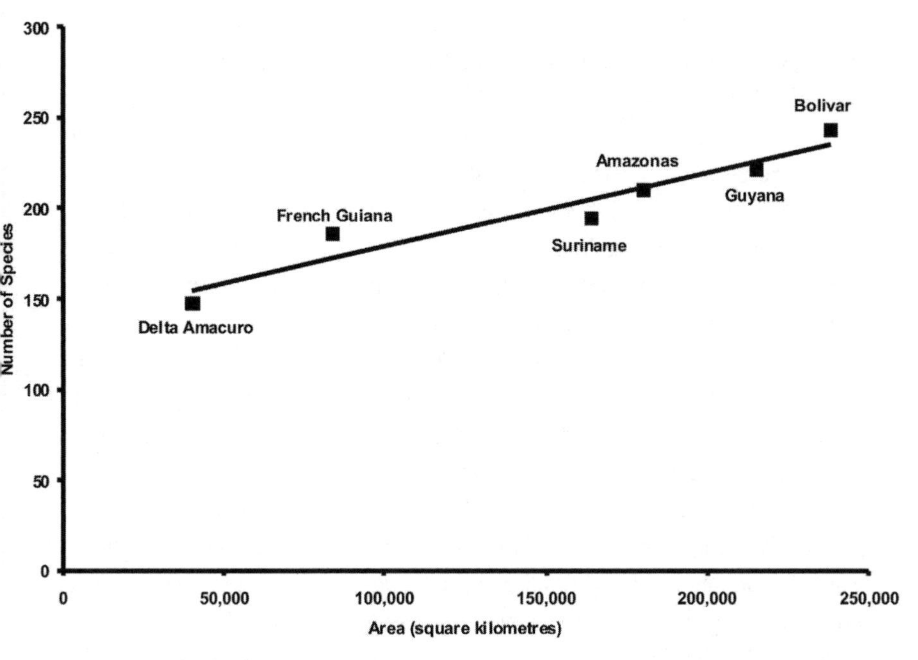

Figure 11.2 Plot of area versus species for the six political units of the Guianas. The two variables have a Pearson product-moment correlation of $r = 0.972$.

and is considered widespread in the region (appendix 11.1). Most of the large-sized mammals are widely distributed with all ungulates, 70% of the carnivores, 67% of the anteaters and sloths, and 50% of the armadillos found in the six political units of the Guianas. The exceptions are two species each of dolphins, manatees, and rabbits that are confined to specific rivers or savannas, and the primates, of which only 20% of the species are widely distributed across the region. For the more speciose small mammals, 40% of the bats, 26% of the marsupials, and 22% of the rodents are found in all the political units.

There are 16 species endemic to the Guianas, of which 9 are rodents. *Sciurus flammifer* is found only in the northern Bolivar state (Linares 1998). *Podoxymys roraimae* is restricted to the top of Mount Roraima in Venezuela (Pérez-Zapata, Lew, Aguilera, and Reig 1992) and Guyana (Anthony 1929); it has not been reported from the Brazilian portion of the tepui but is expected to occur there. *Dasyprocta guamara* occurs only in the Delta Amacuro state (Ochoa G., Bevilacqua, and García 2005). *Isothrix sinnamariensis* is known from lowland areas of Guyana and French Guiana but has not been documented in Suriname (Lim et al. 2006). *Rhipidomys macconnelli* and *R. wetzeli* are known from the Guiana highland areas of Venezuela and Guyana (Lim, Engstrom, and Ochoa 2005b). Finally, there are three undescribed species of *Oecomys* and *Oligoryzomys* from the Caura River area of Bolivar state (J. Ochoa G., pers. comm.). There are five endemic species of bats, including *Lonchorhina fernandezi*, which is found in the Amazonas and Bolivar states (Williams and Genoways 2008). *Platyrrhinus aurarius* is restricted to highland areas >500 m (Lim and Engstrom 2000). *Lasiurus atratus* occurs throughout the lowland regions (Lim et al. 1999). *Molossus barnesi* is known only from French Guiana (Simmons and Voss 1998). Finally, there is an undescribed species of *Artibeus* from the Caura River area of the Bolivar state (J. Ochoa G., pers. comm.). There are two endemic species of marsupials from the Guianas: *Marmosa tyleriana* is known from only the tepuis of Venezuela (Creighton and Gardner 2008), and *Monodelphis reigi* has been documented from Sierra de Lema in the Bolivar state and Mount Ayanganna in Guyana (Lim et al. 2010).

Six of the endemic species (*Marmosa tyleriana, Monodelphis reigi, Platyrrhinus aurarius, Podoxymys roraimae, Rhipidomys macconnelli*, and *R. wetzeli*) occur only in upland areas >500 m, and the other 10 species occur only in lowland areas. Of the lowland endemics, only one species (*Lasiurus atratus*) is widely distributed, whereas two species (*Isothrix sinnamariensis* and *Molossus barnesi*) are restricted to the east and seven species are restricted to the west (*Artibeus sp., Dasyprocta guamara, Lonchorhina fernandezi, Oecomys sp.* 1, *Oecomys sp.* 2, *Oligoryzomys sp.*, and *Sciurus flammifer*).

11.4 Biogeography

11.4.1 ENDEMISM AND DISTRIBUTION

Although the Guianas do not have the highest levels of mammalian biodiversity in the Neotropics—a distinction held by western Amazonia with approximately 215 species in any given area (Voss and Emmons 1996)—its association with the ancient Guiana Shield craton has identified a potential role as a stable core area for biotic diversification (Lim 2008). The Guiana plateau (>500 m) of Venezuela and Guyana is the most prominent topographic feature in the Guianas and has influenced the distributional patterns of mammals in this region both as an area of endemism and as a geographic barrier. However, the remote sandstone formations of the table-topped mountains (tepuis) that reach nearly 3000 m in elevation are in need of detailed study for a better understanding of biodiversity and biogeography in northern South America. The first and most comprehensive study to focus on the nonvolant mammals of the Guianas was the publication of Tate (1939), which I summarize and update with current taxonomy, including the bats.

The mammalian highland or summit fauna (>1500 m) was originally described to comprise nine species from Mounts Duida, Auyantepui, and Roraima (Tate 1939). Besides these species, *Mustela frenata* (Gardner 1990) and *Proechimys hoplomyoides* (Woods and Kilpatrick 2005) should be added to this list. Tate (1939) originally described *P. hoplomyoides* as a subspecies of *P. cayennensis*, based on two specimens from approximately 1400 and 2300 m on Mount Roraima. Three of these 11 highland species are also endemic to the Guiana uplands, including *Marmosa tyleriana*, *Podoxymys roraimae*, and *Rhipidomys macconnelli*. An additional 3 species are endemic to the more inclusive Guiana Shield, including *Didelphis imperfecta*, which was considered as *D. albiventris imperfectus* by Gardner (1990); *Sphiggurrus melanurus*, which was listed as *Coendou* sp. by Tate (1939); and *Proechimys hoplomyoides*. Highland species with Andean affinities include *Necromys urichi* and *Mustela frenata*. More widely distributed species found throughout the Amazon basin include *Micoureus demererae* (which was listed as *M. cinerea* by Gardner 1990), *Tamandua tetradactyla*, and *Nasua nasua*.

Six species were added to the highland fauna by Handley (1976) from Mount Duida and surrounding areas of Venezuela. Species endemic to the Guiana uplands include a series of specimens originally listed as *Rhipidomys fulviventer* but subsequently identified as the new species *R. wetzeli* (Gardner 1990) and *Platyrrhinus aurarius*, the first documented endemic species of bat (Handley and Ferris 1972). The other four species are *Anoura caudifer*, *A. geoffroyi*, and *Rhipidomys nitela* (listed as *R. mastacalis*), each widely distributed in South America, and *Neusticomys venezuelae* (listed as *Daptomys venezuelae*), with Andean affinities.

During studies at Cerro de la Neblina in southern Venezuela, Gardner (1990) reported 11 more species to the highland fauna, including 5 species with Andean affinities and 6 species with wider distributions throughout the Amazon basin. Species of Andean affinity included *Marmosops neblina*, which was originally described as a subspecies of *M. impavidus* by Gardner (1990) and subsequently recognized as a distinct species (Mustrangi and Patton 1997; Patton, da Silva, and Malcolm 2000); *Artibeus bogotensis*, which Handley (1987) considered a subspecies of *A. glaucus*, is a distinct species more closely related to *A. gnomus* (Lim, Engstrom, and Patton et al. 2008b); *Myotis oxyotus*; *Histiotus humboldti*, which was listed as an undescribed species by Gardner (1990) and later described by Handley (1996); and *Lasiurus cinereus*. More widespread species included Rhipi-

domys leucodactylus and *Sturnira tildae* (found also in the Amazon), and *Oecomys trinitatis* (listed as *Oryzomys trinitatis*), *Nyctinomops macrotis*, *Carollia brevicauda*, and *Sturnira lilium* (found even more widely, ranging into Central America).

Ochoa, Molina, and Giner (1993) added three more species to the highland fauna based on an inventory of Parque Nacional Canaima in Venezuela. *Anoura latidens* has Andean affinities, *Tapirus terrestris* is widely distributed in South America and is the only large mammal from the summit fauna, and *Molossus molossus* occurs throughout the Neotropics. An additional Guiana highland endemic is *Monodelphis reigi*, which was recently described from 1400 m at Sierra de Lema in Venezuela (Lew and Pérez-Hernández 2004), and subsequently reported from 1100 m and 2050 m on Mount Ayanganna in Guyana (Lim et al. 2010). There are eight other highland species previously unreported based on my fieldwork from Mounts Ayanganna, Roraima, and Wokomung in the Pakaraima Highlands of Guyana. Two of these species are endemic to the Guiana Shield, *Oecomys rex* and *Lophostoma schulzi*; one species occurs in the eastern Amazon *Hylaeamys megacephalus* (previously included in the genus *Oryzomys*; Weksler, Percequillo, and Voss 2006); one species is found in the Amazon basin, *H. yunganus* (previously included in the genus *Oryzomys*; Weksler, Percequillo, and Voss 2006); one species has affinities to the western Amazon, *Euryoryzomys macconnelli* (previously included in the genus *Oryzomys*; Weksler, Percequillo, and Voss 2006); and three species are widely distributed throughout the Neotropics, *Micronycteris megalotis*, *Myotis riparius*, and *Trachops cirrhosus*.

At present, 40 species of mammals are documented from the highland or summit zone (> 1500 m) of the Guianas, representing 5 opossums, 14 rodents, 17 bats, 2 carnivores, a tamandua, and a tapir. Of these highland species, 6 are endemic to the central plateau region of Guiana (>500 m), 5 are endemic to the more inclusive Guiana Shield, 10 have affinities to the Andes, only 1 has affinities to the eastern Amazon, and 18 are widely distributed in the Amazon and much of South America.

In terms of Tate's (1939) distributional patterns, the 284 species of mammals found in the Guianas (appendix 11.1) can be classified as follows: 97 species are broadly distributed throughout tropical Central America and South America; 37 are distributed from Panama into South America; 34 are distributed throughout the contiguous Amazon forest; 34 are found in the western Amazon and foothills of the Andes; 17 are found in the eastern Amazon; 14 are found primarily in savannas; 12 are found through northwestern South America into Central America; 38 are endemic to the Guiana Shield; none are found in northern South America north of the Amazon (Tate [1939] assigned two species to this category, which is similar to the third); and one is found in northern Venezuela and Colombia. However, some species such as *Enchisthenes hartii* and *Mustela frenata* do not have distributional ranges that match closely with the categories of Tate (1939). In addition, some of the savanna endemics are listed in other categories, such as the Guiana Shield endemic *Lonchorhina fernandezi*. There are twice as many species with western Andean affinities than eastern Amazonian affinities, which was a pattern noted by Tate (1939) and Gardner (1990). Similarly, Voss, Lunde, and Simmons (2001) identified 19 species in southern Venezuela that primarily occur in the western Amazon. However, five of these species (*Bradypus variegatus*, *Scleronycteris ega*, *Rhinophylla fischerae*, *Dactylomys dactylinus*, and *Mesomys hispidus*) also have extensive distribution in the eastern Amazon.

Most Guianan species (87%) are found in the western lowlands, slightly fewer (78%) are found in the eastern lowlands, just over half (56%) are found in the central plateau, and only 38% are also found in Central America (appendix 11.1). However, not all orders of mammals have similar distribution patterns. None of the 15 species of Guianan primates are found in Central America, whereas most (71%) of the carnivores are widely distributed and occur in Central America. Similarly, only a few (7%) of the species of rodents are widely distributed in Central America, whereas over half (55%) of the bats are found in Central America.

11.4.2 FAUNAL SIMILARITY

Similarity of taxonomic composition within circumscribed localities such as national parks or protected areas can be used to assess geographic patterns on a continental basis. There are six lowland forested localities in the Guianas that have been well surveyed for mammals: Paracou (Simmons and Voss 1998; Voss, Lunde, and Simmons 2001) and Arataye (Voss and Emmons 1996) in French Guiana; Brownsberg (Lim et al. 2005) in Suriname; Iwokrama (Lim and Engstrom 2005) in Guyana; and Imataca (Ochoa 1995) and Cunucunuma (Handley 1976; Voss and Emmons 1996) in Venezuela. There are an additional seven lowland forested localities in other regions of South America and Central America that have been well surveyed for mammals (Voss and Emmons 1996): Manaus and Rio Xingu in Brazil; Balta, Manu, and Cuzco Amazonico in Peru; Barro Colorado Island in Panama; and La Selva in Costa Rica. Faunal similarity based on the presence or absence of mammal species among these 13 Neotropical localities was analyzed using Jaccard's coefficient and clustered by the unweighted pair-group method using arithmetic averages (UPGMA), as

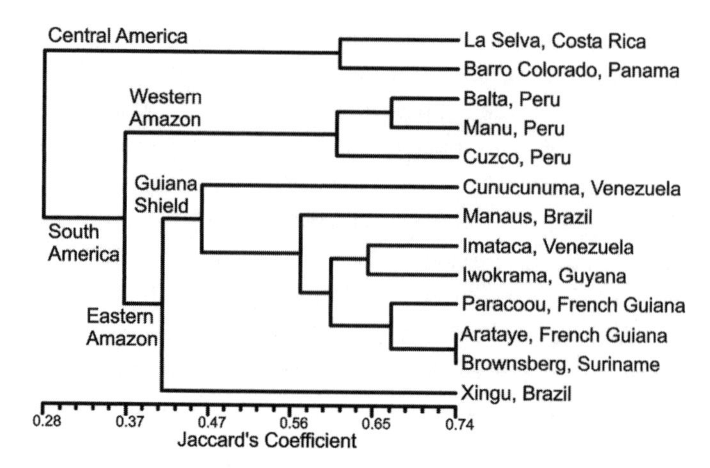

Figure 11.3 Faunal similarity study derived from an unweighted pair-group method with arithmetic mean (UPGMA) cluster analysis of Jaccard's coefficient among 13 Neotropical lowland rain forest localities that have been well surveyed for mammals.

implemented in the computer software program NTSYSpc version 2.11a (Rohlf 2002) and reported earlier by Lim et al. (2005).

There was geographic structuring in the analysis of faunal similarity and localities from the Guianas grouped together (fig. 11.3). The Guianas clustered within eastern Amazonia, which was most similar to western Amazonia and less similar to Central American localities. However, comprehensive surveys of forested localities are currently lacking for Mexico, northern Central America, western versant of the Andes in South America, and the Atlantic Forest, limiting the finer-scale resolution of this analysis.

11.4.3 HISTORICAL BIOGEOGRAPHY

Phylogenies can be used to investigate biogeographic patterns and to hypothesize modes of speciation that account for biotic diversification. Although comprehensive species-level phylogenies for Neotropical mammals are not abundant, there are a few examples that address the historical biogeography of the Guianas. There is a robust phylogeny constructed for New World emballonurid bats (tribe Diclidurini) based on loci from the four genetic transmission systems (mitochondrial, autosomal, X and Y sex chromosomes) found in mammals (Lim, Engstrom, Bickham et al. 2008). Molecular dating with fossil calibration points indicate an episode of biotic diversification in the early Miocene (18–19 Ma) associated with the differentiation of genera (Lim 2007). An analysis of historical biogeography identified the northern Amazon as the ancestral area of most internal nodes of the phylogeny, suggesting within-area speciation events similar to the taxon pulse hypothesis of biotic diversification (Erwin 1979) during expansions and contractions of forest and savanna habitats with the ancient Guiana Shield acting as a stable core area where ancestral species persisted (fig. 11.4; Lim 2008). The more inclusive Guiana highlands, or Pantepui (>1500 m), have been proposed as the species source for surrounding lowland flora during the more recent Quaternary (Rull 2005, 2007). A study of

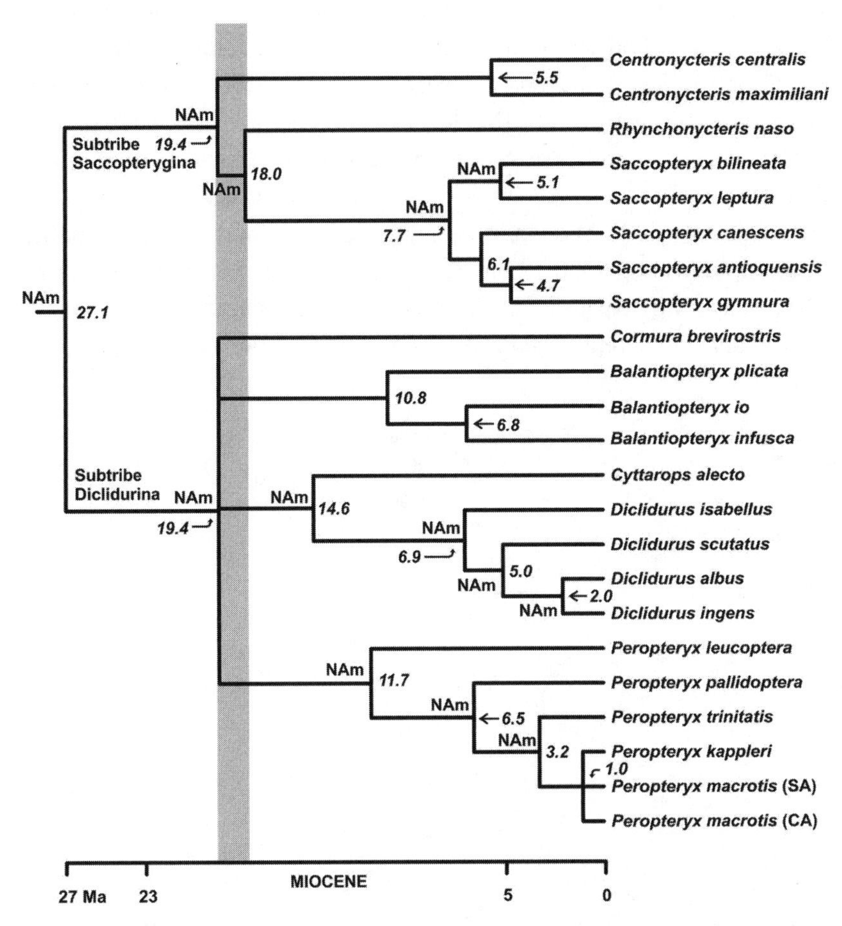

Figure 11.4 Phylogeny for the Bayesian relaxed-clock molecular dating of New World emballonurid bats (Lim 2007). The age of each node is indicated by italicized numbers (Ma) and the northern Amazon (NAm) is indicated as an unequivocal ancestral area. Grey shading identifies an episode of taxon-pulse diversification from a stable core area in the Guiana Shield during the early Miocene (19.4–18.0 Ma) that gave rise to most contemporary genera.

evolutionary patterns of morphology and behavior found a phylogenetic basis to the correlation of ear morphology and echolocation call parameters indicating an adaptive radiation from forest to savanna for the subtribe Diclidurina that corresponded to diversification of genera during the taxon pulse event in the early Miocene (Lim and Dunlop 2008).

A molecular phylogenetic analysis of short-tailed opossums (Lim et al. 2010), including the recently described *Monodelphis reigi* (Lew and Pérez-Hernández 2004), corroborated earlier morphological studies suggesting its close relationship to the *M. adusta* species complex (Solari 2007). As presently known, *M. reigi* is endemic to the highland regions (>1100 m) of the Guiana Shield, and is the only taxon within the *M. adusta* species complex that does not occur in the Andes and contiguous lowlands over 1000 km to the west. The only other short-tailed opossum in the Guianas is the common and widely distributed endemic species *M. brevicaudata* (Voss, Lunde, and Simmons 2001), which is distantly related to *M. reigi*. The biogeographic scenario based on current distributions suggests that the Andes is the ancestral area for the *M. adusta* species complex, which implies that *M. reigi* speciated in isolation in the Guiana highlands after a range expansion of the most recent common ancestor (fig. 11.5). This colonization event from the Andes occurred relatively early in the diver-

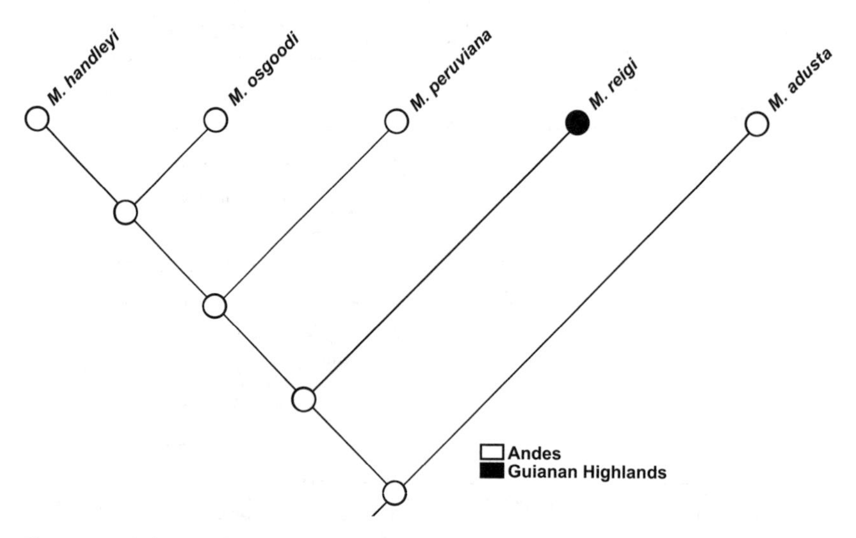

Figure 11.5 Phylogeny of the *Monodelphis adusta* species complex based on molecular phylogenetic analysis of cytochrome-b sequence data (Lim et al. 2010). Optimization of current distributions suggests that the Andes is the ancestral area, implying that a dispersal event to the Guiana highlands 1000 km to the east gave rise to *M. reigi*.

sification of the *M. adusta* species complex, as inferred by the more basal split for *M. reigi*.

Similarly, a recent study of broad-nosed bats (*Platyrrhinus*) found the Guiana upland endemic *P. aurarius* (>500 m) within an Andean clade of seven other species (Velazco and Patterson 2008). However, the speciation of *P. aurarius* and the presumptive sister species *P. infuscus* from their common ancestor probably occurred more recently than the appearance of *Monodelphis reigi*. Velazco and Patterson (2008) suggested that a dispersal event from the Andes to the Guiana plateau in the Pleistocene gave rise to *P. aurarius*.

A molecular phylogenetic analysis of spiny mice (*Neacomys*) found that the three species known from the Guianas did not form a monophyletic clade (Catzeflis and Tilak 2009); sister- group relationships included species found in the Amazon basin. Therefore, Catzeflis and Tilak (2009) hypothesized that multiple faunal exchanges occurred between the Guianan and Amazonian regions. Although not all species of *Neacomys* were included in this study, *N. paracou* from the Guianas was the basal-most species in the phylogeny, which suggests divergence beginning at least in the Pliocene.

Based on low levels of cytochrome-*b* sequence divergence (2.7%), the Guianan endemic *Isothrix sinnamariensis* is presumed to have split from a common ancestor with the eastern Amazonian *I. pagurus* during the Pleistocene (Lim et al. 2006). The recent discovery of the highland Andean endemic *I. barbarabrownae* as a basal lineage sister to the five lowland species indicates a complex biogeographic scenario for brush-tailed rats (Patterson and Velazco 2008). Although the allopatric distribution of species of *Isothrix* is not conducive for unequivocal statements of ancestral areas, the recurrence of the Amazon at internal nodes of the phylogeny during character optimization suggests the possibility of this area's central role in the diversification of species within the genus (fig. 11.6). Incorporating a parsimonious approach based on geographic proximity, I postulate that there were successive range expansions from the western Amazon to the Andes, central Amazon, and eastern Amazon, including more recent dispersal events to the Guianas that gave rise to *I. sinnamariensis*; similar expansions to the Orinoco basin of the Guiana Shield gave rise to *I. orinoci*. An alternative biogeographic scenario at the basal node would be to identify the Andes as the ancestral area with expansion into the adjacent western Amazon (fig. 11.6; Patterson and Velazco 2008).

Phylogeographic studies can also identify divergence of even more recent origin. Two species of mouse opossums that are widely distributed in lowland

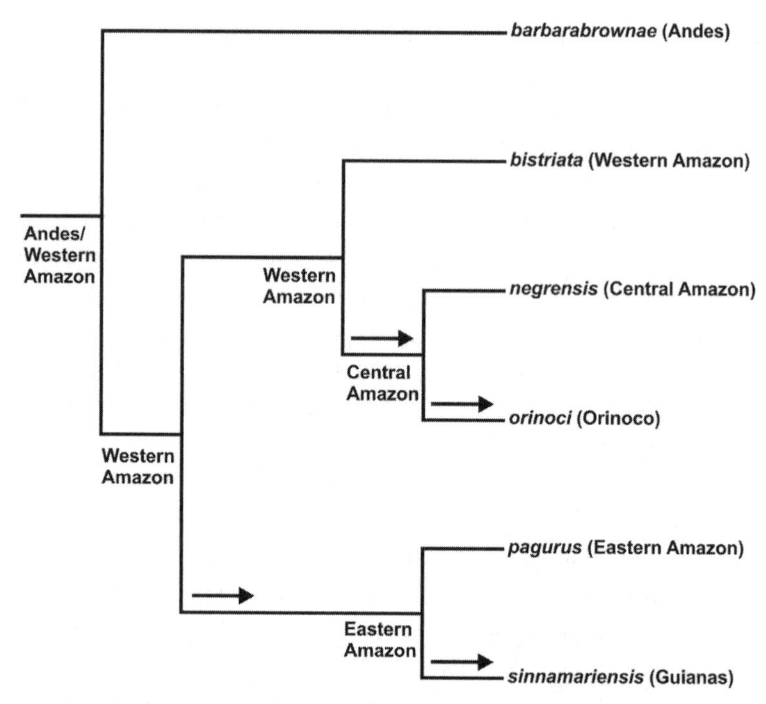

Figure 11.6 Phylogeny of brush-tailed rats *Isothrix* based on the molecular analysis of Patterson and Velazco (2008). Ancestral areas represent a potential parsimonious character optimization at internal nodes of the tree and arrows indicate possible dispersal events. The basal node is equivocal, compatible both with range expansion to the Andes for I. *barbarabrownae* or to the Western Amazon for the most recent common ancestor of all other species of *Isothrix*.

forests of South America each had strong geographic structuring in molecular sequence variation (Steiner and Catzeflis 2004). Populations of *Marmosa murina* and *Micoureus demerarae* from the Guianas appeared more closely related to populations from eastern Amazonia than to those in western Amazonia, suggesting a range expansion from the eastern Amazon into the Guianas. There was further phylogeographic structuring and relatively high average sequence divergence (3%) of *M. demerarae* within the Guianas, but a lack of similar structuring and divergence for *Marmosa murina*. This indicates either a more recent range expansion of *M. murina* into the Guianas, or perhaps a population bottleneck effect during the early Pleistocene. The high average sequence divergence between populations (5.3%) of *Micoureus demerarae* within the Guianas suggests that a closer examination of the species diversity for this taxon is warranted.

11.5 Conclusions

The Guiana Shield is an ancient formation that was relatively stable during times of substantial environmental change in the Miocene (e.g., the uplifting of the northern Andean mountains), and during the Pliocene (e.g., the newly forming Amazon basin). Its location adds significance to its role in the differentiation of the Neotropical biota. While not the hotbed of species diversity that is western Amazon, or of endemism as are the Andes, its biota is a subtle mix of widely distributed species interspersed with species having affinities to adjacent regions. Within the Guianas, the plateau region >500 m in elevation is the most prominent topographic feature. These fractured highlands (tepuis) are both areas of endemism and biogeographic barriers that have influenced distributional patterns. I have given examples for three speciation processes that have contributed to the present-day biodiversity of the Guainas: (1) the Guiana Shield acted as a stable core area for taxon-pulse diversification during changes in the composition of savanna and forest paleoenvironments in the early Miocene, as hypothesized for New World sheath-tailed bats (tribe Diclidurini); (2) the Guiana plateau as an area for range expansion from the Andes during the late Miocene as hypothesized for speciation of short-tailed opossums (*Monodelphis reigi*) and during the early Pleistocene for speciation of broad-nosed bats (*Platyrrhinus aurarius*); and (3) the Guiana lowlands as an area of multiple faunal exchange with the Amazon as seen in spiny mice (*Neacomys*) during the Pliocene, or in the Pleistocene as seen in the speciation of brush-tailed rats (*Isothrix*) in the Orinoco basin (*I. orinoci*) and in the eastern Guianan lowlands (*I. sinnamariensis*) after colonization from the Amazon. Even more recent is the phylogeographic structuring of *Marmosa murina* with range expansion into the Guianas from eastern Amazonia. As is the case with many organismal groups, there is a paucity of comprehensive species phylogenies for New World mammals that are needed to give a better perspective on the historical biogeography of South America that have contributed to the high levels of biodiversity found within this continent.

Acknowledgments

I thank Mark Engstrom and Jose Ochoa for collaborating on the checklist of Guiana Shield mammals (Lim et al. 2005), which formed the basis for this synthesis on the biogeography of the Guianas. Bruce Patterson and Francois Catzeflis gave me suggestions that helped the formulation of ideas that appear in this chapter. My fieldwork and research in the Neotropics has been supported over the past two decades by colleagues in mammalogy at the Royal Ontario Museum, with primary funding generously provided by the Department of Natural History and the ROM Governors.

Literature Cited

Anthony, H. E. 1929. "Two New Genera of Rodents from South America." *American Museum Novitates* 383:1–6.

Bonvicino, C. R., J. P. Boubli, I. B. Otazú, F. C. Almeida, F. F. Nascimento, J. R. Coura, and H. N. Seuánez. 2003. "Morphologic, Karyotypic and Molecular Evidence of a New Form of *Chiropotes* (Primates, Pitheciinae)." *American Journal of Primatology* 61:123–33.

Boubli, J. P., M. N. F. da Silva, M. V. Amado, T. Hrbek, F. B. Pontual, and I. P. Farias. 2008. "A Taxonomic Reassessment of *Cacajao melanocephalus* Humboldt (1811), with the Description of Two New Species." *International Journal of Primatology* 29:723–41.

Casado, F., C. R. Bonvicino, and H. N. Seuánez. 2007. "Phylogeographic Analyses of *Callicebus lugens* (Platyrrhini, Primates)." *Journal of Heredity* 98:88–92.

Catzeflis, F., and M. Tilak. 2009. "Molecular Systematics of Neotropical Spiny Mice (*Neacomys*: Sigmodontinae, Rodentia) from the Guiana Region." *Mammalia* 73:239–47.

Creighton, G. K., and A. L. Gardner. 2008. "Genus *Marmosa*." In *Mammals of South America, Vol. 1: Marsupials, Xenarthrans, Shrews, and Bats*, edited by A. L. Gardner, 51–61. Chicago: University of Chicago Press.

Eger, J. L. 1977. "Systematics of the Genus *Eumops* (Chiroptera: Molossidae)." *Life Sciences Contributions, Royal Ontario Museum* 110:1–69.

———. 2008. "Family Molossidae." In *Mammals of South America, Vol. 1: Marsupials, Xenarthrans, Shrews, and Bats*, edited by A. L. Gardner, 399–439. Chicago: University of Chicago Press.

Erwin, T. L. 1979. "Thoughts on the Evolutionary History of Ground Beetles: Hypotheses Generated from Comparative Faunal Analyses of Lowland Forest Sites in Temperate and Tropical Regions." In *Carabid Beetles: Their Evolution, Natural History, and Classification*, edited by T. L. Erwin, G. E. Ball, and D. R. Whitehead, 539–92. The Hague: Dr. W. Junk.

Gibbs, A. K., and C. N. Barron. 1993. *The Geology of the Guiana Shield*. Oxford: Oxford University Press.

Gardner, A. L. 1990. "Two New Mammals from Southern Venezuela and Comments on the Affinities of the Highland Fauna of Cerro de la Neblina." In *Advances in Neotropical Mammalogy*, edited by K. H. Redford and J. F. Eisenberg, 411–24. Gainesville: Sandhill Crane Press.

Gregorin, R., E. Gonçalves, B. K. Lim, and M. D. Engstrom. 2006. "New Species of Disk-Winged Bats *Thyroptera* and Range Extension of *T. discifera*." *Journal of Mammalogy* 87:238–46.

Handley, C. O., Jr. 1976. "Mammals of the Smithsonian Venezuelan Project." *Brigham Young University Science Bulletin, Biological Series* 20 (5): 1–91, map.

———. 1987. "New Species of Mammals from Northern South America: Fruit-Eating Bats, Genus *Artibeus* Leach." In *Studies in Neotropical Mammalogy: Essays in Honor of Philip Hershkovitz*, edited by B. D. Patterson and R. M. Timm, 163–72. Chicago: Field Museum of Natural History, *Fieldiana: Zoology*, 39:1–506.

———. 1996. "New Species of Mammals from Northern South America: Bats of the Genera *Histiotus* Gervais and *Lasiurus* Gray (Chiroptera: Vespertilionidae)." *Proceedings of the Biological Society of Washington* 109:1–9.

Handley, C. O., Jr., and K. C. Ferris. 1972. "Description of New Bats of the Genus *Vampyrops*." *Proceedings of the Biological Society of Washington* 84:519–24.

Huber, O. 2006. "Herbaceous Ecosystems on the Guayana Shield, A Regional Overview." *Journal of Biogeography* 33:464–75.

Lew, D., and R. Pérez-Hernández. 2004. "Una Nueva Especie del Género *Monodelphis* (Didelphimorphia: Didelphidae) de la Sierra de Lema, Venezuela." *Memorias de la Fundacion La Salle de Ciencias Naturales* 159–160:7–25.

Lew, D., R. Pérez-Hernández, and J. Ventura. 2006. "Two New Species of *Philander* (Didelphimorphia, Didelphidae) from Northern South America." *Journal of Mammalogy* 87:224–37.

Lim, B. K. 2007. "Divergence Times and Origin of Neotropical Sheath-Tailed Bats (Tribe Diclidurini) in South America." *Molecular Phylogenetics and Evolution* 45:777–91.

———. 2008. "Historical Biogeography of New World Emballonurid Bats (Tribe Diclidurini): Taxon Pulse Diversification." *Journal of Biogeography* 35:1385–401.

———. 2009. "Environmental Assessment at the Bakhuis Bauxite Concession: Small-Sized Mammal Diversity and Abundance in the Lowland Humid Forests of Suriname." *Open Biology Journal* 2:42–57.

Lim, B. K., and J. M. Dunlop. 2008. "Evolutionary Patterns of Morphology and Behavior as Inferred from a Molecular Phylogeny of New World Emballonurid Bats (Tribe Diclidurini)." *Journal of Mammalian Evolution* 15:79–121.

Lim, B. K., and M. D. Engstrom. 2000. "Preliminary Survey of Bats from the Upper Mazaruni of Guyana." *Chiroptera Neotropical* 6:119–23.

———. 2005. "Mammals of Iwokrama Forest." *Proceedings of the Academy of Natural Sciences, Philadelphia* 154:71–108.

Lim, B. K., M. D. Engstrom, J. W. Bickham, and J. C. Patton. 2008. "Molecular Phylogeny of New World Emballonurid Bats (Tribe Diclidurini) Based on Loci from the Four Genetic Transmission Systems in Mammals." *Biological Journal of the Linnean Society* 93:189–209.

Lim, B. K., M. D. Engstrom, H. H. Genoways, F. M. Catzeflis, K. A. Fitzgerald, S. L. Peters, M. Djosetro et al. 2005. "Results of the Alcoa Foundation-Suriname Expeditions. XIV. Mammals of Brownsberg Nature Park, Suriname." *Annals of the Carnegie Museum* 74:225–74.

Lim, B. K., M. D. Engstrom, and J. Ochoa G. 2005. "Mammals." In *Checklist of the Terrestrial Vertebrates of the Guiana Shield*, edited by T. Hollowell and R. P. Reynolds, 77–92. *Bulletin of the Biological Society of Washington* 13:1–98.

Lim, B. K., M. D. Engstrom, J. C. Patton, and J. W. Bickham. 2006. "Systematic Relationships of the Guianan Brush-Tailed Rat (*Isothrix sinnamariensis*) and Its First Occurrence in Guyana." *Mammalia* 70:120–25.

———. 2008. "Systematic Review of Small Fruit-Eating Bats (*Artibeus*) from the Guianas, and a Re-evaluation of *A. glaucus bogotensis.*" *Acta Chiropterologica* 10:243–56.

———. 2010. "Molecular Phylogenetics of Reig's Short-Tailed Opossum (*Monodelphis reigi*) and Its Distributional Range Extension into Guyana." *Mammalian Biology* 75:287–93.

Lim, B. K., M. D. Engstrom, R. M. Timm, R. P. Anderson, and L. C. Watson. 1999. "First Records of 10 Bat Species in Guyana and Comments on Diversity of Bats in Iwokrama Forest." *Acta Chiropterologica* 1:179–90.

Linares, O. J. 1998. *Mamíferos de Venezuela.* Caracas: Sociedad Conservacionista Audubon de Venezuela.

Mittermeier, R. A., N. Myers, J. B. Thomsen, G. A. B. da Fonseca, and S. Olivieri. 1998. "Biodiversity Hotspots and Major Tropical Wilderness Areas: Approaches to Setting Conservation Priorities." *Conservation Biology* 12:516–20.

Mustrangi, M. A., and J. L. Patton. 1997. "Phylogeography and Systematics of the Slender Mouse Opossum *Marmosops* (Marsupialia, Didelphidae)." *University of California Publications in Zoology* 130:1–86.

Ochoa G. J. 1995. "Los Mamíferos de la Region de Imataca, Venezuela." *Acta Científica Venezolana* 46:274–87.

Ochoa G., J., M. Bevilacqua, and F. García. 2005. "Evaluación Ecológica Rápida de las Comunidades de Mamíferos en Cinco Localidades del Delta del Orinoco, Venezuela." *Interciencia* 30: 466–75.

Ochoa G. J., C. Molina, and S. Giner. 1993. "Inventario y Estudio Comunitario de los Mamíferos del Parque Nacional Canaima, con una Lista de las Especies Registradas para la Guayana Venezolana." *Acta Científica Venezolana* 44:244–61.

Patterson, B. D., and P. M. Velazco. 2008. "Phylogeny of the Rodent Genus *Isothrix* (Hystricognathi, Echimyidae) and Its Diversification in Amazonia and the Eastern Andes." *Journal of Mammalian Evolution* 15:181–201.

Patton, J. L., and M. N. F. da Silva. 1997. "Definition of Species of Pouched Four-Eyed Opossums (Didelphidae, *Philander*)." *Journal of Mammalogy* 78:90–102.

Patton, J. L., M. N. F. da Silva, and J. R. Malcolm. 2000. "Mammals of the Rio Juruá and the Evolutionary and Ecological Diversification of Amazonia." *Bulletin of the American Museum of Natural History* 244:1–306.

Patton, J. L., and L. H. Emmons. 1985. "A Review of the Genus *Isothrix* (Rodentia, Echimyidae)." *American Museum Novitates* 2817:1–14.

Pérez-Zapata, A., D. Lew, M. Aguilera, and O. A. Reig. 1992. "New Data on the Systematics and Karyology of *Podoxymys roraimae* (Rodentia, Cricetidae)." *Zeitschrift für Säugetierkunde* 57:216–24.

Rohlf, F. J. 2002. *NTSYSpc: Numerical Taxonomy and Multivariate Analysis System, version 2.11a.* Port Jefferson: Applied Biostatistics Inc.

Rull, V. 2005. "Biotic Diversification in the Guayana Highlands: A Proposal." *Journal of Biogeography* 32:921–27.

———.2007. "The Guayana Highlands: A Promised (But Threatened) Land for Ecological and Evolutionary Science." *Biotropica* 39:31–34.

Simmons, N. B., and R. S. Voss. 1998. "The Mammals of Paracou, French Guiana: A Neotropical Lowland Rainforest Fauna, Part 1. Bats." *Bulletin of the American Museum of Natural History* 237:1–219.

Smith, M. F., and J. L. Patton. 1999. "Phylogenetic Relationships and the Radiation of Sigmodontine Rodents in South America: Evidence from Cytochrome b." *Journal of Mammalian Evolution* 6:89–128.

Solari, S. 2007. "New Species of *Monodelphis* (Didelphimorphia: Didelphidae) from Peru, with Notes on *M. adusta* (Thomas, 1897)." *Journal of Mammalogy* 88:319–29.

Steiner, C., and F. M. Catzeflis. 2004. "Genetic Variation and Geographical Structure of Five Mouse-Sized Opossums (Marsupialia, Didelphidae) Throughout the Guiana Region." *Journal of Biogeography* 31:959–73.

Tate, G. H. H. 1939. "Mammals of the Guiana Region." *Bulletin of the American Museum of Natural History* 76:151–229.

van Roosmalen, M. G. M., T. van Roosmalen, and R. A. Mittermeier. 2002. "A Taxonomic Review of the Titi Monkeys, Genus *Callicebus* Thomas, 1903, with the Description of Two New Species, *Callicebus bernhardi* and *Callicebus stephennashi*, from Brazilian Amazonia." *Neotropical Primates* 10 (Suppl.): 1–52.

Velazco, P. M., and B. D. Patterson. 2008. "Phylogenetics and Biogeography of the Broad-Nosed Bats, Genus *Platyrrhinus* (Chiroptera: Phyllostomidae)." *Molecular Phylogenetics and Evolution* 49:749–59.

Voss, R. S., and L. H. Emmons. 1996. "Mammalian Diversity in Neotropical Lowland Rainforest: A Preliminary Assessment." *Bulletin of the American Museum of Natural History* 230:1–115.

Voss, R. S., D. P. Lunde, and N. B. Simmons. 2001. "The Mammals of Paracou, French Guiana: A Neotropical Lowland Rainforest Fauna. Part 2. Nonvolant Species." *Bulletin of the American Museum of Natural History* 263:1–236.

Weksler, M., A. R. Percequillo, and R. S. Voss. 2006. "Ten New Genera of Oryzomyine Rodents (Cricetidae: Sigmodontinae)." *American Museum Novitates* 3537:1–29.

Williams, S. L., and H. H. Genoways. 2008. "Subfamily Phyllostominae." In *Mammals of South America, Vol. 1: Marsupials, Xenarthrans, Shrews, and Bats*, edited by A. L. Gardner, 82–107. Chicago: University of Chicago Press.

Wilson, D. E., and D. M. Reeder, eds. 2005. *Mammal Species of the World: A Taxonomic and Geographic Reference*, 3rd ed. Baltimore: Johns Hopkins University Press.

Woods, C. A., and C. W. Kilpatrick. 2005. "Infraorder Hystricognathi." In *Mammal Species of the World: A Taxonomic and Geographic Reference*, 3rd ed., edited by D. E. Wilson and D. M. Reeder, 1538–600. Baltimore: Johns Hopkins University Press.

Appendix 11.1

Table A11.1 The occurrences of 284 species of mammals documented from the Guianas of northern South America in the eastern lowlands (EL), Guiana highlands (>500 m; GH), western lowlands (WL), and Central America beyond Panama (CA). Geographic distribution is updated based on current taxonomy (Wilson and Reeder 2005) and summarized following the categories of Tate (1939; original category in parentheses) including: (1) widely distributed in Central and South America, (2) widely distributed in South America but does not extend west of Panama, (3) distributed in the Amazon basin including the Guianas, (4) distributed from the Guianas west into Upper Amazonia and the Andean foothills, (5) distributed from the Guianas east into Brazil, (6) occurring primarily in savanna or dry open areas, (7) distributed from the Guianas into Colombia and Panama (and into Central America for several species of bats), (8) endemic to the Guiana Shield, (9) distributed in northern South America north of the Amazon River, and (10) distributed in the northern edge of the Guianas in Venezuela and Colombia. There are 16 species endemic to the Guianas (*). The species list is modified and revised from Lim et al. (2005) including occurrence in Amazonas (A), Bolivar (B), and Delta Amacuro (D) in Venezuela, Guyana (G), Suriname (S), and French Guiana (F).

	CA	WL	GH	EL	Tate	A	B	D	G	S	F
Didelphimorphia (23)	3	17	15	15	12						
Didelphidae (23)	3	17	15	15	12						
Caluromys lanatus		X		X	4 (4)	A	B		G		
Caluromys philander		X	X	X	5 (5)	A	B	D	G	S	F
Chironectes minimus	X	X	X	X	1 (1)	A	B	D	G	S	F
Didelphis imperfecta		X	X	X	8 (1)	A	B		G	S	F
Didelphis marsupialis	X	X	X	X	1	A	B	D	G	S	F
Gracilinanus emiliae		X		X	3		B		G	S	F
Hyladelphys kalinowskii				X	3				G		F
Lutreolina crassicaudata		X	X	X	5 (5)		B	D	G	S	
Marmosa lepida				X	3				G	S	F
Marmosa murina		X	X	X	2 (3)	A	B	D	G	S	F
*Marmosa tyleriana**			X		8 (8)	A	B				
Marmosops neblina		X	X		4	A					
Marmosops parvidens		X	X	X	3 (8)	A	B		G	S	F
Marmosops pinheiroi			X	X	5		B		G	S	F
Metachirus nudicaudatus	X	X	X	X	1 (1)	A	B	D	G	S	F
Micoureus demerarae		X	X	X	2 (1)	A	B	D	G	S	F
Monodelphis brevicaudata		X	X	X	8 (3)	A	B		G	S	F
*Monodelphis reigi**			X		8		B		G		
Monodelphis sp.		X			6		B				
Philander andersoni		X			4	A	B				
Philander deltae		X			7			D			
Philander mondolfii		X			4	A	B				

	CA	WL	GH	EL	Tate	A	B	D	G	S	F
Philander opossum			X	X	5 (1)				G	S	F
Pilosa (6)	**3**	**6**	**5**	**5**	**5**						
Bradypodidae (2)	1	2	1	1	2						
Bradypus tridactylus		X	X	X	5 (8)		B	D	G	S	F
Bradypus variegatus	X	X			1	A	B				
Megalonychidae (1)	0	1	1	1	1						
Choloepus didactylus		X	X	X	3 (1)	A	B	D	G	S	F
Myrmecophagidae (3)	2	3	3	3	3						
Cyclopes didactylus	X	X	X	X	1 (1)	A	B	D	G	S	F
Myrmecophaga tridactyla	X	X	X	X	1 (1)	A	B	D	G	S	F
Tamandua tetradactyla		X	X	X	2 (1)	A	B	D	G	S	F
Cingulata (6)	**1**	**5**	**3**	**5**	**4**						
Dasypodidae (6)	1	5	3	5	4						
Cabassous unicinctus		X	X	X	2 (6)		B	D	G	S	F
Dasypus kappleri		X		X	3 (8)	A	B	D	G	S	F
Dasypus novemcinctus	X	X	X	X	1 (1)	A	B	D	G	S	F
Dasypus sabanicola		X			6		B				
Euphractus sexcinctus				X	6				G	S	
Priodontes maximus		X	X	X	2 (3)	A	B	D	G	S	F
Chiroptera (147)	**81**	**134**	**86**	**126**	**0**						
Emballonuridae (16)	8	15	6	16	0						
Centronycteris maximiliani		X		X	2	A	B		G	S	F
Cormura brevirostris	X	X	X	X	1	A	B	D	G	S	F
Cyttarops alecto	X	X		X	1				G	S	F
Diclidurus albus	X	X	X	X	1	A	B	D	G	S	F
Diclidurus ingens		X	X	X	3	A	B	D	G		F
Diclidurus isabellus		X		X	3	A	B	D	G		
Diclidurus scutatus		X		X	3	A	B		G	S	F
Peropteryx kappleri	X	X	X	X	1		B	D	G	S	F
Peropteryx leucoptera		X		X	2	A			G	S	F
Peropteryx macrotis	X	X	X	X	1	A	B	D	G	S	F
Peropteryx trinitatis		X		X	2	A	B	D			F
Rhynchonycteris naso	X	X	X	X	1	A	B	D	G	S	F
Saccopteryx bilineata	X	X		X	1	A	B	D	G	S	F
Saccopteryx canescens		X		X	3	A	B	D	G	S	F
Saccopteryx gymnura				X	5				G	S	F
Saccopteryx leptura	X	X		X	1	A	B	D	G	S	F
Noctilionidae (2)	2	2	0	2	0						
Noctilio albiventris	X	X		X	1	A	B	D	G	S	F
Noctilio leporinus	X	X		X	1	A	B	D	G	S	F
Mormoopidae (5)	5	5	4	3	0						

(continued)

Table A11.1 *(continued)*

	CA	WL	GH	EL	Tate	A	B	D	G	S	F
Pteronotus davyi	X	X	X		7	A	B	D			
Pteronotus gymnonotus	X	X		X	1	A	B	D	G	S	F
Pteronotus parnellii	X	X	X	X	1	A	B	D	G	S	F
Pteronotus personatus	X	X	X	X	1	A	B	D	G	S	F
Phyllostomidae (74)	39	70	58	63	0						
Phyllostominae (29)	20	27	22	27	0						
Chrotoperus auritus	X	X	X	X	1	A	B	D	G	S	F
Glyphonycteris daviesi	X	X	X	X	1	A	B		G	S	F
Glyphonycteris sylvestris	X	X	X	X	1	A	B		G	S	F
Lampronycteris brachyotis	X	X		X	1	A	B	D	G	S	F
*Lonchorhina fernandezi**		X			8, 6	A	B				
Lonchorhina inusitata		X	X	X	5	A	B		G	S	F
Lonchorhina orinocensis		X			6	A	B				
Lophostoma brasiliense	X	X		X	1	A	B	D	G	S	F
Lophostoma carrikeri		X	X	X	3	A	B		G	S	F
Lophostoma schulzi			X	X	8				G	S	F
Lophostoma silvicolum	X	X	X	X	1	A	B	D	G	S	F
Macrophyllum macrophyllum	X	X		X	1	A	B	D	G	S	F
Micronycteris brosseti				X	3				G		F
Micronycteris hirsuta	X	X	X	X	1	A	B	D	G	S	F
Micronycteris megalotis	X	X	X	X	1	A	B	D	G	S	F
Micronycteris microtis	X	X	X	X	1	A	B	D	G	S	F
Micronycteris minuta	X	X	X	X	1	A	B	D	G	S	F
Micronycteris schmidtorum	X	X		X	1	A	B				F
Mimon bennettii		X	X	X	2	A			G	S	F
Mimon crenulatum	X	X	X	X	1	A	B	D	G	S	F
Phylloderma stenops	X	X	X	X	1	A	B		G	S	F
Phyllostomus discolor	X	X	X	X	1	A	B	D	G	S	F
Phyllostomus elongatus		X	X	X	2	A	B	D	G	S	F
Phyllostomus hastatus	X	X	X	X	1	A	B	D	G	S	F
Phyllostomus latifolius		X	X	X	3		B		G	S	F
Tonatia saurophila	X	X	X	X	1	A	B	D	G	S	F
Trachops cirrhosus	X	X	X	X	1	A	B	D	G	S	F
Trinycteris nicefori	X	X	X	X	1	A	B	D	G	S	F
Vampyrum spectrum	X	X	X	X	1	A	B	D	G	S	F
Glossophaginae (11)	4	11	7	9	0						
Anoura caudifer		X	X	X	2	A	B		G	S	F

	CA	WL	GH	EL	Tate	A	B	D	G	S	F
Anoura geoffroyi	X	X	X	X	1	A	B		G	S	F
Choeroniscus godmani	X	X		X	7		B	D	G	S	
Choeroniscus minor		X	X	X	2	A	B	D	G	S	F
Glossophaga longirostris		X		X	6, 7	A	B	D	G		
Glossophaga soricina	X	X	X	X	1	A	B	D	G	S	F
Lichonycteris obscura		X		X	2		B		G	S	F
Lionycteris spurrelli		X	X	X	2	A	B		G	S	F
Lonchophylla thomasi	X	X	X	X	1	A	B		G	S	F
Scleronycteris ega		X			3	A	B				
Carolliinae (5)	2	5	4	3	0						
Carollia brevicauda		X	X	X	2	A	B	D	G	S	F
Carollia castanea	X	X			7	A					
Carollia perspicillata	X	X	X	X	1	A	B	D	G	S	F
Rhinophylla fischerae		X	X		3	A					
Rhinophylla pumilio		X	X	X	2	A	B	D	G	S	F
Stenodermatinae (27)	11	25	23	22	0						
Ametrida centurio		X	X	X	5	A	B	D	G	S	F
Artibeus amplus		X	X	X	7	A	B		G	S	
Artibeus bogotensis		X	X	X	4	A	B	D	G	S	
Artibeus cinereus		X	X	X	2	A	B	D	G	S	F
Artibeus concolor		X	X	X	5	A	B	D	G	S	F
Artibeus gnomus		X	X	X	2	A	B	D	G	S	F
Artibeus jamaicensis	X	X			7	A	B	D			
Artibeus lituratus	X	X	X	X	1	A	B	D	G	S	F
Artibeus obscurus		X	X	X	2	A	B	D	G	S	F
Artibeus planirostris		X	X	X	2	A	B		G	S	F
*Artibeus sp.**		X			8		B				
Chiroderma trinitatum		X	X	X	2	A	B	D	G	S	F
Chiroderma villosum	X	X	X	X	1	A	B	D	G	S	F
Enchisthenes hartii	X	X	X		4	A					
Mesophylla macconnelli	X	X	X	X	1	A	B	D	G	S	F
*Platyrrhinus aurarius**			X		8	A	B		G	S	
Platyrrhinus brachycephalus		X		X	2		B	D	G	S	F
Platyrrhinus helleri	X	X	X	X	1	A	B	D	G	S	F
Sphaeronycteris toxophyllum		X	X		4	A	B				
Sturnira lilium	X	X	X	X	1	A	B	D	G	S	F
Sturnira tildae		X	X	X	2	A	B	D	G	S	F
Uroderma bilobatum	X	X	X	X	1	A	B	D	G	S	F
Uroderma magnirostrum	X	X	X	X	1	A	B	D	G		

(continued)

Table A11.1 (continued)

	CA	WL	GH	EL	Tate	A	B	D	G	S	F
Vampyressa bidens		X	X	X	3	A	B	D	G	S	F
Vampyressa thyone	X	X	X	X	1	A	B		G		
Vampyrodes caraccioli	X	X	X	X	1	A	B	D	G	S	F
Desmodontinae (2)	2	2	2	2	0						
Desmodus rotundus	X	X	X	X	1	A	B	D	G	S	F
Diaemus youngii	X	X	X	X	1	A	B	D	G	S	F
Natalidae (1)	0	1	1	1	0						
Natalus tumidirostris		X	X	X	6		B		G	S	F
Furipteridae (1)	1	1	0	1	0						
Furipterus horrens	X	X		X	1	A			G	S	F
Thyropteridae (3)	2	2	0	3	0						
Thyroptera discifera	X	X		X	1					S	F
Thyroptera devivoi				X	6				G		
Thyroptera tricolor	X	X		X	1	A	B	D	G	S	F
Vespertilionidae (18)	9	13	8	13	0						
Eptesicus andinus				X	4				G		
Eptesicus brasiliensis	X	X		X	1	A	B	D	G	S	
Eptesicus chiriquinus	X	X		X	1		B	D	G	S	F
Eptesicus diminutus		X			4		B				
Eptesicus furinalis	X	X	X	X	1	A	B	D	G	S	F
Histiotus humboldti			X		4	A	B				
*Lasiurus atratus**		X		X	8		B		G	S	F
Lasiurus blossevillii	X	X	X	X	1	A		D	G	S	F
Lasiurus cinereus		X	X		4	A		D			
Lasiurus ega	X	X	X	X	1	A	B	D	G	S	
Lasiurus egregius				X	5					S	F
Myotis albescens	X	X		X	1	A	B	D	G	S	F
Myotis keaysi		X			4		B				
Myotis nigricans	X	X	X	X	1	A	B	D	G	S	F
Myotis oxyotus			X		4	A	B				
Myotis riparius	X	X	X	X	1	A	B	D	G	S	F
Rhogeessa hussoni				X	5					S	
Rhogeessa io	X	X		X	1	A	B	D	G		
Molossidae (27)	15	25	9	24	0						
Cynomops abrasus		X		X	2		B		G	S	F
Cynomops greenhalli	X	X	X	X	7		B	D		S	F
Cynomops paranus		X		X	3		B		G	S	F
Cynomops planirostris		X	X	X	2	A	B	D	G	S	F
Eumops auripendulus	X	X		X	1	A	B	D	G	S	F
Eumops nanus	X	X		X	7				G		
Eumops dabbenei		X			4		B	D			
Eumops glaucinus	X	X		X	1	A	B		G	S	

	CA	WL	GH	EL	Tate	A	B	D	G	S	F
Eumops hansae	X	X		X	1	A	B	D	G		F
Eumops trumbulli		X	X	X	3		B		G	S	
Molossops neglectus		X		X	2		B		G	S	
Molossops temminckii	X	X		X	1		B		G		
Molossus aztecus	X	X			7		B	D			
*Molossus barnesi**				X	8						F
Molossus coibensis	X	X		X	1	A	B	D	G		
Molossus molossus	X	X	X	X	1	A	B	D	G	S	F
Molossus pretiosus	X	X		X	1		B	D	G		
Molossus rufus	X	X	X	X	1	A	B	D	G	S	F
Molossus sinaloae	X	X		X	1		B	D	G	S	F
Molossus sp.				X	4				G		
Neoplatymops mattogrossensis		X		X	2	A	B		G		
Nyctinomops gracilis		X	X		4	A	B	D			
Nyctinomops laticaudatus	X	X		X	1	A	B		G	S	F
Nyctinomops macrotis	X	X	X	X	1	A	B		G	S	
Promops centralis	X	X	X	X	1	A	B		G	S	F
Promops nasutus		X	X	X	2	A	B		G	S	
Primates (15)	**0**	**12**	**4**	**8**	**11**						
Cebidae (5)	0	4	1	4	4						
Cebus albifrons		X			4 (4)	A	B				
Cebus apella		X		X	3 (9)	A		D	G	S	F
Cebus olivaceus		X	X	X	5	A	B	D	G	S	F
Saguinus midas				X	8 (8)				G	S	F
Saimiri sciureus		X		X	3 (3)	A	B		G	S	F
Atelidae (3)		2	1	2	3						
Alouatta macconnelli		X	X	X	5 (9)	A	B	D	G	S	F
Ateles belzebuth		X			4 (4)	A	B				
Ateles paniscus				X	8 (8)				G	S	F
Aotidae (1)		1	1	0	1						
Aotus trivirgatus		X	X		8 (3)	A	B				
Pitheciidae (6)		5	1	2	3						
Cacajao hosomi		X	X		8	A					
Cacajao melanocephalus		X			4	A					
Callicebus lugens		X			4 (3)	A	B				
Chiropotes chiropotes				X	8 (4)				G	S	F
Chiropotes israelita		X			8	A	B				
Pithecia pithecia		X		X	8 (8)	A	B	D	G	S	F
Carnivora (17)	**12**	**17**	**16**	**16**	**16**						
Canidae	0	2	2	2	2						
Cerdocyon thous		X	X	X	6 (6)	A	B	D	G	S	

(continued)

Table A11.1 (continued)

	CA	WL	GH	EL	Tate	A	B	D	G	S	F
Felidae	6	6	6	6	5						
Leopardus pardalis	X	X	X	X	1 (1)	A	B	D	G	S	F
Leopardus tigrinus	X	X	X	X	1	A	B	D	G	S	F
Leopardus wiedii	X	X	X	X	1 (1)	A	B	D	G	S	F
Panthera onca	X	X	X	X	1 (1)	A	B	D	G	S	F
Puma concolor	X	X	X	X	1 (1)	A	B	D	G	S	F
Puma yagouaroundi	X	X	X	X	1 (1)	A	B	D	G	S	F
Mustelidae	4	5	4	4	5						
Eira barbara	X	X	X	X	1 (1)	A	B	D	G	S	F
Galictis vittata	X	X	X	X	1 (2)	A	B		G	S	F
Mustela frenata	X	X	X		4 (4)	A	B		G		
Lontra longicaudis	X	X	X	X	1 (1)	A	B	D	G	S	F
Pteronura brasiliensis		X		X	2 (3)	A	B	D	G	S	F
Procyonidae	2	4	4	4	3						
Bassaricyon beddardi		X	X	X	8 (8)	A	B		G		
Potos flavus	X	X	X	X	1 (1)	A	B	D	G	S	F
Nasua nasua		X	X	X	2 (8)	A	B	D	G	S	F
Procyon cancrivorus	X	X	X	X	1 (1)	A	B	D	G	S	F
Cetacea (2)	**0**	**2**	**0**	**2**	**0**						
Delphinidae	0	1	0	1	0						
Sotalia fluviatilis		X		X	3		B	D	G	S	F
Platanistidae	0	1	0	1	0						
Inia geoffrensis		X		X	3	A	B	D	G		
Sirenia (2)	**1**	**1**	**0**	**2**	**0**						
Trichecidae	1	1	0	2	0						
Trichecus inunguis				X	3				G		
Trichecus manatus	X	X		X	1		B	D	G	S	F
Perissodactyla (1)	**0**	**1**	**1**	**1**	**1**						
Tapiridae	0	1	1	1	1						
Tapirus terrestris		X	X	X	2 (1)	A	B	D	G	S	F
Cetartiodactyla (5)	**2**	**5**	**5**	**5**	**5**						
Tayassuidae	2	2	2	2	2						
Pecari tajacu	X	X	X	X	1 (1)	A	B	D	G	S	F
Tayassu pecari	X	X	X	X	1 (1)	A	B	D	G	S	F
Cervidae	0	3	3	3	3						
Mazama americana		X	X	X	2 (2)	A	B	D	G	S	F
Mazama gouazoubira		X	X	X	2 (6)	A	B	D	G	S	F
Odocoileus cariacou		X	X	X	6 (6)	A	B	D	G	S	F
Rodentia (58)	**4**	**46**	**27**	**36**	**37**						
Sciuridae (5)	0	4	1	2	5						
Sciurillus pusillus				X	3 (5)				G	S	F

	CA	WL	GH	EL	Tate	A	B	D	G	S	F
Sciurus aestuans		X	X	X	8 (5)	A	B	D	G	S	F
Sciurus gilvigularis		X			4 (5)	A	B				
Sciurus igniventris		X			4 (4)	A	B				
Cricetidae (32)	3	24	19	21	19						
Calomys hummelincki		X			7		B				
Holochilus sciureus		X		X	4 (3)	A	B	D	G	S	F
Neacomys dubosti				X	8					S	F
Neacomys guianae				X	8 (7)	A	B		G	S	
Neacomys paracou			X	X	8		B		G	S	F
Necromys urichi		X	X		4 (8)	A	B		G		
Nectomys melanius		X	X	X	3 (3)	A	B		G	S	F
Nectomys palmipes		X			7		B	D			
Neusticomys oyapocki				X	8						F
Neusticomys venezuelae		X	X	X	4	A	B		G		
Oecomys auyantepui		X	X	X	8 (8)		B		G	S	F
Oecomys bicolor		X	X	X	2 (2)	A	B	D	G	S	F
Oecomys concolor		X	X		4	A	B				
Oecomys rex			X	X	8 (4)		B		G	S	F
Oecomys roberti		X		X	4 (4)	A	B		G		
Oecomys rutilus		X	X	X	8 (4)		B		G	S	F
*Oecomys sp. 1**		X			8		B				
*Oecomys sp. 2**		X			8		B				
Oecomys speciosus		X			6		B				
Oecomys trinitatis	X	X	X	X	1 (10)	A	B		G		
Oligoryzomys fulvescens	X	X	X	X	1 (1)		B	D	G	S	F
*Oligoryzomys sp.**		X			8		B				
Euryoryzomys macconnelli		X	X	X	4 (4)	A	B		G	S	F
Hylaeamys megacephalus		X	X	X	5 (1)	A	B	D	G	S	F
Hylaeamys yunganus		X	X	X	3	A	B	D	G	S	F
*Podoxymys roraimae**			X		8 (8)		B		G		
Rhipidomys leucodactylus		X	X	X	3 (4)	A	B		G		F
*Rhipidomys macconnelli***			X		8 (8)	A	B		G		
Rhipidomys nitela		X	X	X	3 (10)	A	B		G	S	F
*Rhipidomys wetzeli**			X		8	A	B		G		
Sigmodon alstoni		X		X	6 (6)	A	B		G	S	F
Zygodontomys brevicauda	X	X		X	6 (6)	A	B	D	G	S	F
Erethizontidae (2)	0	2	2	2	1						

(continued)

Table A11.1 (continued)

	CA	WL	GH	EL	Tate	A	B	D	G	S	F
Coendou prehensilis		X	X	X	2 (5)	A	B	D	G	S	F
Caviidae (2)	0	2	0	2	2						
Cavia aperea		X		X	6 (6)	A	B		G	S	
Hydrochoeris hydrochaeris		X		X	2 (1)	A	B	D	G	S	F
Dasyproctidae (5)	0	4	1	2	3						
Dasyprocta fuliginosa		X			4 (4)	A	B				
*Dasyprocta guamara**		X			8			D			
Dasyprocta leporina		X	X	X	5 (5)	A	B	D	G	S	F
Myoprocta acouchy		X		X	5 (4)				G	S	F
Myoprocta pratti		X			4	A	B				
Cuniculidae (1)	1	1	1	1	1						
Cuniculus paca	X	X	X	X	1 (1)	A	B	D	G	S	F
Echimyidae (11)	0	9	3	6	7						
Dactylomys dactylinus		X			3	A	B				
Echimys chrysurus				X	5 (5)				G	S	F
Echimys semivillosus		X			10 (10)		B	D			
Isothrix orinoci		X			8 (3)	A	B				
*Isothrix sinnamariensis**				X	8				G		F
Makalata didelphoides		X		X	3 (3)	A	B	D	G	S	F
Mesomys hispidus		X		X	3 (3)	A	B		G	S	F
Proechimys cuvieri		X	X	X	3	A	B		G	S	F
Proechimys guyannensis		X	X	X	3 (2)	A	B	D	G	S	F
Proechimys hoplomyoides		X	X		8 (8)	A	B		G		
Proechimys quadruplicatus		X			4	A					
Lagomorpha (2)	1	2	1	1	1						
Leporidae	1	2	1	1	1						
Sylvilagus brasiliensis		X	X	X	1	A	B			S	
Sylvilagus floridanus	X	X			6 (6)	A	B				
Total	**108**	**248**	**158**	**222**	**92**	**208**	**241**	**147**	**222**	**194**	**186**

12 Speciation in Amazonia
Patterns and Predictions of a Network of Hypotheses

Cibele R. Bonvicino and Marcelo Weksler

Abstract

Many evolutionary and ecological hypotheses have been proposed to account for the biodiversity of Amazonia. Recent work on the phylogeographic patterns of Amazonian organisms and better understanding of the geologic history of the region enable us to specify and test the spatial, temporal, and phylogenetic predictions of these biogeographic models. Four main causal factors have been postulated, and their expectations and predictions are reviewed here: (1) The role of rivers as barriers to species dispersal is evident, but their role as primary agents of speciation is still unclear. Diversification of a few mammalian species in large, blackwater river basins has been associated with rivers, but this role has been rejected for most small mammals of the whitewater Rio Juruá. River characteristics such as size, stream flow, and geology are important in their role as a distributional barrier. (2) Geotectonic changes associated with uplift of the central Andes and increase in global sea levels produced gradual submersion of western Amazonia, creating the wetland system known as Lake Pebas (or Pebas Sea). The effects of marine incursions have been associated with diversification of several organisms, but no mammalian phylogenetic pattern has been effectively linked to Lake Pebas. (3) The genetic structure of rodents and frogs in Western Amazonia is correlated with the location and orientation of the Iquitos Paleoarch, an ancient drainage barrier that survived until the Pliocene. Another paleoarch, the Purus, has also been postulated as an important barrier during Miocene marine incursions. However, some geologists argue that most arches did not greatly influence Cenozoic biological diversification. (4) Several variations of refugia in response to Quaternary climatic fluctuations are not supported by genetic studies of small mammals. New versions of the refugia hypothesis are tied to earlier climatic events in the Tertiary, but this increases the uncertainty of both spatial and temporal predictions. Our knowledge of the biogeographic processes leading to patterns within Amazonia is still incomplete, as the mammalian fauna of this vast region and the spatial and temporal genetic patterns of most of its species remain poorly known.

12.1 Introduction

The Amazonian region (fig. 12.1) harbors one of the richest mammalian faunas of the world, with approximately 140 genera and 430 species (Silva, Rylands, and Fonseca 2005). Although these values probably underestimate the actual biodiversity of the region (de Vivo 1996; Voss and Emmons 1996), Amazonian mammals account for approximately one-third of all South American mammalian diversity (~1260 species; Wilson and Reeder 2005, plus species recognized after 2003). What are the geographic factors and evolutionary processes that generated such biological richness? Alfred R. Wallace (1853) developed the first scientific hypothesis to explain the diversity of primates in the Amazonian basin. Subsequently, several biogeographic models have been proposed to account for the region's plant and animal diversity (table 12.1; see also reviews in Antonelli et al. 2010; Bush 1994; Haffer 1997, 2008; Hall and Harvey 2002; Hoorn et al. 2010; Moritz et al. 2000; Noonan and Wray 2006; Patton and da Silva 2005; Simpson and Haffer 1978).

The historical scenarios proposed to explain genetic differentiation and speciation of Amazonian organisms differ in several respects: modes of speciation (sympatric, parapatric, peripatric, or allopatric; Agapow et al. 2004; Bolnick and Fitzpatrick 2007; Endler 1977); type of physical barriers that control outcrossing and dispersal (topographic, riverine, or ecologic; Hall and Harvey 2002); temporal framework (ranging from late Pleistocene to Miocene; Haffer 2008; Ribas, Gaban-Lima, and Miyaki 2005); and geographic location of speciation events (e.g., Andes; in situ refuges; ecotonal zones; circum-Amazonia; endemism centers). The most compelling hypotheses fall into three major categories based on their causal factors, namely climatic, geomorphic, and ecological, or combinations of these factors (table 12.1). Climatic hypotheses are based on changes in the earth's climate due to Milankovitch cycles (Berger, Loutre, and Dehant 1989), resulting in altered distribution of biomes and vegetation types (Bennett 1990; Haffer 1997). Geomorphic hypotheses are associated with physical structures and paleofeatures of the landscape, such as rivers, mountains, and waterways. The last category of hypotheses rests on speciation models that do not require historical events but instead result from ecological factors, as is the case with gradient (Endler 1977) and taxon-pulse hypotheses (Erwin 1979).

These different scenarios offer predictions on the historical (phylogenetic and coalescent) and spatial structure of the genetic variation of Amazonian species (Aleixo 2004, 2006; Antonelli et al. 2009; Barraclough and Vogler 2000; Lessa, Cook, and Patton 2003). Estimations and statistical tests of the phylogenetic-phylogeographic-population genetics continuum (Arbogast et al. 2002; Carstens et al. 2004; Emerson, Paradis, and Thebaud 2001) can in turn be used to infer genetic configurations like tree topologies (e.g., sister-group

Table 12.1 Major historical and ecological hypotheses explaining mammalian Amazonian diversity. *a*, Some authors argue for older events in the Tertiary (e.g., Haffer 2008); *b*, Depending on river permeability; *c*, Ring species scenario (Marroig and Cerqueira 1997); *d*, Some authors argue for younger events in the Quaternary (e.g., Frailey et al. 1988; Nores 1999).

Scenario	Main causal factors	Barrier to gene flow	Speciation mode	Spatial location	Temporal framework	Primary references
1a. Area Refugia	Climatic	Matrix of nonforested vegetation (surrounding forest refuges)	Allopatric (vicariance among refugia)	In situ (within Amazonia)	Pleistocene Glacial maxima[a]	Haffer 1969, 1997; Simpson and Haffer 1978
1b. River Refugia	Climatic	None (Isolation by Distance along continuous distribution)	Parapatric	Headwaters of tributaries	Pleistocene Glacial maxima	Haffer 1997; Noonan and Wray 2006
2a. Vanishing Refugia	Climatic, Ecological	Ecological	Peripatric	In situ / Peripheral	Continuous during Pleistocene	Vanzolini and Williams 1981
2b. Disturbance-vicariance	Climatic, Geomorphic, Ecological	Matrix of "normal" forest (surrounding "islands of special environment")	Allopatric	Periamazonian, (especially montane areas)	Pleistocene Glacial maxima	Bush 1994; Colinvaux 1998
3. Riverine Barriers	Geomorphic	Rivers	Allopatric or parapatric[b]	In situ	Continuous	Ayres and Clutton-Brock 1992; Wallace 1853

(continued)

Table 12.1 *(continued)*

Scenario	Main causal factors	Barrier to gene flow	Speciation mode	Spatial location	Temporal framework	Primary references
4. Paleoarches/ paleobasins	Geomorphic	Topographic /Environmental; due to tectonic activity either directly or through resulting differences in depositional histories, soils, and eventual vegetation types	Allopatric	Amazon basin	Late Miocene – Early Pliocene	Patton and da Silva 1998; Patton et al. 1997; Patton, da Silva, and Malcolm 2000; da Silva and Patton 1998
5. Marine incursions	Geomorphic, Climatic	Amazon shallow sea	Allopatric, parapatric[c]	Periamazonian (areas above 100m)	Miocene[d]	Frailey et al. 1988; Hoorn and Vonhof 2006; Lovejoy, Albert, and Crampton 2006; Marroig and Cerqueira 1997; Nores 1999
6. Gradient	Ecological	None (steep environmental gradient)	Parapatric	Periamazonian (Ecotones, transition zones)	Continuous	Endler 1977, 1982
7. Taxon pulses	Ecological	None	Sympatric, Parapatric	In situ	Continuous	Antonelli et al. 2010; Erwin 1979; Whinnet et al. 2005

relationships; monophyly versus paraphyly, star patterns; Barraclough and Vogler 2000; Patton and da Silva 1998), historical population size (e.g., recent expansion versus stable population size; Lessa, Cook, and Patton 2003), structuring of genetic diversity (e.g., F_{st} structure; nucleotide divergence among clades), spatial distribution of haplotype groups or haplogroups (e.g., clines, presence of secondary contact or suture zones; Endler 1977), and to date most recent common ancestors or cladogenetic events.

Investigation of the spatial and historical genetic structure of mammalian taxa in Amazonia has started to reveal the complexity of these hypothetical networks. In this chapter we present a brief overview of the most relevant historical and biogeographic models for explaining Amazonian diversity, with some of their genetic consequences and expected patterns. We synthesize the few phylogeographic and phylogenetic reports in Amazonian mammals (and other organisms), and comment on the current geomorphological and palynological evidence for the required landscape associated with each hypothesis.

12.2 Climatic Oscillation: Refugia and Related Hypotheses

The forest refugia hypothesis was progressively developed and refined as a cause of organismal diversity, becoming the dominant explanation for Neotropical diversity during the final decades of the twentieth century (Haffer 1969, 1974, 1981, 1985, 1990; Prance 1973, 1982, 1987; Vanzolini 1970). Allopatric speciation during Pleistocene (or even Tertiary) glacial maxima is the main mechanism proposed for the refugia hypothesis (Haffer 2008). Glacial decreases in temperature and humidity resulted in shrinking islands of tropical rain forests surrounded by less mesic habitats, causing forest-dependent populations to become isolated. During interglacial cycles, differentiated populations dispersed differentially from their restricted ranges as forests expanded; thus, refugia would have maintained their higher diversity. Three sources of evidence pointed to the existence of refugia: (1) palynological records of vegetational changes due to climatic oscillations associated with Quaternary glacial-interglacial intervals—increased aridity during glacial maxima led to open-vegetation formations (llanos and cerrado) in areas now occupied by forest (Boisseau et al. 1996; van der Hammen 1974; van der Hammen and Absy 1994); (2) soil desiccation, suggesting a similar pattern (Ab'Sáber 1982); and (3) congruence of areas of high endemicity of different groups with areas of higher present rainfall (Whitmore and Prance 1987), potentially indicating the location of the past refugia.

There is now a consensus, based on fossil-pollen data, that conditions during parts of the Pleistocene differed from the present because temperature, precipitation, and CO_2 levels were lower and the Amazon basin experienced increased seasonality during glacial maxima (Bush 1994; Colinvaux 1987; Colinvaux et al. 1996; van der Hammen and Hooghiemstra 2000). It is debatable, however, whether open vegetation was extensively distributed throughout the Amazon basin. Reviewing palynological data, van der Hammen and Hooghiemstra (2000) described several open, arid vegetation zones in the lower Rio Branco region (presently central Amazonia), Rondônia (western Amazonian Brazil), and in Guyana and Suriname. Conversely, Haberlea and Masli (1999) and Ron (2000) argued that the Amazon basin was not extensively replaced by savanna vegetation during the last glacial period, as required by traditional refuge theory (Haffer 1969, 1997). It has been suggested that the last glacial cycle resulted in changes in the composition of the Amazonian forest rather than in contraction to small islands (Colinvaux et al. 2001; Lessa, Cook, and Patton 2003). As postulated by Colinvaux (1998, 95), "It is unlikely that ice age climates of the Amazon were sufficiently arid to fragment the forest as required by Haffer's refugia hypothesis. However, glacial Amazonian climates were colder and with reduced CO_2 concentrations which would have been most effective on the biota in the elevated areas stipulated to have been refugia. If local endemicity of butterflies or birds records Pleistocene speciation, this is because glacial climates provided cool, CO_2 starved islands in a sea of continuous forest."

Genetic studies on small mammals in western Amazonia, reviewed by Moritz et al. (2000) and Lessa, Cook, and Patton (2003), attest to a history of population stability of Amazonian forest taxa through Pleistocene glacial cycles, contradicting the refugia hypothesis. In particular, comparisons of population growth parameters between selected Amazonian and Beringian small mammals suggested that the current genetic composition of various Amazonian taxa does not reflect population retractions to small isolates during the last glacial cycle (Lessa, Cook, and Patton 2003). However, the refugia hypothesis is supported by recent studies of ants (Solomon et al. 2008).

Variations of the classic refugia hypothesis include *river refugia* (Haffer 2008) and *vanishing refugia* (Vanzolini and Williams 1981) hypotheses. In the river refugia hypothesis, vegetational changes in northern and southern Amazonia cause mesic vegetation to remain only along the gallery forests of large rivers (Haffer 2008). These conditions would have resulted in a highly linear (unidimensional) distribution of species along river corridors, creating a single axis of gene dispersal and good fit to an isolation-by-distance model, especially for

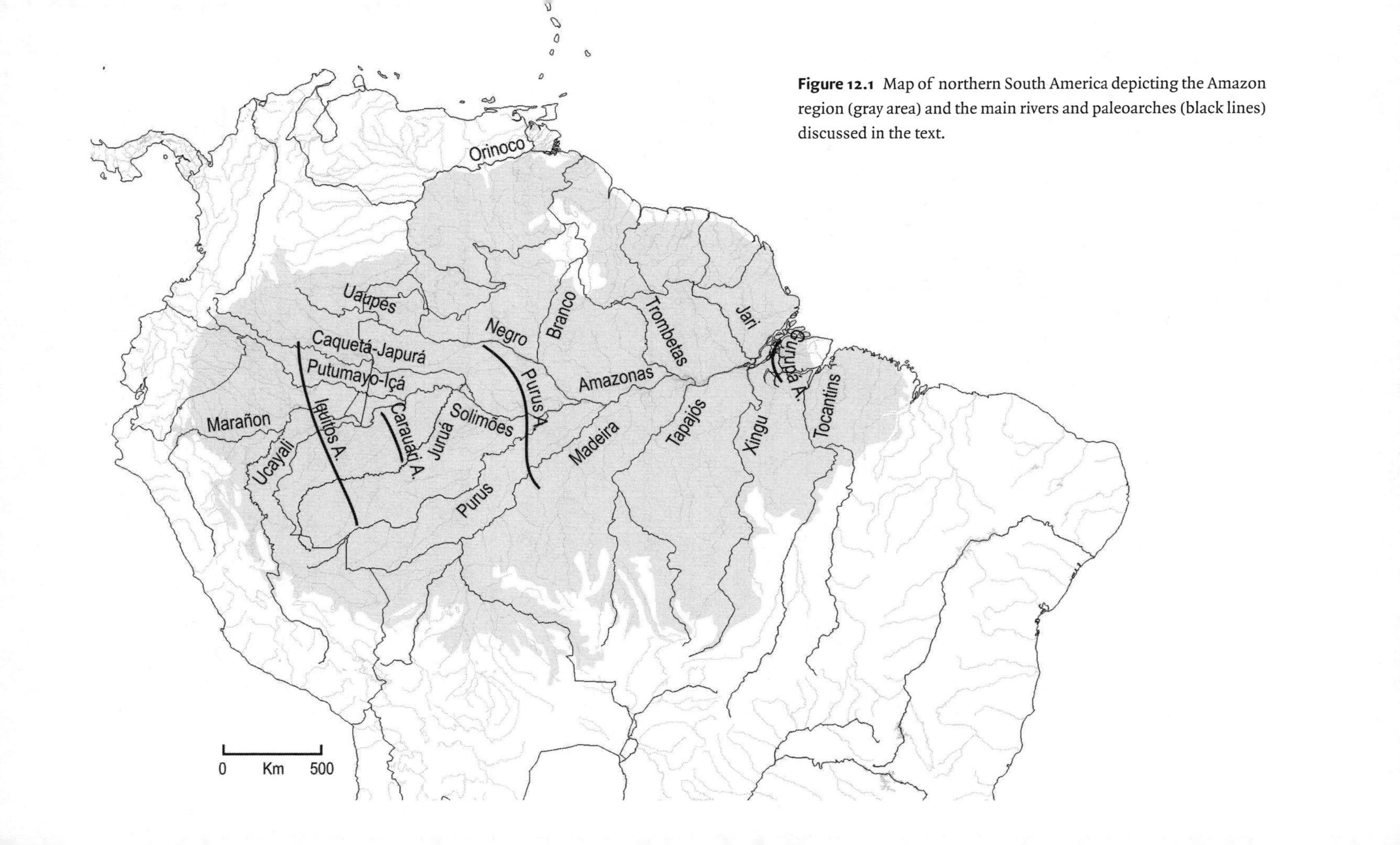

Figure 12.1 Map of northern South America depicting the Amazon region (gray area) and the main rivers and paleoarches (black lines) discussed in the text.

species with deme or metapopulation structure and with limited dispersal ability (Haffer 2008). In the vanishing refuge model (Vanzolini and Williams 1981), some populations differentiated to species via directional selection favoring tolerance of ecotones or dry habitats as rain forest patches became too small to retain viable populations. This hypothesis was based on the observation that sister taxa usually occur in geographically adjacent but different habitats (a pattern also predicted by gradient models; Moritz et al. 2000).

Another hypothesis dependent on climatic oscillations, but coupled with ecological processes, is the *disturbance-vicariance* hypothesis (Bush, 1994; Bush and de Oliveira 2006; Colinvaux 1993, 1998). In this scenario, species became isolated in microrefugia (Rull 2009) in peripheral regions of Amazonia due to invasion and counterinvasion during the cold/warm periods of Pleistocene. In ecotones between the Amazonian forest and open-vegetation biomes (Cerrado and Llanos) or along the elevational transitions of the Andean piedmont and the Guiana Shield (Rull 2004), the drier, colder conditions of glacial maxima might have resulted in regions of *perturbed conditions* (Bush 1994) or of *special forest* (Colinvaux 1998), which became isolated either by topographic landscape (e.g., different mountain systems in the Eastern Andean piedmont) or by ecological conditions (i.e., the *natural forests* of Colinvaux). The disturbance-vicariance hypothesis has received some supporting evidence from studies of dendrobatid frogs (Noonan and Wray 2006), but has not yet been tested using mammals.

12.3 Riverine Hypothesis: Primary Factor or Secondary Barrier?

The role of rivers as barriers to species dispersal has been evident for primates since Wallace (1853) demonstrated the abrupt discontinuity of several primate species across Amazonas tributaries (confirmed by community-composition studies by Ayres and Clutton-Brock 1992; see also Bates 1863; Hershkovitz 1977). However, their role as primary factors causing speciation is not clear. It appears that river size is generally a good predictor of turnover in species composition of faunas on opposite shores (Ayres 1986; Ayres and Clutton-Brock 1992), but it is unclear whether it produces the same effect on haplotype diversity. Rivers might thus be powerful barriers to the range limits of a species, but how do they affect the spatial (geographic) and historical (phylogenetic and coalescent) genetic structure of populations? In addition, do rivers actually lead species to differentiate or simply limit the reexpansion of former allopatric disjuncts (Moritz et al. 2000)?

The riverine hypothesis assumes that animal populations were formed and became effectively isolated from one another following the formation of the current Amazon river network (fig. 12.1) after the late Miocene (Figueiredo et al. 2009). A general model for the riverine diversification hypothesis predicts that populations on one riverbank will be monophyletic with respect to those of the opposite bank (i.e., they will be reciprocally monophyletic), and that the two lineages of opposite-bank populations/taxa will form a clade relative to those from elsewhere within the species range (Patton and da Silva 1998). Thus, three basic patterns of phylogeographic structure can be expected for terrestrial mammals (Gascon et al. 2000; Moritz et al. 2000; Patton and da Silva 1998; Patton, da Silva, and Malcolm 2000): (1) rivers failing to influence genetic structure, in which populations of both banks share haplotypes and the overall phylogeographic structure forms a coalescent tree without riverbank separation (a pattern also expected for Chiroptera; Ditchfield 2000); (2) rivers functioning as barriers to gene flow and representing a primary causal factor for speciation, with reciprocally monophyletic sister groups on each riverbank; and (3) secondary contact of formerly disjunct groups, with clades on each riverbank sharing more recent common ancestors with other populations, to the exclusion of the opposite-bank populations. Genetic distance should also increase, from low estimates in nonbarrier systems to high ones for non-sister clades, while populations undergoing differentiation should show intermediate values.

A review of published phylogeographic patterns of terrestrial mammals with distributions across rivers of the Amazon basin (table 12.1) shows that populations of different riverbanks were reciprocally monophyletic in 27 cases, while there was no phylogenetic (or coalescent) structure between opposite riverbanks in 16 cases. Despite clear indications of rivers as phylogenetic suture zones, populations of different banks formed restricted sister clades in only four taxa: *Callicebus lugens* and *Isothrix negrensis* of the Rio Negro, and *Proechimys echinothrix* and *Proechimys steerei* in the Solimões (and probably *Proechimys echinothrix* in the Juruá and *Bassaricyon* spp. in the Negro). In more than 80% of cases, however, the most recent common ancestor of each bank's sample included populations outside the river system, indicating that other causal factors cannot be discounted.

Variation among rivers was also large. The Negro and Solimões rivers (fig. 12.1) harbor most taxa with reciprocal monophyly between river margins (14 and 7, respectively), but the Juruá constituted a putative barrier for only 3 of 13 taxa. Amazonian rivers show drastic differences regarding sediment suspension and flow (Galvão 1977; Latrubesse et al. 2005). They can be classified as whitewater rivers (with the presence of clay sediments and characterized by the presence of meanders), clean-water rivers (with little clay sediment), and blackwater rivers, due to presence of suspension of organic material and usually without meanders. These differences are important with respect to their role as barriers because rivers with meanders, such as the Juruá, are obviously less efficient geographic barriers than those lacking them. In addition, some rivers, such as the Negro, have maintained their present-day drainage system for much longer than others (Latrubesse and Franzinelli 1998, 2002, 2005), enhancing its effectiveness as a barrier to gene flow long enough to enable speciation.

Among other organisms, the riverine hypothesis has been associated with diversification of terra firme birds in the Rio Negro region (Aleixo 2004); nonpasserine birds in the Amazon, Solimões, and Ucayali rivers (Armenta, Weckstein, and Lane 2005); and birds (Chevirona, Hackett, and Capparella 2005) and dendrobatid frogs (Noonan and Wray 2006) in the Napo and Marañón upper basins. Nevertheless, the role of rivers as primary causal factors or reinforcement for previous differentiation needs fuller examination.

12.4 THE PEBAS SEA AND OTHER WETLAND SCENARIOS

The modern structure of the fluvial megasystem of the Amazon basin is the result of transcurrent faults affecting parts of the South America Plate to the west and the Caribbean Plate to the north. This process began in the Miocene (Franzinelli and Igreja 2002; Figueiredo et al. 2009) and has completely changed the regional landscape. From the time of separation of South America from Africa until the end of the Oligocene, a paleofluvial system—the Paleo-Orinoco—dominated the drainage of northwestern Amazonia and the foreland Andean basins. Rivers drained the periphery of the Amazon Craton and flowed in a northwest (and probably southwest) direction (Hoorn et al. 2010).

In the early Miocene (~23 Ma), geotectonic changes in the Amazonian basin associated with ongoing uplift of the eastern cordillera of the central Andes gradually submerged western Amazonia, creating a long-lived (17–11 Ma) continental water body known as Lake Pebas, or the Pebas Sea (Lundberg et al. 1998; Wesselingh et al. 2002). The Amazon was characterized by a freshwater wetland ecosystem dominated by shallow channels and lakes, floodplain lakes, and swamps supplied with water and sediments from the Andes, and repeatedly influenced by marine incursions when sea levels rose (Hoorn and Vonhof 2006; Lovejoy, Albert, and Crampton 2006). Other versions of the lake model date it to the end of the Pliocene into the Quaternary (Klammer 1984; Marroig and Cerqueira 1997; Sombroek 1966), although there is no available evidence of sediment deposition during these periods (Figueiredo et al. 2009; Hoorn et al. 1995; Hoorn and Vonhof 2006).

The effects of marine incursions have been associated with diversification of freshwater fishes (Lovejoy, Albert, and Crampton 2006), terra firme birds (Aleixo 2004, 2006), ants (Solomon et al. 2008), and both paradoxical (Garda and Cannatella 2007) and poison frogs (Santos et al. 2009). More studies, however, are necessary for understanding the association of mammalian patterns with the marine incursion model, which represents a relevant area of future research for Amazonia mammalian biogeography.

12.5 Paleoarches

Several tectonic elements (geological "ridges" or "paleoarches") cross the Amazon basin (fig. 12.1): the Iquitos (= Jutai) Arch in Peru and Acre; the Carauari Arch across the Negro and Solimões; the Purus Arch (west of Manaus); the Monte Alegre Arch (west of the Tapajós); and the Gurupa (west of Marajó; Patton and da Silva 1998; Roddaz et al. 2005). These arches, connecting the Guianan and Brazilian Shields, subdivided the Amazon basin into several geological sub-basins, and have been implicated as primary factors in biological diversification (Patton et al. 1997; Haffer 2008).

The Paleoarch model postulates that tectonic events associated with Andean orogeny during the Miocene and Pliocene had geological effects in the Arches, in which they in turn influenced the diversification of Amazonian organisms. Under this model, paleobasins between arches form the regional centers for the diversification of organisms (Patton and da Silva 1998). Empirical evidence for the arch as a causal factor comes from correspondence between phylogeographic breaks in genetic data and the position of the arch. The Iquitos Arch is probably the best-known example of association between arch position and phylogeographic breaks. The Iquitos Arch has been postulated as a primary factor in the diversification of rodents (e.g., *Mesomys*, *Proechimys*, *Neacomys*, *Oecomys*) and frogs (Gascon et al. 2000; Patton and da Silva 1998, 2005; Patton et al. 1997; Patton, da Silva, and Malcolm 2000; Symula, Schulte, and Summers 2003). An interesting point is that the haplotype suture zone found in the Juruá River and associated with the Iquitos Arch also corresponded to the limit between "Floresta Ombrofila Aberta" and "Floresta Ombrofila Fechada" (Patton and da Silva 2005), suggesting that by influencing the divergence of forest tree species that comprise mammal habitat, paleoarches may have also helped create ecological variation within Amazonia that is the focus of various ecological hypotheses.

Some geologists, on the other hand, are uncertain of the role of tectonic arches as causal factors of Cenozoic diversification. As Rossetti, Mann de Toledo, and Góes (2005, 86) explained: "[M]ost of these [tectonic arches] are Paleozoic or Mesozoic structures buried under a mantle of Cretaceous and Cenozoic deposits (e.g., Caputo 1991), thus no longer representing elevated elements of the terrain. For instance, the Purus Arch located to the west of Manaus, considered to be an important vicariance feature, is a geologic structure that occurs >1000 m of depth under Cretaceous rocks of the Alter do Chão Formation (Wanderley Filho 1991)." Clearly, additional research on the geology of these arches and their correlation with biogeographic patterns is required to understand their roles on the biological diversification of Amazon mammals.

12.6 Ecological Hypotheses: Gradients and Taxon Pulses

Ecological hypotheses place greater emphasis on the fine-scale processes associated with speciation, whatever its underlying cause (such as geography or climatic oscillations). The principal models in this category are the gradient hypothesis and taxon pulses. The gradient hypothesis is based on parapatric speciation across steep environmental gradients without geographic separation of the populations. The gradient hypothesis does not require isolation of populations by vicariance, but rather postulates that populations can evolve by divergent selection across environmental gradients, like changes in precipitation and elevation, despite ongoing gene flow (Endler 1977; Moritz et al. 2000). Papers testing the gradient hypothesis for Neotropical fauna have both refuted (Patton and Smith 1992; Patton, Myers, and Smith 1990 for small mammals) and supported this hypothesis (Mallet 1993 for butterflies; see Moritz et al. 2000 for a review).

Taxon pulses, as proposed by Erwin (1979, 1981, 1985), are defined as cyclic adaptive shifts over time to different habitats caused by repeated expansion and contraction of taxon ranges. The model assumes that distributional ranges of taxa periodically fluctuate around a more stable, continuously occupied center (Halas, Zamparo, and Brooks 2005). The pulses occur when previous barriers break down and expansion into suitable habitats becomes possible. The expansion may eventually result in fine-scaled allopatric speciation, where the barriers to interbreeding are subtle genetic and environmental differences rather than vicariant obstacles (Antonelli et al. 2010; Lim 2008, 2009). A possible time

Table 12.2 Comparisons of phylogenetic parameters of populations in the Rio Negro and other basins. Listed taxa are present on both banks of the rivers. Unless otherwise noted, the genetic distance is uncorrected nucleotide distance (p%) based on the cytochrome-*b* DNA sequence data. The column "Narrow distribution in each margin" is relevant in case of "Reciprocal monophyly" because rivers acting as barriers could be demonstrated only for taxa with narrow distribution in each margin, while in the case of taxa with widespread distribution other factors could also be acting.

Taxon	Basin	Reciprocal monophyly	Narrow distribution in each margin	Genetic Distance	Number of species	Data source
Didelphimorphia						
Metachirus	Juruá	No		1.78 [b]	1	Patton, da Silva, and Malcolm 2000
Metachirus	Negro	Yes	No	13.02 [b]	2?	da Silva and Patton 1998; Patton, da Silva, and Malcolm 2000
Metachirus	Solimões	Yes	No	8.26 [b]	2?	da Silva and Patton 1998; Patton, da Silva, and Malcolm 2000
Micoureus demerare	Juruá	No		3.91 [b]	1	da Silva and Patton 1998; Patton, da Silva, and Malcolm 2000
Micoureus demerarae vs M. regina	Juruá	Yes	No	16.4 [b]	2	da Silva and Patton 1998; Patton, da Silva, and Malcolm 2000
Micoureus	Negro	Yes	No	7.5 [b]	2	da Silva and Patton 1998; Patton, da Silva, and Malcolm 2000
Micoureus	Solimões	Yes	No	8.3 [b]	2	da Silva and Patton 1998; Patton, da Silva, and Malcolm 2000
Philander	Amazonas	No	—	1.5 [b]	1	Patton and da Silva 1998; Patton, da Silva, and Malcolm 2000
Philander	Juruá	No	—	—	1	Patton, da Silva, and Malcolm 2000
Philander	Negro	Yes	No	6.9 [b]	2?	Patton and da Silva 1998; Patton, da Silva, and Malcolm 2000
Cingulata						
Dasypus novemcinctus	Negro	Yes	No	5.6–8.4	1	Souza 2008
Cetartiodactyla						
Pecari tajacu	Negro	Yes	No	11.7	1	C.R. Bonvicino, unpublished data

(continued)

Table 12.2 *(continued)*

Taxon	Basin	Reciprocal monophyly	Narrow distribution in each margin	Genetic Distance	Number of species	Data source
Carnivora						
Bassarycion	Negro	Yes	Yes?	5.4	2	C.R. Bonvicino, unpublished data
Panthera onca	Amazonas	No	—	—	—	Eizerik et al. 2001
Potos flavus	Negro	Yes	Yes?	1.1	1	C.R. Bonvicino, unpublished data
Primates						
Alouatta	Negro	Yes	No	5.0	2	C.R. Bonvicino, unpublished data
Alouatta macconnelli	Branco	Yes	No	0.4	1	C.R. Bonvicino, unpublished data
Aotus	Negro	Yes	No	6.7	2	Menezes, pers. comm.
Ateles belzebuth	Tapajós	No	—	—	—	Collins and Dubach 2000a,2000b
Callicebus lugens	Negro	Yes	Yes	2.1	1	Casado, Bonvicino, and Seuánez 2007
C. (Cebus)	Negro	Yes	No	3.7	2	C.R. Bonvicino, unpublished data
C. (Sapajus) apella	Negro	No	—	—	1	C.R. Bonvicino, unpublished data
Saguinus fuscicollis	Juruá	Yes	No	2.3 [b]	2?	Patton, da Silva, and Malcolm 2000; Peres, Patton, and da Silva 1996
Saguinus niger complex	Tocantins	Yes	No	8.6 [c]	2	Vallinoto et al. 2006
Rodentia: Caviomorpha						
Dactylomys dactylinus	Amazonas	No	—	2.3 [b]	1	Patton and da Silva 1998; Patton, da Silva, and Malcolm 2000
Echimys sp.	Amazonas	Yes	?	4.2 [b]	?	Patton and da Silva 1998
Isothrix negrensis	Negro	Yes	Yes	4.4	1	Bonvicino, de Menezes, and de Oliveira 2003

Table 12.2 *(continued)*

Taxon	Basin	Reciprocal monophyly	Narrow distribution in each margin	Genetic Distance	Number of species	Data source
Isothrix	Solimões	No	—	5.1 [b]	1?	Patton and da Silva 1998; Patton, da Silva, and Malcolm 2000
Makalata	Branco	Yes	No	10.2	2	C.R. Bonvicino, unpublished data
Makalata	Negro	Yes	No	9.2	2	C.R. Bonvicino, unpublished data
Mesomys hispidus	Juruá	No	—	—	1	Patton, da Silva, and Malcolm 2000
Mesomys hispidus	Solimões	No	—	1.4 [b]	1	Patton and da Silva 1998; Patton, da Silva, and Malcolm 2000
Proechimys echinothrix	Juruá	Yes	No	4.3 [b]	2	Patton, da Silva, and Malcolm 2000
Proechimys echinothrix	Solimões	Yes	Yes	10.8 [b]	2	Patton, da Silva, and Malcolm 2000
Proechimys goeldii group	Negro	Yes	No	10.2 [b]	2	Patton and da Silva 1998; Patton, da Silva, and Malcolm 2000
Proechimys simonsi group	Juruá	No	—	—	1	Matocq, Patton, and da Silva 2000; Patton, da Silva, and Malcolm 2000
Proechimys steerei	Juruá	No	—	—	1	Matocq, da Silva, and Malcolm 2000; Patton, da Silva, and Malcolm 2000
Proechimys steerei	Solimões	Yes	Yes	6.3 [b]	2	Patton and da Silva 1998
Proechimys spp.	Solimões	Yes	No	9.6 [b]	2?	Patton and da Silva 1998
Rodentia: Sigmodontinae						
Euryoryzomys	Solimões	Yes	No	11.2 [b]	2	Patton and da Silva 1998
Hylaeamys megacephalus	Juruá	No	—	—	1	Patton, da Silva, and Malcolm 2000
Hylaeamys	Solimões	Yes	No	15.0 [b]	2	Patton and da Silva 1998; Patton, da Silva, and Malcolm 2000
Neacomys minutus	Juruá	No	—	—	1	Patton, da Silva, and Malcolm 2000
Oecomys bicolor	Juruá	No	—	—	1	Patton, da Silva, and Malcolm 2000
Oecomys sp.	Juruá	Yes	Unknown	7.3 [b]	2	Patton, da Silva, and Malcolm 2000
Oligoryzomys microtis	Juruá	No	—	—	1	Patton, da Silva, and Malcolm 2000

b: Kimura two-parameter distances (%)
c: KHY85+gamma distribution distance, D-loop

framework in South America for taxon pulses is the beginning of Amazon basin formation and increased aridity in northern South America during the Late Miocene; taxon pulses were proposed as an important mechanism in the diversification of emballonurid bats (Lim 2008, 2009).

12.7 Conclusions

Our knowledge of the biogeographic processes leading to patterns within Amazonia is still in its infancy—the mammalian fauna of this vast region is still undersampled and spatial and temporal genetic patterns of most of its organisms are poorly known. A comprehensive biogeography framework for the Amazon basin should be based on robust phylogenetic and phylogeographic structures of multiple groups of organisms with well-known genetic and spatial distributions. Using process-based methods and statistical evaluation of data will allow powerful testing of specific hypotheses, contributing to the development of historical biogeographic reconstructions (Sanmartín, Enghoff, and Ronquist 2001). Assessment of historical demographic changes are possible with coalescent-based methods (e.g., Fu 1997; Kuhner, Yamato, and Felsenstein 1998), which allow for the identification of demographic expansions, dramatic growth from refugia areas, geographic subdivisions, common demographic histories, and stable populations (Lessa, Cook, and Patton 2003). We present here expected values or ranges of phylogenetic and phylogeographic parameters based on the temporal, geographic, and population expectations of the diversity mechanisms (tables 12.1, 12.2).

Acknowledgments

We are grateful to FUNASA for facilitating our fieldwork in Barcelos, Amazonas; to A. Junqueira, J. A. de Oliveira, P. Gonçalvez, V. Pena Firme, P. Borodin, P. Albajar, F. Escarlate, and P. S. D'Andrea for collaborating in fieldwork; to H. N. Seuánez for reading a previous version of this manuscript; and P. Velazco for providing access to unpublished data. Work was supported by CAPES and CNPq (Conselho Nacional para o Desenvolvimento Científio e Tecnológico, Brazil), and fellowship to CRB.

Literature Cited

Ab'Sáber, A. N. 1982. "The Paleoclimate and Paleoecology of Brazilian Amazonia." In *Biological Diversification in the Tropics*, edited by G. T. Prance, 41–59. New York: Columbia University Press.

Agapow, P. M., O. R .P. Bininda-Emonds, K. A. Crandall, J. L. Gittleman, G. M. Mace, J. C.

Marshall, and A. Purvis. 2004. "The Impact of Species Concept on Biodiversity Studies." *Quarterly Review of Biology* 79:161–79.

Aleixo, A. 2004. "Historical Diversification of a Terra-Firme Forest Bird Superspecies: A Phylogeographic Perspective of the Role of Different Hypothesis of Amazonian Diversification." *Evolution* 58:1303–17.

———. 2006. "Historical Diversification of Floodplain Forest Specialist Species in the Amazon: A Case Study with Two Species of the Avian Genus *Xiphorhyncus* (Aves: Dendrocolaptidae)." *Biological Journal of the Linnean Society* 89:383–95.

Antonelli, A., J. A. A. Nylander, C. Pearsson, and I. Sanmartín. 2009. "Tracing the Impact of the Andean Uplift on Neotropical Plant Evolution." *Proceedings of the National Academy of Sciences, USA* 106:9749–54.

Antonelli, A., A. Quijada-Masareñas, A. J. Crawford, J. M. Bates, P. M. Velazco, and W. Wüster. 2010. "Molecular Studies and Phylogeography of Amazonian Tetrapods and Their Relation to Geological and Climatic Models." In *Amazonia, Landscape and Species Evolution*, edited by C. Hoorn and F. P. Wesselingh, 386–404. Chichester: Wiley-Blackwell.

Arbogast, B. S., S. V. Edwards, J. Wakeley, P. Beerli, and J. Slowinski. 2002. "Estimating Divergence Times from Molecular Data on Phylogenetic and Population Genetic Timescales." *Annual Review of Ecology and Systematics* 33:707–40.

Armenta, J. K., J. D. Weckstein, and D. F. Lane. 2005. "Geographic Variation in Mitochondrial DNA Sequences of an Amazonian Nonpasserine: The Black-Spotted Barbet Complex." *Condor* 107:527–36.

Ayres, J. M. 1986. *Uakaris and Amazonian Flooded Forests*. Unpublished PhD diss., University of Cambridge, Cambridge.

Ayres, J. M., and T. H. Clutton-Brock. 1992. "Rivers Boundaries and Species Range Size in Amazonian Primates." *American Naturalist* 140:531–37.

Barraclough, T. G., and A. P. Vogler. 2000. "Detecting the Geographic Pattern of Speciation from Species-Level Phylogenies." *American Naturalist* 155:419–34.

Bates, H. W. 1863. *The Naturalist on the River Amazons*. London: Murray.

Bennett, K. D. 1990. "Milankovitch Cycles and Their Effects on Species in Ecological and Evolutionary Time." *Paleobiology* 16:11–21.

Berger, A., M. F. Loutre, and V. Dehant. 1989. "Pre-Quaternary Milankovitch Frequencies." *Nature* 342:133.

Boisseau, C., J. Vigue, M.-P. Ledru, P. I. S. Braga, F. Soubies, M. Fournier, L. Martin, K. Suguio, and B. Turcq. 1996. "The Last 50,000 Years in the Neotropics (Southern Brazil): Evolution of Vegetation and Climate." *Palaeogeography, Palaeoclimatology, Palaeoecology* 123:239–57.

Bolnick, D. I., and B. M. Fitzpatrick. 2007. "Sympatric Speciation: Models and Empirical Evidence." *Annual Review of Ecology, Evolution, and Systematics* 38:459–87.

Bonvicino, C. R., A. R. de Menezes, and J. A. de Oliveira. 2003. "Molecular and Karyologic Variation in the Genus *Isothrix* (Rodentia, Echimyidae)." *Hereditas* 139:206–11.

Bush, M. B. 1994. "Amazonian Speciation, a Necessarily Complex Model." *Journal of Biogeography* 21:5–17.

Bush, M. B., and P. E. de Oliveira. 2006. "The Rise and Fall of the Refugial Hypoth-

esis of Amazonian Speciation: A Paleoecological Perspective." *Biota Neotropica* 6: bn00106012006.
Caputo, M. V. 1991. "Solimões Megashear: Interplate Tectonics in Northwestern Brazil." *Geology* 19:246–49.
Carstens, B. C., A. L. Stevenson, J. D. Degenhardt, and J. Sullivan. 2004. "Testing Nested Phylogenetic and Phylogeographic Hypotheses in the *Plethodon vandykei* Species Group." *Systematic Biology* 53:781–93.
Casado, F., C. R. Bonvicino, and H. N. Seuánez. 2007. "Phylogeographic Analyses of *Callicebus lugens* (Platyrrhini, Primates)." *Journal of Heredity* 98:88–92.
Chevirona, Z. A., S. J. Hackett, and A. P. Capparella. 2005. "Complex Evolutionary History of a Neotropical Lowland Forest Bird (*Lepidothrix coronata*) and Its Implications for Historical Hypotheses of the Origin of Neotropical Avian Diversity." *Molecular Phylogenetics and Evolution* 36:338–57.
Colinvaux, P. A. 1987. "Amazon Diversity in Light of the Paleoecological Record." *Quaternary Science Reviews* 6:98–114.
———. 1993. "Pleistocene Biogeography and Diversity in Tropical Forests of South America." In *Biological Relationships Between Africa and South America*, edited by P. Goldblatt, 473–99. New Haven: Yale University Press.
———. 1998. "A New Vicariance Model for Amazonian Endemics." *Global Ecology and Biogeography Letters* 7:95–96.
Colinvaux, P. A., P. E. de Oliveira, E. Moreno, M. C. Miller, and M. B. Bush. 1996. "A Long Pollen Record from Lowland Amazonia: Forest and Cooling in Glacial Times." *Science* 274:85–88.
Colinvaux, P. A., G. Irion, M. E. Rosanen, M. B. Bush, and J. A. S. Nunes de Melo. 2001. "A Paradigm to Be Discarded: Geological and Paleogeological Data Falsify the Haffer and Prance Refugia Hypothesis of Amazonian Speciation." *Amazoniana* 16:609–46.
Collins, A. C., and J. M. Dubach. 2000a. "Biogeographic and Ecological Forces Responsible for Speciation in *Ateles*." *International Journal of Primatology* 21:421–43.
———. 2000b. "Phylogenetic Relationships of Spider Monkeys (*Ateles*) Based on Mitochondrial DNA Variation." *International Journal of Primatology* 21:381–420.
Ditchfield, A. D. 2000. "The Comparative Phylogeography of Neotropical Mammals: Patterns of Intraspecific Mitochondrial DNA Variation Among Bats Contrasted to Non Volant Small Mammals." *Molecular Ecology* 9:1307–18.
Eizirik, E., J. Kim, M. Menotti-Raymond Jr., P. G. Crawshaw, and S. J. O'Brien. 2001. "Phylogeography, Population History and Conservation Genetics of Jaguars (*Panthera onca*, Mammalia, Felidae)." *Molecular Ecology* 10:65–79.
Emerson, B. C., E. Paradis, and C. Thebaud. 2001. "Revealing the Demographic Histories of Species Using DNA Sequences." *Trends in Ecology and Evolution* 16:707–16.
Endler, J. A. 1977. *Geographic Variation, Speciation, and Clines*. Princeton: Princeton University Press, *Monographs in Population Biology*, 10.
———. 1982. "Pleistocene Forest Refuges: Fact or Fancy?" In *Biological Diversification in the Tropics*, edited by G. T. Prance, 641–57. New York: Columbia University Press.
Erwin, T. L. 1979. "Thoughts on the Evolutionary History of Ground Beetles: Hypotheses

Generated from Comparative Faunal Analyses of Lowland Forest Sites in Temperate and Tropical Regions." In *Carabid Beetles: Their Evolution, Natural History, and Classification*, edited by T. L. Erwin, G. E. Ball, and D. R. Whitehead, 539–92. The Hague: Dr W. Junk.

———. 1981. "Taxon Pulses, Vicariance, and Dispersal: An Evolutionary Synthesis Illustrated by Carabid Beetles." In *Vicariance Biogeography: A Critique*, edited by G. Nelson and D. E. Rosen, 159–96. New York: Columbia University Press.

———. 1985. "The Taxon Pulse: A General Pattern of Lineage Radiation and Extinction Among Carabid Beetles." In *Taxonomy, Phylogeny and Zoogeography of Beetles and Ants*, edited by G. Ball, 437–72. The Hague: Dr. W. Junk.

Figueiredo, J., C. Hoorn, P. Van der Ven, and E. Soares. 2009. "Late Miocene Onset of the Amazon River and the Amazon Deep-Sea Fan: Evidence from the Foz do Amazonas Basin." *Geology* 37:619–22.

Frailey, C. D., E. L. Lavina, A. Rancy, and J. P. de Souza. 1988. "A Proposed Pleistocene/Holocene Lake in the Amazon Basin and Its Significance to Amazonian Geology and Biogeography." *Acta Amazonica* 18:119–43.

Franzinelli, E., and H. Igreja. 2002. "Modern Sedimentation in the Lower Negro River, Amazonas State, Brazil." *Geomorphology* 44:259–71.

Fu, Y. 1997. "Statistical Tests of Neutrality of Mutations Against Population Growth, Hitchhiking and Background Selection." *Genetics* 147:915–25.

Galvão, M. V. 1977. *Geografia Do Brasil: Região Norte. Fundação Instituto Brasileiro de Geografia e Estatística, Vol.* 1. Rio de Janeiro: Editora Sergraf/IBGE.

Garda, A. A., and D. C. Cannatella. 2007. "Phylogeny and Biogeography of Paradoxical Frogs (Anura, Hylidae, Pseudae) Inferred from 12S and 16S Mitochondrial DNA." *Molecular Phylogenetics and Evolution* 44:104–11.

Gascon, C., J. R. Malcolm, J. L. Patton, M. N. F. da Silva, J. P. Bogart, S. C. Lougheed, C. A. Peres et al. 2000. "Riverine Barriers and the Geographic Distribution of Amazonian Species." *Proceedings of the National Academy of Sciences, USA* 97:13672–77.

Haberlea, S. G., and M. A. Masli. 1999. "Late Quaternary Vegetation and Climate Change in the Amazon Basin Based on a 50,000 Year Pollen Record from the Amazon Fan, ODP Site 932." *Quaternary Research* 15:27–38.

Haffer, J. 1969. "Speciation in Amazonian Forest Birds." *Science* 165:131–37.

———. 1974. "Avian Speciation in Tropical South America." *Publications of the Nuttall Ornithological Club* 14:1–390.

———. 1981. "Aspects of Neotropical Bird Speciation During the Cenozoic." In *Vicariance Biogeography*, edited by G. Nelson and D. E. Rosen, 371–94. New York: Columbia University Press.

———. 1985. "Avian Zoogeography of the Neotropical Lowlands." *Ornithological Monographs* 36:113–46.

———. 1990. "Geoscientific Aspects of Allopatric Speciation." In *Vertebrates in the Tropics: Proceedings of the International Symposium on Vertebrate Biogeography and Systematics in the Tropics, Bonn, June 5–8, 1989*, edited by G. Peters and R. Hutterer, 45–60. Bonn: Alexander Koenig Zoological Research Institute and Zoological Museum.

———. 1997. "Alternative Models of Vertebrate Speciation in Amazonia: An Overview." *Biodiversity and Conservation* 6:451–76.

———. 2008. "Hypotheses to Explain the Origin of Species in Amazonia." *Brazilian Journal of Biology* 68:917–47.

Halas, D., D. Zamparo, and D. R. Brooks. 2005. "A Historical Biogeographical Protocol for Studying Biotic Diversification by Taxon Pulses." *Journal of Biogeography* 31:249–60.

Hall, J. P. W., and D. J. Harvey. 2002. "The Phylogeography of Amazonia Revisited: New Evidence from Riodinid Butterflies." *Evolution* 56:1489–97.

Hershkovitz, P. 1977. *Living New World Monkeys (Platyrrhini), with an Introduction to Primates. Vol. I.* Chicago: University of Chicago Press.

Hoorn, C., J. Guerrero, G. A. Sarmiento, and M. A. Lorente. 1995. "Andean Tectonics as a Cause for Changing Drainage Patterns in Miocene Northern South America." *Geology* 23:237–40.

Hoorn, C., and H. Vonhof. 2006. "Neogene Amazonia: Introduction to the Special Issue." *Journal of South American Earth Sciences* 21:1–4.

Hoorn, C., F. P. Wesselingh, H. ter Steege, M. A. Bermudez, A. Mora, J. Sevink, I. Sanmartin et al. 2010. "Amazonia Through Time: Andean Uplift, Climate Change, Landscape Evolution, and Biodiversity." *Science* 330:927–31.

Klammer, G. 1984. "The Relief of the Extra Andean Amazonian Basin." In *The Amazon: Limnology and Landscape Ecology of a Mighty Tropical River and Its Basin*, edited by H. Sioli, 47–83. Dordrecht: Dr. W. Junk.

Kuhner, M. K., J. Yamato, and J. Felsenstein. 1998. "Maximum Likelihood Estimation of Population Growth Rates Based on the Coalescent." *Genetics* 149:429–34.

Latrubesse, E. M., and E. Franzinelli. 1998. "Late Quaternary Alluvial Sedimentation in the Upper Rio Negro Basin, Amazonia, Brazil: Paleohydrological Implications." In *Paleohydrology and Environmental Change*, edited by G. Benito, V. R. Baker, and K. J. Gregory, 259–71. Chichester: John Wiley & Sons.

———. 2002. "Evidence of Quaternary Paleohydrological Changes in Middle Amazonia: The Aripuanã and Jiparaná Fans Like Systems." *Zeitschrift für Geomorphologie, n.f.* 129:61–62.

———. 2005. "The Late Quaternary Evolution of the Negro River, Amazon, Brazil: Implications for Island and Floodplain Formation in Large Anabranching Tropical Systems." *Geomorphology* 70:372–97.

Latrubesse, E. M., J. C. Stevaux, M. L. Santos, and M. L. Assine. 2005. "Grandes sistemas fluviais: Geologia, geomorfologia e paleohidrologia." In *Quaternário do Brasil*, edited by C. R. G. Souza, K. Suguio, A. M. S. Oliveira, and P. E. Oliveira, 276–97. Ribeirão Preto: Holos Editora.

Lessa, E. P., J. A. Cook, and J. L. Patton. 2003. "Genetic Footprints of Demographic Expansion in North America, But Not Amazonia, During the Late Quaternary." *Proceedings of the National Academy of Sciences, USA* 100:10331–34.

Lim, B. K. 2008. "Historical Biogeography of New World Emballonurid Bats (Tribe Diclidurini): Taxon Pulse Diversification." *Journal of Biogeography* 35:1385–401.

———. 2009. "Review of the Origins and Biogeography of Bats in South America." *Chiroptera Neotropical* 15:391–410.

Lovejoy, N. R., J. S. Albert, and W. G. R. Crampton. 2006. "Miocene Marine Incursions and Marine/Freshwater Transitions: Evidence from Neotropical Fishes." *Journal of South American Earth Sciences* 21:5–13.

Lundberg, J. G., L. G. Marshall, J. Guerrero, B. Horton, M. C. Malabarba, and F. Wesselingh. 1998. "The Stage for Neotropical Fish Diversification: A History of Tropical South American Rivers." In *Phylogeny and Classification of Neotropical Fishes*, edited by L. R. Malabarba, R. E. Reis, R. P. Vari, C. A. S. Lucena, and Z. M. S. Lucena, 13–48. Porto Alegre: Museu de Ciências e Tecnologia, PUCRS.

Mallet, J. 1993. "Speciation, Radiation, and Color Pattern Evolution in *Heliconius* Butterflies: Evidence from Hybrid Zones." In *Hybrid Zones and the Evolutionary Process*, edited by R. G. Harrison, 226–60. Toronto: Oxford University Press.

Marroig, G., and R. Cerqueira. 1997. "Plio-Pleistocene South American History and the Amazon Lagoon Hypothesis: A Piece in the Puzzle of Amazonian Diversification." *Journal of Comparative Biology, Ribeirão Preto* 2:103–19.

Matocq, M. D., J. L. Patton, and M. N. F. da Silva. 2000. "Population Genetic Structure of Two Ecologically Distinct Amazonian Spiny Rats: Separating History and Current Ecology." *Evolution* 54:1423–32.

Moritz, J. L., J. L. Patton, C. J. Schneider, and T. B. Smith. 2000. "Diversification of Rainforest Faunas: An Integrated Molecular Approach." *Annual Review of Ecology and Systematics* 31:533–63.

Noonan, B. P., and K. P. Wray. 2006. "Neotropical Diversification: The Effects of a Complex History on Diversity within the Poison Frog Genus *Dendrobates*." *Journal of Biogeography* 33:1007–20.

Nores, M. 1999. "An Alternative Hypothesis for the Origin of Amazonian Bird Diversity." *Journal of Biogeography* 26:475–85.

Patton, J. L., and M. N. F. da Silva. 1998. "Rivers, Refuges, and Ridges: The Geography of Speciation of Amazonian Mammals." In *Endless Forms: Species and Speciation*, edited by D. J. Howard and S. H. Berlocher, 202–13. Oxford: Oxford University Press.

———. 2005. "The History of Amazonian Mammals: Mechanisms and Timing of Diversification." In *Tropical Rainforests: Past, Present and Future*, edited by C. W. Dick and C. Moritz, 107–26. Chicago: University of Chicago Press.

Patton, J. L., M. N. F. da Silva, M. C. Lara, and M. A. Mustrangi. 1997. "Diversity, Differentiation, and the Historical Biogeography on Nonvolant Small Mammals of the Neotropical Forests." In *Tropical Forest Remnants: Ecology, Management and Conservation of Fragmented Communities*, edited by W. F. Laurance and R. O. Bierregaard, 455–65. Chicago: University of Chicago Press.

Patton, J. L., M. N. F. da Silva, and J. R. Malcolm. 2000. "Mammals of the Rio Juruá and the Evolutionary and Ecological Diversification of Amazonia." *Bulletin of the American Museum of Natural History* 244:1–306.

Patton, J. L., P. Myers, and M. F. Smith. 1990. "Vicariant versus Gradient Models of Diversification: The Small Mammal Fauna of Eastern Andean Slopes of Peru." In *Vertebrates in the Tropics: Proceedings of the International Symposium on Vertebrate Biogeography and Systematics in the Tropics, June 5–8, 1989*, edited by G. Peters and R. Hutterer, 355–71. Bonn: Alexander Koenig Zoological Research Institute and Zoological Museum.

Patton, J. L., and M. F. Smith. 1992. "mtDNA Phylogeny of Andean Mice: A Test of Diversification Across Ecological Gradients." *Evolution* 46:174–83.

Peres, C. A., J. L. Patton, and M. N. F. da Silva. 1996. "Riverine Barriers and Gene Flow in Amazonian Saddle-Back Tamarins." *Folia Primatologica* 67:113–24.

Prance, G. T. 1973. "Phytogeographic Support for the Theory of Pleistocene Forest Refuges in the Amazon Basin, Based on Evidence from Distribution Patterns in Caryocaraceae, Chrysobalanacae, Dichapetalaceae and Lecythidaceae." *Acta Amazonica* 3:5–28.

———, ed. 1982. *Biological Diversification in the Tropics*. New York: Columbia University Press.

———. 1987. "Biogeography of Neotropical Plants." In *Biogeography and Quaternary History in Tropical America*, edited by T. C. Whitmore and G. T. Prance, 46–65. Oxford: Clarendon Press.

Ribas, C. C., R. Gaban-Lima, and C. Y. Miyaki. 2005. "Historical Biogeography and Diversification within the Neotropical Parrot Genus *Pionopsitta* (Aves: Psittacidae)." *Journal of Biogeography* 32:1409–27.

Roddaz, M, P. Baby, S. Brusset, W. Hermoza, and J. M. Darrozes. 2005. "Forebulge Dynamics and Environmental Control in Western Amazonia: The Case Study of the Arch of Iquitos (Peru)." *Tectonophysics* 399:87–108.

Ron, S. R. 2000. "Biogeographic Area Relationships of Lowland Neotropical Rainforest Based on Raw Distributions of Vertebrate Groups." *Biological Journal of the Linnean Society* 71:379–402.

Rossetti, D., P. Mann de Toledo, and A.-M. Góes. 2005. "New Geological Framework for Western Amazonia (Brazil) and Implications for Biogeography and Evolution." *Quaternary Research* 63:78–89.

Rull, V. 2004. "Is the 'Lost World' Really Lost? Palaeoecological Insights into the Origin of the Peculiar Flora of the Guayana Highlands." *Naturwissenschaften* 91:139–42.

———. 2009. "Microrefugia." *Journal of Biogeography* 36:481–84.

Sanmartín, I., H. Enghoff, and F. Ronquist. 2001. "Patterns of Animal Dispersal, Vicariance and Diversification in the Holarctic." *Biological Journal of the Linnean Society* 73:345–90.

Santos, J. C., L. A. Coloma, K. Summers, J. P. Caldwell, R. Ree, and D. C. Cannatella. 2009. "Amazonian Amphibian Diversity is Primarily Derived from Late Miocene Andean Lineages." *PLoS Biology* 7:448–61.

Silva, J. M. C., A. B. Rylands, and G. A. B. Fonseca. 2005. "The Fate of Amazonian Areas of Endemism." *Conservation Biology* 19:689–94.

da Silva, M. N. F., and J. L. Patton. 1993. "Amazonian Phylogeography: mtDNA Sequence Variation in Arboreal Echimyid Rodents (Caviomorpha)." *Molecular Phylogenetics and Evolution* 2:243–55.

da Silva, M. N. F., and J. L. Patton. 1998. "Molecular Phylogeography and the Evolution and Conservation of Amazonian Mammals." *Molecular Ecology* 7:475–86.

Simpson, B. B., and J. Haffer. 1978. "Speciation Patterns in the Amazonian Forest Biota." *Annual Review of Ecology and Systematics* 9:497–518.

Solomon, S. E., M. Bacci Jr., J. Martins Jr., G. G. Vinha, and U. G. Mueller. 2008. "Paleodistributions and Comparative Molecular Phylogeography of Leafcutter Ants (*Atta* spp.) Provide New Insight into the Origins of Amazonian Diversity." *PLoS One* 3:e2738.

Sombroek, W. G. 1966. *Amazon Soils: A Reconnaissance of the Soils of the Brazilian Amazon Region*. Wageningen: Wageningen University.

Souza, W. V. 2008. "Filogenia e Filogeografia em Dasypodidae (Mammalia: Xenarthra) com Ênfase nos Tatus do Gênero *Dasypus*." Unpublished PhD diss. Universidade Estadual do Rio de Janeiro.

Symula, R., R. Schulte, and K. Summers. 2003. "Molecular Systematics and Phylogeography of Amazonian Poison Frogs of the Genus *Dendrobates*." *Molecular Phylogenetics and Evolution* 26:452–75.

Vallinoto, M., J. Araripe, P. S. Rego, C. H. Tagliaro, I. Sampaio, and H. Schneider. 2006. "Tocantins River as an Effective Barrier to Gene Flow in *Saguinus niger* Populations." *Genetics and Molecular Biology* 29:215–19.

van der Hammen, T. 1974. "The Pleistocene Changes of Vegetation and Climate in Tropical South America." *Journal of Biogeography* 1:3–26.

van der Hammen, T., and M. C. Absy. 1994. "Amazonia During the Last Glacial." *Palaeogeography, Palaeoclimatology, Palaeoecology* 109:247–61.

van der Hammen, T., and H. Hooghiemstra. 2000. "Neogene and Quaternary History of Vegetation, Climate, and Plant Diversity in Amazonia." *Quaternary Sciences Review* 19:725–42.

Vanzolini, P. E. 1970. *Zoologia Sistemática, Geografía e a Origem das Espécies*. São Paulo: Instituto Geográfico de São Paulo.

Vanzolini, P. E., and E. E. Williams. 1981. "The Vanishing Refuge: A Mechanism for Ecogeographic Speciation." *Papeis Avulsos de Zoologia* 34:251–55.

Vivo, M. de. 1996. "How Many Species of Mammals Are There in Brazil? Taxonomic Practice and Diversity Evaluation." In *Biodiversity in Brazil: A First Approach*, edited by C. E. M. Bicudo and N. A. Menezes, 313–21. São Paulo: Conselho Nacional de Desenvolvimento Científico e Tecnológico (CNPq).

Voss, R. S., and L. H. Emmons. 1996. "Mammalian Diversity in Neotropical Lowland Rainforests: A Preliminary Assessment." *Bulletin of the American Museum of Natural History* 230:1–115.

Wallace, A. R. 1853. *A Narrative of Travels on the Amazon and Rio Negro: With an Account of the Native Tribes, and Observations of the Climate, Geology, and Natural History of the Amazon Valley*. London: Reeve.

Wanderley Filho, J. R. 1991. "Evolução Estrutural da Bacia do Amazonas e sua Relação com o Embasamento." Unpublished MSc Thesis. Universidade Federal do Pará, Belém.

Wesselingh, F. P., M. E. Räsänen, G. Irion, H. B. Vonhof, R. Kaandorp, W. Renema, L. Romero Pittman, and M. Gingras. 2002. "Lake Pebas: A Palaeoecological Reconstruction of a Miocene, Long-Lived Lake Complex in Western Amazonia." *Cainozoic Research* 1:35–81.

Whinnett, A., M. Zimmermann, K. R. Willmott, N. Herrera, R. Mallarino, F. Simpson, M. Joron et al. 2005. "Strikingly Variable Divergence Times Inferred Across an Amazonian Butterfly 'Suture Zone.'" *Proceedings of the Royal Society of London, B, Biological Sciences* 272:2525–33.

Whitmore, R. T., and G. T. Prance, eds. 1987. *Biogeography and Quaternary History in Tropical America. Oxford Monographs on Biogeography*, 3. Oxford: Clarendon Press.

Wilson, D. E., and D. M. Reeder, eds. 2005. *Mammal Species of the World: A Taxonomic and Geographic Reference*, 3rd ed. Baltimore: Johns Hopkins University Press.

13 Historical Fragmentation Shaping Vertebrate Diversification in the Atlantic Forest Biodiversity Hotspot

Leonora P. Costa and Yuri L. R. Leite

Abstract

The Atlantic Forest is one of the world's 32 biodiversity hotspots, reflecting its high levels of plant and animal richness, endemism, and threats. Traditional biogeographic studies, based mainly on plants and other vertebrate groups, such as lizards, have long pointed out the singularities of this area, as well as provided indications of a marked distinction between north and south components in the Atlantic Forest. This usually overlooked distinction has recently attracted more attention via the results of historical biogeographic studies using molecular markers. In this chapter we use information from our own work and from the literature to locate the limits and better understand this biogeographic partition. Phylogeographic studies have documented and reinforced the general pattern of southern and northern components for many taxonomic groups, including rodents, marsupials, carnivores, and xenarthrans, as well as for other vertebrates, such as lizards, birds, snakes, and frogs. There are striking phylogeographic breaks within the Atlantic Forest, most of them represented by northern and southern components that converge at 20°S latitude, suggesting a common vicariant event. However, it remains to determine when and where the Atlantic Forest phylogroups were isolated, what events were responsible for their isolation, and how the different groups responded to such events.

13.1 The Atlantic Forest: Diversity, Endemism and Affinities

The Atlantic Forest stretches mainly along Brazil's Atlantic coast, extending inland to the Oriental Region of Paraguay and the Province of Misiones in northeastern Argentina, and exhibiting a variety of vegetation types (fig. 13.1). The remaining vegetation represents only 7% to 8% of the 1 to 1.5 million km^2 forest that Europeans found in 1500 AD (Galindo-Leal and Câmara 2003). This reduction of Atlantic Forest habitat is mainly due to tree-cutting for timber, sugar cane, coffee and soy plantations, cattle pastures, roads, and the growth of cities (Dean 1995). Brazil's first centers of development were located in the

Atlantic Forest, which now houses the country's industrial center where 61% of its 190 million people live in dense urban settlements (Fundação SOS Mata Atlântica and INPE 2009). The situation in Paraguay is similar: 98% of its population lives in the Oriental Region, which represents 40% of the country's land area (D'Elía et al. 2008), and two-thirds of its Atlantic Forest was lost between 1973 and 2000 (Huang et al. 2007). Although Atlantic Forest covers only about 1% of Argentina, where it is known as the Misiones or Paraná Forest, it is one of the country's most diverse ecoregions, exhibiting a high concentration of endemic species. Despite severe habitat destruction, the Misiones Forest retains some of the most continuous and least-altered portions of the Interior Atlantic Forest (Giraudo et al. 2003).

Despite extreme reduction and fragmentation of forest cover (fig. 13.1) and ongoing threats, the Atlantic Forest still exhibits high levels of diversity and endemism, and is consequently one of the top biodiversity hotspots for conservation (Myers et al. 2000). Following Fonseca's (1985) pioneering work, the biological importance of the Atlantic Forest has been highlighted in several scientific volumes (e.g., Metzger 2009; Morellato and Haddad 2000). Nevertheless, the processes that generated its striking diversity are still unknown, the ideas so far presented to explain it are controversial, and the knowledge about patterns of endemism is poor compared to some other areas (see Moritz et al. 2000).

The Atlantic Forest houses approximately 8000 endemic plants and 567 vertebrate species (Myers et al. 2000). Mittermeier et al. (2005) recorded nearly 270 species of mammals in the Atlantic Forest, of which 72 are endemic. Updated records expand this to 88 endemic species, with rodents and primates accounting for 82% of the endemics, with 54 and 19 species, respectively (appendix 13.1). In South America, Atlantic Forest mammal diversity is surpassed only by that of the Tropical Andes and the Amazon (Costa et al. 2000). Its extraordinary levels of diversity and endemism have been traditionally explained by the isolation of the Atlantic Forest from other forested biomes in South America (e.g., Patton et al. 1997). Although this offers a partial explanation, it does not represent a comprehensive story.

The Atlantic Forest appears widely isolated from other humid forests today, but there are indications of former connections through intervening regions now occupied by open and dry habitats. The palynological record of the Quaternary shows that between 33,000–25,000 BP, the central Brazilian region was moister than today and covered by rain forests (Ledru 1993), and during the last glacial maximum (18,000–12,000 BP), the present-day corridor of xeric vegetation was covered by extensive woodland (Prado and Gibbs 1993). These findings

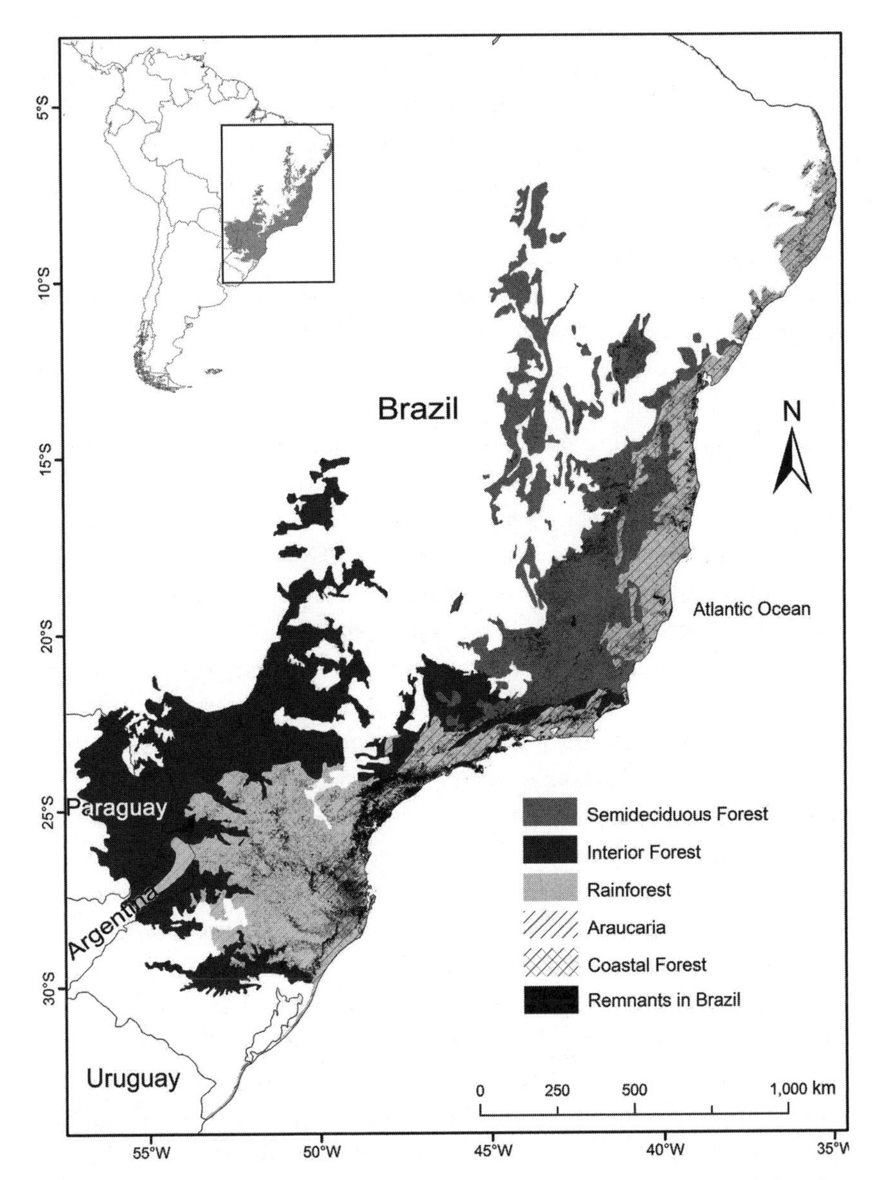

Figure 13.1 Vegetation types within the Atlantic Forest and forest remnants in Brazil (based on Fundação SOS Mata Atlântica and INPE, 2009).

indicate the predominance of seasonal forest and woodland vegetation during most of the Pleistocene. For example, a palynological profile of the latest Pleistocene (10,990–10,540 BP) from the Caatinga region revealed pollen of taxa found in present-day Amazonian and Atlantic forests. The pollen concentration is high, probably reflecting a large and well-drained watershed under a climate conducive to dense forest cover (Oliveira, Franca-Barreto, and Suguio 1999). Based on speleothems and travertine deposits, Auler et al. (2004) and Wang et al. (2004) showed that the currently semi-arid Caatinga experienced relatively frequent and abrupt changes in rainfall during the Pleistocene, suggesting that intermittent wetter conditions probably allowed the expansion and extension of gallery forests, affected rain forest distribution, and opened forest corridors connecting Amazonian and Atlantic rain forests.

There are gallery forests and other kinds of mesic enclaves in the open intervening habitats that separate the Atlantic Forest from other humid biomes. Botanical studies of these mesic enclaves and coastal forests in the northeastern region reveal floristic connections to either the Atlantic Forest or the Amazon (Bigarella, Andrade-Lima, and Riehs 1975; Coimbra-Filho and Câmara 1996; Rizzini 1967). The humid forests support forest-adapted animals. For example, species of small mammals typical of rain forests, such as the slender mouse opossum *Marmosops incanus* and the climbing rat *Rhipidomys mastacalis*, are currently found in the humid forest patches called *brejos* in the states of Ceará and Bahia in northeastern Brazil (see Carmignotto, de Vivo, and Langguth, chapter 14, this volume). Costa (2003) presented many examples of phylogeographic connections between Atlantic forest and Amazonian small mammals. Her data suggested that the Atlantic Forest is not historically isolated, either from the Amazon or from other humid forested habitats dominated by dry forests and other open-habitat vegetation. Indeed, Costa's (2003) results showed that the Atlantic Forest clades were either not reciprocally monophyletic or were the sister group to all the other clades included in the analyses; in many cases, the Atlantic Forest unit was deeply nested in a more inclusive clade, whereas in other cases it showed up as a composite area.

Another striking and unexpected relationship between the Atlantic Forest and other distant forested habitats is provided by a small number of vertebrate taxa including frogs, birds, and rodents. In contrast with the examples given earlier, the mammal taxa exhibiting this singular historical pattern of relationship are not widespread throughout South America but instead are disjunct taxa, endemic to the Atlantic Forest and to the Andes. *Rhagomys rufescens* is a sigmodontine rodent that occurs in the eastern Atlantic Forest (Percequillo, Gonçalves, and de Oliveira 2004, Steiner-Souza et al. 2008). *Rhagomys* was considered a monotypic genus, endemic to the Atlantic Forest and inhabiting lowland and montane humid forests, until the recent description of *R. longilingua* from the eastern Andean slopes of Peru and Bolivia (Luna and Patterson 2003; Villalpando, Vargas, and Salazer-Bravo 2006). Another sigmodontine rodent exhibiting an Atlantic Forest-Andean connection is *Drymoreomys albimaculatus* (Percequillo, Weksler, and Costa 2011), a newly described genus and species endemic to the lower montane areas (400–1000 m) in the Atlantic Forest of eastern Brazil, which was recovered as sister of *Eremoryzomys polius*, an Andean endemic (760–2100 m) from the valley of the lower Río Marañón in northern Peru. Additionally, the two taxa are both restricted to well-known Neotropical centers of endemism (Cracraft 1985; Müller 1973). *Drymoreomys albimaculatus* occupies the "Serra do Mar" center, while *E. polius* is distributed in the "Marañón" center. Other sigmodontine rodents also share the pattern described here for thomasomyines and oryzomyines: phylogenetic studies of the genus *Akodon* found that Atlantic Forest species of the *cursor* group were sister to the Andean *Akodon boliviensis* group (Gonçalves et al. 2007; Smith and Patton 2007).

Other examples of phylogenetic relationships between Atlantic Forest and Andean groups can be found among hylid frogs (Faivovich et al. 2004; Faivovich et al. 2005) and parrots of the genus *Pionus* (Ribas et al. 2007). As noted by Percequillo, Weksler, and Costa (2011), the general congruence of biogeographic patterns among birds, frogs, and rodents likely highlights a historical connection between the Atlantic Forest and the Andean life zones; their terrestrial small vertebrate communities subsequently diverged into present-day lineages.

13.2 Biogeographical Patterns within the Atlantic Forest

Floristic and phytogeographic studies have long pointed out the distinction between northern and southern components in the Atlantic Forest, specifically to the north and south of Espírito Santo and southern Bahia (Coimbra-Filho and Câmara 1996; Mori, Boom, and Prance 1981; Rizzini 1967). Regarding the Atlantic Forest fauna, classic biogeographical studies provided examples of a north-to-south partition of the biota, such as the breaks in distributions identified for amphibians (Lynch 1979; Müller 1973) and lizards (Vanzolini 1988). Atlantic Forest mammal species typically range either from northeastern to southeastern Brazil (e.g., *Chaetomys* and *Leontopithecus*; Oliver and Santos 1991) or from southeastern to southern Brazil, eastern Paraguay, and Misiones, Argentina (e.g., *Abrawayaomys*, Pardiñas, Teta, and D'Elía 2009; *Juliomys*, de la Sancha et al. 2009). Another example of this faunal differentiation concerns the distributional limits of closely related taxa ranging to the north and south of the Rio Doce, which runs eastward through the central part of the state of Espiríto Santo in eastern Brazil (fig. 13.2). The tamarins *Callithrix flaviceps*

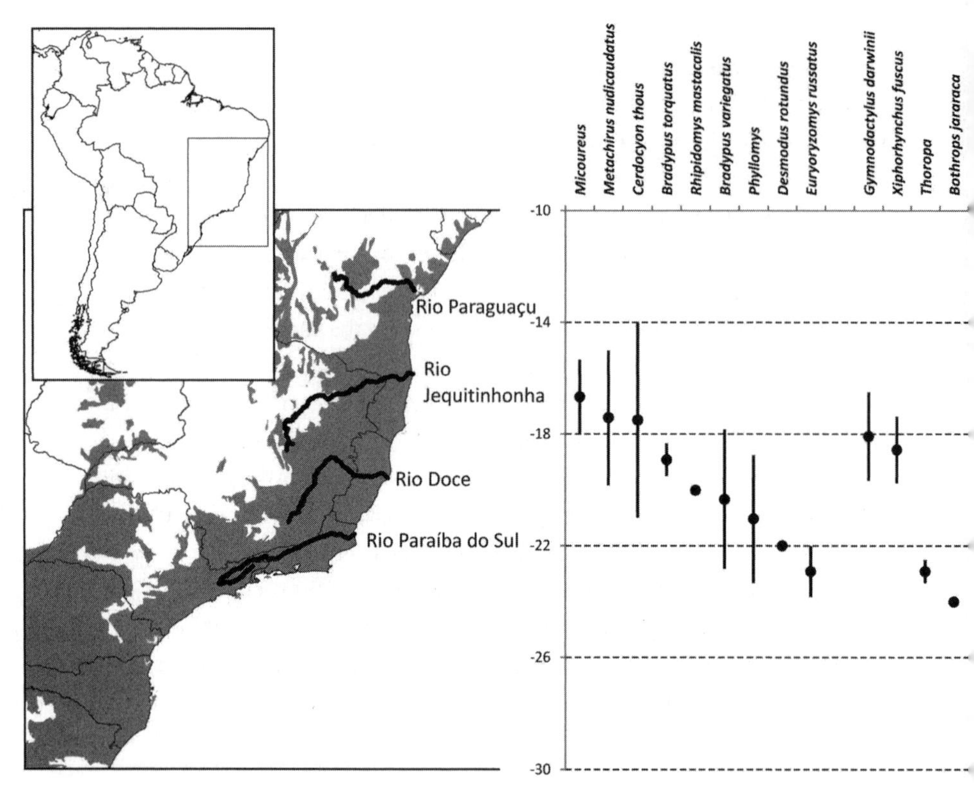

Figure 13.2 Extent of the Atlantic Forest domain in eastern Brazil and approximate latitudinal position of the phylogeographic break between northern and southern clades of mammals and other vertebrate taxa. Numbers represent latitudes, vertical lines represent the range of the phylogeographic break (i.e., from the southernmost record in the northern clade to the northernmost record in the southern clade), and dots represent median values. Data sources for each taxon: *Micoureus* (Costa 2003), *Metachirus nudicaudatus* (Costa 2003), *Cerdocyon thous* (Tchaika et al. 2007), *Bradypus torquatus* (Lara-Ruiz et al. 2008), *Rhipidomys mastacalis* (Costa 2003), *Bradypus variegatus* (Moraes-Barros et al. 2006), *Phyllomys* (Leite 2003), *Desmodus rotundus* (Martins et al. 2007), *Euryoryzomys russatus* (Miranda et al. 2007), *Gymnodactylus darwinii* (Pellegrino et al. 2005), *Xiphorhynchus fuscus* (Cabanne et al. 2007), *Thoropa* (Fitzpatrick et al. 2009), and *Bothrops jararaca* (Grazziotin et al. 2006).

and *C. aurita*, for example, only occur south of the Rio Doce, while *C. geoffroyi*, *C. kuhlii*, and *C. jacchus* are distributed to the north of this river (Mendes 1997). Similarly, the distributional limits of the capuchin monkeys *Cebus nigritus* and *C. robustus* meet at the Rio Doce, the former occurring to the south and the latter to the north (Rylands et al. 1997).

Biogeographic studies using cladistic approaches have subsequently corroborated this general trend. In one of the first, Cracraft and Prum (1988) found the Atlantic Forest to be a composite area of endemism for birds, one that supports lineages with two different histories. Subsequently, Bates, Hackett, and Cracraft (1998) used raw geographic distributions for Neotropical species and subspecies of perching birds (order Passeriformes) to present a hypothesis of area-relationships for twelve areas of endemism in the lowland Neotropics. Their results corroborated Cracraft and Prum's (1988) findings, and the Atlantic Forest always appeared as a composite area with separate, although closely related, northern and southern clusters. In some cases, the northern Atlantic Forest clustered with eastern Amazonia. Both Cracraft and Prum (1988) and Bates, Hackett, and Cracraft (1998) presented evidence suggesting that the Atlantic Forest contains some taxa that have recently diverged from Amazonian forms and others that are the result of older in situ speciation (Bates, Hackett, and Cracraft 1998).

The first study to show the pattern of northern and southern components in the Atlantic Forest using areas of endemism for mammals was presented by Costa et al. (2000). Using Parsimony Analysis of Endemism on distributions of marsupials, rodents, and primates, we identified two related regions in the Atlantic Forest—whose limits meet between the mouths of the rivers Doce and Jequitinhonha—along the southeastern coast of Brazil (Costa et al. 2000).

After the emergence of phylogeography as a field for biogeographic analysis (Avise et al. 1987), several studies corroborated the same pattern of northern and southern components in the Atlantic Forest, using different methods and a variety of taxa. In a study of the biogeographic relationships between the Amazon and the Atlantic Forest, Costa (2003) called attention to northern and southern components in the Atlantic Forest for didelphid marsupials and sigmodontine rodents (fig. 13.2). The phylogeographic breaks in the distributions of species and clades of both groups were similar, probably resulting from common historic events. Leite (2003) examined the biogeography of Atlantic Forest tree rats of the genus *Phyllomys* (Echimyidae), which are endemic to this biome. He found that the break between the northeastern and southern clades lies at about 19°S latitude, with the highest diversity (8 species) found 22°–24°S.

Subsequently, additional phylogeographic studies with more vagile mammals, as well as other vertebrate groups, have shown consistently congruent results, even within small and restricted assemblages, such as intraspecific groups. In a study of comparative phylogeography of the Atlantic Forest endemic maned sloth (*Bradypus torquatus*) and the widespread three-toed sloth

(*Bradypus variegatus*), Moraes-Barros et al. (2006) inferred two main phylogeographic groups in the Atlantic Forest, representing north and south lineages. Despite the low levels of genetic diversity identified, *B. torquatus* had at least two divergent mitochondrial phylogroups: southeastern Bahia and Santa Teresa (in Espírito Santo), north and south of the Rio Doce, respectively. For the more widespread three-toed sloth, nested clade analysis showed that the clade representing São Paulo (south of Rio Doce) is divergent from the clades found in Bahia and Minas Gerais (north of Rio Doce). In an extension of the work by Moraes-Barros et al. (2006), Lara-Ruiz, Chiarello, and Santos (2008) attempted to characterize the geographic structure and to produce estimates of genetic diversity on the endemic maned sloth based on broader sampling from the largest remnant populations in the states of Bahia, Espírito Santo, and Rio de Janeiro. Their phylogeographic analysis revealed very divergent genetic clusters specific to different geographical regions, indicating that at least two independent evolutionary units are present in the Atlantic Forest. Their analyses have conservation implications, as their results support the existence of two evolutionarily significant units and several management units (Lara-Ruiz, Chiarello, and Santos 2008).

Studying the phylogeography and population history of the crab-eating fox, *Cerdocyon thous*, Tchaicka et al. (2007) found a strong phylogeographic partition between northeastern Brazil and other portions of its distribution, with complete separation between southern and northern components in the Atlantic Forest. Comparing the phylogeographic patterns for the fox to those observed for other Neotropical vertebrates, they identified a major north–south demographic discontinuity that seems to have marked the history of the Atlantic Forest biota, suggesting the occurrence of a shared, large-scale historical fragmentation event in that region. They concluded that the apparent unity of present-day biomes may be misleading, and that a complex history may underlie the formation and biogeographic interactions of these ecosystems (Tchaicka et al. 2007).

Martins et al. (2007) found marked population structure in the common vampire bat (*Desmodus rotundus*), which ranges from Mexico to southern South America. On the basis of mitochondrial sequence analyses, they described four geographically circumscribed clades in Brazil that included southern and northern Atlantic Forest clades. The pattern described for this bat is similar to those described for other mammals, suggesting geographical congruence between historical vicariant processes, including likely vicariant events between north and south Atlantic Forest (Martins et al. 2007). In this case, however, the phylogeographic break is situated farther south (22°S) than most others (19°S–20°S; fig. 13.2).

Miranda et al. (2007) investigated geographic patterns of genetic variation using the mitochondrial cytochrome-*b* and nuclear IRBP genes, and found the recurrent northern-southern phylogeographic break in one sigmodontine rodent species (*Euryoryzomys russatus*) but not in two others (*Hylaeamys megacephalus* and *Sooretamys angouya*). In both phylogenetic and network analyses, specimens of *E. russatus* formed three groups: one from southern Brazil, in the Pampa-Atlantic Forest transitional area (29°–30°S); a second from southeastern Brazil south of the Paraíba do Sul river (23°–29°S), constituting what they called the southern Atlantic Forest group; and a third, including sites in mountainous areas of Rio de Janeiro and Espírito Santo and coastal Bahia (13°–23°S), corresponding to what they called the northern Atlantic Forest group. Similar to the pattern found for *D. rotundus* (Martins et al. 2007), the break occurs at ~22°S.

Other vertebrates also show phylogeographic patterns similar to those of Atlantic Forest mammals, including lizards (Pellegrino et al. 2005), snakes (Grazziotin et al. 2006), birds (Cabanne, Santos, and Miyaki 2007), and frogs (Fitzpatrick et al. 2009). Pellegrino et al. (2005) recovered three distinctive lizard phylogroups in the *Gymnodactylus darwinii* complex: a southeastern clade and two clades from northern regions. The genetic structure of the major clades coincided with the presence of river systems in the Atlantic Forest with the deepest phylogeographical break coincident with the Rio Doce system.

Grazziotin et al. (2006) presented results of a large-scale survey of variation in the mitochondrial cytochrome-*b* gene of the pit viper *Bothrops jararaca* and two closely related insular species, *Bothrops insularis* and *Bothrops alcatraz*, all endemic to the Atlantic Forest. Phylogenetic and network analyses identified *B. jararaca* as a highly heterogenous taxon, composed of two geographically structured phylogroups with southern and northern distributions, respectively, separated by a deep divergence. These genetic results corroborate earlier morphological data (Hoge, Belluomi, and Fernandes 1977) that revealed variation in southern and northern populations of *B. jararaca*, with intermediate morphologies found in populations from São Paulo. Together, the genetic and morphological evidence points to existence of a hybrid zone between the southern and northern populations. In this case, however, rivers do not seem to be effective present-day dispersal barriers to *B. jararaca*, as other rivers of comparable size cross the range of the species without inducing genetic differentiation (Grazziotin et al. 2006).

Cabanne, Santos, and Miyaki (2007) described three main mitochondrial lineages of *Xiphorhynchus fuscus* (Aves: Dendrocolaptidae), an endemic passeriform from the southern Atlantic Forest in Brazil and Argentina. One lineage was associated with the subspecies *X. f. tenuirostris* and the other two with *X. f. fuscus*, which meet at the Rio Doce. The authors hypothesized a contact zone between these two subspecies along the Rio Doce in Espírito Santo, and another between two clades of *X. f. fuscus*, located to the southwest of the Paraíba do Sul river valley.

Using mitochondrial and nuclear sequences, Fitzpatrick et al. (2009) found significant diversity among frog lineages of *Thoropa miliaris* and *T. taophora* in northern and southern regions of the Atlantic Forest, as well as evidence that populations in the two regions have had very different evolutionary histories. Their data revealed that northern and southern regions of the Atlantic Forest carry the genetic signal of very different historical demographic processes, and the discontinuity between the two species seems to occur at the Paraíba do Sul river valley, the same place where sister clades of rodents (Miranda et al. 2007) and birds (Cabanne, Santos, and Miyaki 2007) come in contact with each other.

13.3 Suture Zones, Rivers, and Refuges

Only a few studies have addressed the evolutionary origins of biodiversity in the megadiverse Atlantic Forest (e.g., Cabanne, Santos, and Miyaki 2007; Carnaval 2002; Carnaval and Bates 2007; Carnaval et al. 2009; Fitzpatrick et al. 2009; Grazziotin et al. 2006; Moritz et al. 2000; Pellegrino et al. 2005). Phylogeographic patterns shared by numerous co-occurring taxa in a region are likely the result of common historical events. Therefore, phylogeographic studies can be used to understand poorly known regions and biomes of conservation importance. However, detailed phylogeographic studies of Atlantic Forest organisms are few in number and limited in taxonomic coverage, as exemplified by this review. Nevertheless, the few studies available reveal a pattern of concordant phylogeographic breaks within this biome, dividing it into northern and southern components.

Remington (1968, 322) defined a suture zone as "a band of geographic overlap between major biotic assemblages, including some pairs of species or semispecies which hybridize in the zone." Thus, according to the traditional view, suture zones are clusters of hybrid zones that form: (1) between glacial refugia, as a result of biotic expansion from each of the refuges in interglacial periods; (2) at the bases of mountains, as a result of dispersal through mountain passages during periods of warm climate; or (3) in regions that have suffered significant anthropogenic impact during the last hundred years (see Swenson and Howard 2005). These patterns were used to reconstruct the historical events that produced the current regional patterns of diversity. In the past few years, there has been increasing interest in suture zones (Swenson and Howard 2004) resulting from the identification of recurrent phylogeographic breaks at specific localities in temperate regions (Guillaume et al. 2000; Hewitt 2001; Redenbach and Taylor 2002). These research results led Swenson and Howard (2004) to expand the definition of suture zones to include phylogeographic breaks and clusters of contact zones of species that fail to hybridize. Some suture zones form in regions where taxa with northern and southern distributions meet. Hewitt (1999) believed these zones form in response to expansion processes related to the retraction of Pleistocene glaciers, creating clusters called "north/south contact zones." Thus far, phylogeographic studies of suture zones have been concentrated in temperate regions and biomes, with a few exceptions such as the one Patton, da Silva, and Malcolm (2000) described on the Juruá River in the Brazilian Amazon. Since the data accumulated for mammals and other vertebrates in the Atlantic Forest appear to reflect the same patterns predicted for suture zones in temperate biomes, an effort should be made to investigate such tropical suture zones. Because of its complexity, phylogeography and historical demographic studies across a broad range of taxa will be necessary to uncover large-scale patterns underlying Atlantic Forest diversity and to evaluate potential scenarios of diversification in eastern Brazil (Rull 2008).

In many of the aforementioned studies, phylogeographic breaks within the Atlantic Forest are located between latitudes 14° and 21°S in the region between the state of Espírito Santo and southern Bahia, recognized as the Atlantic Forest Central Corridor (Galindo-Leal and Câmara 2003). In 8 out of the 13 studies reviewed here, the break falls between 19° and 21°S latitude, roughly coincident with the Rio Doce drainage at 20°S (fig. 13.2). Four studies revealed a break situated farther south, roughly coinciding with the Paraíba do Sul river drainage. Although historical and more recent works have addressed the importance of rivers as barriers and important factors in the diversification of many zoological groups in the Amazon (Wallace 1852; Patton and da Silva 1998; Patton, da Silva, and Malcolm 2000; Gascon et al. 2000; see Bonvicino and Weksler, chapter 12, this volume), the role of Atlantic Forest rivers as barriers has not yet been rigorously tested. Among the studies reviewed here, one reported evidence of rivers acting as possible barriers for vertebrates in the Atlantic Forest (Pellegrino et al. 2005), while another was inconclusive on this topic (Grazziotin et al. 2006). The study of Carnaval and Moritz (2008) showed that at least two rivers (Doce and São Francisco) are located at the boundaries of forest refugia suggested by their climatic models, thus providing an alternative view to the riverine-barrier hypothesis.

Earlier authors have pointed out alternative scenarios to explain the northern and southern components in the Atlantic Forest; most cited climatic and vegetation changes that occurred from the end of the Pliocene through the Quaternary as possible explanations. Such changes could have fragmented the Brazilian Atlantic Forest into distinct patches isolated by savanna habitats, with one of these splits separating the northeastern and southeastern areas of Brazil (Auler and Smart 2001; Behling and Lichte 1997; Ledru et al. 1998; Lichte and Behling 1999; Turcq, Pressinotti, and Martin 1997; de Vivo and Carmignotto 2004; Whitmore and Prance 1987). These events could have been responsible for the remarkable divergence between northern and southern Atlantic Forest populations, especially for taxa with arboreal habits or otherwise restricted to humid forest habitats (e.g., see Lara-Ruiz, Chiarello, and Santos 2008; Moraes-Barros et al. 2006). Miranda et al. (2007) noted that the distribution of the southern Atlantic Forest clade of the rice rat *Euryoryzomys russatus* corresponds to one of the three Atlantic Forest centers of evolution proposed by Por, Imperatriz-Fonseca, and Neto (2005). This center is said to be located south of the Paraíba do Sul river delta and ends in the Serra Geral mountain range (in southern Brazil, bordering the Pampas). Moreover, the remaining northern Atlantic Forest group consisted of samples collected from localities in the mountainous region of the Atlantic Forest, a possible Pleistocene forest refuge (Vanzolini and Williams 1981).

However, other studies do not support relating the diversification of southern and northern lineages in the Atlantic Forest to Pleistocene refuges. Costa (2003) found major differences in times of lineage diversification for Neotropical small mammals. For those with a major phylogeographic break in the Atlantic Forest (e.g., the climbing rat *Rhipidomys* and the woolly mouse opossum *Micoureus*), times of divergence are mostly pre-Pleistocene, although recent intraspecific lineage splits congruent with a Pleistocene age are known (e.g., the brown four-eyed opossum *Metachirus nudicaudatus*). These results are in agreement with those for other old vertebrate lineages, such as the pit viper *B. jararaca*. Grazziotin et al. (2006) concluded that neither Pleistocene forest refuges nor Pliocene orogeny alone could explain the phylogeography of

B. jararaca in the Atlantic Forest. They proposed that the divergence of the two main lineages of *B. jararaca* is associated with an old fragmentation event in the Brazilian Atlantic Forest during the Pliocene, but that genetic and geographic variation in the subpopulations may have been shaped by late Pleistocene climatic oscillations.

In any case, there is recent interest in the environmental history of the southern and northern parts of the Atlantic Forest throughout the Quaternary. Some suggest that Pleistocene climatic fluctuations had a greater impact in the southern portion of the biome (Carnaval and Moritz 2008; Carnaval et al. 2009; Por 1992). Other climate models (Carnaval and Moritz 2008; Carnaval et al. 2009) predict the presence of major stable refugial areas over most of the central corridor of the Atlantic rain forest, and relative instability of forested habitats south of the Rio Doce, suggesting strikingly different Quaternary histories for these regions. These predictions are in agreement with the demographic studies presented for bats (Martins et al. 2007), birds (Cabanne, Santos, and Miyaki 2007), and frogs (Carnaval et al. 2009), jointly suggesting that higher latitudes probably experienced more intensely the effect of Pleistocene climatic oscillations, leading to forest contraction and fragmentation, and subsequent advances into southern areas of the Atlantic Forest biome in response to late Quaternary warming. But for other small mammals of relatively ancient lineages, such as the slender mouse opossum *Marmosops incanus* and the tree rat *Phyllomys pattoni*, the molecular phylogeography conflicts with model predictions, showing higher molecular diversity in southern populations (Carnaval and Moritz 2008). Additional studies including multiple species occurring in this region should help establish the generality of phylogeographic patterns and their connections to Pleistocene climatic oscillations.

As mentioned, most phylogeographic breaks are situated between 19°–21°S, in the Atlantic Forest Central Corridor, a region traversed by the Rio Doce. This region extends from the state of Espírito Santo into the southern portion of Bahia, an area covering more than 8.5 million hectares and containing high biological diversity that includes many endangered species and others with severely restricted ranges (Aguiar et al. 2005; Ayres et al. 2005; IPEMA 2005). The Central Corridor encompasses two centers of endemism, defined on the basis of vertebrate distributions (Costa et al. 2000; Kinzey 1982; Müller 1973; Silva, Sousa, and Casteleti 2004), butterflies (Tyler, Brown, and Wilson 1994), and plants (Prance 1982; Soderstrom, Judziewicz, and Clark 1988). There we find many physiognomies, including the broad-leaf evergreen forest, semideciduous forests, “restingas” (sand dune habitats), and mangroves. Southernmost

Bahia and northern Espírito Santo constitute the areas of Tabuleiro Forests, which lie on Tertiary sediments of the Barreiras series. Compared to other Neotropical rain forest formations, the Tabuleiro Forests are uncommon, due to their high levels of species diversity and elevated density of vines and epiphytes (Peixoto and Gentry 1990; Thomas 2008). Given the inadequate state of current knowledge, this region deserves intensive biological inventories in order to generate refined biogeographical analyses, given its potential to help elucidate the patterns of animal diversification in the highly rich and endemic Atlantic Forest.

This review of Atlantic Forest phylogeography suggests a picture of great diversity and historical complexity in the Atlantic Forest Central Corridor. It indicates the existence of a Neotropical suture zone and provides data for a comprehensive explanation of the diversity patterns we observe today. There are striking phylogeographic breaks within the Atlantic Forest, most represented by northern and southern components that converge at roughly 20°S, suggesting a common vicariant event. However, we still do not know when and where the Atlantic Forest phylogroups were isolated and what events could have been responsible for this isolation, or how the different groups responded to such events. This understanding can only come with a more extensive sampling across the geographic distribution of many species across this area. We are still recovering the patterns, and our knowledge regarding the processes behind them is still very limited.

Acknowledgments

We would like to thank Bruce Patterson for organizing the symposia on Historical Biogeography of Neotropical Mammals during the 10th International Mammalogical Congress (IMC), and Ricardo Ojeda, president of the 10th IMC held in Mendoza, Argentina. James L. Patton and Guillermo D'Elía provided insightful comments that improved the quality of this manuscript. We thank the Critical Ecosystem Partnership Fund (CEPF), Conselho Nacional de Desenvolvimento Cientifico e Tecnológico (CNPq), Fundação de Apoio à Ciência e Tecnologia do Espírito Santo (FAPES), and Fundo de Apoio à Ciência e Tecnologia do Município de Vitória (FACITEC) for continuous grant support.

Literature Cited

Aguiar, A. P., A. G. Chiarello, S. L. Mendes, and E. N. Matos. 2005. "Os Corredores Central e da Serra do Mar na Mata Atlântica Brasileira." In *Mata Atlântica: Biodiversidade, Ameaças e Perspectivas*, edited by C. Galindo-Leal and I. G. Câmara, 119–32. Belo Horizonte: Fundação SOS Mata Atlântica e Conservação Internacional.

Auler, A. S., and P. L. Smart. 2001. "Late Quaternary Paleoclimate in Semi-Arid Northeastern Brazil from U-series Dating of Travertine and Water Table Speleothems." *Quaternary Research* 55:159–67.

Auler, A. S., X. R. Wang, L. Edwards, H. Cheng, P. S. Cristalli, P. L. Smart, and D. A. Richards. 2004. "Quaternary Ecological and Geomorphic Changes Associated with Rainfall Events in Presently Semi-Arid Northeastern Brazil." *Journal of Quaternary Science* 19:693–701.

Avise, J. C., J. Arnold, R. M. Ball, E. Bermingham, T. Lamb, J. E. Neigel, C. A. Reeb, and N. C. Saunders. 1987. "Intraspecific Phylogeography: The Mitochondrial DNA Bridge between Population Genetics and Systematics." *Annual Review of Ecology and Systematics* 18:489–522.

Ayres, A. P., G. A. B. Fonseca, A. B. Rylands, H. L. Queiroz, L. P. Pinto, D. Mastreson, and R. B. Cavalcanti. 2005. *Os Corredores Ecológicos das Florestas Tropicais do Brasil*. Belém: Sociedade Civil Mamirauá.

Bates, J. M., S. J. Hackett, and J. Cracraft. 1998. "Area-Relationships in the Neotropical Lowlands: A Hypothesis Based on Raw Distributions of Passerine Birds." *Journal of Biogeography* 25:783–93.

Behling, H., and M. Lichte. 1997. "Evidence of Dry and Cold Climatic Conditions at Glacial Times in Tropical Southeastern Brazil." *Quaternary Research* 48:348–58.

Bigarella, J. J., D. Andrade-Lima, and P. J. Riehs. 1975. "Considerações a Respeito das Mudanças Paleoambientais na Distribuição de Algumas Espécies Vegetais e Animais no Brasil." *Anais da Academia Brasileira de Ciências* 47:411–64.

Cabanne, G. S., F. R. Santos, and C. Y. Miyaki. 2007. "Phylogeography of *Xiphorhynchus fuscus* (Passeriformes, Dendrocolaptidae): Vicariance and Recent Demographic Expansion in Southern Atlantic Forest." *Biological Journal of the Linnean Society* 91:73–84.

Carnaval, A. C. 2002. "Phylogeography of Four Frog Species in Forest Fragments of Northeastern Brazil—A Preliminary Study." *Integrative and Comparative Biology* 42:913–21.

Carnaval, A. C., and J. M. Bates. 2007. "Amphibian DNA Shows Marked Genetic Structure and Tracks Pleistocene Climate Change in Northeastern Brazil." *Evolution* 61:2942–57.

Carnaval, A. C., M. J. Hickerson, C. F. B. Haddad, M. T. Rodrigues, and C. Moritz. 2009. "Stability Predicts Genetic Diversity in the Brazilian Atlantic Forest Hotspot." *Science* 323:785–89.

Carnaval, A. C., and C. M. Moritz. 2008. "Historical Climate Modeling Predicts Patterns of Current Biodiversity in the Brazilian Atlantic Forest." *Journal of Biogeography* 35:1187–201.

Coimbra-Filho, A. F., and I. G. Câmara. 1996. *Os Limites Originais do Bioma Mata Atlântica na Região Nordeste do Brasil*. Rio de Janeiro: Fundação Brasileira para a Conservação da Natureza.

Costa, L. P. 2003. "The Historical Bridge Between the Amazon and the Atlantic Forest of Brazil: A Study of Molecular Phylogeography with Small Mammals." *Journal of Biogeography* 30:71–86.

Costa, L. P., Y. L. R. Leite, G. A. B. Fonseca, and M. T. Fonseca. 2000. "Biogeography of

South American Forest Mammals: Endemism and Diversity in the Atlantic Forest." *Biotropica* 32:872–81.

Cracraft, J. 1985. "Historical Biogeography and Patterns of Differentiation within the South American Avifauna: Areas of Endemism." In *Neotropical Ornithology*, edited by P. A. Buckley, M. S. Foster, E. S. Morton, R. S. Ridgely, and F. G. Buckley, 49–84. Lawrence, KS: American Ornithologists Union, *Ornithological Monographs* 36.

Cracraft, J., and R. O. Prum. 1988. "Patterns and Processes of Diversification: Speciation and Historical Congruence in Some Neotropical Birds." *Evolution* 42:603–20.

Dean, W. 1995. *With Broadax and Firebrand: The Destruction of the Brazilian Atlantic Forest*. Berkeley: University of California Press.

D'Elía, G., I. Mora, P. Myers, and R. D. Owen. 2008. "New and Noteworthy Records of Rodentia (Erethizontidae, Sciuridae, and Cricetidae) from Paraguay." *Zootaxa* 1784:39–57.

De la Sancha, N., G. D'Elía, F. Netto, P. Pérez, and J. Salazar-Bravo. 2009. "Discovery of *Juliomys* (Rodentia, Sigmodontinae) in Paraguay, a New Genus of Sigmodontinae for the Country's Atlantic Forest." *Mammalia* 73:162–67.

Faivovich, J., P. C. A. Garcia, F. Ananias, L. Lanari, N. G. Basso, and W. C. Wheeler. 2004. "A Molecular Perspective on the Phylogeny of the *Hyla pulchella* Species Group (Anura, Hylidae)." *Molecular Phylogenetics and Evolution* 32:938–50.

Faivovich, J., C. F. B. Haddad, P. C. A. Garcia, D. R. Frost, J. A. Campbell, and W. C. Wheeler. 2005. "Systematic Review of the Frog Family Hylidae, with Special Reference to Hylinae: Phylogenetic Analysis and Taxonomic Revision." *Bulletin of the American Museum of Natural History* 294:1–240.

Fitzpatrick, S. W., C. A. Brasileiro, C. F. B. Haddad, and K. R. Zamudio. 2009. "Geographical Variation in Genetic Structure of an Atlantic Coastal Forest Frog Reveals Regional Differences in Habitat Stability." *Molecular Ecology* 18:2877–96.

Fonseca, G. A. B. 1985. "The Vanishing Brazilian Atlantic Forest." *Biological Conservation* 34:17–34.

Fundação SOS Mata Atlântica and INPE (Instituto Nacional de Pesquisas Espaciais). 2009. *Atlas dos Remanescentes Florestais da Mata Atlântica: Período 2005*□*2008. Relatório Parcial, São Paulo.* http://mapas.sosma.org.br/site_media/download/atlas%20mata%20atlantica-relatorio2005-2008.pdf.

Galindo-Leal, C., and I. G. Câmara. 2003. "Atlantic Forest Hotspot Status: An Overview." In *The Atlantic Forest of South America: Biodiversity Status, Threats and Outlook*, edited by C. Galindo-Leal and I. G. Câmara, 3–11. Washington: Island Press.

Gascon, C. S., J. R. Malcolm, J. L. Patton, M. N. F. da Silva, J. P. Bogart, S. C. Lougheed, C. A. Peres et al. 2000. "Riverine Barriers and the Geographic Distribution of Amazonian Species." *Proceedings of the National Academy of Sciences, USA* 97:13672–77.

Giraudo, A. R., H. Povedano, M. J. Belgrano, U. Pardiñas, A. Miquelarena, D. Ligier, E. Krauczuk et al. 2003. "Biodiversity Status of the Interior Atlantic Forest of Argentina." In *The Atlantic Forest of South America: Biodiversity Status, Threats, and Outlook*, edited by C. Galindo-Leal and I. G. Câmara, 160–80. Washington: Island Press.

Gonçalves, P. R., P. Myers, J. F. Vilela, and J. A. de Oliveira. 2007. "Systematics of Species of the Genus *Akodon* (Rodentia: Sigmodontinae) in Southeastern Brazil and Implications for the Biogeography of the Campos de Altitude." *Miscellaneous Publications, Museum of Zoology, University of Michigan* 197:1–24.

Grazziotin, F. G., M. Monzel, S. Echeverrigaray, and S. L. Bonatto. 2006. "Phylogeography of the *Bothrops jararaca* Complex (Serpentes: Viperidae): Past Fragmentation and Island Colonization in the Brazilian Atlantic Forest." *Molecular Ecology* 15:3969–82.

Guillaume, C. P., B. Heulin, M. J. Arrayago, A. Bea, and F. Brana. 2000. "Refuge Areas and Suture Zones in the Pyrenean and Cantabrian Regions: Geographic Variation of the Female MPI Sex-Linked Alleles Among Oviparous Populations of the Lizard *Lacerta (Zootoca) vivipara*." *Ecography* 23:3–10.

Hewitt, G. M. 1999. "Post-Glacial Re-Colonization of European Biota." *Biological Journal of the Linnean Society* 68:87–112.

———. 2001. "Speciation, Hybrid Zones and Phylogeography—Or Seeing Genes in Space and Time." *Molecular Ecology* 10:537–49.

Hoge, A. R., H. E. Belluomini, and W. Fernandes. 1977. "Variação do Número de Placas Ventrais de *Bothrops jararaca* em Função dos Climas (Viperidae, Crotalinae)." *Memórias do Instituto Butantan* 40/41:11–17.

Huang, C., S. Kim, A. Altstatt, J. R. G. Townshend, P. Davis, K. Song, C. J. Tucker et al. 2007. "Rapid Loss of Paraguay's Atlantic Forest and the Status of Protected Areas: A Landsat Assessment." *Remote Sensing of Environment* 106:460–66.

IPEMA (Instituto de Pesquisas da Mata Atlântica). 2005. *Conservação da Mata Atlântica no Espírito Santo: Cobertura Florestal e Unidades de Conservação*. Vitória: IPEMA and Conservação Internacional.

Kinzey, W. G. 1982. "Distribution of Primates and Forest Refuges." In *Biological Diversification in the Tropics*, edited by G. T. Prance, 455–82. New York: Columbia University Press.

Lara-Ruiz, P., A. G. Chiarello, and F. R. Santos. 2008. "Extreme Population Divergence and Conservation Implications for the Rare Endangered Atlantic Forest Sloth, *Bradypus torquatus* (Pilosa: Bradypodidae)." *Biological Conservation* 141:1332–42.

Ledru, M.-P. 1993. "Late Quaternary Environmental and Climatic Changes in Central Brazil." *Quaternary Research* 39:90–98.

Ledru, M. P., J. Bertaux, A. Sifeddine, and K. Suguio. 1998. "Absence of Last Glacial Records in Lowland Tropical Forests." *Quaternary Research* 49:233–37.

Leite, Y. L. R. 2003. "Evolution and Systematics of the Atlantic Tree Rats, Genus *Phyllomys* (Rodentia. Echimyidae), with Description of Two New Species." *University of California Publications in Zoology* 132:1–118.

Lichte, M., and H. Behling. 1999. "Dry and Cold Climatic Conditions in the Formation of the Present Landscape in Southeastern Brazil: An Interdisciplinary Approach to a Controversial Topic." *Zeitschrift für Geomorphologie* 43:341–58.

Luna, L., and B. D. Patterson. 2003. "A Remarkable New Mouse (Muridae: Sigmodontinae) from Southeastern Peru, with Comments on the Affinities of *Rhagomys rufescens* (Thomas, 1886)." *Fieldiana, Zoology, n.s.* 101:1–24.

Lynch, J. D. 1979. "The Amphibians of the Lowland Tropical Forests." In *The South American Herpetofauna: Its Origin, Evolution and Dispersal*, edited by W. E. Duellman, 189–216. Lawrence: Museum of Natural History, University of Kansas.

Martins, F. M., A. D. Ditchfield, D. Meyer, and J. S. Morgante. 2007. "Mitochondrial DNA Phylogeography Reveals Marked Population Structure in the Common Vampire Bat, *Desmodus rotundus* (Phyllostomidae)." *Journal of Zoological Systematics and Evolutionary Research* 45:372–78.

Mendes, S. L. 1997. "Padrões Biogeográficos e Vocais em *Callithrix* (Primates, Callitrichidae) do Grupo Jacchus." Unpublished PhD diss. Universidade Estadual de Campinas, Sao Paulo.

Metzger, J. P. 2009. "Conservation Issues in the Brazilian Atlantic Forest." *Biological Conservation* 142:1138–40.

Miranda, G. B., J. Andrades-Miranda, J. F. B. Oliveira, A. Langguth, and M. S. Mattevi. 2007. "Geographic Patterns of Genetic Variation and Conservation Consequences in Three South American Rodents." *Biochemical Genetics* 45:839–56.

Mittermeier, R. A., P. R. Gil, M. Hoffman, J. Pilgrim, T. Brooks, C. G. Mittermeier, J. Lamoreux, and G. A. B. Fonseca. 2005. *Hotspots Revisited: Earth's Biologically Richest and Most Endangered Terrestrial Ecoregions*. Washington: CEMEX and Conservation International.

Moraes-Barros, N., J. A. B. Silva, C. Y. Myiaki, and J. S. Morgante. 2006. "Comparative Phylogeography of the Atlantic Forest Endemic Sloth (*Bradypus torquatus*) and the Widespread Three-Toed Sloth (*Bradypus variegatus*) (Bradypodidae, Xenarthra)." *Genetica* 126:189–98.

Morellato, L. P. C., and C. F. B. Haddad. 2000. "Introduction: The Brazilian Atlantic Forest." *Biotropica* 32:786–92.

Mori, S., B. M. Boom, and G. T. Prance. 1981. "Distribution Patterns and Conservation of Eastern Brazilian Coastal Forest Tree Species." *Brittonia* 33:233–45.

Moritz, C., J. L. Patton, C. J. Schneider, and T. B. Smith. 2000. "Diversification of Rainforest Faunas: An Integrated Molecular Approach." *Annual Review of Ecology and Systematics* 31:533–63.

Müller, P. 1973. *Dispersal Centres of Terrestrial Vertebrates in the Neotropical Realm. Biogeographica*, 2. The Hague: Dr. W. Junk.

Myers, N., R. A. Mittermeier, C. G. Mittermeier, G. A. B. Fonseca, and J. Ken. 2000. "Biodiversity Hotspots for Conservation Priorities." *Nature* 403:853–58.

Oliveira, P. E., A. M. Franca-Barreto, and K. Suguio. 1999. "Late Pleistocene/ Holocene Climatic and Vegetational History of the Brazilian Caatinga: The Fossil Dunes of the Middle São Francisco River." *Palaeogeography, Palaeoclimatology, Palaeoecology* 152:319–37.

Oliver, W. L. R., and I. B. Santos. 1991. "Threatened Endemic Mammals of the Atlantic Forest Region of South-East Brazil." *Special Scientific Report of Jersey Wildlife Preservation Trust* 4:1–125.

Paglia, A. P., G. A. B. Fonseca, A. B. Rylands, G. Herrmann, L. Aguiar, A. G. Chiarello, Y. L. R. Leite et al. Forthcoming. *Annotated Checklist of Brazilian Mammals*, 2nd ed. Washington: Conservation International, *Occasional Papers in Conservation Biology*.

Pardiñas, U. F. J., P. Teta, and G. D'Elía. 2009. "Taxonomy and Distribution of *Abrawayaomys* (Rodentia: Cricetidae), an Atlantic Forest Endemic with the Description of a New Species." *Zootaxa* 2128:39–60.

Patton, J. L., and M. N. F. da Silva. 1998. "Rivers, Refuges and Ridges: The Geography of Speciation of Amazonian Mammals." In *Endless Forms: Species and Speciation*, edited by D. J. Howard and S. H. Berlocher, 202–12. New York: Oxford University Press.

Patton, J. L., M. N. F. da Silva, M. C. Lara, and M. A. Mustrangi. 1997. "Diversity, Differentiation, and the Historical Biogeography of Non-Volant Small Mammals of the Neotropical Forests." In *Tropical Forest Remnants: Ecology, Management, and Conservation of Fragmented Communities*, edited by W. F. Laurance and R. O. Bierregaard Jr., 455–65. Chicago: University of Chicago Press.

Patton, J. L., M. N. F. da Silva, and J. R. Malcolm. 2000. "Mammals of the Rio Juruá and the Evolutionary and Ecological Diversification of Amazonia." *Bulletin of the American Museum of Natural History* 244:1–306.

Peixoto, A. L., and A. Gentry. 1990. "Diversidade e Composição Florística da Mata de Tabuleiro na Reserva Florestal de Linhares (Espírito Santo, Brasil)." *Revista Brasileira de Botânica* 13:19–25.

Pellegrino, K. C. M., M. T. Rodrigues, A. N. Waite, M. Morando, Y. Y. Yassuda, and J. W. Sites Jr. 2005. "Phylogeography and Species Limits in the *Gymnodactylus darwinii* Complex (Gekkonidae, Squamata): Genetic Structure Coincides with River Systems in the Brazilian Atlantic Forest." *Biological Journal of the Linnean Society* 85:13–26.

Percequillo, A. R., P. R. Gonçalves, and J. A. de Oliveira. 2004. "The Rediscovery of *Rhagomys rufescens* (Thomas, 1886), with a Morphological Redescription and Comments on its Systematic Relationships Based on Morphological and Molecular (Cytochrome *b*) Characters." *Mammalian Biology* 69:238–57.

Percequillo, A. R., M. Weksler, and L. P. Costa. 2011. "A New Genus and Species of Rodent from the Brazilian Atlantic Forest (Rodentia, Cricetidae, Sigmodontinae, Oryzomyini), with Comments on Oryzomyine Biogeography." *Zoological Journal of the Linnean Society* 161:357–90.

Por, F. D. 1992. *Sooretama: The Atlantic Rain Forest of Brazil*. The Hague: SPB Academic Publishing.

Por, F. D., V. L. Imperatriz-Fonseca, and F. L. Neto. 2005. *Biomes of Brazil: An Illustrated Natural History*. Sofia: Pensoft Publishers.

Prado, D. E., and P. E. Gibbs. 1993. "Patterns of Species Distributions in the Dry Seasonal Forests of South America." *Annals of the Missouri Botanical Garden* 80:902–27.

Prance, G. T. 1982. "Forest Refuges: Evidence from Woody Angiosperms." In *Biological Diversification in the Tropics*, edited by G. T. Prance, 137–58. New York: Columbia University Press.

Redenbach, Z., and E. B. Taylor. 2002. "Evidence for Historical Introgression Along a Contact Zone between Two Species of Char (Pisces: Salmonidae) in Northwestern North America." *Evolution* 56:1021–35.

Remington, C. L. 1968. "Suture-Zones of Hybrid Interaction between Recently Joined Biotas." In *Evolutionary Biology*, vol. 2, edited by T. Dobzhansky, M. K. Hecht, and W. C. Steere, 321–428. New York: Plenum.

Ribas, C. C., R. G. Moyle, M. Y. Miyaki, and J. Cracraft. 2007. "The Assembly of Montane Biotas: Linking Andean Tectonics and Climatic Oscillations to Independent Regimes of Diversification in *Pionus* Parrots." *Proceedings of the Royal Society B, Biological Sciences* 274:2399–408.

Rizzini, C. T. 1967. "Delimitação, Caracterização e Relações da Flora Silvestre Hiléiana." *Atlas Simpósio Biota Amazônica* 4:13–36.

Rull, V. 2008. "Speciation Timing and Neotropical Biodiversity: The Tertiary-Quaternary Debate in the Light of Molecular Phylogenetic Evidence." *Molecular Ecology* 17:2722–29.

Rylands, A. B., G. A. B. Fonseca, Y. L. R. Leite, and R. A. Mittermeier. 1997. "Primates of the Atlantic Forest: Origin, Endemism, Distribution, and Communities." In *Adaptive Radiations of Neotropical Primates*, edited by M. Norconk, A. Rosenberger, and P. Garber, 21–51. New York: Plenum Press.

Silva, J. M. C., M. C. Sousa, and C. H. M. Casteleti. 2004. "Areas of Endemism for Passerine Birds in the Atlantic Forest." *Global Ecology and Biogeography* 13:85–92.

Smith, M. F., and J. L. Patton. 2007. "Molecular Phylogenetics and Diversification of South American Grass Mice, Genus *Akodon*." In *The Quintessential Naturalist: Honoring the Life and Legacy of Oliver P. Pearson*, edited by D. A. Kelt, E. P. Lessa, J. Salazar-Bravo, and J. L. Patton, 827–58. Berkeley: University of California Press, *University of California Publications in Zoology* 134.

Soderstrom, T. R., E. J. Judziewicz, and L. G. Clark. 1988. "Distribution Patterns of Neotropical Bamboos." In *Proceedings of a Workshop on Neotropical Distribution Patterns*, edited by P. E. Vanzolini and W. R. Heyer, 121–57. Rio de Janeiro: Academia Brasileira de Ciências.

Steiner-Souza, F., P. Cordeiro-Estrela, A. R. Percequillo, A. F. Testoni, and S. L. Althoff. 2008. "New Records of *Rhagomys rufescens* (Rodentia: Sigmodontinae) in the Atlantic Forest of Brazil." *Zootaxa* 1824:28–34.

Swenson, N. G., and D. J. Howard. 2004. "Do Suture Zones Exist?" *Evolution* 58:2391–97.

———. 2005. "Clustering of Contact Zones, Hybrid Zones, and Phylogeographic Breaks in North America." *American Naturalist* 166:581–91.

Tchaicka, L., E. Eizirik, T. G. Oliveira, J. F. Candido, and T. R. O. Freitas. 2007. "Phylogeography and Population History of the Crab-Eating Fox (*Cerdocyon thous*)." *Molecular Ecology* 16:819–38.

Thomas, W. W., ed. 2008. *The Atlantic Coastal Forest of Northeastern Brazil*. New York: New York Botanical Garden, *Memoirs of the New York Botanical Garden* 100.

Turcq, B., M. M. N. Pressinotti, and L. Martin. 1997. "Paleohydrology and Paleoclimate of the Past 33,000 Years at the Tamanduá River, Central Brazil." *Quaternary Research* 47:284–94.

Tyler, H., K. S. Brown Jr., and K. Wilson. 1994. *Swallowtail Butterflies of the Americas: A Study in Biological Dynamics, Ecological Diversity, Biosystematics and Conservation*. Gainesville: Scientific Publishers.

Vanzolini, P. E. 1988. "Distributional Patterns of South American Lizards." In *Proceedings of a Workshop on Neotropical Distribution Patterns*, edited by P. E. Vanzolini and W. R. Heyer, 317–42. Rio de Janeiro: Academia Brasileira de Ciências.

Vanzolini, P. E., and Williams, E. E. 1981. "The Vanishing Refuge: A Mechanism for Ecogeographic Speciation." *Papéis Avulsos de Zoologia* 34:251–55.

Villalpando, G., J. Vargas, and J. Salazar-Bravo. 2006. "First Record of *Rhagomys* (Mammalia: Sigmodontinae) in Bolivia." *Mastozoologia Neotropical* 13:143–49.

Vivo, M. de, and A. P. Carmignotto. 2004. "Holocene Vegetation Change and the Mammal Faunas of South America and Africa." *Journal of Biogeography* 31:943–57.

Wallace, A. R. 1852. "On the Monkeys of the Amazon." *Proceedings of the Zoological Society of London* 20:107–10.

Wang, X., A. S. Auler, R. L. Edwards, H. Cheng, P. S. Cristalli, P. L. Smart, D. A. Richards, and C.-C. Shen. 2004. "Wet Periods in Northeastern Brazil Over the Past 210 kyr Linked to Distant Climate Anomalies." *Nature* 432:740–43.

Whitmore, T. C., and G. T. Prance, eds. 1987. *Biogeography and Quaternary History in Tropical America*. Oxford: Oxford University Press, *Oxford Biogeography Series* 3.

Appendix 13.1

Table A13.1 Taxonomic list of mammal species endemic to the Atlantic Forest, grouped by genus, family, and order (source: Paglia et al., forthcoming).

DIDELPHIMORPHIA
- Didelphidae
  - *Cryptonanus*
    - *C. guahybae* (Tate, 1931)
  - *Didelphis*
    - *D. aurita* (Wied-Neuwied, 1826)
  - *Gracilinanus*
    - *G. microtarsus* (Wagner, 1842)
  - *Marmosops*
    - *M. paulensis* (Tate, 1931)
  - *Monodelphis*
    - *M. iheringi* (Thomas, 1888)
    - *M. rubida* (Thomas, 1899)
    - *M. scalops* (Thomas, 1888)
    - *M. theresa* Thomas, 1921

PILOSA
- Bradypodidae
  - *Bradypus*
    - *B. torquatus* Illiger, 1811

CETARTIODACTYLA
- Cervidae
  - *Mazama*
    - *M. bororo* Duarte, 1996

PRIMATES
- Atelidae
  - *Alouatta*
    - *A. guariba* (Humboldt, 1812)

(continued)

Table A13.1 *(continued)*

Brachyteles
B. arachnoides (É. Geoffroy, 1806)
B. hypoxanthus (Kuhl, 1820)
Callitrichidae
Callithrix
C. aurita (É. Geoffroy, 1812)
C. flaviceps (Thomas, 1903)
C. geoffroyi (Humboldt, 1812)
C. jacchus (Linnaeus, 1758)
C. kuhlii Coimbra-Filho, 1985
Leontopithecus
L. caissara Lorini and Persson, 1990
L. chrysomelas (Kuhl, 1820)
L. chrysopygus (Mikan, 1823)
L. rosalia (Linnaeus, 1766)
Cebidae
Cebus
C. nigritus (Goldfuss, 1809)
C. robustus Kuhl, 1820
C. xanthosternos (Wied-Neuwied, 1826)
Pitheciidae
Callicebus
C. coimbrai Kobayashi and Langguth, 1999
C. melanochir (Wied-Neuwied, 1820)
C. nigrifrons (Spix, 1823)
C. personatus (É. Geoffroy, 1812)
CHIROPTERA
Molossidae
Nyctinomops
N. aurispinosus (Peale, 1848)
Vespertilionidae
Eptesicus
E. taddeii Miranda, Bernardi, and Passos, 2006
Histiotus
H. alienus Thomas, 1916
Lasiurus
L. ebenus Fazzolari-Corrêa, 1994
Rhogeessa
R. hussoni Genoways and Baker, 1996
RODENTIA
Caviidae
Cavia
C. intermedia Cherem, Olimpio, and Ximenez, 1999

Cricetidae

Abrawayaomys

A. ruschii Cunha and Cruz, 1979

Akodon

A. mystax Hershkovitz, 1998

A. sanctipaulensis Hershkovitz, 1990

A. serrensis Thomas, 1902

Bibimys

B. labiosus (Winge, 1887)

Blarinomys

B. breviceps (Winge, 1887)

Brucepattersonius

B. griserufescens Hershkovitz, 1998

B. igniventris Hershkovitz, 1998

B. iheringi (Thomas, 1896)

B. soricinus Hershkovitz, 1998

Delomys

D. collinus Thomas, 1917

D. dorsalis (Hensel, 1873)

D. sublineatus (Thomas, 1903)

Euryoryzomys

E. russatus (Wagner, 1848)

Hylaeamys

H. laticeps (Lund, 1840)

H. oniscus (Thomas, 1904)

Juliomys

J. ossitenuis Costa, Pavan, Leite, and Fagundes, 2007

J. pictipes (Osgood, 1933)

J. rimofrons Oliveira and Bonvicino, 2002

Oxymycterus

O. caparaoe Hershkovitz, 1998

O. dasytrichus (Schinz, 1821)

O. judex Thomas, 1909

O. quaestor Thomas, 1903

O. rufus (G. Fischer, 1814)

Phaenomys

P. ferrugineus (Thomas, 1894)

Rhagomys

R. rufescens (Thomas, 1886)

Sooretamys

S. angouya (G. Fischer, 1814)

Thaptomys

T. nigrita (Lichtenstein, 1829)

(continued)

Table A13.1 *(continued)*

Wilfredomys
W. oenax (Thomas, 1928)
Dasyproctidae
Dasyprocta
D. catrinae (Thomas, 1917)
Echimyidae
Callistomys
C. pictus (Pictet, 1841)
Phyllomys
P. sulinus Leite, Christoff, and Fagundes, 2008
P. dasythrix Hensel, 1872
P. kerri (Moojen, 1950)
P. lundi Leite, 2003
P. mantiqueirensis Leite, 2003
P. medius (Thomas, 1909)
P. nigrispinus (Wagner, 1842)
P. pattoni Emmons, Leite, Kock, and Costa, 2002
P. thomasi (Ihering, 1871)
P. unicolor (Wagner, 1842)
Trinomys
T. bonafidei (Moojen, 1948)
T. dimidiatus (Günther, 1877)
T. elegans (Lund, 1841)
T. eliasi (Pessôa and Reis, 1993)
T. iheringi (Thomas, 1911)
T. mirapitanga Lara, Patton, and Hingst-Zaher, 2002
T. panema (Moojen, 1948)
T. paratus (Moojen, 1948)
T. setosus (Desmarest, 1817)
Erethizontidae
Chaetomys
C. subspinosus (Olfers, 1818)
Sphiggurus
S. villosus (F. Cuvier, 1823)
Sciuridae
Guerlinguetus
G. ingrami (Thomas, 1901)

Mammals of the Cerrado and Caatinga
Distribution Patterns of the Tropical Open Biomes of Central South America

Ana Paula Carmignotto, Mario de Vivo, and Alfredo Langguth

Abstract

Cerrado and Caatinga are neighboring open biomes of tropical South America, but their vegetation, soil, and climate characteristics render them distinctive formations. Despite these differences, their mammal faunas are largely shared. Once thought to be impoverished faunas derived from forest biomes, Cerrado and Caatinga possess rich assemblages of mammals made up of species strongly associated with open areas, including a significant endemic component mostly represented by small rodents associated with open formations. Because the Cerrado and the Caatinga are characterized by the presence of both forested and open-vegetation habitats, their mammal faunas are a composite, including species that also occur in the Amazon, the Atlantic Rain Forest, the Chaco, and open biomes of the temperate zone. The gallery, semideciduous, and deciduous forests contribute importantly to the species richness of these faunas, which together with the open habitats such as grasslands, savannas, and shrubby caatinga, enhance the overall species richness and high regional diversity of these formations. However, distinct histories of the open and forested formations are evident in mammal species distributions. The disjunct distribution of some arboreal mammals in the eastern Amazon and northern Atlantic rain forests, together with Pleistocene records of arboreal mammals in present-day areas of the Caatinga, reveal that this biome was not always an open, semiarid area but once constituted a forested formation. On the other hand, the higher number of endemic species of the Cerrado together with species shared with the Chaco (contrasted with the lower richness and endemism of the Caatinga) suggests the long-term persistence of open habitats in present-day areas of the Cerrado.

14.1 Introduction

When Alfred Russel Wallace (1876) published his treatise of the world's zoogeography, he divided the South American continent in two faunistic regions: a tropical Brazilian subregion and a temperate Chilean subregion. This division influenced biogeographers concerned with the evolution of mammals (e.g., Hershkovitz 1972), and was perceived almost as synonymous with forest (tropical subregion) and open (temperate subregion) biomes. However, Burmeister (1854) was able to distinguish Atlantic and Amazonian rain forest mammal faunas within the tropical subregion as well as an open-vegetation assemblage that he called the fauna of the "sertão." The concept of what was then called sertão has been greatly refined through the characterization of its distinct biomes, the Cerrado and the Caatinga (e.g., Eva et al. 2002; Hueck 1972; Rizzini 1997). This refinement led zoologists to endeavor tracing the ecological and evolutionary affinities of its fauna. One of the main results of this effort was an unfortunate emphasis on the forested nature of these open-biome mammal faunas, which have been generally regarded as impoverished and forest-derived (Cerqueira 1982; Mares 1992; Redford and Fonseca 1986).

Questions that concern us here are: What mammal species inhabit the Cerrado and Caatinga? How different are these faunas from each other? What are the relationships of these faunas to other tropical formations of Central South America?

We begin by characterizing both biomes and providing updated lists of their mammal faunas. Then we analyze their diversity by comparing endemic and shared components of their faunas. Finally, we discuss the evolution of these biomes and their mammal faunas since the Cenozoic, noting important patterns of diversification.

14.2 Characterization of the Biomes

Tropical South American biomes south of the equator and east of the Andes include two major rain forests, Amazonian and Atlantic, separated by a series of distinctive, more or less open vegetation types, the Caatinga, Cerrado, and Chaco (fig. 14.1A). Each of these biomes is defined by characteristic phytophysiognomies (Hueck and Seibert 1981). Additionally, where biomes meet, spatial patterns of floral substitution are complex, sometimes forming transitional belts and other times mosaics. Within their core areas and at their periphery, they contain sizable pockets of other vegetation types. Thus, no biome is homogeneous, and all show the influence of neighboring formations (Ab'Saber 1977).

The Cerrado is mostly a savanna, and its phytophysiognomy varies from large tracts of grassy fields to dense woodlands. Its core area comprises 2 million km^2 in the center of tropical South America where it borders the Chaco, Amazonian, and Atlantic rain forests, and the Caatinga. Enclaves of

Cerrado vegetation and physiognomy exist within neighboring biomes such as the Amazonian and Atlantic rain forests, ranging in size from a few to thousands of square kilometers (Ratter, Ribeiro, and Bridgewater 1997). Although climatically seasonal (fig. 14.1B), the Cerrado is irrigated by large perennial rivers as well as by numerous smaller watercourses, and therefore it is not dry, but rather constitutes a humid savanna (Coutinho 2006). The numerous

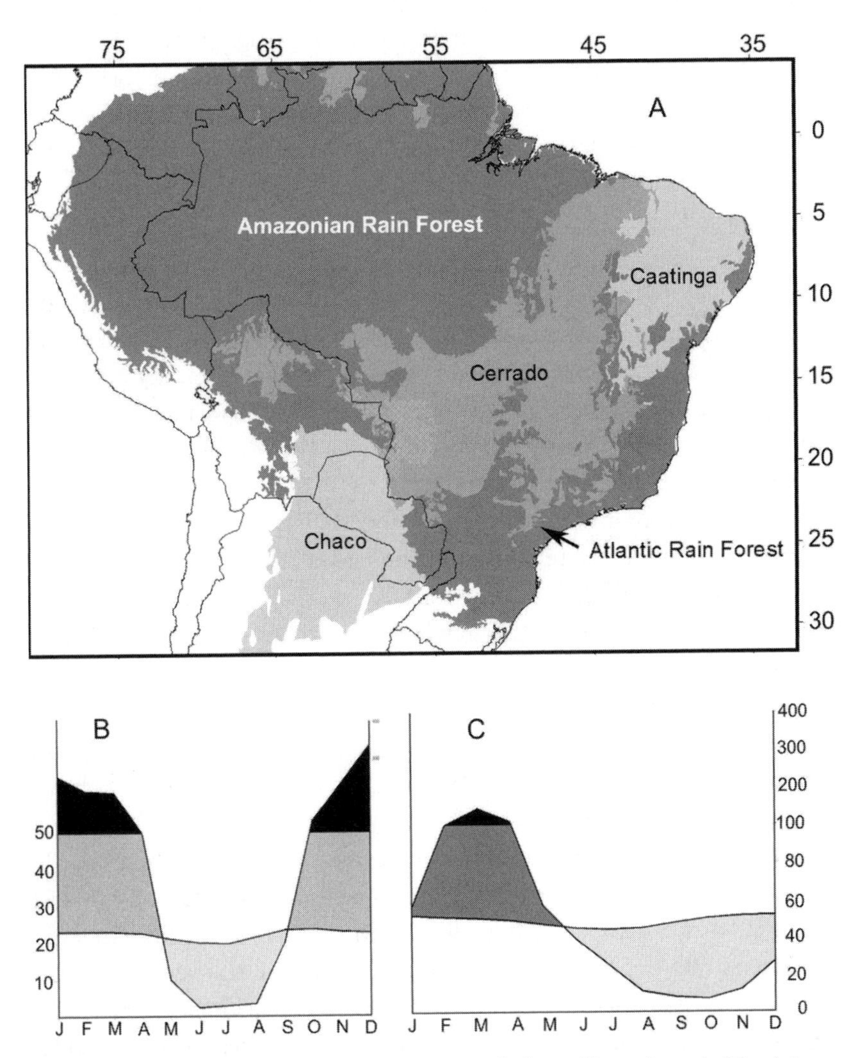

Figure 14.1 (A) Distribution of Cerrado and Caatinga and adjacent biomes in tropical South America based on Dinnerstein et al. (1995); (B) Walter's (1986) climatic diagram for the central region of the Cerrado; (C) climatic diagram for the central region of the Caatinga. Diagrams show rainy season in medium gray and black, dry season is in light gray (climate data from Leemans and Cramer 1991).

watercourses make gallery forests and seasonally flooded plains important components of Cerrado landscapes (fig. 14.2). Typically, the dry season lasts approximately four months, and annual precipitation is at least 1200 mm. In some places, mainly in the vicinity of the Amazonian rain forest, precipitation can reach 2000 mm, but most of the Cerrado receives ~1500 mm, with a mean annual temperature of ~22°C (Nimer 1989).

Core areas of Cerrado are located in the highlands of central Brazil, 1000 to 1300 m elevation, where old, Precambrian terrains predominate, and are now covered by laterites and Cenozoic sediments (Parada and Andrade 1977). Red and yellow latosoils and sand dominate, covering 65% of the Cerrado region, being very deep, well-drained, and rich in organic matter and minerals, especially aluminum and iron (Furley and Ratter 1988). Characteristics of climate, relief, and soil, together with the influence of fire, pasture, and humans, distinguish the major habitats present in Cerrado today (Oliveira-Filho and Ratter 2002). The vegetation is comprised of a complex mosaic (fig. 14.2) that ranges from open formations, such as dry ("*campo limpo*" and "*campo sujo*") and humid grasslands ("*campo úmido*" and "*veredas*"), to dense savannas ("*campo cerrado*" and "*cerrado*" sensu stricto), to truly forested ones, such as woodlands ("*cerradão*") and prominent gallery forests. Semideciduous and deciduous forests are present as pockets of variable size (Eiten 1972).

The Caatinga in northeastern Brazil is a deciduous dry forest and shrub formation covering ~0.8 million km^2 (Prado 2008); Coutinho (2006) considered it a dry savanna. Situated between a narrow strip of Atlantic rain forest and the eastern extent of the Amazonian rain forest, it meets the Cerrado over a large front (fig. 14.1A). In places, the contact between these biomes is clear-cut, with typical phytophysiognomies occurring side by side, while in others it includes vast tracts of transitional areas (Ab'Sáber 1974). The climate is hot and dry (fig. 14.1C), with a marked seasonal rainfall regime. The dry season lasts roughly seven months per year, followed by five months of precipitation averaging 800 mm, and a mean annual temperature ~27°C. An important climatic characteristic is that drier years are frequent and not predictable (Nimer 1989). Most of the rivers that drain the Caatinga are seasonal, and cacti are a prominent feature of the landscape, while trees and shrubs are deciduous, and herbs and grasses grow only during the rainy season. Caatinga resembles the Chaco biome of Bolivia, Paraguay, and northwestern Argentina in its deciduous physiognomy as well as its abundance of cacti (Hueck 1972; Rizzini 1997), although its floristic affinities to that biome are not strong (Prado 2008). Unlike the Cerrado, gallery forests are not important, and dry forests mostly cover

Figure 14.2 Habitat types found in the Cerrado: (A) gallery forest and flooded grassland; (B) woodland; (C) flooded grassland (foreground) and in the background the "vereda" (on the left) and a savanna (on the right); (D) dense and sparse savanna configurations; (E) "campo sujo" (shrubby grassland); and (F) "campo limpo" (dry grassland). Photos (A), (B), (C), (E), and (F) by Ana Paula Carmignotto; photo (D) by Fernando Nadal Junqueira Villela.

Figure 14.3 Habitat types found in the Caatinga: (A) rocky outcrop with shrubs; (B) shrubby caatinga; (C) mesic orographic forest enclave ("brejo de altitude"); (D) shrubby caatinga with cacti; and (E) aerial view of woody caatinga. Photos (A), (B), (D), and (E) by Miguel T. Rodrigues; photo (C) by Alfredo Langguth.

mountain ranges and interplanaltic depressions. Most of the vegetation has been destroyed by man during the last 500 years. Some forested areas receive orographic precipitation in excess of typical rainfall (the so-called "brejos de altitude," which are enclaves of Atlantic rain forest within the Caatinga). The Caatinga differs from Cerrado in relief and soils. Large extensions of the Caatinga biome occupy interplanaltic depressions covered by Tertiary sediments that lie over Precambrian crystalline basal formations. Soils are shallow and rocky, with numerous rock outcrops (Ab'Saber 1974). Major habitats (fig. 14.3) include tall dry forests ("*caatinga arbórea*") with trees up to 15–20 m in height occurring in humid areas with richer soils; shrubby vegetation ("*caatinga arbustiva*") in areas of rocky soils, with a high density of cacti and bromeliads; more mesic habitats ("*brejos*") in areas with orographic rainfall, and, finally, gallery forests along perennial rivers (Andrade-Lima 1981).

14.3 Cerrado and Caatinga Mammals

Cerrado mammals have attracted serious attention only in the last half century (Alho 1981; Alho, Pereira, and Paula 1986; Ávila-Pires 1960, 1966, 1972; Fonseca and Redford 1984; Hershkovitz 1990a, 1990b, 1993; Lacher and Alho 1989, 2001; Lacher, Mares, and Alho 1989; Mares, Ernest, and Gettinger 1986; Mares, Braun, and Gettinger 1989; Mello 1980; Mello and Moojen 1979; Valle et al. 1982). One of the first lists of mammal species from the Cerrado was that of Redford and Fonseca (1986), which contained 67 genera and 100 terrestrial mammal species and was considered a rich and diverse assemblage of mammals (Mares, Braun, and Gettinger 1989). However, the number of endemics was judged to be very low since most species were derived from mesic formations, were shared with adjacent biomes, and/or were widespread within the Cerrado (Ávila-Pires 1966; Mello-Leitão 1946; Redford and Fonseca 1986).

Regarding the Caatinga mammals, two important expeditions took place, one between 1952 and 1955 conducted by the "Serviço Nacional de Peste"—which collected mammals in 40 distinct localities of Brazilian Northeast (Freitas 1957)—and the other to the Exu-Bodocó region in Pernambuco between 1967 and 1971 (Karimi, de Almeida, and Petter 1976). Both expeditions obtained mainly, but not only, marsupials and small rodents. Additionally, other important series of specimens were collected by Carnegie Museum of Natural History researchers supported by the Brazilian Academy of Sciences from 1975 to 1978 in the "Chapada do Araripe" region, which resulted in a number of articles related to the ecology, systematics, and geographic distribution of Caatinga mammals (Lacher 1981; Lacher and Mares 1986; Mares and Lacher 1987; Mares et al. 1981; Mares, Willig, and Lacher 1985; Streilein 1982a, 1982b, 1982c, 1982d, 1982e, 1982f; Willig 1983, 1985, 1986; Willig and Mares 1989). Finally, from 1976 on, mammal collections were started in the federal universities of Pernambuco and Paraíba, which jointly maintain the most representative collections of Caatinga mammals today.

The first checklists of Caatinga mammals (Mares et al. 1981; Willig and Mares 1989) included 8 orders, 23 families, 63 genera, and 86 species. These authors concluded that the Caatinga mammal fauna was species-poor, included few endemics, and its species were not well-adapted to semiarid environments but instead were characteristic of mesic formations. They regarded Caatinga mammals as a subset of Cerrado mammals (Mares, Willig, and Lacher 1985; Willig 1983), essentially corroborating the findings of Sick (1965) and Vanzolini (1976) for birds and reptiles, respectively.

We have developed a revised list of the Cerrado and Caatinga mammals (appendix 14.1) to reevaluate their zoogeographical patterns. Mammal distribution data were compiled from several general publications (Anderson 1997; Bonvicino, de Oliveira, and D'Andrea 2008; Eisenberg and Redford 1999; Gardner 2008; IUCN 2009; Marinho-Filho, Rodrigues, and Juarez 2002; de Oliveira, Gonçalves, and Bonvicino 2008; Redford and Eisenberg 1992; Reis et al. 2006). Many additional publications and not-yet-published theses were also used, because we trusted their accuracy for the taxa under consideration over more general sources (Aires 2008; Bezerra and de Oliveira 2010; Bezerra et al. 2007; Black-Décima et al. 2010; Cáceres and Carmignotto 2006; Cáceres, Ferreira, and Carmignotto 2007; Campos 2009; Carmignotto 2005; Carmignotto and Monfort 2006; Gregorin 2006; Gregorin and Taddei 2002; Gregorin, Carmignotto, and Percequillo 2008; Gurgel-Filho 2010; Iack-Ximenes 1999, 2005; Kasper et al. 2009; Leite 2003; Percequillo 1998; Percequillo, Hingst-Zaher, and Bonvicino 2008; Rossi 2005; Silva 2010; Silva 2001; Varela et al. 2010; de Vivo 1991, unpubl. data; Voss and Jansa 2009; Weksler, Percequillo, and Voss 2006). We have followed the nomenclature and taxonomic arrangement of contributors to Wilson and Reeder (2005) and Gardner (2008), except where these were modified by the aforementioned works. Species published but not named (such as "species a" or "species b") have not been counted.

Listing the mammals for entire biomes required a number of decisions regarding questionable identifications. Another problem was how to deal with regions not immediately classifiable as belonging to a given biome, such as the

Pantanal wetlands and floodplains and the transitional area known as "Zona dos Cocais" (palm tree region). The Pantanal is a vast flood- plain located between the Brazilian states of Mato Grosso and Mato Grosso do Sul, southeastern Bolivia, and northeastern Paraguay. We included it within the Cerrado biome because it supports several mammal species that occur in the Cerrado, and the two biomes constitute a single zoogeographical unit (Fonseca, Herrmann, and Leite 1999). The Zona dos Cocais is a large area with dense concentrations of Babassu palms in the central and northern portion of the Brazilian state of Maranhão; it lies adjacent to typical Cerrado, Amazonian rain forest, and Caatinga. We have excluded the Zona dos Cocais because it is not very well understood, we have little data on the distribution of its mammals, and because authors have classified it either as a transitional vegetation belt or alternatively allocated it to one or another of its neighbor biomes (Prado 2008; Rizzini 1997). In addition, the Zona dos Cocais may be of anthropogenic origin (Fernandes and Bezerra 1990).

Another important step was to develop criteria by which a species should or should not be associated with a given biome. The majority of cases were reasonably straightforward: we tallied as belonging to a biome all species that occurred in sizable areas of that biome, even when their preferred habitat is atypical for that biome. For example, mammals occupying forested habitats inside the open biomes of the Cerrado and Caatinga have been counted as belonging to the faunas of those biomes because we regard those biomes as fundamentally heterogeneous. We did not count any species as belonging to a biome if its occurrence in that biome was deemed marginal. For instance, in the northern portion of Mato Grosso, some typical Amazonian primates, such as species of *Ateles* and *Aotus*, occur in broad gallery forests. These primates do not range far into the Cerrado and were thus omitted from the lists.

Finally, some species characteristics of open formations have been recorded in areas far inside forested biomes. This type of historic range expansion is even more common in heavily disturbed regions of the Atlantic rain forest. In these cases, we decided to infer stricter biome affiliations and ignore the putative range expansion (for instance, we left out species of *Thrichomys* that now are found within the Atlantic rain forest of northeastern Brazil).

14.3.1 PATTERNS OF RICHNESS

Cerrado mammals comprise 126 genera and 227 species, while the Caatinga supports 101 genera and 153 species. The number of species for both biomes is substantially higher than previously known (table 14.1), indicating the growth

Table 14.1 Number of genera and species (in parentheses) of mammals registered by distinct authors for the Cerrado and Caatinga.

Cerrado	Caatinga	Reference
–	63 (86)	Mares et al. (1981)
67 (100) nonvolant	–	Redford and Fonseca (1986)
109 (159)	74 (102)	Fonseca, Herrmann, and Leite (1999)
–	95 (143)	de Oliveira, Gonçalves, and Bonvicino (2008)
121 (194)	–	Marinho-Filho, Rodrigues, and Juarez (2002)
126 (227)	101 (153)	Present study

of taxonomic and distributional knowledge in recent years. This has resulted from an increased number of inventories, especially in areas that had not previously been surveyed (Carmignotto 2005; Gregorin, Carmignotto, and Percequillo 2008; Marinho-Filho and Veríssimo 1997; Oliveira and Langguth 2004) and continued systematic studies (Basile 2003; Bonvicino 2003; Bonvicino and Almeida 2000; Bonvicino and Weksler 1998; Bonvicino, Lima, and Almeida 2003; Braggio and Bonvicino 2004; Gregorin 2006; Langguth and Bonvicino 2002; Oliveira 1998; Percequillo, Hingst-Zaher, and Bonvicino 2008; Tribe 2005). However, mammal inventories in both biomes are still incomplete. In the Caatinga, surveys have been mainly concentrated in mesic and transitional regions, and vast areas remain completely unknown (de Oliveira, Gerude, and de Silva 2008). In the Cerrado, the northern and western regions are the least studied, and relatively few species have been recorded there (Cáceres et al. 2008; Carmignotto 2005; de Oliveira, Gerude, and de Silva 2008).

Fonseca, Hermann, and Leite (1999) analyzed the relationship between the number of species and the area of the main tropical biomes of South America. They employed a regression analysis in which the Atlantic rain forest fell well above the regression line, the Amazonian rain forest exactly on it, and the Cerrado and the Caatinga were distinctly below expectation, implying that the latter biomes had fewer species than would be expected by their area alone. On the other hand, for each 100,000 km^2 of area, the Cerrado supports 11.4 species and the Caatinga 19.2 species, while (using the dated estimates of Fonseca, Herrmann, and Leite 1999) the Amazonian rain forest supports only 5.5 species. The explanation for these paradoxical patterns lies in proportions of species shared between tropical South America biomes, as explained beyond.

The higher species richness of the Cerrado relative to the Caatinga can be explained by its area, which is more than twice that of the Caatinga. Bats and rodents are the most diverse orders in both biomes, a typical Neotropical pattern (Patterson 2000, 2001); carnivores and marsupials come next, with the remaining orders represented by fewer species (table 14.2). This pattern had already been demonstrated for Cerrado and Caatinga, but the greater number of xenarthrans (Cingulata and Pilosa), especially in Caatinga, is not evident in our data (Fonseca, Herrmann, and Leite 1999; Marinho-Filho, Rodrigues, and Juarez 2002).

The Caatinga bat fauna is unexpectedly rich, with numbers of species and genera almost as high as the Cerrado. In fact, bats comprise half of all Caatinga mammals, while in the Cerrado they amount to slightly more than one-third of the fauna (see table 14.2). Bat faunal composition within the Caatinga is quite heterogeneous. Drier and treeless habitats support impoverished bat communities, and most of the richness is associated with more mesic habitats (Willig 1983). This elevated richness seems to be related to the latitudinal diversity gradient, which is strongly developed among New World bats (Stevens 2004; Willig and Selcer 1989).

Removing bats from the analysis, rodents, carnivores, and marsupials are the most diverse groups, and all remaining orders have fewer species and can be lumped together (fig. 14.4). At the generic level, the two biomes present similar proportions for all of the groups. At the species level, the Caatinga has proportionally more medium- and large-bodied mammals (such as carnivores and primates), while the Cerrado has more marsupials and rodents.

14.3.2 ADJACENT BIOMES AND SHARED MAMMALS

The distributions of mammals in the Cerrado, Caatinga, Chaco, and adjacent Amazonian and Atlantic rain forests are distinctly hierarchical. Cerrado and Caatinga share 120 mammal species, representing 78% of the Caatinga and 59% of the Cerrado mammal fauna. Shared species include 62 bats (52%); excluding bats, 22 (38%) are rodents, 14 (24%) carnivores, 7 (12%) marsupials, and 15 (26%) represent the remaining orders. However, most of these species are also shared with other biomes. In fact, the largest component is a fauna shared by all biomes considered here—or all except the Chaco (see the last categories of table 14.3 and fig. 14.5). A total of 79 species are shared (52% of the Caatinga and 35% of the Cerrado mammal fauna), of which 52 are bats—the remainder are mostly large species with broad geographic distributions (11 carnivores, 5 xenarthrans, 3 ungulates, and 3 large rodents: porcupine, capybara, and paca). This is noteworthy, as most previous discussions of the faunal similarities of Caatinga and Cerrado did not distinguish between exclusively shared faunal components and taxa more broadly distributed over tropical South America (Vanzolini 1963,

Table 14.2 Richness and endemism of genera and species (in parentheses) for each mammalian order that occurs in the Cerrado (CE) and Caatinga (CA) biomes. Values include simple counts and percentages of the total mammalian faunas.

	Richness		Endemics		Richness		Endemics	
ORDER	CE	CE%	CE	CE%	CA	CA%	CA	CA%
Didelphimorphia	12 (23)	9.5 (10.1)	0 (2)	0 (8.0)	7 (8)	6.9 (5.2)	0 (1)	0 (12.5)
Cingulata	5 (8)	4.0 (3.5)	0	0	4 (5)	4.0 (3.3)	0	0
Pilosa	3 (3)	2.4 (1.3)	0	0	3 (3)	3.0 (2.0)	0	0
Primates	5 (8)	4.0 (3.5)	0 (1)	0 (4.0)	4 (6)	4.0 (3.9)	0 (1)	0 (12.5)
Rodentia	35 (78)	27.8 (34.4)	5 (21)	100 (84.0)	23 (35)	22.8 (22.9)	0 (5)	0 (62.5)
Lagomorpha	1 (1)	0.8 (0.4)	0	0	1 (1)	1.0 (0.7)	0	0
Chiroptera	45 (80)	35.7 (35.2)	0 (1)	0 (4.0)	44 (77)	43.6 (50.3)	1 (1)	100 (12.5)
Carnivora	14 (19)	11.1 (8.4)	0	0	11 (14)	10.9 (9.2)	0	0
Perissodactyla	1 (1)	0.8 (0.4)	0	0	1 (1)	1.0 (0.7)	0	0
Cetartiodactyla	5 (6)	4.0 (2.6)	0	0	3 (3)	3.0 (2.0)	0	0
TOTAL	126 (227)		5 (25)		101 (153)		1 (8)	

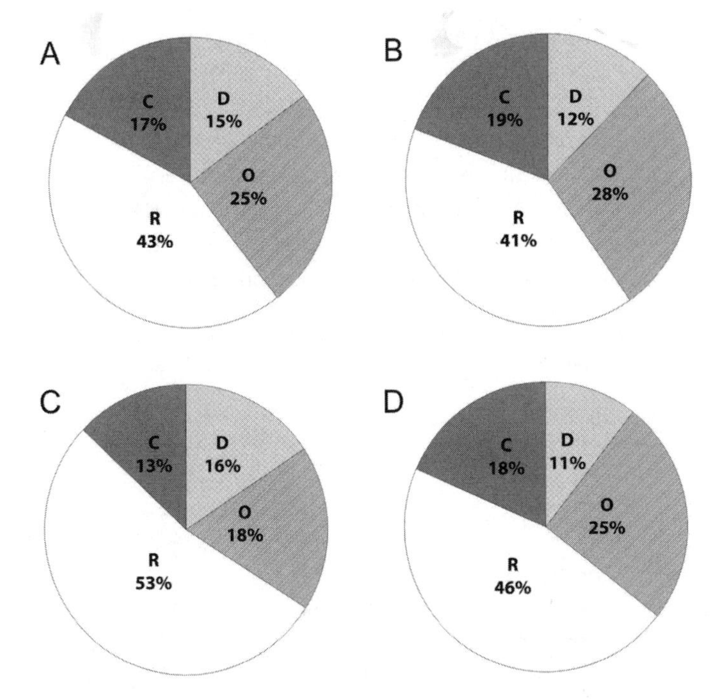

Figure 14.4 Richness of nonvolant mammal genera and species for Cerrado and Caatinga. (A) percentage of genera for the Cerrado; (B) percentage of genera for the Caatinga; (C) percentage of species for the Cerrado; (D) percentage of species for the Caatinga. Abbreviations: C, Carnivora; D, Didelphimorphia; R, Rodentia; O, others.

1976; Vitt 1991 for reptiles; Sick 1965, 1966 for birds; and Ávila-Pires 1966; Fonseca, Herrmann, and Leite 1999; Redford and Fonseca 1986 for mammals). Only 15 species (mostly belonging to small-sized taxa) are exclusively shared between Cerrado and Caatinga, and 5 additional species (two marsupials and three rodents) are also shared with the Chaco (table 14.3, fig. 14.5).

The largest part of the Cerrado mammalian fauna (50 species, 22%) is composed by species also present in Amazonia, Atlantic rain forest, the Chaco, and Caatinga, followed by a number of species also shared with Amazonian and Atlantic rain forests and Caatinga (29, 13%; see table 14.3 and fig. 14.5). Again, broadly distributed mammals, such as bats and medium- and large-sized species, are responsible for this pattern. Endemics represent the third largest component of Cerrado mammal fauna (25 species, 11%), almost all of which are rodents (see appendix 14.1 and table 14.2). By comparison, the Caatinga has only 8 endemic species (5% of the total, mostly rodents) and it shares more species with adjacent biomes than the Cerrado (33% with all biomes consid-

Table 14.3 Unique and shared elements of the Cerrado and Caatinga mammal faunas. Values are the absolute number of species and percentages of their respective faunas. Tropical South America biomes are represented by the following abbreviations: AF, Atlantic Forest; AM, Amazon Forest; CA, Caatinga; CE, Cerrado; CH, Chaco; "Other" refers to any open biome not contiguous with either Cerrado or Caatinga.

Comparisons	Cerrado	Caatinga
Endemics	25 (11.0%)	8 (5.2%)
CH	25 (11.0%)	0
AM	21 (9.3%)	6 (3.9%)
CE–CA	15 (6.6%)	15 (9.8%)
AF	10 (4.4%)	10 (6.5%)
Other	2 (0.9%)	0
AM–AF	7 (3.1%)	6 (3.9%)
CE–CA–CH	5 (2.2%)	5 (3.3%)
CE–CA–AF	5 (2.2%)	5 (3.3%)
CE–CA–AM	5 (2.2%)	5 (3.3%)
CH–AM	5 (2.2%)	0
CH–AF	2 (0.9%)	1 (0.7%)
CH–AM–AF	10 (4.4%)	2 (1.3%)
CE–CA–CH–AF	6 (2.6%)	6 (3.9%)
CE–CA–CH–AM	5 (2.2%)	5 (3.3%)
CE–CA–AM–AF	29 (12.8%)	29 (19.0%)
CE–CA–CH–AM–AF	50 (22.0%)	50 (32.7%)
Total	227 (100%)	153 (100%)

ered in the analysis and 19% with Amazonian and Atlantic rain forests and Cerrado). Another important distinction is the number of species shared with the Chaco, which is quite high for Cerrado (25 species, 11%, involving rodents, marsupials, primates, and some artiodactyls and carnivores), but nonexistent for Caatinga (table 14.3 and fig. 14.5).

Ignoring ubiquitous species in tropical central South America, different patterns are evident. For the Cerrado, the number of endemic species (25), and the ones shared with a single adjacent biome become the most important categories (25 species shared with Chaco, 21 species with Amazon rain forest, 15 species with Caatinga, and 10 species with Atlantic rain forest). For Caatinga, the most

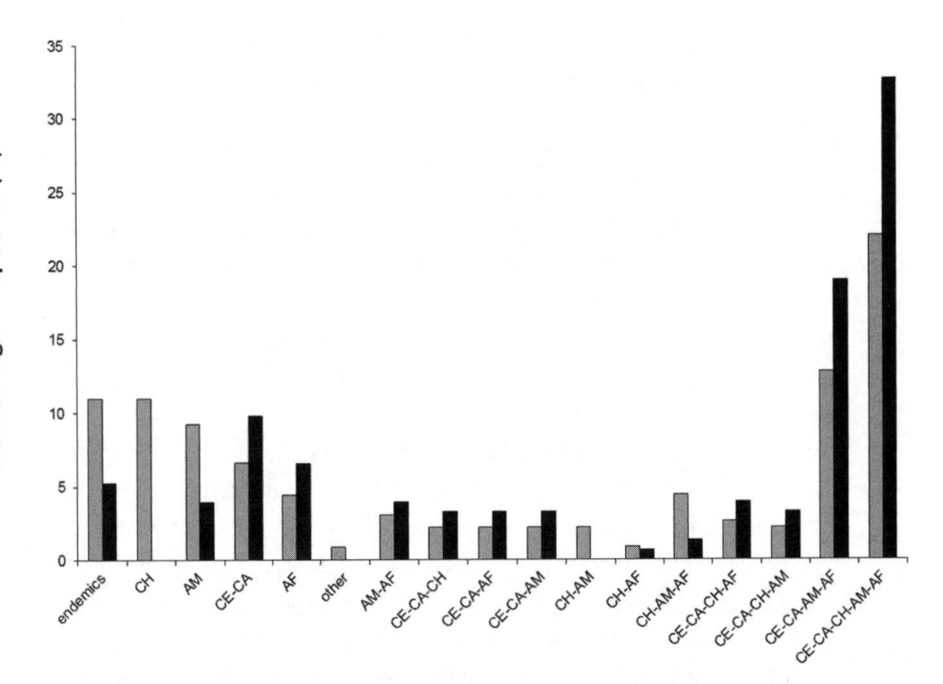

Figure 14.5 Unique and shared components of the Cerrado and Caatinga mammal faunas. Values are percentages of their total faunas, gray bars represent the Cerrado, and black bars the Caatinga. Tropical South America biomes are represented by the following abbreviations: AF, Atlantic rain forest; AM, Amazonian rain forest; CA, Caatinga; CE, Cerrado; CH, Chaco. Except for endemics, each pair of bars compares the percentage of species shared by the Cerrado and the Caatinga with other sets of South American biomes. In cases where a histogram is lacking, there are no shared species between the biomes considered.

important relationships are with the Cerrado and the Atlantic rain forest (15 and 10 shared species, respectively), followed by endemics (8 species), and species shared with Amazon rain forest (6). The mere existence of so many significant categories of shared species between Cerrado, Caatinga, and adjacent biomes is itself an important faunal pattern.

The highly heterogeneous nature of these biomes, reflecting a mosaic of both forested and open-vegetation habitats, is responsible for this complex network of relationships. However, Cerrado and Caatinga differ in habitat heterogeneity. The contribution of forest formations to the overall diversity of Cerrado and Caatinga has long been appreciated (Fonseca, Herrmann, and Leite 1999; Johnson, Saraiva, and Coelho 1999; Mares, Willig, and Lacher 1985; Mares, Ernest, and Gettinger 1986; Mares, Braun, and Gettinger 1989; Marinho-Filho and Reis 1989). The gallery forests function as mesic corridors that allow range extensions of rain forest species within the open biomes (Bishop 1974; Cerqueira 1982; Vanzolini 1963); they also provide refuge, food, and water resources for nonforest species (Mares, Willig, and Lacher 1985; Redford and Fonseca 1986). The wide-ranging mammal species, mainly bats and medium- and large-sized species, are mostly habitat generalists, occupying both the open and forest habitats in Cerrado and Caatinga (Fonseca, Herrmann, and Leite 1999; Marinho-Filho, Rodrigues, and Juarez 2002). On the other hand, species shared with Amazon and Atlantic rain forests represent forest dwellers (30% of the combined fauna of Cerrado and Caatinga), mostly represented by marsupial, primate, rodent, and bat species with restricted distributions in open biomes.

Forest habitats are organized differently in Cerrado and Caatinga. In the Cerrado, gallery forests extend far into the biome along major river systems. These rivers originate in the Brazilian highlands and drain into alternate major river basins, thereby linking all major South America tropical biomes via their gallery forests. The deciduous and semideciduous forests of the Cerrado occur in both small and large pockets and are maintained not only by substantial precipitation during the rainy season but also by favorable soil conditions. In contrast, the Caatinga presents only two main rivers, the São Francisco and Parnaíba, and the pockets of dry forest within the biome today reflect the distribution of orographic precipitation and topographic relief. The continuity of forests in the Cerrado, versus their patchy distribution in the Caatinga, may explain the higher number of Amazonian species shared with the Cerrado (21 species) versus the Caatinga (6 species).

Cerrado mammals are unequally shared with Amazonian and Atlantic rain forests—21 species are shared with the Amazon rain forest and only 10 with the Atlantic rain forest. A possible explanation for this fact could be that the Cerrado borders the Amazonian rain forest from northeastern Bolivia to northeastern Brazil, and this border lacks any major relief. In contrast, Atlantic rain forest is separated from the Cerrado by extensive mountain ranges. In general, the gallery forests play an important role in increasing the regional diversity of the Cerrado, as most forest species do not range between Amazon and Atlantic rain forests across the entire Cerrado (Carmignotto 2005).

Similar distributional patterns have been observed for plants, butterflies, and birds from Cerrado (Brown and Gifford 2002; Oliveira-Filho and Ratter 1995; da Silva 1996). Analyzing phylogeographic data from small rodents and marsupials, Costa (2003) showed that Amazon-Cerrado and Atlantic-Cerrado relationships are closer than relationships between Amazon and Atlantic rain forest species. The influence of Amazonian and Atlantic rain forests on the distributions of Caatinga mammals is less clear (de Oliveira, Gonçalves, and Bonvicino 2008; Rodrigues 2008; Silva et al. 2008). However, the contribution of forest habitats to the mammal diversity of the Cerrado and Caatinga has been overemphasized (Fonseca, Herrmann, and Leite 1999; Mares, Willig, and Lacher 1985; Redford and Fonseca 1986); determining how many and which species of mammals inhabit the open facies of these biomes has received less attention (Carmignotto 2005). In fact, open-habitat specialists include 80 species (33 species endemic to Cerrado and Caatinga, 15 shared between Cerrado and Caatinga, 25 species shared between Cerrado and Chaco, and 7 species shared between Cerrado, Caatinga, and other open formations). Thus, the open-habitat fauna represents 31% of the combined fauna, equaling the number of forest specialists (30%). The remaining mammals (39%) are habitat generalists, occurring broadly across tropical South America. This assemblage is in fact the background fauna of the entire South American tropics.

There is a strong Chaco influence in the southwestern Cerrado, which is reflected in its distinctive mammal fauna (Cáceres, Ferreira, and Carmignotto 2007; Cáceres et al. 2008). Despite the close affinity of the Caatinga vegetation to the Chaco (Andrade-Lima 1981; Prance 1987), this relationship is not evident in the mammal fauna. Mammal species shared between Cerrado and Caatinga biomes are either widely distributed or have more restricted distributions in the general region where the two biomes meet (Carmignotto 2005; this study).

Another interesting faunal pattern is the low number of species shared with other temperate open biomes of South America (see appendix 14.1). Only 24 species (8 bats, 5 carnivores, 5 rodents, 3 marsupials, 2 deer, and 1 armadillo) of 260 species that occur in Cerrado and Caatinga range into the Chilean (Wallace 1876) or Patagonian subregion (Hershkovitz 1972).

14.3.3 ENDEMIC COMPONENTS OF TROPICAL OPEN FORMATIONS

The Cerrado has 5 endemic genera and 25 endemic species, whereas the Caatinga has only 1 genus and 8 species. The group of endemics most representative for both biomes is the rodents, with marsupials, bats, and primates also contributing (see table 14.2). The most striking difference between these open formations is shown by the sigmodontines, which are represented by 13 endemic species in the Cerrado but only one (*Rhipidomys cariri*) in the Caatinga. Collectively, the Cerrado and the Caatinga support 48 endemic species (18.5% of the combined fauna); excluding bats, the proportion of endemism increases to 26.1% for these open country formations.

In terms of distributional patterns, most of the endemics present restricted distributions within the Cerrado or the Caatinga, and some regions can be recognized based on the occurrence of endemics alone, such as the following: dry forests in the eastern portion of the Cerrado, which support endemic plants, birds, and mammals (Carmignotto 2005; Oliveira-Filho and Ratter 1995; da Silva and Bates 2002); the high-elevation grasslands, which support numerous endemic plants, lizards, birds, and mammals (Giullieti and Pirani 1988; Rodrigues 1988; da Silva and Bates 2002; Weksler and Bonvicino 2005); the "brejos" and sand dunes in the Caatinga, with lizards and mammals restricted to these habitats (Borges-Nojosa and Caramaschi 2008; Leite 2003; Rodrigues 2008); and the central Brazilian highlands, where most of the endemic mammals of the Cerrado occur (Carmignotto 2005).

There are two principal patterns of endemism for mammals of the tropical open formations of Central South America. The first is of open-formation species in genera that diversified in the rain forest biomes of Amazonia or eastern Brazil, while the second includes taxa derived from lineages that diversified in the open biomes.

The first kind of endemism is well-represented among the primates. The Cerrado and the Caatinga have a number of endemic primate species, such as *Callithrix penicillata*, *Cebus libidinosus*, *Callicebus barbarabrownae*, and *Alouatta ululata*, all of which belong to genera distributed in the neighboring rain forests. The marsupial *Monodelphis umbristriata* is an open-biome species amid a host of forest congeners. Among the rodents, *Urosciurus urucumus* is a squirrel endemic to the dry forests of the Pantanal that has clear Amazonian affinities. The sigmodontines *Akodon lindberghi*, *Euryoryzomys lamia*, *Oligoryzomys moojeni*, *O. rupestris*, *O. stramineus*, *Rhipidomys cariri*, and *R. macrurus*, as well as the hystricognaths *Phyllomys blainvillii*, *Trinomys minor*, *T. yonenagae*, and *Dasyprocta nigriclunis* are vicariant species of genera well-diversified in the Amazon or Atlantic rain forests. The same can be said for the bats *Lonchophylla dekeyseri*, *Micronycteris sanborni*, and *Thyroptera devivoi*.

Phylogenetic information is lacking for many of these genera, but the few phylogenies available are suggestive. Gregorin (2006) published a phylogeny of *Alouatta* in which the Central American *A. palliata* is the sister of all South American species, and within this South American clade, *A. ululata*, the species from the Cerrado and Caatinga, is sister to *A. belzebul*, which is distributed in disjunct fashion in the Amazon and Atlantic rain forest. Weksler (2006) analyzed the phylogenetic relationships among oryzomyine rodents and found that Cerrado endemic *Euryoryzomys lamia* is the sister of *E. russatus*, an Atlantic rain forest inhabitant, and that *Oligoryzomys stramineus* of the Caatinga and *O. nigripes*, a species widely distributed in both open and forest biomes of South America, are closely related. In the genus *Phyllomys*, the Caatinga endemic *P. blainvillii* is sister to the clade composed by *P. lamarum* and *P. brasiliensis* (Leite 2003), both Atlantic rain forest inhabitants from northern Minas Gerais. Cerrado endemic *Dasyprocta nigriclunis* appears most closely related to an agouti species from the dry forests of Pantanal and thus in turn belongs to a group of species distributed in Caatinga and Amazon rain forests (Iack-Ximenes 1999). The available phylogenetic information indicates that open-formation vicars of forest lineages may have evolved early and in response to historical events during the Late Tertiary, predating the climatic fluctuations of the Pleistocene (Costa 2003; Lara and Patton 2000; Leite 2003).

The other pattern of endemism is of Cerrado and Caatinga endemics belonging to lineages long associated with open biomes. Most mammals endemic to the Cerrado and Caatinga belong to this category. The marsupials include *Cryptonanus* and *Thylamys*, as essentially open-vegetation forms. They are well-represented in tropical and temperate open formations of South America (Carmignotto and Monfort 2006; Palma et al. 2002; Voss, Lunde, and Jansa 2005), and their affinities place them in quite different positions in the phylogeny of South American marsupials. They belong to a clade that includes both open (*Chacodelphys*, *Lestodelphys*, and *Thylamys*) and forest (*Gracilinanus*) forms, and the two were not found to be closely related (Voss and Jansa 2009; Voss, Lunde, and Jansa 2005). Therefore, these open-biome marsupial genera seem to have evolved independently. The only phylogeny that included Brazilian representatives is that of *Thylamys* (Giarla, Voss, and Jansa 2010), which places both *T. velutinus* and *T. karimii* as basal taxa to other extant species, and not as sister groups. Again, this points to their independent differentiation in the tropical open biomes, likely predating the temperate diversification of this group (Braun et al. 2005; Palma et al. 2002).

Several lineages of sigmodontine rodents have genera endemic to open formations: akodontines (*Juscelinomys*, *Necromys*, *Thalpomys*), scapteromyines (*Kunsia*, *Scapteromys*), phyllotines (*Calomys*), and wiedomyines (*Wiedomys*) (D'Elia 2003; D'Elia et al. 2006; Weksler 2003). The open-vegetation oryzomyines *Cerradomys* and *Microakodontomys* group with a clade that includes other open-vegetation genera such as *Holochilus*, *Lundomys*, and *Pseudoryzomys* (Weksler 2003). The differentiation of these clades, estimated as 5–9 Ma (Engel et al. 1998) or 10–14 Ma (Smith and Patton 1999), is thought to predate the Pleistocene.

Hystricognath genera are also endemic to the open formations, including *Carterodon*, *Clyomys* and *Thrichomys* (Echimyidae: Eumysopinae), *Ctenomys* (Ctenomyidae), and *Galea* and *Kerodon* (Caviidae). The Eumysopinae comprise open-country (*Carterodon*, *Clyomys*, *Thrichomys*) and forest inhabitants (*Euryzygomatomys*, *Hoplomys*, *Proechimys*, *Trinomys*), with *Thrichomys* radiating among these genera, presenting an unresolved phylogenetic relationship. Indeed, the monophyly of Eumysopinae is open to question due to the star phylogeny obtained among Echimyidae genera (Galewski et al. 2005; Leite and Patton 2002). The family Ctenomyidae is the sister group of the Octodontidae, all members strongly associated with open-vegetation formations (Honeycutt, Rowe, and Gallardo 2003; Lessa and Cook 1998). *Galea* is sister to the clade composed by *Cavia* and *Microcavia* (comprising the subfamily Caviinae), which in turn is sister to the members of the subfamily Hydrochoerinae, with *Hydrochoerus* and *Kerodon* as sister taxa, all being Caviidae strongly associated with open formations (Rowe et al. 2010; Woods and Kilpatrick 2005). Both molecular and fossil estimates for the diversification of these genera range from the middle to late Miocene (Galewski et al. 2005; Honeycutt, Rowe, and Gallardo 2003; Rowe et al. 2010) to the early Pleistocene (Lessa and Cook 1998).

14.4 A History of Ideas and Distribution Patterns

The arrival of European naturalists in South America from the late eighteenth century fostered the recognition of biotic provinces based on the flora and fauna. As early as 1837, Von Martius recognized the Cerrado and Caatinga as major vegetation physiognomies (Romariz 1968), and Burmeister (1854) distinguished three main regions based on mammals: the Atlantic and Amazonian rain forests and the sertão, corresponding to the Cerrado and Caatinga. Wallace (1876) divided the South American continent into two main areas, an Andean-Patagonian (or Chilean) temperate province and a Brazilian province, of tropical character. Wallace's regions were still very much valid a century later, being the basis for Hershkovitz (1972) division of the continent in mammalian faunal regions. All of these studies were based on faunal lists, and the introduction of phylogenetic considerations in interpreting these patterns has been relatively recent (e.g., Barros et al. 2009).

It is important to point out that authors focusing on the descriptions of biogeographical patterns of Cerrado and Caatinga have treated the subject at two very distinct scales: from the point of view of South American open formations as a whole, and more restrictively, comparing the Cerrado and Caatinga outside a continental context. Aiming to integrate these views, our discussion will encompass all modern ideas on the subject, independently of scale.

Hershkovitz (1962, 1972) offered the first serious, modern attempt to trace patterns of ecological differentiation among South American mammal lineages. He generally saw the continent's forest mammals as possessing "generalized," primitive morphological features and "pastoral" or open-vegetation forms as being more derived and specialized. He went further in his analysis of open-vegetation Sigmodontinae—several genera of open-country rodents were derived from forest ancestors, while others were forms derived from already-pastoral ancestors. Additionally, many hystricognath rodents, rabbits, deer, and a few other taxa belong to this group of mammals with ancient connections to open vegetation. Here we have shown that the endemic mammals of the Cerrado and Caatinga include lineages that are highly diversified in primarily open biomes as well as forest vicariants, which agrees coarsely with Hershkovitz's ideas. The composition of Hershkovitz's monophyletic groups has changed with our growing knowledge of phylogenetic relationships, and some of his assessments of sigmodontine relationships are no longer valid. Among the rodents, the phyllotine and akodont sigmodontines, and several hystricognaths (such as the eumysopine Echimyidae, Ctenomyidae, and Caviidae) are essentially open-vegetation forms and form monophyletic groups (see D'Elia et al. 2006 for the sigmodonts and Galewski et al. 2005 and Rowe et al. 2010 for the hystricognaths).

Another line of inquiry began with the assertion by Vanzolini (1963) that the Caatinga had only a subset of the Cerrado lizard fauna; later (Vanzolini 1988), he maintained that despite the description of species endemic to these biomes, Caatinga and Cerrado lizard faunas were indistinguishable. We believe that this discussion is now settled: limited but definite patterns of endemism have been demonstrated for several vertebrate groups for the Cerrado and the Caatinga (Carmignotto 2005; Colli, Bastos, and Araújo 2002; Marinho-Filho, Rodrigues, and Juarez 2002; Oliveira, Gonçalves, and Bonvicino 2008; Rodrigues 2008; da Silva and Bates 2002; this chapter). The fact that the two biomes share a large faunal component with each other and their neighbors does not invalidate the reality of the unique open-vegetation faunas within their limits.

Mares, Willig, and Lacher (1985) concluded that the Caatinga mammal fauna did not present advanced adaptations to semiarid environments, and that the very existence of "tropical" faunas in this environment depended on mesic enclaves (brejos and arboreal Caatinga). Similarly, Redford and Fonseca (1986) considered the Cerrado mammal fauna as mostly composed of versatile, mesic-derived forms. We believe that these assessments are misleading, because most of the endemic mammals of the Cerrado and Caatinga belong to lineages long associated with open environments (examples include phyllotine, akodontine, and eumysopine rodents). Further, these groups occur widely over the continent and are not restricted to the tropics. Although the number of endemic mammals in the Caatinga is small, so is its area. Relative to area, the Caatinga is not particularly poor (Caatinga = 1.92 species/10,000 km^2 and Cerrado = 1.14 species /10,000 km^2). Understanding the origin of the endemic fauna of either of the two biomes is impossible without considering their relationships with the entire complex of open vegetation mammalian faunas of South America. We would summarize the Caatinga and the Cerrado mammal fauna as follows: (1) both are open-vegetation biomes that exclusively share a substantial number of mammal species; (2) both support a number of forest mammal species derived from and shared with the Amazonian and Atlantic rain forests; (3) most endemic species of both biomes belong to lineages that have long been associated with open formations, either temperate or Andean.

14.5 A Summary of Patterns and Their Evolution

Research on the history of Cerrado and Caatinga biomes has taken place under two quite different perspectives and scales. Generally, researchers from outside South America have been interested in unveiling the history of open formations on the continent as a whole, and have frequently sought to place this history in a global perspective. On the other hand, South American scholars have looked more particularly at them and neighboring biomes to understand their histories. Both approaches have shed light on the Cerrado and Caatinga mammal faunas, because they can be analyzed both as part of the South American open-formation faunas and, alternatively, as biome-specific faunas.

Janis (2003) proposed that during the Paleocene and the Eocene, the world was mostly covered by tropical or "tropical-like" forests, with small areas occupied by savanna formations. In South America, savannas would be an Eocene innovation, lying largely south of the Cerrado and Caatinga today, extending from southern Brazil and Paraguay to the tip of the continent. Burnham and Johnson (2004) confirmed the existence of closed canopy rain forest in South America by the Eocene, but the extent of these forests remains is uncertain. Thus "savannas" during the Eocene might have been restricted to the southern portion of South America, as proposed by Janis, or be more broadly or narrowly distributed. However, the Paleogene mammals of South America appear exceptional, because hypsodonty (a dental syndrome commonly related to grass diets) appeared in the fossil record by mid-Eocene (Patterson and Pascual 1972), leading these authors to believe that extensive grassy landscapes already were present in South America by the Eocene.

As the world cooled from the end of the Eocene to the Oligocene, it is generally believed that grasslands and grassy savannas expanded widely across the globe, bringing about the evolution of specialized grazing lineages, a trend that continued throughout the Neogene (Flynn, Charrier, Croft and Wyss, chapter 4, this volume; de Vivo 2008). Given the existence of a Miocene primate fauna (see Rosenberger 2002) and echimyid rodents (Kramarz and Bellosi 2005) in Patagonia, it is reasonable to suppose that mesic savannas extended throughout much of South America, and that the much drier, open vegetation of the Southern Cone today did not exist then, at least in its present location.

If more mesic conditions were prevalent throughout South America in the Miocene, what paleoenvironments would have characterized the regions now occupied by the Cerrado and Caatinga biomes? Kay and Madden (1997) suggested three distinct climatic regimes for the region now corresponding to eastern and central Brazil south to Uruguay: extreme southern Brazil and Uruguay with precipitation up to 1000 mm per year; a strip crossing central to the east coast of Brazil with precipitation in the range of the modern Cerrado biome (i.e., between 1000 and 2000 mm per year); and a third section encompassing the northern parts of central Brazil and the entire northeast region of the country, with precipitation at least 2000 mm. If these estimates are correct, dry and humid savanna and dry forests could have existed in the southern range and rain forests to the north. Part of the region with a savanna climate coincides with the present area of distribution of the south and western parts of the Cerrado, but the entire area of the modern Caatinga would be covered by rain forests, not the dry forest and shrub of today. This reconstruction is similar to the one presented more recently by Willis and McElwain (2002).

Botanists working in Brazil have long devoted their attention to the age of the Cerrado and Caatinga formations. Rizzini (1997) believed that both biomes are relatively modern features in their present configurations and geographical location, but it was unclear to him whether Cerrado vegetation was long associated with the region it now occupies. Ledru (2002), working with palynological records from central Brazil, agreed with Rizzini and gave 32,000 years as the earliest record of pollen of Cerrado plants. She concluded that modern-looking

Cerrado vegetation did not exist before 7000 to 10,000 years ago. There are, however, scholars who believe the Cerrado to be very old because of the number of endemics and the characteristic adaptations of its flora (Furley 1999; Oliveira-Filho and Ratter 2002).

Prado (2008) reviewed the literature on the Caatinga and showed that most authors had placed its origins in the Quaternary, as corroborated by de Queiroz (2006), who suggested that climate fluctuations beginning in the Neogene may have had a role in creating the climatic conditions for the establishment of the modern Caatinga. Based on molecular estimates of divergence times of most of the endemic taxa present in the Cerrado and Caatinga today, which date as far back as middle to late Miocene, the open-vegetation formations should have already been well-developed (Barros et al. 2009; D'Elia et al. 2006; Honeycutt, Rowe, and Gallardo 2003; Leite 2003; Smith and Patton 1999).

There are Pleistocene records of typical open-vegetation inhabitants in the Cerrado and Caatinga (Cartelle 1999; Rancy 1999). De Vivo (1997) inferred that the Caatinga was once mostly covered by humid forests because of the fossil primates they once contained, as well as the disjunct distributions of extant forest species in the Amazon and Atlantic rain forests. This suggests that Quaternary climatic fluctuations may have played a role either in the diversification of some of the lineages present today in the Cerrado and Caatinga, or at least in shaping their modern distributions.

14.6 Conclusions

Cerrado and Caatinga mammals are highly diverse, with nearly 20% of the species being endemic to both biomes. One of the more notable faunal patterns is that most of these 48 endemic species are strongly associated with the open habitats in these biomes, and most represent lineages that diversified in open-country formations. Additional endemic taxa consist of open-formation vicars of genera that are diverse in the Amazonian and Atlantic rain forests, showing the complex history of faunal evolution in these tropical open biomes.

Another significant pattern is the high proportion of species shared with neighboring biomes. This pattern is based in the mosaic composition of the biomes and contributes importantly to the regional diversity of Cerrado and Caatinga—distinct mammal communities occur in distinct regions within these biomes. The shared components are not only distributed in the adjacent rain forest biomes, but also in other open tropical and temperate biomes of South America.

Acknowledgments

We would like to thank the editors of this volume for inviting us to contribute. APC also thanks them for the invitation to participate in the Mendoza symposium. We also would like to thank two referees whose criticisms have greatly improved our manuscript. APC and MDV would like to thank FAPESP (Fundação de Amparo à Pesquisa do Estado de São Paulo, grants # 98/05075-7 and # 00/06642-4). AL thanks CNPq (Conselho Nacional de Desenvolvimento Cientifico e Tecnológico) for research fellowship.

Literature Cited

Ab'Sáber, A. N. 1974. "O Domínio Morfoclimático Semi-árido das Caatingas Brasileiras." *Geomorfologia* 53:1–19.

———. 1977. "Os Domínios Morfoclimáticos na América do Sul—Primeira Aproximação." *Geomorfologia* 52:1–22.

Aires, C. C. 2008. "Caracterização das Espécies Brasileiras de *Myotis* Kaup, 1829 (Chiroptera: Vespertilionidae) e Ensaio Sobre a Filogeografia de *Myotis nigricans* (Schinz, 1821) e *Myotis riparius* Handley, 1960." Unpublished PhD diss., Universidade de São Paulo.

Alho, C. J. R. 1981. "Small Mammal Populations of Brazilian Cerrado: The Dependence of Abundance and Diversity on Habitat Complexity." *Revista Brasileira de Biologia* 41:223–30.

Alho, C. J. R., L. A. Pereira, and A. C. Paula. 1986. "Patterns of Habitat Utilization By Small Mammal Populations in Cerrado Biome of Central Brazil." *Mammalia* 50:447–60.

Anderson, S. 1997. "Mammals of Bolivia: Taxonomy and Distribution." *Bulletin of the American Museum of Natural History* 231:1–652.

Andrade-Lima, D. 1981. "The Caatingas Dominium." *Revista Brasileira de Botânica* 4:149–63.

Ávila-Pires, F. D. 1960. "Roedores Colecionados na Região de Lagoa Santa, Minas Gerais, Brasil." *Arquivos do Museu Nacional, Rio de Janeiro* 50: 25–45.

———. 1966. "Observações Gerais Sobre a Mastozoologia do Cerrado." *Anais da Academia Brasileira de Ciências* 38(suplemento): 331–40.

———. 1972. "A New Subspecies of *Kunsia fronto* (Winge, 1888) from Brazil (Rodentia, Cricetidae)." *Revista Brasileira de Biologia* 32:419–22.

Barros, M. C., I. Sampaio, H. Schneider, and A. Langguth. 2009. "Molecular Phylogenies, Chromosomes and Dispersion in Brazilian Akodontines (Rodentia, Sigmodontinae)." *Iheringia, Série Zoologia* 99:373–80.

Basile, P. 2003. "Taxonomia de *Thrichomys* Trouessart, 1880 (Rodentia, Echimyidae)." Unpublished MSc thesis. Departamento de Biologia, Faculdade de Filosofia, Universidade de São Paulo, Ribeirão Preto.

Bezerra, A. M. R., A. P. Carmignotto, A. P. Nunes, and F. H. G. Rodrigues. 2007. "New Data on the Distribution, Natural History, and Morphology of *Kunsia tomentosus* (Lichtenstein, 1830) (Rodentia: Cricetidae: Sigmodontinae)." *Zootaxa* 1505:1–18.

Bezerra, A. M. R., and J. A. de Oliveira. 2010. "Taxonomic Implications of Cranial Morphometric Variation in the Genus *Clyomys* Thomas, 1916 (Rodentia: Echimyidae)." *Journal of Mammalogy* 91:260–72.

Bishop, I. R. 1974. "An Annotated List of Caviomorph Rodents Collected in North-Eastern Mato Grosso, Brazil." *Mammalia* 38:489–502.

Black-Décima, P., R. V. Rossi, A. Vogliotti, J. L. Cartes, L. Maffei, J. M. B. Duarte, S. González, and J. P. Juliá. 2010. "Brown Brocket Deer *Mazama gouazoubira* (Fischer, 1814)." In *Neotropical Cervidology*, edited by J. M. B. Duarte and S. Gonzalez, 190–201. Gland: IUCN/FUNEP.

Bonvicino, C. R. 2003. "A New Species of *Oryzomys* (Rodentia, Sigmodontinae) of the *subflavus* Group from the Cerrado of Central Brazil." *Mammalian Biology* 68: 78–90.

Bonvicino, C. R., and F. C. Almeida. 2000. "Karyotype, Morphology and Taxonomic Status of *Calomys expulsus* (Rodentia: Sigmodontinae)." *Mammalia* 64:339–51.

Bonvicino, C. R., J. A. de Oliveira, and P. S. D'Andrea. 2008. *Guia dos Roedores do Brasil, com Chaves para Gêneros Baseadas em Caracteres Externos*. Rio de Janeiro: Centro Pan-Americano de Febre Aftosa—OPAS/OMS. *Série de Manuais Técnicos* 11.

Bonvicino, C. R., J. F. S. Lima, and F. C. Almeida. 2003. "A New Species of *Calomys* Waterhouse (Rodentia, Sigmodontinae) from the Cerrado of Central Brazil." *Revista Brasileira de Zoologia* 20:301–07.

Bonvicino, C. R., and M. Weksler. 1998. "A New Species of *Oligoryzomys* (Rodentia, Sigmodontinae) from Northeastern and Central Brazil." *Zeitschrift für Säugetierkunde* 63: 90–103.

Borges-Nojosa, D. M., and U. Caramaschi. 2008. "Composição e Análise Comparativa da Diversidade e das Afinidades Biogeográficas dos Lagartos e Anfisbenídeos (Squamata) dos Brejos Nordestinos." In *Ecologia e Conservação da Caatinga*, 3rd ed., edited by I. R. Leal, M. Tabarelli, and J. M. C. da Silva, 463–514. Recife: Editora Universitária, Universidade Federal de Pernambuco.

Braggio, E., and C. R. Bonvicino. 2004. "Molecular Divergence in the Genus *Thrichomys* (Rodentia, Echimyidae)." *Journal of Mammalogy* 85:316–20.

Braun, J. K., R. A. Van den Bussche, P. K. Morton, and M. A. Mares. 2005. "Phylogenetic and Biogeographic Relationships of Mouse Opossums *Thylamys* (Didelphimorphia, Didelphidae) in Southern South America." *Journal of Mammalogy* 86:147–59.

Brown, K., Jr., and D. R. Gifford. 2002. "Lepidoptera in the Cerrado Landscape and the Conservation of Vegetation, Soil, and Topographical Mosaics." In *The Cerrados of Brazil: Ecology and Natural History of a Neotropical Savanna*, edited by P. S. Oliveira and R. J. Marquis, 201–22. New York: Columbia University Press.

Burmeister, H.1854. *Systematische Uebersicht der Thiere Brasiliens, Welche Während einer Reise Durch die Provinzen von Rio de Janeiro und Minas Geraës Gesammelt Oder Beobachtet Wurden von Dr. Hermann Burmeister. Säugethiere (Mammalia)*, I. Berlin: Georg Reimer.

Burnham, R. J., and K. R. Johnson. 2004. "South American Palaeobotany and the Origins of Neotropical Rainforests." *Philosophical Transactions of the Royal Society of London, series* B 359:1595–610.

Cáceres, N. C., and A. P. Carmignotto. 2006. *Caluromys lanatus*. *Mammalian Species* 803:1–6.

Cáceres, N. C., A. P. Carmignotto, E. Fischer, and C. F. Santos. 2008. "Mammals from Mato Grosso do Sul, Brazil." *CheckList* 4:321–35.

Cáceres, N. C., V. L. Ferreira, and A. P. Carmignotto. 2007. "The Occurrence of the Mouse Opossum *Marmosops ocellatus* (Marsupialia, Didelphidae) in Western Brazil." *Mammalian Biology* 72:45–48.

Campos, B. T. P. 2009. "Filogeografia de *Rhipidomys* (Muridae: Sigmodontinae) nos Brejos de Altitude do Nordeste Brasileiro." Unpublished MSc thesis. Programa de Pós-Graduação em Ciências Biológicas (Zoologia), Universidade Federal da Paraíba, João Pessoa.

Carmignotto, A. P. 2005. "Pequenos Mamíferos Terrestres do Bioma Cerrado: Padrões Faunísticos Locais e Regionais." Unpublished PhD diss., Universidade de São Paulo.

Carmignotto, A. P., and T. Monfort. 2006. "Taxonomy and Distribution of the Brazilian Species of *Thylamys* (Didelphimorphia: Didelphidae)." *Mammalia* 70:126–44.

Cartelle, C. 1999. "Pleistocene Mammals of the Cerrado and Caatinga of Brazil." In *Mammals of the Neotropics, Vol. 3, The Central Neotropics: Ecuador, Peru, Bolivia, Brazil*, edited by J. F. Eisenberg and K. H. Redford, 27–46. Chicago: University of Chicago Press.

Cerqueira, R. 1982. "South American Landscapes and their Mammals." In *Mammalian Biology in South America*, edited by M. A. Mares and H. H. Genoways. Pittsburgh: University of Pittsburgh, *Special Publication Series, Pymatuning Laboratory of Ecology* 6.

Colli, G. R., R. P. Bastos, and A. F. B. Araújo. 2002. "The Character and Dynamics of the Cerrado Herpetofauna." In *The Cerrados of Brazil: Ecology and Natural History of a Neotropical Savanna*, edited by P. S. Oliveira and R. J. Marquis, 223–41. New York: Columbia University Press.

Costa, L. P. 2003. "The Historical Bridge Between the Amazon and the Atlantic Forests of Brazil: A Study of Molecular Phylogeography with Small Mammals." *Journal of Biogeography* 30:71–86.

Coutinho, L. M. 2006. "O Conceito de Bioma." *Acta Botánica Brasileira* 20:13–23.

D'Elia, G. 2003. "Phylogenetics of Sigmodontinae (Rodentia, Muroidea, Cricetidae), with Special Reference to the Akodont Group, and with Additional Comments on Historical Biogeography." *Cladistics* 19:307–23.

D'Elia, G., L. Luna, E. M. González, and B. D. Patterson. 2006. "On the Sigmodontinae Radiation (Rodentia, Cricetidae): An Appraisal of the Phylogenetic Position of *Rhagomys*." *Molecular Phylogenetics and Evolution* 38:558–64.

Dinnerstein, E., D. M. Olson, D. J. Graham, A. L. Webster, S. A. Primm, M. P. Bookbinder, and G. Ledec. 1995. *A Conservation Assessment of the Terrestrial Ecoregions of Latin America and the Caribbean*. Washington: World Wildlife Foundation, World Bank.

Eisenberg, J. F., and K. H. Redford, eds. 1999. *Mammals of the Neotropics, Vol. 3. The Central Neotropics: Ecuador, Peru, Bolivia, Brazil*. Chicago: University of Chicago Press.

Eiten, G. 1972. "The Cerrado Vegetation of Brazil." *The Botanical Review* 38:201–341.

Engel, S. R., K. M. Hogan, J. F. Taylor, and S. K. Davis. 1998. "Molecular Systematics and Paleobiogeography of the South American Sigmodontine Rodents." *Molecular Biology and Evolution* 15:35–49.

Eva, H. D., E. E. de Miranda, C. M. Di Bella, V. Gond, O. Huber, M. Sgrenzaroli, S. Jones et al. 2002. *A Vegetation Map of South America*. Luxembourg: European Commission, Joint Research Centre.

Fernandes, A., and P. Bezerra. 1990. *Estudo Fitogeográfico do Brasil*. Fortaleza: Stylus Comunicações.

Fonseca, G. A. B., G. Herrmann, and Y. L. R. Leite. 1999. "Macrogeography of Brazilian Mammals." In *Mammals of the Neotropics, Vol. 3. The Central Neotropics: Ecuador, Peru, Bolivia, Brazil*, edited by J. F. Eisenberg and K. H. Redford, 549–63. London: University of Chicago Press.

Fonseca, G. A. B., and K. H. Redford. 1984. "The Mammals of IBGE's Ecological Reserve, Brasília, and an Analysis of the Role of Gallery Forests in Increasing Diversity." *Revista Brasileira de Biologia* 44:517–23.

Freitas, C. A. de. 1957. "Notícia Sobre a Peste no Nordeste." *Revista Brasileira de Malariologia e Doenças Tropicais* 9:123–33.

Furley, P. A. 1999. "The Nature and Diversity of Neotropical Savanna Vegetation with Particular Reference to the Brazilian Cerrados." *Global Ecology and Biogeography* 8:223–41.

Furley, P. A., and J. A. Ratter. 1988. "Soil Resources and Plant Communities of the Central Brazilian Cerrado and Their Development." *Journal of Biogeography* 15:97–108.

Galewski, T., J.-F. Mauffrey, Y. L. R. Leite, J. L. Patton, and E. J. P. Douzery. 2005. "Ecomorphological Diversification Among South American Spiny Rats (Rodentia; Echimyidae): A Phylogenetic and Chronological Approach." *Molecular Phylogenetics and Evolution* 34:601–15.

Gardner, A. L., ed. 2008. *Mammals of South America, Vol. 1. Marsupials, Xenarthrans, Shrews, and Bats*. Chicago: University of Chicago Press.

Giarla, T. C., R. S. Voss, and S. A. Jansa. 2010. "Species Limits and Phylogenetic Relationships in the Didelphid Marsupial Genus *Thylamys* Based on Mitochondrial DNA Sequences and Morphology." *Bulletin of the American Museum of Natural History* 346:1–67.

Giullieti, A. M., and J. R. Pirani. 1988. "Patterns of Geographic Distribution of Some Plant Species from the Espinhaço Range, Minas Gerais and Bahia, Brazil." In *Proceedings of a Workshop on Neotropical Distribution Patterns*, edited by P. E. Vanzolini and W. R. Heyer, 39–69. Rio de Janerio: Academia Brasileira de Ciências.

Gregorin, R. 2006. "Taxonomia e Variação Geográfica das Espécies do Gênero *Alouatta* Lacépède (Primates, Atelidae) no Brasil." *Revista Brasileira de Zoologia* 23:64–144.

Gregorin, R., A. P. Carmignotto, and A. R. Percequillo. 2008. "Quirópteros do Parque Nacional da Serra das Confusões, Piauí, Nordeste do Brasil." *Chiroptera Neotropical* 14:366–83.

Gregorin, R., and V. A. Taddei. 2002. "Chave Artificial para a Identificação de Molossídeos Brasileiros (Mammalia, Chiroptera)." *Mastozoologia Neotropical* 9:13–32.

Gurgel-Filho, N. M. 2010. "Pequenos Mamíferos do Ceará: Didelphimorphia (Didelphidae), Chiroptera (Microchiroptera) e Rodentia (Sigmodontinae)." Unpublished MSc thesis, Universidade Federal da Paraíba, João Pessoa.

Hershkovitz, P. 1962. "Evolution of Neotropical Cricetine Rodents (Muridae), with Special Reference to the Phyllotine Group." *Fieldiana Zoology* 46:1–524.

———. 1972. "The Recent Mammals of the Neotropical Region: A Zoogeographic and

Ecological Review." In *Evolution, Mammals and Southern Continents*, edited by A. Keast, F. C. Erk, and B. Glass, 311–41. Albany: State University of New York Press.

———. 1990a. "The Brazilian Rodent Genus *Thalpomys* (Sigmodontinae, Cricetidae) with a Description of a New Species." *Journal of Natural History* 24: 763–83.

———. 1990b. "Mice of the *Akodon boliviensis* Size Class (Sigmodontinae, Cricetidae), with the Description of Two New Species from Brazil." *Fieldiana Zoology, n.s.* 57:1–35.

———. 1993. "A New Central Brazilian Genus and Species of Sigmodontine Rodent (Sigmodontinae) Transitional Between Akodonts and Oryzomyines, with a Discussion of Muroid Molar Morphology and Evolution." *Fieldiana Zoology, n.s.* 75:1–18.

Honeycutt, R. L., D. L. Rowe, and M. H. Gallardo. 2003. "Molecular Systematics of the South American Caviomorph Rodents: Relationships Among Species and Genera in the Family Octodontidae." *Molecular Phylogenetics and Evolution* 26:476–89.

Hueck, K. 1972. *As Florestas da América do Sul: Ecologia, Composição e Importância Econômica*. São Paulo: Editora da Universidade de Brasília, Editora Polígono.

Hueck, K., and P. Seibert. 1981. *Vegetationskarte von Südamerika*. Stuttgart: Gustav Fischer Verlag.

Iack-Ximenes, G. E. 1999. "Sistemática da Familia Dasyproctidae Bonaparte, 1838 (Rodentia, Hystricognathi) no Brasil." Unpublished MSc thesis. Universidade de São Paulo.

———. 2005. "Revisão de *Trinomys* Thomas, 1921 (Rodentia:Echimyidae)." Unpublished PhD diss., Universidade de São Paulo.

International Union for Conservation of Nature and Natural Resources (IUCN). 2009. *The IUCN Red List of Threatened Species*. Available at www.iucnredlist.org.

Janis, C. 2003. "Tectonics, Climate Change, and the Evolution of Mammalian Ecosystems." In *Evolution on Planet Earth: Impact of the Physical Environment*, Vol. 1, edited by L. Rothschild and A. Lister, 319–38. London: Academic Press.

Johnson, M. A., P. M. Saraiva, and D. Coelho. 1999. "The Role of Gallery Forests in the Distribution of Cerrado Mammals." *Revista Brasileira de Biologia* 59:421–27.

Karimi, Y., C. R. de Almeida, and F. Petter. 1976. "Note sur les Rongeurs du Nord-Est du Brésil." *Mammalia* 40:257–66.

Kasper, C. B., M. L. da Fontoura-Rodrigues, G. N. Cavalcanti, T. R. O. de Freitas, F. H. G. Rodrigues, T. G. de Oliveira, and E. Eizirik. 2009. "Recent Advances in the Knowledge of Molina's Hog-Nosed Skunk *Conepatus chinga* and Striped Hog-Nosed Skunk *C. semistriatus* in South America." *Small Carnivore Conservation* 41:25–28.

Kay, R., and R. Madden. 1997. "Paleogeography and Paleoecology." In *Vertebrate Paleontology in the Neotropics: The Miocene Fauna of La Venta, Colombia*, edited by R. F. Kay, R. H. Madden, R. L. Cifelli, and J. J. Flynn, 520–50. Washington: Smithsonian Institution Press.

Kramarz, A. G., and E. S. Bellosi. 2005. "Hystricognath Rodents from the Pinturas Formation, Early-Middle Miocene of Patagonia: Biostratigraphic and Paleoenvironmental Implications." *Journal of South American Earth Sciences* 18:199–212.

Lacher, T. E., Jr. 1981. "The Comparative Social Behavior of *Kerodon rupestris* and *Galea spixii* and the Evolution of Behavior in the Caviidae." *Bulletin of the Carnegie Museum* 17:1–71.

Lacher, T. E., Jr., and C. J. R. Alho. 1989. "Microhabitat Use Among Small Mammals in the Brazilian Pantanal." *Journal of Mammalogy* 70:396–401.

———. 2001. "Terrestrial Small Mammal Richness and Habitat Associations in an Amazonian-Cerrado Contact Zone." *Biotropica* 33:171–81.

Lacher, T. E, Jr., and M. A. Mares. 1986. "The Structure of Neotropical Mammal Communities: An Appraisal of Current Knowledge." *Revista Chilena de Historia Natural* 59:121–34.

Lacher, T. E., Jr., M. A. Mares, and C. J. R. Alho. 1989. "The Structure of a Small Mammal Community in a Central Brazilian Savanna." In *Advances in Neotropical Mammalogy*, edited by K. H. Redford and J. F. Eisenberg, 137–62. Gainesville: Sandhill Crane Press.

Langguth, A., and C. R. Bonvicino. 2002. "The *Oryzomys subflavus* Species Group, with Description of Two New Species (Rodentia, Muridae, Sigmodontinae)." *Arquivos do Museu Nacional, Rio de Janeiro* 60:285–94.

Lara, M. C., and J. L. Patton. 2000. "Evolutionary Diversification of Spiny Rats (Genus *Trinomys*, Rodentia: Echimyidae) in the Atlantic Forest of Brazil." *Zoological Journal of the Linnean Society* 130:661–86.

Ledru, M.-P. 2002. "Late Quaternary History and Evolution of the Cerrados as Revealed by Palynological Records." In *The Cerrados of Brazil: Ecology and Natural History of a Neotropical Savanna*, edited by P. S. Oliveira and R. J. Marquis, 33–50. New York: Columbia University Press.

Leemans, R., and W. P. Cramer. 1991. *The IIASA Database for Mean Monthly Values of Temperature, Precipitation, and Cloudiness on a Global Terrestrial Grid*. RR-91–18. Laxenburg: International Institute for Applied Systems Analysis.

Leite, Y. L. R. 2003. "Evolution and Systematics of the Atlantic Tree Rats, Genus *Phyllomys* (Rodentia, Echimyidae), with Description of Two New Species." *University of California Publications, Zoology* 132:1–118.

Leite, Y. L. R., and J. L. Patton. 2002. "Evolution of South American Spiny Rats (Rodentia, Echimyidae): The Star-Phylogeny Hypothesis Revisited." *Molecular Phylogenetics and Evolution* 25:455–64.

Lessa, E. P., and J. A. Cook. 1998. "The Molecular Phylogenetics of Tuco-Tucos (genus *Ctenomys*, Rodentia: Octodontidae) Suggests an Early Burst of Speciation." *Molecular Phylogenetics and Evolution* 9:88–99.

Mares, M. A. 1992. "Neotropical Mammals and the Myth of Amazonian Biodiversity." *Science* 255:976–79.

Mares, M. A., J. K. Braun, and D. Gettinger. 1989. "Observations on the Distribution and Ecology of the Mammals of the Cerrado Grasslands of Central Brazil." *Annals of the Carnegie Museum* 58:1–60.

Mares, M. A., K. A. Ernest, and D. Gettinger. 1986. "Small Mammal Community Structure and Composition in the Cerrado Province of Central Brazil." *Journal of Tropical Ecology* 2:289–300.

Mares, M. A., and T. Lacher. 1987. "Ecological, Morphological, and Behavioral Convergence in Rock-Dwelling Mammals." In *Current Mammalogy*, vol. 1, edited by H. H. Genoways, 308–48. New York: Plenum Press.

Mares, M. A., M. R., Willig, and T. E. Lacher. 1985. "The Brazilian Caatinga in South American Zoogeography: Tropical Mammals in a Dry Region." *Journal of Biogeography* 12:57–69.

Mares, M. A., M. R. Willig, K. Streilein, and T. E. Lacher Jr. 1981. "The Mammals of Northeastern Brazil: A Preliminary Assessment." *Annals of the Carnegie Museum of Natural History* 50:80–137.

Marinho-Filho, J., and M. L. Reis. 1989. *A Fauna de Mamíferos Associada às Matas de Galeria.* In *Anais do Simpósio sobre Mata Ciliar*, coordinated by L. M. Barbosa, 43–61. Campinas: Fundação Cargill.

Marinho-Filho, J., F. H. G. Rodrigues, and K. M. Juarez. 2002. "The Cerrado Mammals: Diversity, Ecology, and Natural History." In *The Cerrados of Brazil: Ecology and Natural History of a Neotropical Savanna*, edited by P. S. Oliveira and R. J. Marquis, 266–86. New York: Columbia University Press.

Marinho-Filho, J., and E. W. Veríssimo. 1997. "The Rediscovery of *Callicebus personatus barbarabrownae* in Northeastern Brazil with a New Western Limit for Its Distribution." *Primates* 38:429–33.

Mello, D. A. 1980. "Estudo Populacional de Algumas Espécies de Roedores do Cerrado (Norte do Município de Formosa, Goiás)." *Revista Brasileira de Biologia* 40:843–60.

Mello, D. A., and L. E. Moojen. 1979. "Nota Sobre uma Coleção de Roedores e Marsupiais de Algumas Regiões do Cerrado do Brasil Central." *Revista Brasileira de Pesquisas Médicas e Biológicas* 12:287–91.

Mello-Leitão, C. 1946. "As Zonas de Fauna da América Tropical." *Revista Brasileira de Geografia* 8:71–118.

Nimer, E. 1989. *Climatologia do Brasil*, 2nd ed. Rio de Janeiro: Instituto Brasileiro de Geografia e Estatística.

Oliveira, F. F., and A. Langguth. 2004. "Pequenos Mamíferos (Didelphimorphia e Rodentia) de Paraíba e Pernambuco, Brasil." *Revista Nordestina de Biologia* 18:19–86.

Oliveira, J. A. de. 1998. "Morphometric Assessment of Species Groups in the South American Rodent Genus *Oxymycterus* (Sigmodontinae), with Taxonomic Notes Based on the Analysis of Type Material." Unpublished PhD diss. Texas Tech University, Lubbock.

Oliveira, J. A. de., P. R. Gonçalves, and C. R. Bonvicino. 2008. "Mamíferos da Caatinga." In *Ecologia e Conservação da Caatinga*, 3rd ed., edited by I. R. Leal, M. Tabarelli, and J. M. C. Silva, 275–302. Recife: Editora Universitária, Universidade Federal de Pernambuco.

Oliveira, T. G. de, R. G. Gerude, and J. de S. e Silva Jr. 2008. "Unexpected Mammalian Records in the State of Maranhão." *Boletim Museu Paraense Emilio Goeldi, Ciências Naturais* 2:23–32.

Oliveira-Filho, A. T., and J. A. Ratter. 1995. "A Study of the Origin of Central Brazilian Forests by the Analysis of Plant Species Distribution Patterns." *Edinburgh Journal of Botany* 52:141–94.

———. 2002. "Vegetation Physiognomies and Woody Flora of the Cerrado Biome." In *The Cerrados of Brazil: Ecology and Natural History of a Neotropical Savanna*, edited by P. S. Oliveira and R. J. Marquis, 91–120. New York: Columbia University Press.

Palma, R. E., E. Rivera-Milla, T. L. Yates, P. A. Marquet, and A. P. Meynard. 2002. "Phylogenetic and Biogeographic Relationships of the Mouse Opossum *Thylamys* (Didelphimorphia, Didelphidae) in Southern South America." *Molecular Phylogenetics and Evolution* 25:245–53.

Parada, J. M., and S. M. Andrade. 1977. "Cerrados: Recursos Minerais." In *IV Simpósio sobre o Cerrado: Bases para utilização agropecuária*, edited by M. G. Ferri, 195–209. São Paulo: Editora da Universidade de São Paulo.

Patterson, B., and R. Pascual. 1972. "The Fossil Mammal Fauna of South America." In *Evolution, Mammals, and Southern Continents*, edited by A. Keast, F. C. Erk, and B. Glass, 247–309. Albany: State University of New York Press.

Patterson, B. D. 2000. "Patterns and Trends in the Discovery of New Neotropical Mammals." *Diversity and Distributions* 6:145–51.

———. 2001. "Fathoming Tropical Biodiversity: The Continuing Discovery of Neotropical Mammals." *Diversity and Distributions* 7:191–96.

Percequillo, A. R. 1998. "Sistemática de *Oryzomys* Baird, 1858 do leste do Brasil (Muroidea, Sigmodontinae)." Unpublished MSc thesis. Departamento de Zoologia, Universidade de São Paulo.

Percequillo, A. R., E. Hingst-Zaher, and C. R. Bonvicino. 2008. "Systematic Review of Genus *Cerradomys* Weksler, Percequillo and Voss, 2006 (Rodentia: Cricetidae: Sigmodontinae: Oryzomyini), with Description of Two New Species from Eastern Brazil." *American Museum Novitates* 3622:1–46.

Prado, D. E. 2008. "As Caatingas da América do Sul." In *Ecologia e Conservação da Caatinga*, 3rd ed., edited by I. R. Leal, M. Tabarelli, and J. M. C. Silva, 3–73. Recife: Editora Universitária, Universidade Federal de Pernambuco.

Prance, G. T. 1987. "Vegetation." In *Biogeography and Quaternary History in Tropical America*, edited by T. C. Whitmore and G. T. Prance, 28–45. Oxford: Oxford Science Publications.

Queiroz, L. P. de. 2006. "The Brazilian Caatinga: Phytogeographical Patterns Inferred from Distribution Data of the Leguminosae. In *Neotropical Savannas and Dry Forests: Plant Diversity, Biogeography, and Conservation*, edited by R. T. Pennington, G. P. Lewis, and J. A. Ratter, 113–49. Boca Raton: CRC Press, Systematics Association Special Volume, 69.

Rancy, A. 1999. "Fossil Mammals of the Amazon as a Portrait of a Pleistocene Environment." In *Mammals of the Neotropics, Vol. 3, The Central Neotropics: Ecuador, Peru, Bolivia, Brazil*, edited by J. F. Eisenberg and K. H. Redford, 20–26. London: University of Chicago Press.

Ratter, J. A., J. F. Ribeiro, and S. Bridgewater. 1997. "The Brazilian Cerrado Vegetation and Threats to Its Biodiversity." *Annals of Botany* 80:223–30.

Redford, K. H., and G. A. B. da Fonseca. 1986. "The Role of Gallery Forests in the Zoogeography of the Cerrado's Non-Volant Mammalian Fauna." *Biotropica* 18:126–35.

Redford, K. H., and J. F. Eisenberg. 1992. *Mammals of the Neotropics, Vol. 2. The Southern Cone: Chile, Argentina, Uruguay, Paraguay*. London: University of Chicago Press.

Reis, N. R., A. L. Peracchi, W. A. Pedro, and I. P. Lima. 2006. *Mamíferos do Brasil*. Paraná: Editora Universidade Estadual de Londrina.

Rizzini, C. T. 1997. *Tratado de Fitogeografia do Brasil. Aspectos Ecológicos, Sociológicos e Florísticos.* Rio de Janeiro: Âmbito Cultural Edições.

Rodrigues, M. T. 1988. "Distribution of Lizards of the Genus *Tropidurus* in Brazil (Sauria, Iguanidae)." In *Proceedings of a Workshop on Neotropical Distribution Patterns*, edited by P. E. Vanzolini and W. R. Heyer, 305–16. Rio de Janeiro: Academia Brasileira de Ciências.

———. 2008. "Herpetofauna da Caatinga." In *Ecologia e Conservação da Caatinga*, 3rd ed., edited by I. R. Leal, M. Tabarelli, and J. M. C. Silva, 181–236. Recife: Editora Universitária, Universidade Federal de Pernambuco.

Romariz, D. A. 1968. "A vegetação." In *Brasil: A Terra e o Homem*, edited by A. Azevedo, 419–562. São Paulo: Companhia Editora Nacional.

Rosenberger, A. L. 2002. "Platyrrhine Paleontology and Systematics: The Paradigm Shifts." In *The Primate Fossil Record*, edited by W. C. Hartwig, 151–59. Cambridge: Cambridge University Press.

Rossi, V. R. 2005. "Revisão taxonômica de *Marmosa* Gray, 1821 (Didelphimorphia, Didelphidae)." Unpublished PhD diss. Universidade de São Paulo.

Rowe, D. L., K. A. Dunn, R. M. Adkins, and R. L. Honeycutt. 2010. "Molecular Clocks Keep Dispersal Hypotheses Afloat: Evidence for Trans-Atlantic Rafting by Rodents." *Journal of Biogeography* 37:305–24.

Sick, H. 1965. "A fauna do Cerrado." *Arquivos de Zoologia (São Paulo)* 12:71–93.

———. 1966. "A fauna do Cerrado como fauna arborícola." *Anais Academia Brasileira de Ciências* 38:355–63.

Silva, J. M. C. da. 1996. "Distribution of Amazonian and Atlantic Birds in Gallery Forests of the Cerrado Region, South America." *Ornitología Neotropical* 7:1–18.

Silva, J. M. C. da, and J. M. Bates. 2002. "Biogeographic Patterns and Conservation in the South American Cerrado: A Tropical Savanna Hotspot." *Bioscience* 52:225–33.

Silva, J. M. C. da, M. A. de Souza, A. G. D. Bieber, and C. J. Carlos. 2008. "Aves da Caatinga: Status, Uso do Habitat e Sensitividade." In *Ecologia e Conservação da Caatinga*, 3rd ed., edited by I. R. Leal, M. Tabarelli, and J. M. C. Silva, 237–74. Recife: Editora Universitária, Universidade Federal de Pernambuco.

Silva Jr., J. S. 2001. "Especiação nos Macacos-Prego e Caiararas, gênero *Cebus* Erxleben, 1777 (Primates, Cebidae)." Unpublished PhD diss. Universidade Federal do Rio de Janeiro.

Silva, T. C. F. 2010. "Estudo da Variação da Pelagem e da Distribuição Geográfica em *Cebus flavius* e *Cebus libidinosus* do Nordeste do Brasil." Unpublished MSc thesis. Universidade Federal da Paraíba, João Pessoa.

Smith, M. F., and J. L. Patton. 1999. "Phylogenetic Relationships and the Radiation of Sigmodontine Rodents in South America: Evidence from Cytochrome b." *Journal of Mammalian Evolution* 6:89–128.

Stevens, R. D. 2004. "Untangling Latitudinal Richness Gradients at Higher Taxonomic Levels: Familial Perspectives on the Diversity of New World Bat Communities." *Journal of Biogeography* 31:665–74.

Streilein, K. E. 1982a. "Behavior, Ecology, and Distribution of South American Marsupials." In *Mammalian Biology in South America*, edited by M. A. Mares and H. H. Genoways,

231–50. Pittsburgh: University of Pittsburgh, *Special Publication Series, Pymatuning Laboratory of Ecology*, Vol. 6.

———. 1982b. "Ecology of Small Mammals in the Semiarid Brazilian Caatinga. I. Climate and Faunal Composition." *Annals of the Carnegie Museum* 51:79–107.

———. 1982c. "Ecology of Small Mammals in the Semiarid Brazilian Caatinga. II. Water Relations." *Annals of the Carnegie Museum* 51:109–26.

———. 1982d. "Ecology of Small Mammals in the Semiarid Brazilian Caatinga. III. Reproductive Biology and Population Ecology." *Annals of the Carnegie Museum* 51:251–69.

———. 1982e. "Ecology of Small Mammals in the Semiarid Brazilian Caatinga. IV. Habitat Selection." *Annals of the Carnegie Museum* 51:331–43.

———. 1982f. "Ecology of Small Mammals in the Semiarid Brazilian Caatinga. V. Agonistic Behavior and Overview." *Annals of the Carnegie Museum* 51:345–69.

Tribe, C. J. 2005. "A New Species of *Rhipidomys* (Rodentia, Muroidea) from North-Eastern Brazil." *Arquivos do Museu Nacional, Rio de Janeiro* 63:131–46.

Valle, C. M. de C., M. C. Alves, I. B. Santos, and J. B. M. Varejão. 1982. "Observações Sobre Dinâmica de População de *Zygodontomys lasiurus* (Lund, 1841), *Calomys expulsus* (Lund, 1841) e *Oryzomys subflavus* (Wagner, 1842) em Vegetação de Cerrado no Vale do Rio das Velhas (Prudente de Morais, Minas Gerais, Brazil) Rodentia: Cricetidae." *Lundiana* 2:71–83.

Vanzolini, P. E. 1963. "Problemas faunísticos do cerrado." In *Simpósio Sobre o Cerrado I*, edited by M. G. Ferri, 305–22. São Paulo: Editora da Universidade de São Paulo.

———. 1976. "On the Lizards of a Cerrado-Caatinga Contact: Evolutionary and Zoogeographical Implications. (Sauria)." *Papeís Avulsos de Zoologia, São Paulo* 29:111–19.

———. 1988. "Distributional Patterns of South American Lizards." In *Proceedings of a Workshop on Neotropical Distribution*, edited by P. E. Vanzolini and W. R. Heyer, 317–42. Rio de Janeiro: Academia Brasileira de Ciências.

Varela, D. M., R. G. Trovati, K. R. Guzmán, R. V. Rossi, and J. M. B. Duarte. 2010. "Red Brocket Deer *Mazama americana* (Erxleben 1777)." In *Neotropical Cervidology*, edited by J. M. B. Duarte and S. Gonzalez, 151–59. Gland: IUCN/FUNEP.

Vitt, L. J. 1991. "An Introduction to the Ecology of Cerrado Lizards." *Journal of Herpetology* 25:79–90.

de Vivo, M. 1991. *Taxonomia de Callithrix Erxleben, 1977 (Callitrichidae, Primates)*. Belo Horizonte: Fundação Biodiversitas.

———. 1997. "Mammalian Evidence of Historical Ecological Change in the Caatinga Semiarid Vegetation of Northeastern Brazil." *Journal of Comparative Biology* 2:65–73.

———. 2008. "Mamíferos e Mudanças Climáticas." In Biologia e Mudanças Climáticas no Brasil, edited by M. S. Buckeridge, 207–23. São Paulo: Rima Editora.

Voss, R. S., and S. A. Jansa. 2009. "Phylogenetic Relationships and Classification of Didelphid Marsupials, an Extant Radiation of New World Metatherian Mammals." *Bulletin of the American Museum of Natural History* 322:1–177.

Voss, R. S., D. P. Lunde, and S. A. Jansa. 2005. "On the Contents of *Gracilinanus* Gardner and Creighton, 1989, with the Description of a Previously Unrecognized Clade of Small Didelphid Marsupials." *American Museum Novitates* 3482:1–34.

Wallace, A. R. 1876. *The Geographical Distribution of Animals*. London: Macmillan.

Walter, H. 1986. *Vegetação e Zonas Climáticas: Tratado de Ecologia Global*. São Paulo: EPU.
Weksler, M. 2003. "Phylogeny of Neotropical Oryzomyine Rodents (Muridae: Sigmodontinae) Based on the Nuclear IRPB Exon." *Molecular Phylogenetics and Evolution* 29:331–49.
———. 2006. "Phylogenetic Relationships of Oryzomyine Rodents (Muroidea: Sigmodontinae): Separate and Combined Analyses of Morphological and Molecular Data." *Bulletin of the American Museum of Natural History* 196:1–149.
Weksler, M., and C. R. Bonvicino. 2005. "Taxonomy of Pygmy Rice Rats (genus *Oligoryzomys*, Rodentia: Sigmodontinae) of the Brazilian Cerrado, with the Description of Two New Species." *Arquivos do Museu Nacional, Rio de Janeiro* 63:113–30.
Weksler, M., A. R. Percequillo, and R. S. Voss. 2006. "Ten New Genera of Oryzomyine Rodents (Cricetidae: Sigmodontinae)." *American Museum Novitates* 3537:1–29.
Willig, M. R. 1983. "Composition, Microgeographic Variation, and Sexual Dimorphism in Caatingas and Cerrado Bat Communities from Northeast Brazil." *Bulletin of Carnegie Museum of Natural History* 23:1–131.
———. 1985. "Reproductive Patterns of Bats from Caatingas and Cerrado Biomes in Northeastern Brazil." *Journal of Mammalogy* 66:668–81.
———. 1986. "Bat Community Structure in South America: A Tenacious Chimera." *Revista Chilena de Historia Natural* 59:151–68.
Willig, M. R., and M. A. Mares. 1989. "Mammals from the Caatinga: An Updated List and Summary of Recent Research." *Revista Brasileira de Biologia* 49:361–67.
Willig, M. R., and K. W. Selcer. 1989. "Bat Species Density Gradients in the New World: A Statistical Assessment." *Journal of Biogeography* 16:189–95.
Willis, K. J., and J. C. McElwain. 2002. *The Evolution of Plants*. Oxford: Oxford University Press.
Wilson, D. E., and D. M. Reeder, eds. 2005. *Mammal Species of the World: A Taxonomic and Geographic Reference*, 3rd ed. Baltimore: Johns Hopkins University Press.
Woods, C. A., and C. W. Kilpatrick. 2005. "Infraorder Hystricognathi." In *Mammal Species of the World: A Taxonomic and Geographic Reference*, 3rd ed., edited by D. E. Wilson and D. M. Reeder, 1538–600. Baltimore: Johns Hopkins University Press.

Appendix 14.1

Table A14.1 An updated list of the Cerrado and Caatinga mammals, showing their occurrences in adjacent biomes and in other formations of South America. CH, Chaco; AM, Amazon Forest; AF, Atlantic Forest; "Other" refers to any open biome not contiguous with either Cerrado or Caatinga, including the Llanos, Pampas, Monte, and others.

Taxa	Cerrado	Caatinga	CH	AM	AF	Other
DIDELPHIMORPHIA						
DIDELPHIDAE						
Caluromyinae						
Caluromys lanatus	X			X	X	

(Continued)

Table A14.1 (*continued*)

Taxa	Cerrado	Caatinga	CH	AM	AF	Other
Caluromys philander	X			X	X	
Didelphinae						
Chironectes minimus	X			X	X	
Cryptonanus agricolai		X				
Cryptonanus chacoensis	X		X			X
Didelphis albiventris	X	X	X			X
Gracilinanus agilis	X	X	X	X		
Lutreolina crassicaudata	X		X	X	X	X
Marmosa murina	X	X		X	X	
Marmosops incanus	X				X	
Marmosops noctivagus	X			X		
Marmosops ocellatus	X		X			
Micoureus constantiae	X		X	X		
Micoureus demerarae	X	X		X	X	
Micoureus paraguayanus	X				X	
Monodelphis americana	X	X		X	X	
Monodelphis domestica	X	X	X			
Monodelphis kunsi	X		X			
Monodelphis umbristriata	X					
Philander frenatus	X				X	
Philander opossum	X		X	X		
Thylamys karimii	X	X				
Thylamys macrurus	X		X			
Thylamys velutinus	X					
CINGULATA						
DASYPODIDAE						
Dasypodinae						
Dasypus novemcinctus	X	X	X	X	X	
Dasypus septemcinctus	X	X	X		X	
Euphractinae						
Euphractus sexcinctus	X	X	X	X	X	
Tolypeutinae						
Cabassous tatouay	X				X	
Cabassous unicinctus	X	X		X		
Priodontes maximus	X		X	X	X	
Tolypeutes matacus	X		X			X
Tolypeutes tricinctus	X	X				
PILOSA						
BRADYPODIDAE						
Bradypus variegatus	X	X		X	X	

Table A14.1 (continued)

Taxa	Cerrado	Caatinga	CH	AM	AF	Other
MYRMECOPHAGIDAE						
Myrmecophaga tridactyla	X	X	X	X	X	
Tamandua tetradactyla	X	X	X	X	X	
PRIMATES						
CEBIDAE						
Callitrichinae						
Callithrix jacchus		X			X	
Callithrix melanura	X		X	X		
Callithrix penicillata	X					
Cebinae						
Cebus cay	X		X			
Cebus libidinosus	X	X				
Cebus xanthosternos		X			X	
AOTIDAE						
Aotus azarae	X		X	X		
PITHECIIDAE						
Callicebinae						
Callicebus barbarabrownae		X				
Callicebus melanochir		X			X	
Callicebus pallescens	X		X			
ATELIDAE						
Alouattinae						
Alouatta caraya	X		X			
Alouatta ululata	X	X				
RODENTIA						
SCIURIDAE						
Sciurinae						
Guerlinguetus alphonsei		X		X	X	
Guerlinguetus ingrami		X			X	
Urosciurus urucumus	X					
CRICETIDAE						
Sigmodontinae						
Akodon cursor	X	X			X	
Akodon lindberghi	X					
Akodon montensis	X		X		X	
Akodon toba	X		X			
Calomys callosus	X		X			X
Calomys expulsus	X	X				
Calomys tener	X	X			X	
Calomys tocantinsi	X					

(Continued)

Table A14.1 *(continued)*

Taxa	Cerrado	Caatinga	CH	AM	AF	Other
Cerradomys langguthi		X			X	
Cerradomys maracajuensis	X		X			
Cerradomys marinhus	X					
Cerradomys scotti	X		X			
Cerradomys subflavus	X				X	
Cerradomys vivoi		X			X	
Euryoryzomys lamia	X					
Euryoryzomys nitidus	X			X		
Euryoryzomys russatus	X				X	
Holochilus brasiliensis		X	X		X	X
Holochilus chacarius	X		X			
Holochilus sciureus	X	X		X		
Hylaeamys megacephalus	X			X		
Hylaeamys yunganus	X			X		
Juscelinomys candango	X					
Juscelinomys guaporensis	X					
Juscelinomys huanchacae	X					
Kunsia fronto	X		X			
Kunsia tomentosus	X					
Microakodontomys transitorius	X					
Neacomys spinosus	X			X		
Necromys lasiurus	X	X	X	X	X	
Nectomys rattus	X	X	X	X		
Nectomys squamipes	X				X	
Oecomys bicolor	X			X		
Oecomys catherinae	X				X	
Oecomys mamorae	X			X		
Oecomys paricola	X			X		
Oecomys rex	X			X		
Oecomys roberti	X			X		
Oecomys trinitatis	X			X		
Oligoryzomys chacoensis	X		X			
Oligoryzomys fornesi	X	X	X			
Oligoryzomys moojeni	X					
Oligoryzomys nigripes	X	X	X		X	X
Oligoryzomys rupestris	X	X				
Oligoryzomys stramineus	X	X				
Oxymycterus dasytrichus	X	X			X	
Oxymycterus delator	X		X			
Pseudoryzomys simplex	X	X	X			

Table A14.1 (continued)

Taxa	Cerrado	Caatinga	CH	AM	AF	Other
Rhipidomys cariri		X				
Rhipidomys emiliae	X			X		
Rhipidomys macrurus	X					
Rhipidomys mastacalis	X	X			X	
Rhipidomys nitela	X			X		
Thalpomys cerradensis	X					
Thalpomys lasiotis	X					
Wiedomys cerradensis	X					
Wiedomys pyrrhorhinos	X	X				
ERETHIZONTIDAE						
Erethizontinae						
Coendou prehensilis	X	X	X	X	X	
Sphiggurus insidiosus		X			X	
CAVIIDAE						
Caviinae						
Cavia aperea	X	X	X			X
Galea flavidens	X					
Galea spixii	X	X				
Hydrochoerinae						
Hydrochoerus hydrochaeris	X	X	X	X	X	X
Kerodon acrobata	X					
Kerodon rupestris	X	X				
DASYPROCTIDAE						
Dasyprocta azarae	X			X	X	
Dasyprocta nigriclunis	X					
Dasyprocta prymnolopha		X		X	X	
Dasyprocta punctata	X		X			
CUNICULIDAE						
Cuniculus paca	X	X	X	X	X	
CTENOMYIDAE						
Ctenomys boliviensis	X		X			
Ctenomys brasiliensis	X					
Ctenomys nattereri	X					
ECHIMYIDAE						
Dactylomyinae						
Dactylomys dactylinus	X			X		
Echimyinae						
Makalata didelphoides	X			X		
Phyllomys blainvillii		X				
Phyllomys lamarum		X			X	

(Continued)

Table A14.1 (continued)

Taxa	Cerrado	Caatinga	CH	AM	AF	Other
Eumysopinae						
Carterodon sulcidens	X					
Clyomys laticeps	X		X			
Proechimys longicaudatus	X		X			
Proechimys roberti	X			X		
Thrichomys apereoides	X					
Thrichomys inermis	X	X				
Thrichomys laurentius		X				
Thrichomys pachyurus	X		X			
Trinomys albispinus	X	X			X	
Trinomys minor		X				
Trinomys yonenagae		X				
LAGOMORPHA						
LEPORIDAE						
Sylvilagus brasiliensis	X	X	X	X	X	
CHIROPTERA						
EMBALLONURIDAE						
Diclidurinae						
Diclidurus albus		X		X	X	
Emballonurinae						
Peropteryx macrotis	X	X	X	X	X	
Peropteryx trinitatis		X		X		
Rhynchonycteris naso	X	X		X	X	
Saccopteryx bilineata		X		X	X	
Saccopteryx leptura		X		X	X	
PHYLLOSTOMIDAE						
Carolliinae						
Carollia brevicauda	X	X		X	X	
Carollia perspicillata	X	X	X	X	X	
Desmodontinae						
Desmodus rotundus	X	X	X	X	X	X
Diaemus youngii	X		X	X	X	
Diphylla ecaudata	X	X		X	X	
Glossophaginae						
Anoura caudifer	X		X	X	X	
Anoura geoffroyi	X	X		X	X	
Glossophaga soricina	X	X	X	X	X	
Lonchophyllinae						
Lionycteris spurrelli	X	X	X	X	X	
Lonchophylla bokermanni		X			X	

Table A14.1 (continued)

Taxa	Cerrado	Caatinga	CH	AM	AF	Other
Lonchophylla dekeyseri	X					
Lonchophylla mordax	X	X		X	X	
Xeronycteris vieirai		X				
Phyllostominae						
Chrotopterus auritus	X	X	X	X	X	
Glyphonycteris behnii	X			X		
Lonchorhina aurita	X	X		X	X	
Lophostoma brasiliense	X	X	X	X	X	
Lophostoma carrikeri		X		X		
Lophostoma silvicolum		X	X	X	X	
Macrophyllum macrophyllum	X		X	X	X	
Micronycteris megalotis	X	X		X	X	
Micronycteris minuta	X	X		X	X	
Micronycteris sanborni	X	X				
Micronycteris schmidtorum	X	X		X	X	
Mimon bennettii	X	X		X	X	
Mimon crenulatum	X	X		X	X	
Phylloderma stenops	X	X		X	X	
Phyllostomus discolor	X	X	X	X	X	
Phyllostomus hastatus	X	X		X	X	
Tonatia bidens	X	X	X		X	
Tonatia saurophila		X		X		
Trachops cirrhosus	X	X		X	X	
Vampyrum spectrum		X		X		
Stenodermatinae						
Artibeus cinereus	X	X		X	X	
Artibeus concolor	X	X		X		
Artibeus lituratus	X	X	X	X	X	
Artibeus obscurus		X		X	X	
Artibeus planirostris	X	X	X	X	X	
Chiroderma doriae	X				X	
Chiroderma villosum	X	X		X	X	
Platyrrhinus brachycephalus	X			X		
Platyrrhinus helleri	X			X		
Platyrrhinus lineatus	X	X	X		X	
Platyrrhinus recifinus		X			X	
Sturnira lilium	X	X	X	X	X	
Sturnira tildae	X			X	X	
Uroderma bilobatum	X	X		X	X	
Uroderma magnirostrum	X	X		X	X	

(Continued)

Table A14.1 *(continued)*

Taxa	Cerrado	Caatinga	CH	AM	AF	Other
Vampyressa pusilla	X				X	
MORMOOPIDAE						
Pteronotus gymnonotus	X	X		X		
Pteronotus parnellii	X	X		X		
Pteronotus personatus		X		X		
NOCTILIONIDAE						
Noctilio albiventris	X	X	X	X		
Noctilio leporinus	X	X	X	X	X	
FURIPTERIDAE						
Furipterus horrens	X	X		X	X	
THYROPTERIDAE						
Thyroptera devivoi	X					X
Thyroptera discifera	X			X	X	
NATALIDAE						
Natalus stramineus	X	X		X	X	
MOLOSSIDAE						
Molossinae						
Cynomops abrasus	X	X	X	X	X	
Cynomops greenhalli		X		X		
Cynomops planirostris	X	X	X	X	X	
Eumops auripendulus	X	X	X	X	X	
Eumops delticus	X	X		X	X	
Eumops glaucinus	X		X	X	X	
Eumops hansae	X			X	X	
Eumops perotis		X	X	X	X	
Molossops temminckii	X	X	X	X		
Molossus molossus	X	X	X	X	X	
Molossus rufus	X	X	X	X	X	
Neoplatymops mattogrossensis	X	X		X	X	
Nyctinomops aurispinosus	X	X	X	X		
Nyctinomops laticaudatus	X	X	X	X	X	
Nyctinomops macrotis	X		X	X		
Promops nasutus	X	X	X	X	X	
Tadarida brasiliensis	X	X	X	X	X	X
VESPERTILIONIDAE						
Eptesicus andinus	X			X		
Eptesicus brasiliensis	X	X	X	X	X	
Eptesicus diminutus	X	X	X	X	X	
Eptesicus furinalis	X	X	X	X	X	
Histiotus velatus	X	X		X	X	

Table A14.1 (continued)

Taxa	Cerrado	Caatinga	CH	AM	AF	Other
Lasiurus blossevillii	X	X	X	X	X	X
Lasiurus cinereus	X		X	X	X	X
Lasiurus ega	X	X	X	X	X	X
Lasiurus egregius	X	X		X	X	
Myotis albescens	X	X	X	X	X	X
Myotis nigricans	X	X	X	X	X	
Myotis riparius	X	X	X	X	X	
Rhogeessa io	X			X		
Rhogeessa hussoni	X	X				X
CARNIVORA						
FELIDAE						
Felinae						
Leopardus braccatus	X		X			
Leopardus pardalis	X	X	X	X	X	
Leopardus tigrinus	X	X	X	X	X	
Leopardus wiedii	X	X	X	X	X	
Puma concolor	X	X	X	X	X	X
Puma yagouaroundi	X	X	X	X	X	X
Pantherinae						
Panthera onca	X	X	X	X	X	
CANIDAE						
Cerdocyon thous	X	X	X		X	X
Chrysocyon brachyurus	X		X			
Lycalopex vetulus	X	X				
Speothos venaticus	X		X	X	X	
MUSTELIDAE						
Lutrinae						
Lontra longicaudis	X		X	X	X	
Pteronura brasiliensis	X	X	X	X	X	
Mustelinae						
Eira barbara	X	X	X	X	X	
Galictis cuja	X		X		X	X
Galictis vittata	X	X		X	X	
MEPHITIDAE						
Conepatus semistriatus	X	X				X
PROCYONIDAE						
Nasua nasua	X	X	X	X	X	
Procyon cancrivorus	X	X	X	X	X	
PERISSODACTYLA						
TAPIRIDAE						
Tapirus terrestris	X	X	X	X	X	

(Continued)

Table A14.1 *(continued)*

Taxa	Cerrado	Caatinga	CH	AM	AF	Other
ARTIODACTYLA						
TAYASSUIDAE						
Pecari tajacu	X	X	X	X	X	
Tayassu pecari	X	X	X	X	X	
CERVIDAE						
Capreolinae						
Blastocerus dichotomus	X		X			
Mazama americana	X		X	X	X	
Mazama gouazoubira	X	X	X		X	X
Ozotoceros bezoarticus	X		X			X
Total	227	153	111	146	138	25

15

The Role of the Andes in the Diversification and Biogeography of Neotropical Mammals

Bruce D. Patterson, Sergio Solari,
and Paúl M. Velazco

Abstract

The Andes are the world's longest mountain chain, simultaneously presenting dispersal corridors to montane species and dispersal barriers for lowland forms. Steep environmental and climatic gradients on Andean slopes, especially along the Eastern Versant, cause most montane distributions to be far longer (N-S) than they are wide (E-W). The development of the Andes has indelibly marked the divergence of many tropical lowland taxa with basal splits into trans-Andean and cis-Andean components. Besides defining the limits of various lowland centers of endemism, the Andes house several very distinctive biotas of their own—the arid Western Slope, alpine communities including páramo, jalca, and puna extending onto the Altiplano, and moist forested communities on the Eastern Versant of the northern and central Andes reaching into northwestern Argentina. These biotas seem to be historically as well as ecologically distinctive. Speciation appears to have been recent and rapid on the Eastern Versant and in the Altiplano, while the Western Slope biota has a more relictual character. In some cases, Andean radiations are rooted in the tropical lowlands while in others, lowland radiations seem to be derived from Andean (or proto-Andean) ancestry. Few Andean mammals are well studied, but residents of the middle elevations of the Eastern Slope are especially poorly known. Middle-elevation faunas are highly diverse, show substantial degrees of endemism, and their species may be critical to developing accurate historical reconstructions of groups with widespread Neotropical distributions. Sampling these habitats should therefore be a high priority for future surveys.

15.1 Introduction

Areas of endemism—places where co-occurring taxa evolve together and in some degree of isolation from other such areas—transcend taxonomic limits, because they represent discrete areas that share common environments and histories. In practice, however, areas of endemism are shaped by the vagility, age, and rate of diversification of the taxa that occupy them. These differences lead to discordances among groups, in both centers and the distributional limits. Because mammal groups vary so dramatically in vagility and generation time, and because different groups have resided in South America for vastly different time periods, regions of endemism for bats do not coincide with those of carnivores (which are of comparable geographic scale but more recent origin) or of rodents (typically much finer scale but of comparable age, at least for some groups). Thus, the most distinctive region of endemism for South American bats is the Western Slope of the Andes, where 39% of bat species are endemic (cf. Koopman 1982), whereas this same area is more poorly defined for rodents and carnivores. On the other hand, bats are only modestly differentiated in the temperate Valdivian forests of Chile and Argentina, where the generic endemism of rodents and marsupials rivals that of New Guinea (Patterson 1992).

Despite such differences, there are commonalities to biotic distributions that afford a common framework for analysis and discussion. The Andes are unquestionably the most important biogeographic feature in South America, dwarfing the effects of major rivers or biome boundaries in limiting the distributions of lowland and highland species alike. Despite differences in their diversity and endemism, virtually all taxa show the pronounced influence of the Andes between the Caribbean and 25–30°S latitude (fig. 15.1). Temperate climates in the south permit highland species to extend their ranges to lower elevations, eroding the Andean character of distributions in Patagonia (Pardiñas et al. 2003). In the Southern Cone, many taxa occur from the Pacific to the Atlantic and cannot be considered properly Andean, although a number of forms (e.g., *Lama*, *Abrothrix*, *Euneomys*) found near sea level in Tierra del Fuego occur at high elevations in the southern and central Andes.

Our goal in this review is to identify new information on the effects of the Andes on both Andean and lowland mammal faunas. We do not attempt to enumerate Andean diversity (q.v., Solari, Velazco, and Patterson, chapter 8, this volume) or to develop a comprehensive theory on their role in continental radiations (Reig 1981, 1986). Instead, we discuss a number of groups, events, and area relationships in which the Andes seem to have played a decisive role. The Andes variously represent an arena for the diversification and dispersal of the richest biotas in the Neotropics, as well as distributional barriers, vicariant events, and ultimately the climatic drivers for adjacent lowland faunas.

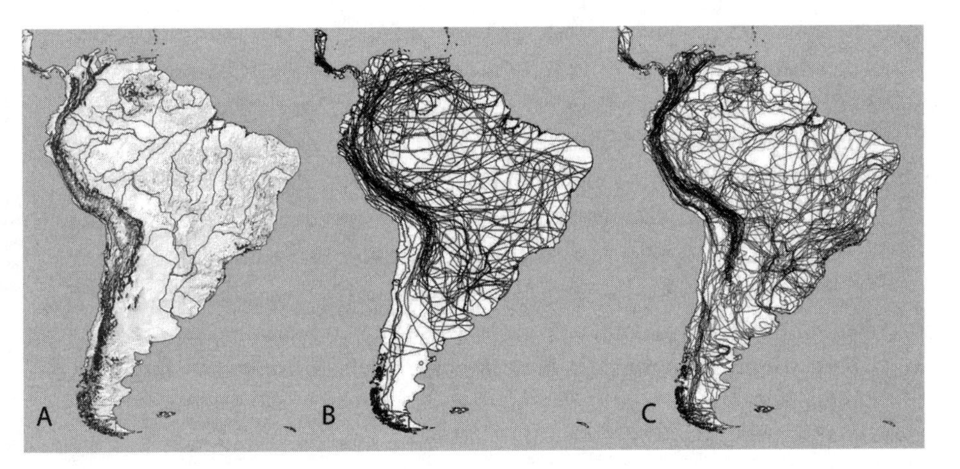

Figure 15.1 The Andes and areography of Neotropical mammals: (A) digital elevation model, with 1500 m a.s.l. contour superimposed; (B) range limits for Neotropical bat species, exclusive of Phyllostomidae (8 families); and (C) range limits for cricetid rodents. Distributions derived from the 2008 Global Mammal Assessment (http://www.iucnredlist.org).

15.2 Provincialism

The major centers of endemism that are largely or exclusively Andean include the Western Slope, Northern and Central Andean forests (including the Eastern Versant and yungas), and various alpine communities. *Western Slope* faunas include mainly xeric forest and scrub inhabitants in southwestern Ecuador and western Peru (Patterson, Pacheco, and Ashley 1992). Its districts include the upper Río Marañón and its tributaries, the Tumbesian or Ecuadorean subcenter, and the Pacific Slope of Peru and northernmost Chile (Chapman 1926; Cracraft 1985). The union of the distributions of the bats *Amorphochilus schnablii*, *Platalina genovensium*, and *Artibeus fraterculus*, all endemics to the region, fairly delimits it. The region is remarkable in terms of bat endemism (Koopman 1982), but is also marked for some rodents: *Eremoryzomys*, *Paralomys*, and *Aegialomys xanthaeolus* are likewise restricted to this region. Some endemic taxa reach low elevations, even sea level, along the arid Pacific coast (Zamora and Zeballos 2009).

Humid forests in the *Northern and Central Andes* are home to rich endemic faunas. Many Andean species of both plants and animals have geographic ranges much longer than they are wide (e.g., Graves 1988), giving rise to a striking succession of ecological communities as one moves up or down along the cordillera's steep slopes. The region is bounded on the north by the Atrato-

San Juan drainages in northwestern Colombia (Alberico 1990) and the Orinoco watershed in central Venezuela (Voss 1988). Andean faunas become attenuated as they extend onto the Sierra Nevada de Santa Marta, Sierra de Mérida, variously disjunct front ranges (e.g., Vilcabamba in Peru), and in the *yungas* south of Tarija, Bolivia—moist forest associations become scattered between 25–30°S. The Andes also support dry *Polylepis* forests at very high elevations (up to 4500 m), with special adaptations to desiccation and cold. Although each of these units has its own endemics and therefore constitutes an independent theater of mammalian evolution, inadequate sampling and limited understanding of mammalian diversification in them prevents us from organizing them hierarchically.

Andean *alpine environments* tend to be wetter and more fragmented in the north, from northern Peru to Venezuela, where they are called "*páramos*." From southern Peru to central Argentina, the high-elevation communities are termed "*puna*" and are more xeric and continuous, extending across western Bolivia and neighboring countries in a vast tableland termed the Altiplano. Intermediate in terms of both moisture and latitude, transitional communities above 3000 m in central Peru are known as "*jalca*" (Soejima et al. 2008). Reig (1986) quantified the substantial endemism of rodent species in páramo and puna (48% and 68%, respectively) and noted the low faunal resemblance (33%) of these areas.

Indisputably Andean biotas are flanked by regions that extend from the Andean piedmont into the lowlands, including the Chocó of Colombia and northwestern Ecuador, Caribbean lowlands of Colombia and Venezuela, Amazonia, Cerrado, Chaco, Monte, Patagonian steppe, Magellanic and Valdivian Forests, and the arid Pacific lowlands. In some cases, Andean orogeny contributed to the development of these other biomes. For example, Verzi (2001) argued that the caviomorph rodents *Tympanoctomys* and *Octomys*, together with the late Pliocene *Abalosia*, constitute an octodontid clade of desert specialists that differentiated in western Argentina in the developing rain shadow of the emergent Andes. On the fringes of the Andes, some faunas appear to be composites. The Guianan *tepuis* seem to be populated by Andean clades of bats (e.g., Lim et al. 2008; Velazco and Patterson 2008) and birds (Perez-Eman 2005), but their terrestrial faunas may have lowland affinities (Gardner 1990; but see Lim et al. 2010). Some Patagonian lineages also appear to be derived from Andean stocks, such as the sigmodontines *Abrothrix* and *Loxodontomys*, while others (e.g., *Reithrodon* and *Eligmodontia*) may have Chacoan, Monte, or Pampean ancestry (Lessa, D'Elia, and Pardiñas 2010; chapter 16, this volume). And there has been a dynamic interplay between the richest regions of endemism in the Neotropics, the tropical Andes and Amazonia, in generating their remarkable species diversity, as discussed in the following.

15.3 Orogenic Processes

The Andean orogeny is deeply rooted in time, traceable to conditions that predated the formation of Pangaea (Ramos 2008). Even in the southern Andes of Argentina, there is evidence of precordillaran terranes that date to the Ordovician (Thomas and Astini 2007). From Jurassic to mid-Cretaceous times, the Andean continental margin constituted a fringing island arc that resembled today's Sunda Shelf, being perforated by continental seaways that only closed as Andean orogeny progressed (Lomize 2008). Orogeny was triggered when western movements of the South American plate caused the Andes, already moderately thick and anchored in the mantle, to serve as a stop that generated compression and produced further continental growth.

By the middle Eocene, sufficient crust had accumulated and been compressed to initiate the present orogenic phase. Compression of the Central Andes by 250 km increased the crust's thickness from 35–75 km. The eventual involvement of the Nazca Plate to the north and the Juan Fernandez Plate to the south in subduction 10–12 Ma triggered the onset of the uplift and growth of the Andes (Lomize 2008). Volcanism played an important role in Cenozoic orogeny in sections of the cordillera where the angle of the oceanic slab subducting beneath the continental margin was high, typically 25–30° (Orme 2007b). However, variation in the angle of subduction along the length of the Andes, reaching as little as 5° in places, produces extensive regions where volcanism is practically nonexistent; these alternate with sections having frequent volcanic activity. Concise and readable summaries of both the tectonic history of South America and its relationship to developing climates and landscapes are presented by Orme (2007a, 2007b).

In places, both Andean habitats and those of surrounding regions became transformed by mountain-building. This is especially apparent in what is now the Valdivian province of Patagonia, which now boasts elevated degrees of endemism but supported a more continental biota 30 Ma (Flynn et al. 2003). The same orogeny that isolated Valdivian forest and matorral scrub also generated the xeric formations that developed in the Andes' eastern rain shadow. This evolving landscape shaped the Pliocene and subsequent development of several groups of arid-adapted rodents, including the sigmodontine *Eligmodontia* (Lanzone, Ojeda, and Gallardo 2007).

15.4 Cis- and Trans-Andean Distributions

The lowlands of northwestern Colombia between the Andes and the Panamanian isthmus are rich in species, many belonging to lowland groups that are endemic to South America proper. Some of these result from recent dispersal across the Andes, while many belong to groups that were widespread prior to Andean orogeny and became isolated by the developing mountain system. These groups often show basal dichotomies into cis- and trans-Andean components, being found exclusively on the east and west sides of the Andean cordillera, respectively. Albert, Lovejoy, and Crampton (2006) associated cis- and trans-disjunctions of freshwater fishes with the initial rise of the Eastern Cordillera ~12 Ma. Emergence of the Sierra Nevada de Santa Marta ~8 Ma would have subdivided the trans-Andean fauna into Magdalena and Pacific Slope components, while the coeval rise of the Mérida Andes isolated the Maracaibo and Orinoco Basins (Albert, Lovejoy, and Crampton 2006). The montane features that constituted barriers for lowland animals would have served as corridors for colonizers from the Central Andes or (once the Northern Andes became occupied) for interchange with the Darién of Panama, Santa Marta, and the Mérida Andes.

Many mammals reflect this east-west pattern, although the association of this diversification with the initial Andean orogeny is unclear in some cases. Analyses of mitochondrial sequences (Cortes-Ortiz et al. 2003) indicated that the primary dichotomy among howler monkeys separates the Central American species (*Alouatta pigra* and *A. palliata*) from those in South America (*A. seniculus*, *A. sara*, *A. macconnelli*, *A. caraya*, *A. belzebul*, and *A. guariba*). Their molecular-clock estimates dated this divergence to the late Miocene and Pliocene, coincident with Andean orogeny. Among marsupials, Gutiérrez, Jansa, and Voss (2010) showed that a major split in the mouse opossums *Marmosa* (including *Micoureus*) involved a trans-Andean clade of species found mainly in Central America and western South America (*mexicana*, *zeledoni*, *isthmica*, *robinsoni*, *xerophila*, *simonsi*, and *rubra*) and a cis-Andean group (*murina*, *tyleriana*, *waterhousei*, *macrotarsus*, *constantiae*, *regina*, *demerarae*, *paraguayana*, and *lepida*). Although Gutiérrez et al. did not date this node in their tree, they clearly associated it with Andean orogeny, pointing out that similar distribution patterns in pimelodid catfishes and parrots date to 6–8 Ma.

Molecular analyses of fruit bats of the genus *Artibeus* (Phyllostomidae) have documented a primary division between *A. fraterculus*, *A. inopinatus*, and *A. hirsutus* on the one hand and *A. jamaicensis*, *A. lituratus*, *A. obscurus*, *A. planirostris*, and *A. schwartzi* on the other (Hoofer et al. 2008; Larsen et al. 2007; Redondo et al. 2008). The former group is trans-Andean: *A. fraterculus* is found on the Western Slope of Peru and Ecuador, *A. inopinatus* in the arid Pacific forests of Central America, and *A. hirsutus* in Pacific Mexico. The other clade is exclusively cis-Andean, including the Greater and Lesser Antilles. A related group of fruit bats, *Dermanura*, shows similar phylogeographic divisions (Solari et al. 2009). The initial diversification of this clade began roughly 11–12 Ma (Redondo et al. 2008).

Short-tailed fruit bats of the genus *Carollia* echo this pattern, although the pattern is somewhat complicated by modern-day gene flow (Hoffmann and Baker 2003). Three lineages (*C. brevicauda*, *C. perspicillata*, and the *C. castanea* complex) each show the closer relationship of western Ecuadorean samples to Central American populations than to nearby populations on the other side of the Andes. That the genus did in fact cross, and perhaps recross, the Northern Andes can be adduced from the derivation of *C. brevicauda* and *C. perspicillata* from the Central American endemic *C. sowelli* (the sister to this trio of species is another Central American endemic, *C. subrufa*), their broad distributions in South America, and their reciprocal monophyly. The authors assigned diversification of *Carollia* to the last 4.5 Ma (Hoffmann and Baker 2003).

Despite their towering heights, the Andean cordillera has not completely eliminated gene flow and dispersal. Pacheco and Patterson (1992) documented allozyme variation in four Andean species of the New World epauletted fruit bats *Sturnira*. Surprisingly, genetic distances among populations on the Eastern Versant, in the Marañón valley, and on the Western Slope fit an isolation-by-distance model, although the fit of the model was improved if straight-line distances between sites were replaced by distances along inhabited elevation contours. Hoffmann and Baker (2003) showed that differentiation of *Carollia* species (short-tailed fruit bats) on opposite sides of the Andes depends on their upper range limits. *C. castanea* is restricted to lower elevations on the west; populations on the east side of the Andes show a whopping 7.8% cytochrome-*b* sequence divergence, and are now recognized as the sister species *C. benkeithi* (Solari and Baker 2006). On the other hand, *C. brevicauda* and *C. perspicillata*, which range into cloud forest on both sides of the Andes, show more modest differentiation (1.1–1.8%) between populations on the east and west sides of the Andes, consistent with at least some recurring gene flow.

15.5 Great American Biotic Interchange

The Pliocene formation of the Panamanian isthmus brought an end to South America's Tertiary isolation and triggered a massive and well-documented exchange of continental biotas termed the Great American Biotic Interchange (GABI; Stehli and Webb 1985). The isthmus was the gateway for colonizers moving both northward and southward. Prior to its formation, the South American element in Central America would likely have resembled the Neotropical portion of the Antillean fauna, comprised of a handful of vagile taxa (bats, sloths, monkeys, and rodents) successful in overwater dispersal (Koopman 1982).

The genus *Ateles* (spider monkeys) supposedly originated in South America, and basal branches include *A. chamek* (Medeiros et al. 1997) and *A. paniscus* (Collins and Dubach 2000b). A subsequent dichotomy separates *A. belzebuth*, *A. hybridus*, and *A. marginatus* in South America east of the Andes from a trans-Andean lineage that includes *A. fusciceps* in the Chocó and eastern Panama and *A. geoffroyi* in the remainder of Central America. Most nodes in this phylogeny apparently date to the middle to late Pliocene (Collins and Dubach 2000a), after establishment of the land bridge.

The tent-making bat *Uroderma bilobatum* also appears to have originated in South America. Three chromosomal races are known: a $2n = 42$ form in South America east of the Andes, a $2n = 38$ karyotype from eastern Central America and NW South America, and a $2n = 44$ race from NW Central America. In an analysis of intraspecific sequence variation in cytochrome-*b* (mt-DNA), the oldest divergences within this species were found within the South American race (Hoffmann, Owen, and Baker 2003). Genetic distances between races ranged from 2.5–2.9%, whereas the within-race variation was more modest: 1.7% ($2n = 42$ race), 0.9% ($2n = 38$), and 0.5% ($2n = 44$). Other instances of colonization of Central America by South American autochthons, with subsequent differentiation on opposite sides of the Andes, include the pouched opossums *Philander* and *Didelphis* (Patton et al. 1997).

Contemporaneously, North and Central American lineages were invading South America. Steppan, Adkins, and Anderson (2004) analyzed phylogenetic relationships among sigmodontine rodents (Cricetidae) to determine when they colonized and diversified in South America. They hypothesized that three major lineages (some Sigmodontini, some Ichthyomyini, and the ancestor of the Oryzomyalia—the remaining New World complex-penis mice) colonized South America by 6 Ma, long before a continuous land bridge had formed. Cooler temperatures and lower sea levels at that time would have increased the extent of montane forests and reduced overwater dispersal distances, facilitating both colonization of the Antilles and movement across the narrowing straits that would become the Panamanian isthmus (Miller, Bermingham, and Ricklefs 2007). However, there is a sizable gap between the estimated molecular divergence dates and the fossil record (D'Elía 2000; Pardiñas, D'Elía, and Ortiz 2002), and conflict among character sets in reconstructions of the basal radiation of Sigmodontinae (e.g., Voss [1988] argued that the ichthyomyines initially radiated in South America). Better resolution, dating, and stability in the basal topology of Sigmodontinae will be needed before this group can fully inform historical reconstructions.

An example involving a far more modest South American radiation is presented by small-eared shrews, genus *Cryptotis*, whose South American range is restricted to the northern Andes. The Colombian small-eared shrew, *Cryptotis colombiana*, is recorded from the Colombian Andes and belongs to the *C. nigrescens* group, which otherwise occurs in Central America. All other known South American shrews belong to the *C. thomasi* group, which appears to be monophyletic (Woodman, Cuartes-Calle, and Delgado-V. 2003). Shrews clearly invaded South America two or more times.

The Andes constituted a colonization corridor for many of these northern invaders. Wang and Carranza-Castaneda (2008, fig. 13) hypothesized that the skunk *Conepatus* colonized the length of South America via the Andes. From a North American origin, it underwent successive derivations of *C. semistriatus* (Central and northern South America), *C. chinga* (Central Andes, Chaco, and subtropical Brazil), and *C. humboldti* in Patagonia and the Southern Cone. They dated the South American clade to 4–5 Ma, indicating that this group was an early participant in the GABI. Their proposed phylogeny invites genetic corroboration (see Eizirik, chapter 7, this volume).

The colilargos (Sigmodontinae: *Oligoryzomys*) are monophyletic (Weksler 2006) and found from southern Mexico to Tierra del Fuego. The genus apparently contains 2 species groups, one designated as the "Amazon–Cerrado" assemblage, which includes Central American species, and the other the "Pampa–Andean" clade, containing species restricted to the Andes or the

Southern Cone (Miranda et al. 2009). The derived position of the latter group (which includes *O. magellanicus*) in a phylogeny subtended by *O. moojeni* and *O. fornesi* (which are Cerrado species) generates a north-to-south geographic pattern, supporting the hypothesis that the genus radiated from north to south, and at least part of that radiation took place in the Andes. Because *Oligoryzomys* is widespread, diversified, and abundant, it is an excellent model for recovering continental and intercontinental relationships, and sampling more taxa and areas should improve the resolution and stability of the topology, which at present remains tentative.

The Andes also served as a corridor for northward dispersal of austral forms. The tribe Abrotrichini includes five sigmodontine genera that are all found in the Southern Andes and Patagonia and possibly autochonous there (D'Elía et al. 2007). The range of one genus (*Abrothrix*) extends through Andean habitats into northern Chile and Argentina all the way to the *jalca* of central Peru (Patterson, Teta, and Smith, forthcoming). Another example is provided by the mountain viscachas, *Lagidium* (Chinchillidae), whose distribution coincides with that of *Abrothrix* in Patagonia and the Andes as far north as central Peru. Recently, a new species of *Lagidium* was described from montane scrub and forest at 2000 m in western Ecuador (Ledesma et al. 2009). Although the phylogenetic relationships of *L. ahuacaense* to other *Lagidium* remain uncertain, the remaining extant genera in Chinchillidae (*Chinchilla* and *Lagostomus*) have austral distributions, enhancing the likelihood that the genus originated in southern South America.

15.6 Differentiation within the Northern and Central Andes

Geologists often subdivide the Andes near the Gulf of Guayaquil in Ecuador because of changes in the orientation of subducting slabs, but biologists typically use another boundary for Andean centers of endemism. The Huancabamba Deflection and Marañón Valley in northern Peru mark the southern limit of many northern species and northern limit of southern ones (Chapman 1926). Nevertheless, many mammals transcend these boundaries without apparent differentiation (Carleton and Musser 1989; Lunde and Pacheco 2003).

Phylogenetic analyses are still lacking for many mammals distributed over this region. However, the northern Andes are very young and strictly montane forms should have colonized them from the Central Andes. Picard, Sempere, and Plantard (2008) studied *Globodera pallida*, a nematode parasite of potatoes that requires cool temperatures and thrives above 2000–2500 m in Peru. Phylogeographic analysis shows a clear evolutionary pattern with deeper, older lineages in southern Peru and shallower ones occurring progressively northwards, showing genetic divergence as populations progressively colonized highland areas as they appeared to the north. Molecular dating indicates that the 2000–2500 m threshold was reached with the northward propagation of crustal thickening by the Early Miocene in southernmost Peru, in the Middle Miocene near Abancay (~13°38'S), and in the latest Miocene in central and northern Peru. The adaptive radiation and diversification of some plants (Paranepheliinae: Asteraceae) apparently followed a similar timetable and progression (Soejima et al. 2008).

Few Andean rodents have been studied adequately to resolve their intrageneric relationships, yet their diverse phylogenies and restricted distributions should offer refined resolution of area relationships. Preliminary findings from several doctoral dissertations on diverse, widespread rodent groups promise great insights. Tribe (1996) assessed taxonomic variation among climbing rats of the genus *Rhipidomys* (Cricetidae), grouping species based on cranial traits and overall similarity. He divided species into three sections, with the *fulviventer* section mostly distributed over the northern Andes in Colombia, Venezuela, and the Guiana Shield. Three species occupied the main cordilleras; *caucensis* in the Western, *latimanus* in the Central, and *fulviventer* in the Eastern Cordillera, apparently reflecting the vicariant effect of the Cauca and Magdalena valleys between these primary ranges. Molecular analyses of these taxa could corroborate the relationships and indicate the time frame of these events, but have not yet been conducted.

Nephelomys rice rats (previously the *albigularis* group of *Oryzomys*; Cricetidae) inhabit montane and cloud forests of the Andes, from Bolivia to Venezuela and Colombia, as well as into Panama and Costa Rica. Percequillo (2003) parsed geographic variation in the group to reveal 14 species-level taxa whose distributions neatly coincide with proposed avian and anuran centers of endemism. He attributed their diversification patterns to Andean uplift during the Neogene and to Quaternary climate change.

Pacheco's (2003) phylogenetic analysis of morphological characters for mice of the genus *Thomasomys* (Cricetidae) recovered a monophyletic Andean thomasomyine group composed of the genera *Thomasomys*, *Rhipidomys*, *Chilomys*, and *Aepeomys*, which is successively sister to lowland thomasomyines, with the closest being *Rhagomys*. His topology also showed some geographic division among members of the group, including the pairs *T. bombycinus* and *T. cinereiventer*, which are separated by the Cauca Valley of the northern Andes, as well as *T. contradictus* and *T. dispar*, and *T. emeritus* and *T. laniger*. In molecular analyses of IRBP sequence variation with poorer taxon sampling (i.e., lacking *Chilomys*), *Rhagomys* joins with *Thomasomys*, *Rhipidomys*, and *Aepeomys* (D'Elía et al. 2005), linking the Andean region of endemism with montane forests in southeastern Brazil (see also Percequillo, Weksler, and Costa 2011).

Voss, Gómez-Laverde, and Pacheco (2002) observed that all records of the mouse genus *Handleyomys* come from localities in the Western and Central cordilleras, despite energetic collecting efforts elsewhere. They also cited distributions of other nonvolant vertebrates that suggest that these mountain ranges were connected more recently than either was with the Eastern cordillera.

15.7 The Altiplano

Phylogeographic evidence indicates that the Altiplano has been subject to subdivision in the recent past. Populations of the camelid *Vicugna* from the northern Altiplano (18–22°S) show evidence of demographic expansion associated with the last major glacial event of the Pleistocene. On the other hand, scattered populations in the extremely arid belt known as the "Dry Diagonal" (to 29°S) show the genetic signature of persistence and demographic isolation, hallmarks of former refugia (Marín et al. 2007). Analyses of guanaco populations (Marín et al. 2008) likewise showed striking variation among Peruvian and northern Chilean populations and very modest differentiation among Patagonian populations, reflecting the antiquity of the Altiplano center and their evident demographic expansion and recent bottlenecks in Patagonia.

The antiquity of the southern Altiplano center is also evident in the radiations of phyllotine mice (Cricetidae), a group adapted to generally xeric environments. Altiplano phyllotines tend to exhibit ancestral character states (telocentric chromosomes, high diploid numbers) and basal positions in phylogenies based on protein and cytochrome-*b* gene sequences, whereas lowland taxa both northwards and southwards tend to exhibit derived character states (Spotorno et al. 2001). In *Eligmodontia*, the northern species have chromosomes with arm sizes <9% of the total karyotype, while the southern species *E. typus* and *E. morgani* exhibit longer arms, probably derived via tandem fusions. Thus, the Patagonian species are derived, probably from a northern ancestor with 2n = 50 and FN = 48 (Spotorno, Sufan-Catalan, and Walker 1994). That interpretation is supported by evidence for poleward demographic expansion in this and nine other lineages of sigmodontine rodent (Lessa, D'Elía, and Pardiñas 2010; this volume).

15.8 The Guianas

A biogeographical connection between the Andes and the Guianas has been proposed for various groups of plants and animals, particularly between the Guiana Shield and the sandstone outcrops on mountain ranges of southeastern Ecuador and northeastern Peru (e.g., Cisneros-Heredia and McDiarmid 2006). The affinities of *Monodelphis reigi* in Venezuela and Guyana with a clade of mostly Andean short-tailed opossums suggested a Miocene-aged connection between these landforms (Lim et al. 2010). However, the Guianan endemic *Platyrrhinus aurarius* belongs to a well-supported Andean clade that dates to more recent times (Velazco and Patterson 2008), suggesting that this ancient connection has been overwritten by more recent dispersal. *Rhipidomys wetzeli* and *Podoxymys roraimae* also appear to show this same pattern (Gardner 1990; Pérez-Zapata et al. 1992).

15.9 Atlantic Forest

Surprisingly, biogeographic connections also exist between the Andes and the Atlantic Forest, specifically with montane forests of the Serra do Mar. Luna and Patterson (2003) described a new species of *Rhagomys* from cloud forest in southeastern Peru bearing numerous morphological synapomorphies with a mouse then known only from nineteenth-century collections near Rio de Janeiro. Their new species, *R. longilingua*, was subsequently recorded from Bolivia (Villalpando, Vargas, and Salazar-Bravo 2006), even as *R. rufescens* was rediscovered in Brazil, and was found to range as far south as Santa Catarina (Percequillo, Gongalves, and Olneira 2004; Steiner-Souza et al. 2008). Very recently, Percequillo, Weksler, and Costa (2011) described *Drymoreomys albimaculatus*, a new genus and species of rice rat from humid montane forests in the Serra do Mar; this rodent was robustly recovered in analyses of morphology and mt-DNA as sister to *Eremoryzomys polius* of the Marañón subdivision of the Western Slope. The authors offered detailed discussion on this novel biogeographic association. Another group that may show the same distributional track is the prehensile-tailed bamboo rats (Echimyidae): *Olallamys* of the northern Andes, *Dactylomys* of the central Andes and Amazonia, and *Kannabateomys* of Atlantic Forest.

15.10 Modes of Speciation

The Andes present a broad spectrum of environmental conditions in close proximity, including elevation-related variation in numerous factors affecting fitness and distribution. Such conditions are essential for "ecological speciation" or more properly parapatric speciation, an alternative to the more usual allopatric model as a general explanation of organismal diversification. Despite its inherent plausibility, the richness of Andean biotas, and the uncontested sharpness of its gradients, there is no compelling evidence for parapatric speciation in Andean mammals, and little for it in other vertebrates.

Patton and Smith (1992) examined small mammals distributed over a 3 km elevational gradient in southern Peru to test the applicability of the parapatric speciation model. Phylogenetic analyses compared the relationships of highland and lowland species of mice in the genus *Akodon* (Cricetidae): mice in different river drainages at the same elevation were compared to their elevational replacements within the same river drainage. Parapatric speciation would link species within a drainage basin, while allopatric speciation would cause the members of a given drainage to be polyphyletic. Analyses of *Akodon torques*, *A. mimus*, and *A. aerosus* falsify the applicability of the gradient model (Patton and Smith 1992). Interestingly, all three taxa form a clade sister to *A. subfuscus*, which is distributed atop the adjacent Altiplano.

Phylogenetic analyses of *Nephelomys* rice rats offer another example. The species *N. keaysi* and *N. levipes* have ribbon-like distributions limited to the Eastern Versant of southern Peru and northern Bolivia. Although they overlap throughout the cloud-forest zone in southeastern Peru (Solari et al. 2006), they are generally elevational replacements, with *N. keaysi* at lower elevations and *N. levipes* above. Despite the apparent division of these ecological replacements into discrete drainage basins, the two appear to be sister taxa, each reciprocally monophyletic (Patton, Myers, and Smith 1990).

Sadly, the straightforward test developed by Patton and Smith has not been widely applied. This is undoubtedly a consequence of its rigorous sampling requirements: sampling two or more valleys at two or more elevations, in situations involving elevational species replacements. However, indications from other taxa also suggest that parapatric speciation is rarely responsible for the diversity and distributions of Andean vertebrates (Arctander and Fjeldsa 1994; Dingle et al. 2006).

15.11 Museums, Cradles, and Diversity Pumps

The major hypotheses for high species diversity in the tropics invoke either lowered extinction rates or elevated speciation rates relative to more temperate regions (Moritz et al. 2000), or invoke both plus increased immigration (Jablonski, Roy, and Valentine 2006). The plausibility of lowered extinction rates is tied to the great age and climatic stability of the tropics.

Some advocates for elevated speciation rates in the tropics point to climatic oscillations during the Pleistocene, which are thought to have triggered the contraction of tropical forest habitats into glacial-episode refugia, where allopatric differentiation and speciation could take place (Haffer 1969; Vanzolini and Williams 1970). The reexpansion of rain forests with the return of warm wet climates during interglacials would have brought these newly divergent forms back into contact, and repetitions of this process would constitute a diversity pump.

Amazonia and the Andes are the two richest regions of endemism in the Neotropics. They have different species pools, different environmental conditions, and have undergone different biogeographic histories; it is therefore reasonable to suppose that they were populated differently and at different times. Average species age and rates of speciation are obvious variables of interest in such comparisons. Unfortunately, there has been no systematic comparison of rates of Andean and Amazonian diversification in mammals; however, studies on birds and frogs offer an approach toward a general answer, and parallel analyses of mammals would be most illuminating, given their ecological and evolutionary diversity.

Poison frogs of the genus *Epipedobates* (Dendrobatidae) exhibit the majority of their phenotypic and species diversity in the Andean foothills and are represented by both highland and lowland endemics. Phylogenetic analysis showed that highland species are monophyletic and derived from a population of lowland ancestry in northern Peru, save for a widespread Amazonian species, *Epipedobates trivittatus*, which is a member of the highland clade that reinvaded the lowlands (Roberts et al. 2006). Comparative analyses of coloration revealed that divergence among populations and species in the highlands has been accelerated relative to the lowlands. Selection was implicated in the divergence of coloration among populations and species (Roberts et al. 2006).

Weir (2006) assembled datable molecular phylogenies for 27 avian taxa to contrast the timing and rates of diversification for lowland and highland faunas in the Neotropics. In terms of rate changes over time, lowland taxa exhibited decreasing speciation rates while highland taxa showed no trend. Combining rates of different taxa into million-year bins, he showed that fauna-wide diversification rates in the lowlands were highest in the late Miocene and declined toward the present. In the highlands, rates varied little until the last million years when they spiked following the onset of severe glacial conditions. Apparently, habitat shifts with changing climates were insufficient to increase the rate of diversification in lowland faunas, but direct fragmentation of habitats by glaciers and elevational shifts of vegetation zones appear to have resulted in late-Pleistocene increases in highland diversification rates. Fully one-third of the highland bird species sampled arose during the last one million years (Weir 2006).

15.12 Amazonian Roots of Andean Diversity

The antiquity of lowland habitats in South America and their stability over time made them ready colonization sources for newly emergent montane areas. A broad literature addresses the Amazonian roots of Andean diversity, but sadly little of this focuses on mammals. Relationships among slender mouse opossums (*Marmosops*) may eventually prove illuminating. Voss, Tarifa, and Yensen (2004) analyzed cytochrome-*b* sequences and showed that *M. noctivagus*, *M. impavidus*, *M. ocellatus*, and *M. creightoni*—all distributed on the middle and lower slopes of the Eastern Versant of Peru and Bolivia—appear to be sister taxa. This clade is sister to a group of exclusively lowland taxa. On the other hand, nuclear (IRBP) sequences showed that Andean taxa (*noctivagus* and *impavidus*) form a clade with the Atlantic Forest endemic *M. incanus*, and that this group is sister to an Amazonian group containing *M. parvidens* (Voss and Jansa 2003). Fuller taxon sampling and total evidence can be expected to clarify these relationships. Within a small radiation of short-tailed opossums (*Monodelphis*), Solari (2007) found that *M. adusta* from the northern Amazonian lowlands was basal to a radiation having at least two species with mostly montane distributions, *M. peruviana* and *M. osgoodi*. However, at least in one case, the montane species *M. osgoodi* is sister to a lowland form, *M. handleyi*, lending ambiguity to these area relationships.

Both allozymes and restriction-fragment-length polymorphisms of mtDNA suggested the same phylogenetic hypothesis for *Leptopogon* flycatchers: the lowland, tropical species was recovered as the basal member of the clade, and the upper-tropical-zone species was sister to the two upper-elevation species. These data are consistent with the diversification of this genus into successively higher-elevation habitats in the Andes (Bates and Zink 1994). Colonization and subsequent Andean diversification is also reported for *Thamnophlius* antshrikes (Brumfield and Edwards 2007), *Pionus* parrots (Ribas et al. 2007), *Cyanolyca* jays (Bonaccorso 2009), hummingbirds (McGuire et al. 2007), and *Elaenia* flycatchers (Rheindt, Christidis, and Norman 2008). Diversification within the Andes via Pleistocene climatic oscillations and their large-scale effects on habitat was also thought to be responsible for explosive radiations of the plant *Lupinus* (Hughes and Eastwood 2006).

15.13 Andean Roots of Lowland Diversity

Although the Andes are relatively young, their habitats are full of species, and over time, some of these might have successfully colonized the Amazon basin. Patterson and Velazco (2006) described a new species of echimyid rodent (Caviomorpha) from cloud forests in southeastern Peru, assigning the taxon to the bottle-brush rat genus *Isothrix*, which is otherwise confined to lowland forests. Their phylogenetic analyses of cytochrome-*b* sequence variation show that the five lowland species of *Isothrix* all form a clade that is sister to *I. barbarabrownae*, the lone Andean endemic (Patterson and Velazco 2008). Because *Isothrix* has no clear sister genus, it is impossible to determine its ancestral distribution, whether Andean or Western Amazonian (see Lim, chapter 11, this volume, fig. 11.6). In any case, the basal split of *Isothrix* into Andean and lowland clades, confirmed in combined mitochondrial and nuclear analyses, identifies the Andes as a theater of its early evolution.

A similar topology is apparent for the bamboo rats *Dactylomys* (Echimyidae). Like *Isothrix barbarabrownae*, *D. peruanus* is distributed in Andean cloud forests, 1000–3000 m in elevation, while its congeners, *D. dactylinus* and *D. boliviensis*, occur in the piedmont and Amazon basin. Analyses of cytochrome-*b* sequence showed that the Andean species is sister to the lowland pair (Patterson and Velazco 2008). If the sister group of *Dactylomys* proves to be Andean (dentition suggests that it is likely *Olallamys*, a northern Andean endemic), this radiation would be clearly rooted in the Andes.

Both *Isothrix* and *Dactylomys* are echimyid rodents, and studies of echimyid phylogeny (Galewski et al. 2005; Lara, Patton, and da Silva 1996; Leite and Patton 2002) have suggested the rapid radiation of most modern genera in the middle Miocene. By that time (9–15 Ma), a proto-Andean block was already in place, creating the semiarid-to-arid climate that characterized the Central Andes for most of the Miocene and Pliocene. This landform and associated climate changes helped to trigger hypsodonty in Cenozoic faunas (Croft 2001) and would have been a major landscape feature during the period in which echimyids diversified. Patterson and Velazco (2008) hypothesized that the Andean components of both *Isothrix* and *Dactylomys* evolved not in the Andes per se but rather in this proto-Andean block, identifying the Andean region as a source pool for Amazonian diversity. Santos and colleagues (2009) developed a similar argument with broad phylogenetic support for the Andean derivation of many poison frog lineages now extant in Amazonia.

Other groups old enough to antedate the latest episodes of Andean orogeny may also present elements of this pattern. The arboreal woolly opossums *Marmosa* (*Micoureus*) were shown to present a phylogenetic pattern that is rooted in the Andes (Patton, da Silva, and Malcolm 2000). In analyses of 630 bp of cytochrome-*b*, the Andean foothill and western Amazonian species *M. regina* is sister to remaining members of the genus, including the Central American *M. alstoni*, Atlantic Forest *M. limae*, and Amazonian and Guianan *M. demerarae*. Voss and Jansa (2003) presented a combined phylogeny of morphological and nuclear sequence characters that showed the Andean foothill genus *Caluromysiops* as sister to the lowland forest taxa *Caluromys lanatus* and *C. philander*. The phyllotine rodent *Calomys* (Cricetidae) can be separated into two large clades (Haag et al. 2007). One clade (*C. musculinus*, *C. lepidus*, and *C. sorellus*) is restricted to the highlands, where they have undergone some local differentiation. However, members of the second clade invaded the lowlands, including nonforested biomes, where they have undergone substantial radiations.

15.14 Future Challenges

The Andes constitute one of the world's hottest hotspots of diversity because of the huge variety of habitats and wealth of species they support as well as the diverse impacts and threats of human developments. Nowhere is this truer than for the tropical and subtropical biomes on the eastern slopes of the Andes, which are some of the richest and at the same time most poorly known ecosystems on Earth (Young 1992). Understanding this complexity before it is obliterated by anthropogenic effects is a monumental challenge.

As one example, Manu Biosphere Reserve in southeastern Peru has attracted repeated biological sampling by many research teams, including national and foreign participants, whose efforts have shown it to be incredibly diverse. It currently ranks as the world's richest protected area for both birds and mammals (Patterson, Stotz, and Solari 2006). However, recent sampling (1999–2001) uncovered at least a dozen species new to science, including 1 marsupial, 6 bats, and 5 rodents. Besides documenting the remarkable incompleteness of vertebrate surveys in Manu, the distributions of new species also document their vulnerability. Despite declining species richness at higher elevations, and many fewer species sampled there, three-quarters of the newly discovered species have distributions that are limited to montane habitats (fig. 15.2). As noted by Voss (2003) for Ecuadorean faunas, mid-elevation faunas of the Eastern Versant also show remarkable beta diversity, so there is substantial species replacement over short distances, both horizontally and vertically. High beta diversity impedes efforts to generalize and extrapolate the findings of the few

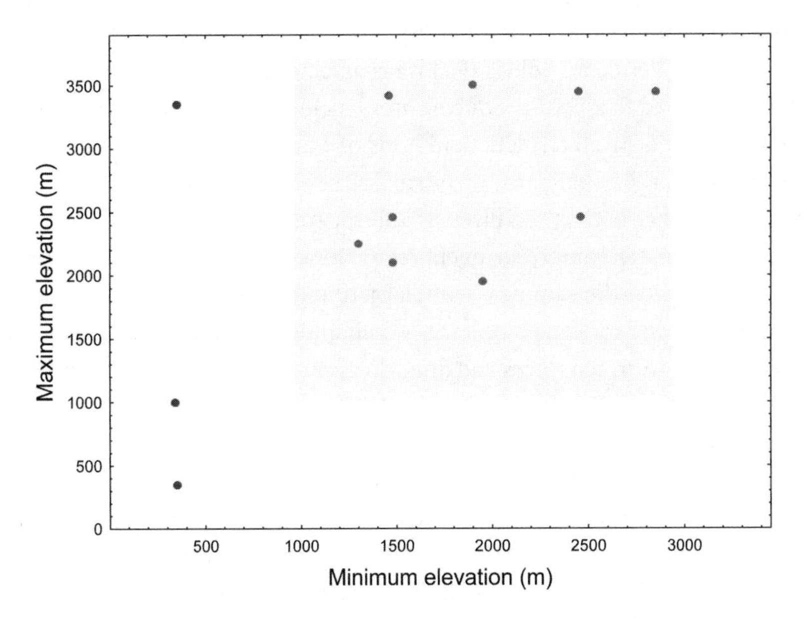

Figure 15.2 Plot of upper and lower range limits documented for 12 new mammal species discovered during recent biological surveys of Manu Biosphere Reserve (Solari et al. 2006). Three-quarters of these new species are restricted to montane forests above 1000 m, despite a strongly negative relationship between mammalian species richness and elevation. This reflects the greater endemism and poorer understanding of montane forest communities. Andean cloud forest and elfin forest are also highly threatened by development (Young and Valencia 1992), making them especially "hot" in terms of biodiversity.

thorough surveys thus far accomplished. The wealth of undiscovered taxa at these elevations, such as *Rhagomys longilingua* and *Isothrix barbarabrownae*, and their crucial importance in deciphering and recasting group phylogenies and biogeographies (D'Elía et al. 2005; Patterson and Velazco 2008), make surveys of the Eastern versant among the highest priorities for fostering understanding of Andean diversity, evolution, and biogeography.

Finally, political instability, insecurity, and xenophobia limit scientific understanding in the Andes as elsewhere in the developing world. Nearly a third of Colombia, which holds the world's richest avifauna, remains mostly unsampled owing to the activities of FARC (the "Revolutionary Armed Forces of Colombia") and other antigovernment guerilla organizations (Paynter and Traylor 1981). Similarly, terrorist acts by Peru's Sendero Luminoso group closed whole regions of the Andes to resident and foreign scientists alike. In addition to political violence, drug wars for control of areas for basic production or transport of drugs (e.g., coca, marijuana, or opium) hampers access to and reduces natural cover of montane forests along the eastern versant of the Andes in Colombia, Peru, and Bolivia (Fjeldså et al. 2005). Using 1° x 1° cells, those authors showed that coca-producing areas include at least 20% of the distributions of 67% of all resident land birds of South America! We believe this percentage may be similar for mammals, as several of these areas are also of high importance in terms of diversity and endemism.

But progressively more stringent restrictions on permits to study, collect, and export scientific samples—often instituted with concerns to enhance conservation and forestall biopiracy—limit understanding just as surely and effectively as do armed forces and drug violence. Early this year, signatories to the Convention on Biological Diversity (CBD) began discussion of new measurements to protect national biodiversity, which will include exploitation of genetic resources (Gilbert 2010). Indiscriminate use of restrictions aimed to protect national resources has resulted in a severe slowing of biological research in some Neotropical countries, like Brazil and Colombia; this is because this legislation does not discriminate between basic (i.e., genetic profile of species, including identification through DNA barcodes) and commercial research (i.e., bioprospecting for new drugs). This lack of clarity represents a major threat to international funding and collaboration with megadiverse developing countries. Devising effective and efficient means to catalyze multinational collaborations will be necessary to solve the larger remaining questions in Andean (and tropical) biogeography.

Acknowledgments

We thank the organizers of IMC-10, especially Ricardo Ojeda, for the invitation to present this work. Fieldwork in Manu was supported by NSF (DEB-9870191), the Field Museum's Barbara E. Brown, Marshall Field III, H. B. Conover, and Street Expeditionary Funds, as well as by the Bertha LeBus Charitable Trust and the family of George Jacobus. For insights about their study groups and access to unpublished materials, the authors thank E. Gutierrez, A. R. Percequillo, S. J. Steppan, and R. S. Voss. Jim Patton and Marcelo Weksler provided very helpful comments on an earlier draft of the manuscript and Leonora Costa handled the editorial chores.

Literature Cited

Alberico, M. S. 1990. "A New Species of Pocket Gopher (Rodentia: Geomyidae) from South America and Its Biogeographic Significance." In *Vertebrates in the Tropics: Proceedings of the International Symposium on Vertebrate Biogeography and Systematics in the Tropics, Bonn, June 5–8, 1989*, edited by G. Peters and R. Hutterer, 103–11. Bonn: Museum Alexander Koenig.

Albert, J. S., T. E. Lovejoy, and W. G. R. Crampton. 2006. "Miocene Tectonism and the Separation of Cis- and Trans-Andean River Basins: Evidence from Neotropical Fishes." *Journal of South American Earth Sciences* 21:14–27.

Arctander, P., and J. Fjeldsa. 1994. "Andean Tapaculos of the Genus *Scytalopus* (Aves, Rhinocryptidae): A Study of Speciation Using DNA Sequence Data." *Conservation Genetics* 68:205–25.

Bates, J. M., and R. M. Zink. 1994. "Evolution into the Andes: Molecular Evidence for Species Relationships in the Genus *Leptopogon*." *Auk* 111:507–15.

Bonaccorso, E. 2009. "Historical Biogeography and Speciation in the Neotropical Highlands: Molecular Phylogenetics of the Jay Genus *Cyanolyca*." *Molecular Phylogenetics and Evolution* 50:618–32.

Brumfield, R. T., and S. V. Edwards. 2007. "Evolution into and out of the Andes: A Bayesian Analysis of Historical Diversification in *Thamnophilus* Antshrikes." *Evolution* 61:346–67.

Carleton, M. D., and G. G. Musser. 1989. "Systematic Studies of Oryzomyine Rodents (Muridae, Sigmodontinae): A Synopsis of *Microryzomys*." *Bulletin of the American Museum of Natural History* 191:1–83.

Chapman, F. M. 1926. "The Distribution of Bird Life in Ecuador." *Bulletin of the American Museum of Natural History* 55:1–784.

Cisneros-Heredia, D. F., and R. W. McDiarmid. 2006. "A New Species of the Genus *Centrolene* (Amphibia : Anura : Centrolenidae) from Ecuador with Comments on the Taxonomy and Biogeography of Glassfrogs." *Zootaxa* 1244:1–32.

Collins, A. C., and J. M. Dubach. 2000a. "Biogeographic and Ecological Forces Responsible for Speciation in *Ateles*." *International Journal of Primatology* 21:421–44.

———. 2000b. "Phylogenetic Relationships of Spider Monkeys (*Ateles*) Based on Mitochondrial DNA Variation." *International Journal of Primatology* 21:381–420.

Cortes-Ortiz, L., E. Bermingham, C. Rico, E. Rodriguez-Luna, I. Sampaio, and M. Ruiz-Garcia. 2003. "Molecular Systematics and Biogeography of the Neotropical Monkey Genus, *Alouatta*." *Molecular Phylogenetics and Evolution* 26:64–81.

Cracraft, J. 1985. "Historical Biogeography and Patterns of Differentiation within the South American Avifauna: Areas of Endemism." In *Neotropical Ornithology*, edited by P. A. Buckley, M. S. Foster, E. S. Morton, R. S. Ridgely, and F. G. Buckley, 49–84. Washington: American Ornithologists Union, *Ornithological Monographs* 36.

Croft, D. A. 2001. "Cenozoic Environmental Change in South America As Indicated by Mammalian Body Size Distributions (Cenograms)." *Diversity and Distributions* 7:271–87.

D'Elía, G. 2000. "Comments on Recent Advances in Understanding Sigmodontine Phylogeny and Evolution." *Mastozoología Neotropical* 7:47–54.

D'Elía, G., L. Luna, E. M. González, and B. D. Patterson. 2005. "On the Structure of the Sigmodontine Radiation (Rodentia, Cricetidae): An Appraisal of the Phylogenetic Position of *Rhagomys*." *Molecular Phylogenetics and Evolution* 38:558–64.

D'Elía, G., U. F. J. Pardiñas, P. Teta, and J. L. Patton. 2007. "Definition and Diagnosis of a New Tribe of Sigmodontine Rodents (Cricetidae: Sigmodontinae), and a Revised Classification of the Subfamily." *Gayana* 71:187–94.

Dingle, C., I. J. Lovette, C. Canaday, and T. B. Smith. 2006. "Elevational Zonation and the Phylogenetic Relationships of the *Henicorhina* Wood-Wrens." *Auk* 123:119–34.

Fjeldså, J., M. D. Alvarez, J. M. Lazcano, and B. Leon. 2005. "Illicit Crops and Armed Conflict as Constraints on Biodiversity Conservation in the Andes Region." *Ambio* 34: 205–11.

Flynn, J. J., A. R. Wyss, D. A. Croft, and R. Charrier. 2003. "The Tinguiririca Fauna, Chile: Biochronology, Paleoecology, Biogeography, and a New Earliest Oligocene South American Land Mammal 'Age.'" *Palaeogeography, Palaeoclimatology, Palaeoecology* 195:229–59.

Galewski, T., J.-F. Mauffrey, Y. L. R. Leite, J. L. Patton, and E. J. P. Douzery. 2005. "Ecomorphological Diversification Among South American Spiny Rats (Rodentia; Echimyidae): A Phylogenetic and Chronological Approach." *Molecular Phylogenetics and Evolution* 34:601–15.

Gardner, A. L. 1990. "Two New Mammals from Southern Venezuela and Comments on the Affinities of the Highland Fauna of Cerro de la Neblina." In *Advances in Neotropical Mammalogy*, edited by K. H. Redford and J. F. Eisenberg, 411–24. Gainesville: Sandhill Crane Press.

Gilbert, N. 2010. "Biodiversity Hope Faces Extinction." *Nature* 467:764.

Graves, G. R. 1988. "Linearity of Geographic Range and Its Possible Effect on the Population Structure of Andean Birds." *Auk* 105:47–52.

Gutiérrez, E. E., S. A. Jansa, and R. S. Voss. 2010. "Molecular Systematics of Mouse Opossums (Didelphidae: Marmosa): Assessing Species Limits Using Mitochondrial DNA Sequences, with Comments on Phylogenetic Relationships and Biogeography." *American Museum Novitates* 3692:1–22.

Haag, T., V. C. Muschner, L. B. Freitas, L. F. B. Oliveira, A. R. Langguth, and M. S. Mattevi. 2007. "Phylogenetic Relationships Among Species of the Genus *Calomys* with Emphasis on South American Lowland Taxa." *Journal of Mammalogy* 88:769–76.

Haffer, J. 1969. "Speciation in Amazonian Forest Birds." *Science* 165:131–37.

Hoffmann, F. G., and R. J. Baker. 2003. "Comparative Phylogeography of Short-Tailed Bats (*Carollia*: Phyllostomidae)." *Molecular Ecology* 12:3403–14.

Hoffmann, F. G., J. G. Owen, and R. J. Baker. 2003. "mtDNA Perspective of Chromosomal Diversification and Hybridization in Peters' Tent-Making Bat (*Uroderma bilobatum*: Phyllostomidae)." *Molecular Ecology* 12:2981–93.

Hoofer, S. R., S. Solari, P. A. Larsen, R. D. Bradley, and R. J. Baker. 2008. "Phylogenetics of the Fruit-Eating Bats (Phyllostomidae: Artibeina) Inferred from Mitochondrial DNA Sequences." *Occasional Papers Museum of Texas Tech University* 277:1–15.

Hughes, C., and R. Eastwood. 2006. "Island Radiation on a Continental Scale: Exceptional Rates of Plant Diversification After Uplift of the Andes." *Proceedings of the National Academy of Sciences, USA* 103:10334–39.

Jablonski, D., K. Roy, and J. Valentine. 2006. "Out of the Tropics: Evolutionary Dynamics of the Latitudinal Diversity Gradient." *Science* 314:102–06.

Koopman, K. F. 1982. "Biogeography of the Bats of South America." In *Mammalian Biology in South America*, edited by M. A. Mares and H. H. Genoways, 273–302. Pittsburgh: Pymatuning Symposia in Ecology. University of Pittsburgh.

Lanzone, C., R. A. Ojeda, and M. H. Gallardo. 2007. "Integrative Taxonomy, Systematics and Distribution of the Genus *Eligmodontia* (Rodentia, Cricetidae, Sigmodontinae) in the Temperate Monte Desert of Argentina." *Mammalian Biology* 72:299–312.

Lara, M. C., J. L. Patton, and M. N. F. da Silva. 1996. "The Simultaneous Diversification of South American Echimyid Rodents (Hystricognathi) Based on Complete Cytochrome b Sequences." *Molecular Phylogenetics and Evolution* 5:403–13.

Larsen, P. A., S. R. Hoofer, M. C. Bozeman, S. C. Pedersen, H. H. Genoways, C. J. Phillips, D. E. Pumo et al. 2007. "Phylogenetics and Phylogeography of the *Artibeus jamaicensis* Complex Based on Cytochrome-b DNA Sequences." *Journal of Mammalogy* 88:712–27.

Ledesma, K. J., F. A. Werner, A. E. Spotorno, and L. Albuja V. 2009. "A New Species of Mountain Viscacha (Chinchillidae: *Lagidium* Meyen) from the Ecuadorean Andes." *Zootaxa* 2126:41–57.

Leite, Y. L. R., and J. L. Patton. 2002. "Evolution of South American Spiny Rats (Rodentia, Echimyidae): The Star-Phylogeny Hypothesis Revisited." *Molecular Phylogenetics and Evolution* 25:455–64.

Lessa, E. P., G. D'Elía, and U. F. J. Pardiñas. 2010. "Genetic Footprints of Late Quaternary Climate Change in the Diversity of Patagonian-Fueguian Rodents." *Molecular Ecology* 19:3031–37.

Lim, B. K., M. D. Engstrom, J. C. Patton, and J. W. Bickham. 2008. "Systematic Review of Small Fruit-Eating Bats (*Artibeus*) from the Guianas, and a Re-Evaluation of *A. glaucus bogotensis*." *Acta Chiropterologica* 10:243–56.

———. 2010. "Molecular Phylogenetics of Reig's Short-Tailed Opossum (*Monodelphis reigi*) and Its Distributional Range Extension into Guyana." *Mammalian Biology* 75:287–93.

Lomize, M. G. 2008. "The Andes As a Peripheral Orogen of the Breaking-Up Pangea." *Geotectonics* 42:206–24.

Luna, L., and B. D. Patterson. 2003. "A Remarkable New Mouse (Muridae: Sigmodontinae) from Southeastern Peru, with Comments on the Affinities of *Rhagomys rufescens* (Thomas, 1886)." *Fieldiana: Zoology, n.s.* 101:1–24.

Lunde, D. P., and V. Pacheco. 2003. "Shrew Opossums (Paucituberculata: *Caenolestes*) from the Huancabamba Region of East Andean Peru." *Mammal Study* 28:145–48.

Marín, J. C., C. S. Casey, M. Kadwell, K. Yaya, D. Hoces, J. Olazabal, R. Rosadio et al. 2007. "Mitochondrial Phylogeography and Demographic History of the Vicuña: Implications for Conservation." *Heredity* 99:70–80.

Marín, J. C., A. E. Spotorno, B. A. González, C. Bonacic, J. C. Wheeler, C. S. Casey, M. W. Bruford et al. 2008. "Mitochondrial DNA Variation and Systematics of the Guanaco (*Lama guanicoe*, Artiodactyla: Camelidae)." *Journal of Mammalogy* 89:269–81.

McGuire, J. A., C. C. Witt, D. L. Altshuler, and J. V. Remsen. 2007. "Phylogenetic Systematics and Biogeography of Hummingbirds: Bayesian and Maximum Likelihood Analyses of Partitioned Data and Selection of an Appropriate Partitioning Strategy." *Systematic Biology* 56:837–56.

Medeiros, M. A., R. M. S. Barros, J. C. Pieczarka, C. Y. Nagamachi, M. Ponsa, M. Garcia, F. Garcia et al. 1997. "Radiation and Speciation of Spider Monkeys, Genus *Ateles*, from the Cytogenetic Viewpoint." *American Journal of Primatology* 42:167–78.

Miller, M. J., E. Bermingham, and R. E. Ricklefs. 2007. "Historical Biogeography of the New World Solitaires (*Myadestes* spp.)." *Auk* 124:868–85.

Miranda, G. B., L. F. B. Oliveira, J. Andrades-Miranda, A. Langguth, S. M. Callegari-Jacques, and M. S. Mattevi. 2009. "Phylogenetic and Phylogeographic Patterns in Sigmodontine Rodents of the Genus *Oligoryzomys*." *Journal of Heredity* 100:309–21.

Moritz, C., J. L. Patton, C. J. Schneider, and T. B. Smith. 2000. "Diversification of Rainforest Faunas: An Integrated Molecular Approach." *Annual Review of Ecology and Systematics* 31:533–63.

Orme, A. R. 2007a. "Tectonism, Climate, and Landscape Change." In *The Physical Geography of South America*, edited by T. T. Veblen, K. R. Young, and A. R. Orme, 23–44. New York: Oxford University Press.

———. 2007b. "The Tectonic Framework of South America." In *The Physical Geography of South America*, edited by T. T. Veblen, K. R. Young, and A. R. Orme, 3–22. New York: Oxford University Press.

Pacheco, V. R. 2003. "Phylogenetic Analyses of the Thomasomyini (Muroidea: Sigmodontinae) Based on Morphological Data." Unpublished PhD diss. City University of New York.

Pacheco, V., and B. D. Patterson. 1992. "Systematics and Biogeographic Analysis of Four Species of *Sturnira* (Chiroptera: Phyllostomidae), with Emphasis on Peruvian Forms." In *Biogeografía, Ecología y Conservación del Bosque Montano en el Perú*, edited by K. R. Young and N. Valencia, 57–81. Lima: Memorias del Museo de Historia Natural. Universidad Nacional Mayor de San Marcos.

Pardiñas, U. F. J., G. D'Elía, and P. E. Ortiz. 2002. "Sigmodontinos Fósiles (Rodentia,

Muroidea, Sigmodontinae) de América del Sur: Estado Actual de su Conocimiento y Prospectiva." *Mastozoología Neotropical* 9:209–52.

Pardiñas, U. F. J., P. Teta, S. Cirignoli, and D. H. Posdestá. 2003. "Micromamíferos (Didelphimorphia y Rodentia) de Norpatagonia Extra-Andina, Argentina: Taxonomía Alfa y Biogeografía." *Mastozoología Neotropical* 10:69–113.

Patterson, B. D. 1992. "A New Genus and Species of Long-Clawed Mouse (Rodentia: Muridae) from Temperate Rainforests of Chile." *Zoological Journal of the Linnean Society* 106:127–45.

Patterson, B. D., V. Pacheco, and M. V. Ashley. 1992. "On the Origins of the Western Slope Region of Endemism: Systematics of Fig-Eating Bats, Genus *Artibeus*." In *Biogeografía, Ecología y Conservación del Bosque Montano en el Perú*, edited by K. R. Young and N. Valencia, 189–205. Lima: Universidad Nacional Mayor de San Marcos, *Memorias del Museo de Historia Natural*.

Patterson, B. D., D. F. Stotz, and S. Solari, eds. 2006. "Mammals and Birds of the Manu Biosphere Reserve, Peru." *Fieldiana: Zoology, n.s.* 110:1–49.

Patterson, B. D., P. Teta, and M. F. Smith. Forthcoming. "*Abrothrix*." In *Mammals of South America*, vol. 2, edited by J. L. Patton. Chicago: University of Chicago Press.

Patterson, B. D., and P. M. Velazco. 2006. "A Distinctive New Cloud-Forest Rodent (Hystricognathi: Echimyidae) from the Manu Biosphere Reserve, Peru." *Mastozoología Neotropical* 13:175–91.

———. 2008. "Phylogeny of the Rodent Genus *Isothrix* (Hystricognathi, Echimyidae) and Its Diversification in Amazonia and the Eastern Andes." *Journal of Mammalian Evolution* 15:181–201.

Patton, J. L., M. N. F. da Silva, M. C. Lara, and M. A. Mustrangi. 1997. "Diversity, Differentiation, and the Historical Biogeography of Nonvolant Small Mammals of the Neotropical Forests." In *Tropical Forest Remnants: Ecology, Management, and Conservation of Fragmented Communities*, edited by W. F. Laurance and R. O. Bierregaard Jr., 455–64. Chicago: University of Chicago Press.

Patton, J. L., M. N. F. da Silva, and J. R. Malcolm. 2000. "Mammals of the Rio Juruá and the Evolutionary and Ecological Diversification of Amazonia." *Bulletin of the American Museum of Natural History* 244:1–306.

Patton, J. L., P. Myers, and M. F. Smith. 1990. "Vicariant Versus Gradient Models of Diversification: The Small Mammal Fauna of Eastern Andean Slopes of Peru." In *Biogeography and Systematics in the Tropics, Bonn, June 5–8 1989*, edited by G. Peters and R. Hutterer, 355–71. Bonn: Alexander Koenig Zoological Research Institute and Zoological Museum.

Patton, J. L., and M. F. Smith. 1992. "MtDNA Phylogeny of Andean Mice: A Test of Diversification Across Ecological Gradients." *Evolution* 46:174–83.

Paynter, R. A., Jr., and M. A. Traylor Jr. 1981. *Ornithological Gazetteer of Colombia*. Cambridge: Bird Department, Museum of Comparative Zoology, Harvard University.

Percequillo, A. R. 2003. "Sistemática de *Oryzomys* Baird, 1858: Definição dos Grupos de Espécie e Revisão Taxonômica do Grupo *albigularis* (Rodentia, Sigmodontinae)." Unpublished PhD diss. Universidade de São Paulo.

Percequillo, A. R., P. R. Gonçalves, and J. A. de Oliveira. 2004. "The Rediscovery of *Rhagomys rufescens* (Thomas, 1886), with a Morphological Redescription and Comments on Its Systematic Relationships Based on Morphological and Molecular (Cytochrome *b*) Characters." *Mammalian Biology* 69:238–57.

Percequillo, A. R., M. Weksler, and L. P. Costa. 2011. "A New Genus and Species of Rodent from the Brazilian Atlantic Forest (Rodentia, Cricetidae, Sigmodontinae, Oryzomyini), with Comments on Oryzomyine Biogeography." *Zoological Journal of the Linnean Society* 161:357–90.

Perez-Eman, J. L. 2005. "Molecular Phylogenetics and Biogeography of the Neotropical Redstarts (*Myioborus*; Aves, Parulinae)." *Molecular Phylogenetics and Evolution* 37:511–28.

Pérez-Zapata, A., D. Lew, M. Aguilera, and O. A. Reig. 1992. "New Data on the Systematics and Caryology of *Podoxymys roraimae* (Rodentia, Cricetidae)." *Zeitschrift für Säugetierkunde* 57:216–24.

Picard, D., T. Sempere, and O. Plantard. 2008. "Direction and Timing of Uplift Propagation in the Peruvian Andes Deduced from Molecular Phylogenetics of Highland Biotaxa." *Earth and Planetary Science Letters* 271:326–36.

Ramos, V. A. 2008. "The Basement of the Central Andes: The Arequipa and Related Terranes." *Annual Review of Earth and Planetary Sciences* 36:289–324.

Redondo, R. A. F., L. P. S. Brina, R. F. Silva, A. D. Ditchfield, and F. R. Santos. 2008. "Molecular Systematics of the Genus *Artibeus* (Chiroptera: Phyllostomidae)." *Molecular Phylogenetics and Evolution* 49:44–58.

Reig, O. A. 1981. "Teoría del Origen y Desarrollo de la Fauna de Mamíferos de America del Sur." *Monographiae Naturae, Museo Municipal de Ciencias Naturales "Lorenzo Scaglia"* 1:1–162.

———. 1986. "Diversity Patterns and Differentiation of High Andean Rodents." In *High Altitude Tropical Biogeography*, edited by F. Vuilleumier and M. Monasterio, 404–39. New York: Oxford University Press.

Rheindt, F. E., L. Christidis, and J. A. Norman. 2008. "Habitat Shifts in the Evolutionary History of a Neotropical Flycatcher Lineage from Forest and Open Landscapes." *BMC Evolutionary Biology* 8:193.

Ribas, C. C., R. G. Moyle, C. Y. Miyaki, and J. Cracraft. 2007. "The Assembly of Montane Biotas: Linking Andean Tectonics and Climatic Oscillations to Independent Regimes of Diversification in *Pionus* Parrots." *Proceedings of the Royal Society B, Biological Sciences* 274:2399–408.

Roberts, J. L., J. L. Brown, R. von May, W. Arizabal, R. Schulte, and K. Summers. 2006. "Genetic Divergence and Speciation in Lowland and Montane Peruvian Poison Frogs." *Molecular Phylogenetics and Evolution* 41:149–64.

Santos, J. C., L. A. Coloma, K. Summers, J. P. Caldwell, R. Ree, and D. C. Cannatella. 2009. "Amazonian Amphibian Diversity is Primarily Derived from Late Miocene Andean Lineages." *PLoS Biology* 7:e1000056.

Soejima, A., J. Wen, M. Zapata, and M. O. Dillon. 2008. "Phylogeny and Putative Hybridization in the Subtribe Paranepheliinae (Liabeae, Asteraceae): Implications for Classification, Biogeography, and Andean Orogeny." *Journal of Systematics and Evolution* 46:375–90.

Solari, S. 2007. "New Species of *Monodelphis* (Didelphimorphia: Didelphidae) from Peru, with Notes on *M. adusta* (Thomas, 1897)." *Journal of Mammalogy* 88:319–29.

Solari, S., and R. J. Baker. 2006. "Mitochondrial DNA Sequence, Karyotypic, and Morphological Variation in the *Carollia castanea* Species Complex (Chiroptera: Phyllostomidae) with Description of a New Species." *Occasional Papers, Museum of Texas Tech University* 254:1–16.

Solari, S., S. R. Hoofer, P. A. Larsen, A. D. Brown, R. J. Bull, J. A. Guerrero, J. Ortega et al. 2009. "Operational Criteria for Genetically Defined Species: Analysis of the Diversification of the Small Fruit-Eating Bats, *Dermanura* (Phyllostomidae: Stenodermatinae)." *Acta Chiropterologica* 11:279–88.

Solari, S., V. Pacheco, L. Luna, P. M. Velazco, and B. D. Patterson. 2006. "Mammals of the Manu Biosphere Reserve." In *Mammals and Birds of the Manu Biosphere Reserve, Peru*, edited by B. D. Patterson, D. F. Stotz, and S. Solari, 13–22. Fieldiana: Zoology, n.s., 110. Chicago: Field Museum of Natural History.

Spotorno, A. E., J. Sufan-Catalan, and L. I. Walker. 1994. "Cytogenetic Diversity and Evolution of Andean Species of *Eligmodontia* (Rodentia, Muridae)." *Zeitschrift für Säugetierkunde* 59:299–308.

Spotorno, A. E., L. I. Walker, S. V. Flores, M. Yevenes, J. C. Marin, and C. Zuleta. 2001. "Evolution of Phyllotines (Rodentia, Muridae) in the Southern Andes." *Revista Chilena de Historia Natural* 74:151–66.

Stehli, F. G., and S. D. Webb, eds. 1985. *The Great American Biotic Interchange*. New York: Plenum Press.

Steiner-Souza, F., P. Cordeiro-Estrela, A. R. Percequillo, A. F. Testoni, and S. L. Althoff. 2008. "New Records of *Rhagomys rufescens* (Rodentia : Sigmodontinae) in the Atlantic Forest of Brazil." *Zootaxa* 1824:28–34.

Steppan, S. J., R. M. Adkins, and J. Anderson. 2004. "Phylogeny and Divergence-Date Estimates of Rapid Radiations in Muroid Rodents Based on Multiple Nuclear Genes." *Systematic Biology* 53:533–53.

Thomas, W. A., and R. A. Astini. 2007. "Vestiges of an Ordovician West-Vergent Thin-Skinned Ocloyic Thrust Belt in the Argentine Precordillera, Southern Central Andes." *Journal of Structural Geology* 29:1369–85.

Tribe, C. J. 1996. "The Neotropical Rodent Genus *Rhipidomys* (Cricetidae: Sigmodontinae): A Taxonomic Revision." Unpublished PhD diss. University College, London.

Vanzolini, P. E., and E. E. Williams. 1970. "South American Anoles: The Geographic Differentiation and Evolution of the *Anolis chrysolepis* Species Group (Sauria, Iguanidae)." *Arquivos de Zoologia (Estado do São Paulo)* 19:1–298.

Velazco, P. M., and B. D. Patterson. 2008. "Phylogenetics and Biogeography of the Broad-Nosed Bats, Genus *Platyrrhinus* (Chiroptera: Phyllostomidae)." *Molecular Phylogenetics and Evolution* 49:749–59.

Verzi, D. H. 2001. "Phylogenetic Position of *Abalosia* and the Evolution of the Extant Octodontinae (Rodentia, Caviomorpha, Octodontidae)." *Acta Theriologica* 46:243–68.

Villalpando, G., J. Vargas, and J. Salazar-Bravo. 2006. "First Record of *Rhagomys* (Mammalia: Sigmodontinae) in Bolivia." *Mastozoología Neotropical* 13:143–49.

Voss, R. S. 1988. "Systematics and Ecology of Ichthyomine Rodents (Muroidea): Patterns of Morphological Evolution in a Small Adaptive Radiation." *Bulletin of the American Museum of Natural History* 188:259–493.

———. 2003. "A New Species of *Thomasomys* (Rodentia: Muridae) from Eastern Ecuador, with Remarks on Mammalian Diversity and Biogeography in the Cordillera Oriental." *American Museum Novitates* 3421:1–47.

Voss, R. S., M. Gómez-Laverde, and V. Pacheco. 2002. "A New Genus for *Aepeomys fuscatus* Allen, 1912, and *Oryzomys intectus* Thomas, 1921: Enigmatic Murid Rodents from Andean Cloud Forests." *American Museum Novitates* 3373:1–42.

Voss, R. S., and S. A. Jansa. 2003. "Phylogenetic Studies on Didelphid Marsupials II. Nonmolecular Data and New IRBP Sequences: Separate and Combined Analyses of Didelphine Relationships with Denser Taxon Sampling." *Bulletin of the American Museum of Natural History* 276:1–82.

Voss, R. S., T. Tarifa, and E. Yensen. 2004. "An Introduction to *Marmosops* (Marsupialia: Didelphidae), with the Description of a New Species from Bolivia and Notes on the Taxonomy and Distribution of Other Bolivian Forms." *American Museum Novitates* 3466:1–40.

Wang, X. M., and O. Carranza-Castaneda. 2008. "Earliest Hog-Nosed Skunk, *Conepatus* (Mephitidae, Carnivora), from the Early Pliocene of Guanajuato, Mexico and Origin of South American Skunks." *Zoological Journal of the Linnean Society* 154:386–407.

Weir, J. T. 2006. "Divergent Timing and Patterns of Species Accumulation in Lowland and Highland Neotropical Birds." *Evolution* 60:842–55.

Weksler, M. 2006. "Phylogenetic Relationships of Oryzomine [*sic*] Rodents (Muroidea, Sigmodontinae): Separate and Combined Analyses of Morphological and Molecular Data." *Bulletin of the American Museum of Natural History* 296:1–149.

Woodman, N., C. A. Cuartas-Calle, and C. A. Delgado-V. 2003. "The Humerus of *Cryptotis colombiana* and Its Bearing on the Species' Phylogenetic Relationships (Soricomorpha: Soricidae)." *Journal of Mammalogy* 84:832–39.

Young, K. R. 1992. "Biogeography of the Montane Forest Zone of the Eastern Slopes of Peru." In *Biogeografía, Ecología y Conservación del Bosque Montano en el Perú*, edited by K. R. Young and N. Valencia, 119–54. Lima: Universidad Nacional Mayor de San Marcos, *Memorias del Museo de Historia Natural*.

Young, K. R., and N. Valencia, eds. 1992. *Biogeografía, Ecología y Conservación del Bosque Montano en el Perú*. Lima: *Memorias del Museo de Historia Natural*, 21. Universidad Nacional Mayor de San Marcos.

Zamora, H. T., and H. Zeballos. 2009. "Distribución de los Murciélagos de la Costa Desértica y Vertientes Occidentales de Perú." *Journadas, 4th Congreso Boliviano de Mastozoología*. Cochabamba, Bolivia: ABIMA.

16 Mammalian Biogeography of Patagonia and Tierra del Fuego

Enrique P. Lessa, Guillermo D'Elía,
and Ulyses F. J. Pardiñas

Abstract

The Patagonian-Fuegian region, located in southern South America roughly south of 40°S, comprises areas of Argentinean monte, Patagonian steppe and grasslands, and Valdivian temperate and Magellanic subpolar forests. Although the area was affected by the glacial cycles of the Neogene, glacial sheets were typically much more limited in South America than in northern continents. In this context, we review distributional, phylogenetic, phylogeographic, and population genetic information on the composition and historical biogeography of mammals in the region. Although many species are likely relatively recent colonizers of the region, distributional and phylogenetic data provide several examples of endemic species and others that likely resulted from local diversification. Phylogeographic analyses provide additional indications of differentiation within the region. Phylogeographic breaks divide species distributions by latitude rather than between major habitats. Population genetic analyses reveal several cases of demographic expansion, all of which can be assigned to the late Pleistocene (i.e., the last 500,000 years). However, very few of these can be attributed to events postdating the Last Glacial Maximum (~21,000 BP). Overall, the current mammalian fauna of Patagonia and Tierra del Fuego is the result of a complex mix of local fragmentation, differentiation, and colonization from lower latitudes.

16.1 Introduction

The Patagonian-Fuegian region comprises the southern end of South America, and is politically divided between Chile and Argentina. It is a large area with a small human population. The term Patagonia is broadly used to encompass continental areas roughly south of 40°S, although other uses (e.g., equating Patagonia to the Patagonian steppe) are also found in the literature. Tierra del Fuego is an archipelago that includes "Isla Grande" and surrounding islands. Three major biomes are represented in the region. The monte is found in northeastern Patagonia. The forested areas are associated with the southern Andes, and can be further subdivided into the Valdivian temperate rain forest and the Magellanic subpolar forest. Finally, the Patagonian steppe sensu lato, which includes a more mesic section located in southern Santa Cruz and Isla Grande, is often described as Patagonian grasslands. These habitats comprise most of the open continental areas east of the Andes, as well as the northeastern area of Isla Grande de Tierra del Fuego (fig. 16.1).

In this review, we consider the biogeography of terrestrial mammals roughly south of 40°S. However, because all three major biomes and many mammalian species extend into lower latitudes, we broaden our scope when necessary. Our focus is on extant species, but we first provide an overview of the geological and climatic history of the region.

16.2 Geographical and Historical Context

Besides Antarctica, South America is the only continent that extends significantly into latitudes beyond 40°S, thus providing a unique opportunity to examine the biogeography of high latitudes in the Southern Hemisphere (e.g., Oesterheld et al. 1998; Pardiñas et al. 2003; Soriano et al. 1983).

The transition from the last glacial maximum (LGM) ~21,000 BP to the current climate in the region has received much attention, and has been documented by geological (Rodbell, Smith, and Mark 2009) and, to a lesser extent, paleodistribution modeling of grass species (Jakob, Martinez-Meyer, and Blattner 2009).

The absolute chronology of the southern South American glaciations is one of the most studied in the world and probably the best documented for the Southern Hemisphere outside Antarctica. This is due, in part, to the ample application of absolute dating techniques (e.g., 40Ar/39Ar dating of volcanic rocks associated with glacial deposits; Rabassa et al. 2005). Using these methods, it was established that some events were in phase with those of the Northern Hemisphere (e.g., Heusser and Heusser 2006), while some others were not (e.g., Schaefer et al. 2006). The earliest registered glaciation occurred in the late Miocene-early Pliocene ~7–5 Ma. Subsequently, in the middle and late Pliocene, at least eight glaciations occurred. The largest Patagonian glaciation, known as the Great Patagonian Glaciation (GPG), occurred 1.2–1.0 Ma in the early Pleistocene. During the GPG, ice tongues reached the Atlantic in the continental area for the first time in the Cenozoic, south of the Río Gallegos (about 52°S, southern Santa Cruz province, Argentina; Clapperton 1993; Rabassa et al.

2000). After the GPG, there were 14–16 less intense glacial periods (see review in Rabassa et al. 2005).

The last glaciation reached its maximum, known as the Last Glacial Maximum (LGM), 25,000–17,000 BP, in the late Pleistocene. It is estimated that during the LGM the ice sheet covered large areas west and east of the Andes, with a total extension of about 480,000 km^2 and a volume of about 500,000 km^3, equivalent to 1.2 m of the water column of all oceans (Glasser et al. 2004; Hulton et al. 2002; McCulloch et al. 2000). The ice sheet extended from 38°–55°S, reaching the Pacific Ocean south of 43°S (Hulton et al. 2002). Large ice sheets centered on the southern Andes developed during glacial phases of Plio-Pleistocene glacial cycles (Clapperton 1993). However, glaciations in South America differed from those in the Northern Hemisphere in several respects. First, rather than covering extensive continental areas, they were generally much more restricted, usually limited to areas along the southern Andes (Rabassa 2008). In particular, the last glaciation did not extend much beyond the southern Andean foothills (fig. 16.1; Hulton et al. 2002; McCulloch et al. 2000). The Great Patagonian Glaciation (GPG) that reached the Atlantic coast in the Province of Santa Cruz, Argentina, ended ~1 Ma (Rabassa, Coronato, and Salemme 2005). Second, the South American continental shelf is extensive in the southern Atlantic, so that large, currently submerged areas were exposed during glacial phases (Patterson 2010, fig. 1). During glaciations subsequent to the GPG, the increases in land area made available due to lowered sea level may have exceeded the areas covered by ice sheets.

Compared to present climatic conditions in the Patagonian-Fuegian region, during the LGM ~21,000 years ago summer rains were reduced to a third, winters were 3–6°C cooler, and there was a wider range of temperature variation between seasons (Markgraf and McGlone 2005). The transition from the LGM to the region's current climate has received much attention, and has been documented by geological (Rodbell, Smith, and Mark 2009) and, to a lesser extent, paleodistribution modeling of selected species of grasses (Jakob, Martinez-Meyer, and Blattner 2009). Multiple lines of evidence (reviewed in McCulloch et al. 2000) indicate that deglaciation happened very quickly after ~17,000 BP, especially in the north, where contemporary conditions were quickly reached (Moreno 1997; Moreno et al. 1999). The paleoclimatic record indicates that areas east of the Andes at high and middle latitudes warmed uniformly (Glasser et al. 2004). Since then, conditions have remained relatively stable (Heusser and Streeter 1980; Rabassa and Clapperton 1990). However, deglaciation was

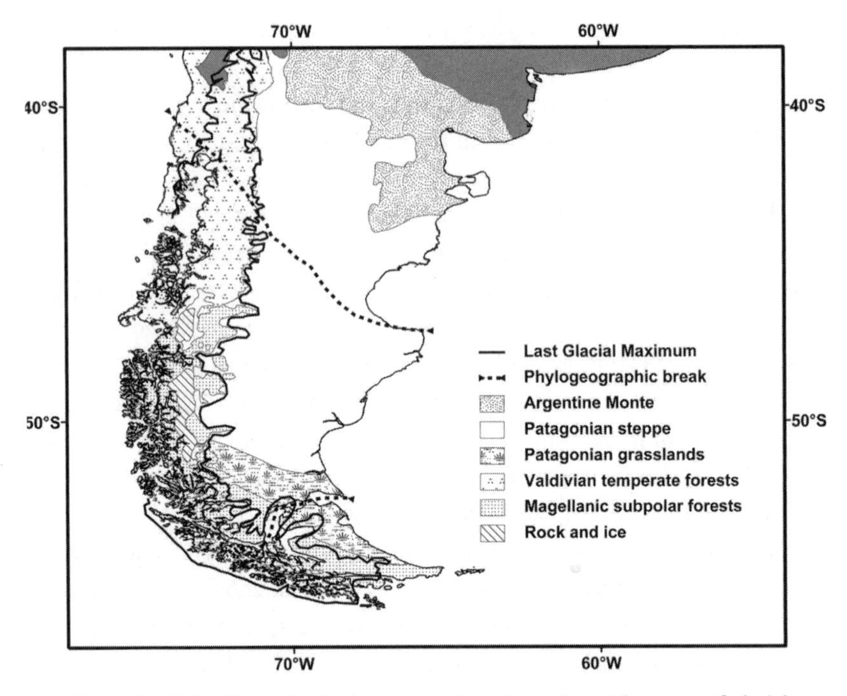

Figure 16.1 Major biomes in the Patagonian-Fueguian regions. The extent of glacial sheets at last glacial maximum (LGM) ~21,000 BP is outlined. Dotted lines represent the approximate location of phylogeographic breaks in some species of sigmodontine rodents (Lessa, D'Elía, and Pardiñas 2010).

interrupted twice by glacial advances of limited intensity (Rabassa, Coronato, and Salemme 2005). In the north they were of small magnitude (Ariztegui et al. 1997), while in the south they were of greater intensity, especially during the Antarctic Cold Reversal (Fogwill and Kubik 2005; McCulloch et al. 2000).

This geographical and historical scenario provides a few constraints on the possible evolution of the mammalian fauna of the Patagonian-Fuegian regions, but is generally insufficient to produce detailed, predictive hypotheses:

Effects of latitude: The most general, but least novel, expectation is for a decline in species numbers with increasing latitude, especially given that the continent progressively narrows.

Insularity: A drop in standing diversity is expected between the continent and Isla Grande de Tierra del Fuego; more generally, habitable area is expected to affect standing diversity. Much of the coastal archipelago (with the noteworthy exception of northeastern Isla Grande) was covered by ice at the

LGM, a factor that must have imposed additional restrictions on current species diversity.

Changes in habitat distributions, continuity, and availability: Because much of the current distribution of southern Andean forest was covered by ice during the last glaciation, species limited to or strongly associated with this habitat must have experienced the greatest distributional perturbations. Unfortunately, it is difficult at present to offer more detailed predictions. In the case of steppe, the limited available reconstructions of past distributions suggest much greater stability (Jakob, Martinez-Meyer, and Blattner 2009). So does the substantial local differentiation of *Liolaemus* lizards (Morando et al. 2007, and references therein), suggesting that steppe environments have remained relatively stable over time.

During glaciations, including the LGM, all pollen records south of 37°S, both west and east of the Andes, show substantial reductions of trees, in some cases their total absence. Extreme aridity is reported for southern South America during the LGM in those ice-free areas north of 36°S, south of the 43°S, and east of the Andes (Markgraf et al. 1992). Forest—mixed with steppe species—would have persisted during the LGM to the west of the Andes from 36°S to 43°S; however, these forests were less diverse than contemporary ones (Heusser et al. 1996; Moreno 1997). The palynological evidence indicates that these communities would be similar to current ecotonal zones of Argentinean Patagonia where forest, shrubs, and grasses mix (Markgraf et al. 2002). Records in these areas of large herbivorous mammals, including mastodons, horses, camelids, and sloths (Moreno et al. 1994; Nuñez, Grosjean, and Cartajena 2001), also indicate the existence of relatively open environments.

At the same time, the effect of Quaternary glaciations on populations of individual plant species have also been studied by analyses of genetic variation, mostly assessed in the form of random amplified polymorphic DNA and protein electrophoresis, evidence that in general is not assessed with an explicit historical approach. In addition, most of these studies have limited geographic coverage (but see Marchelli and Gallo 2006). Some of the most iconic tree species of the regions, such as *alerce* (*Fitzroya cupressoides*), monkey-puzzle tree (*Araucaria araucana*), *ciprés de las Guiatecas* (*Pilgerodendron uviferum*), Chilean cedar (*Austrocedrus chilensis*), and southern beeches (*Nothofagus* spp.), as well as a catsear (*Hypochaeris* sp.), have been studied with these approaches (Allnutt et al. 1999; Bekessy et al. 2002; Marchelli and Gallo 2004; Mathiasen and Premoli 2010; Muellner et al. 2005; Pastorino and Gallo 2002; Pastorino, Gallo, and Hattemer 2004; Premoli, Kitzberger, and Veblen 2000; Premoli et al. 2002). These studies indicate the existence of multiple refugia, both west and east of the Andes, including places that were not recognized as such from the palynological record. The nature of these refugia was variable—in some cases, they were located to the north of present-day distributions (e.g., *Austrocedrus chilensis*; Pastorino and Gallo 2002), but in others to the south (e.g., *Fitzroya cupressoides*; Premoli, Kitzberger, and Veblen 2000), and in still others to both the south and north (e.g., *Nothofagus nervosa*, Marchelli and Gallo 2004, 2006; *N. pumilio*, Mathiasen and Premoli 2010).

16.3 Species Richness, Endemicity, and Phylogenetic Relationships

Species richness tends to decrease from the tropics to the poles; mammals in general and those of the Patagonian-Fuegian region in particular are no exception (Osgood 1943; Schipper et al. 2008). Table 16.1 provides a summary of the numbers of terrestrial mammalian species found on three latitudinal segments of the region, in turn subdivided by major habitats. The overall pattern is well illustrated by mammals present in the steppe, decreasing from 40 in northern Patagonia to 37 in southern Patagonia (31 of which reach the Magellanic Straits), and dropping to 10 on Isla Grande. A more detailed latitudinal breakdown based on distributional records of sigmodontine rodents (table 16.2) indicates that the latitudinal decay in numbers of species is already considerable at the level of the Magellanic Straits, and that numbers are further reduced in Tierra del Fuego (Osgood 1943; Pardiñas, Udrizar Sauthier, and Teta 2009).

Endemics of the region include three marsupials (*Rhyncholestes raphanurus*, *Lestodelphys halli*, and *Dromiciops gliroides*, although the latter two extend further

Table 16.1 Number of terrestrial species of mammals in the Patagonian-Fueguian region, subdivided into three geographical segments—Northern Patagonia (40°–46°S), Southern Patagonia (46°S–Magellanic Straits), and Isla Grande de Tierra del Fuego—each subdivided into its major habitats.

	Andean forest	Steppe	Monte
Northern Patagonia	29	40	24
Southern Patagonia	22	37	
Isla Grande	11	10	

Table 16.2 Numbers of species of sigmodontine rodents found at several latitudes in Patagonia and Tierra del Fuego.

	Forest	Steppe	Monte	Total[a]
Patagonia				
40°–42° S	10	15	9	23
42°–46° S	8	14	8	18
46°–49° S	8	16		18
N margin, Magellanic Strait	7	6		11
Tierra del Fuego	4	4		7

[a] The total is not a marginal sum because a number of species are shared among major habitats.

north), several species of sigmodontine rodents mainly in the Abrotrichini (D'Elía et al. 2007) and in the genera *Irenomys* and *Euneomys* (D'Elía, Teta, and Pardiñas 2006), several species of *Ctenomys* (Parada 2007), the southern vizcacha (*Lagidium wolffsohni*), the pudu (*Pudu puda*), the zorro chilote (*Lycalopex fulvipes*), and the bats *Histiotus magellanicus*, *Lasiurus varius*, and *Myotis aelleni*. Some of these endemics are restricted to forest habitats, but others are found in or limited to the steppe.

Information about species' geographical distributions and phylogenies may be combined to make inferences about geographical patterns of speciation. For example, Cheviron, Capparella, and Vuilleumier (2005) used a phylogenetic approach to ask whether two species of Patagonian-Fuegian *Geositta* (Furnaridae) were sister species and, therefore, might represent a case of recent (e.g., Pleistocene) speciation within the region. The fact that these two birds were not sister to each other and were genetically divergent argued against that scenario and pointed to possible vicariant events between Patagonia-Tierra del Fuego, and other lower latitude areas. Another alternative is that the Patagonian-Fuegian species of interest colonized the region after having originated elsewhere. Indeed, colonization has been a major contributor to the current diversity of mammals at high latitudes in the Northern Hemisphere (Hewitt 2004; Waltari et al. 2007).

The dual requirement of sister relationships and recent divergence is a reasonable approach to assessing whether species pairs represent possible speciation events within the Patagonian-Fuegian region. However, the phylogenetic argument may be extended further by examination of clade endemicity and, more generally, the geographical distribution of species within and outside the focal region in a phylogeny (Wiens and Donoghue 2004). Thus, a case in which several species within a region may form a paraphyletic group relative to

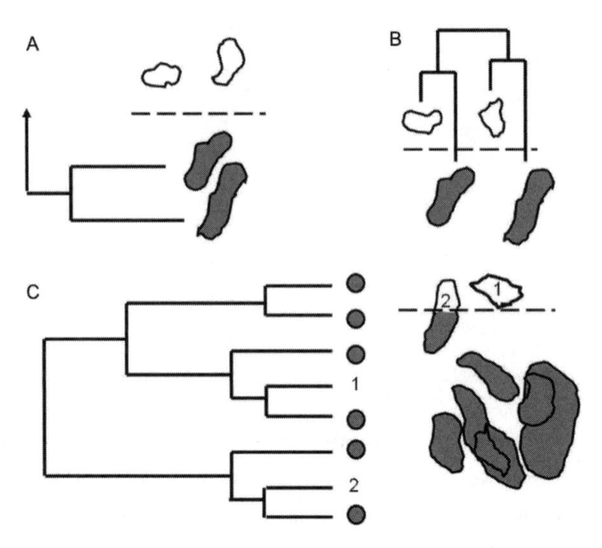

Figure 16.2 Selected cases of patterns of geographical and phylogenetic differentiation: (A) pairs of closely related, sister species are found within a biogeographical region, suggesting speciation within it; (B) each species of a genus within a biogeographical region has its sister species outside the region, suggesting either speciation or colonization across regions; (C) species within a biogeographical region form a paraphyletic group relative to one or a few species that extend beyond, or are exclusively found outside the region, suggesting that much of the diversification of the clade took place within the region of interest.

one or to a few relatives outside the region suggests that the latter are derived from emigrants of the former region (e.g., Stevens 2006). Figure 16.2 outlines the phylogenetic consequences (and tests) of some of these biogeographic scenarios.

Some mammals have their closest relatives outside Patagonia; this is the case of the predominantly Patagonian "monito del monte" (*Dromiciops gliroides*), which is most closely related to marsupials from the Australasian region (Kirsch et al. 1991, Nilsson et al., 2003). Many other species of the region have large distributions that extend well beyond the region (e.g., *Galictis cuja*, *Lama guanicoe*, *Oligoryzomys longicaudatus*, *Puma concolor*, *Tadarida brasiliensis*). For instance, the mouse *Calomys musculinus* is found mostly in the northern and eastern part of Patagonia but has a large distribution outside the region. As the remaining species of the genus are distributed outside Patagonia (Musser and Carleton 2005), and *C. musculinus* is nested within the genus (i.e., is not sister to a clade formed by the remaining species of the genus; Salazar-Bravo et al. 2001), it is safe to assume that this species originated outside Patagonia and secondarily invaded it.

Among rodents, species such as *Abrothrix olivaceus* are widespread in the region. *Abrothrix olivaceus* extends substantially into lower latitudes, particularly west of the Andes, and its sister species appears to be *A. andina*, which is found farther north (Feijoo et al. 2010; Palma et al. 2005). *Akodon iniscatus* and *Phyllotis xanthopygus* are Patagonian species that are nested well within their respective genera (Smith and Patton 2007; Steppan et al. 2007). Two species of silky mice (*Eligmodontia typus* and *E. morgani*) are widespread in Patagonia and extend to lower latitudes. Recent phylogenetic analyses, however, suggest that they are not sister to each other. Rather, each is sister to a different species in northern Argentina (Mares et al. 2008). In these cases, it is not possible to rule out a role of the Patagonian-Fuegian region in the differentiation of these species, but speciation within the region is, in the face of present evidence, unlikely.

In contrast with the cases just outlined, two clades appear to have fostered a significant part of their diversity within the Patagonian region. First, Parada (2007) identified a distinct clade of tuco-tucos (*Ctenomys*) that includes several Patagonian-Fuegian species: *C. magellanicus*, *C. sericeus*, *C. colburni*, *C. fodax*, *C. coyhaiquensis*, and *C. haigi*. Although species limits within this clade need to be clarified, the validity of several species is supported by the mitochondrial DNA data of Parada (2007), and his results strongly suggest several speciation events within the region. Second, the tribe Abrotrichini (D'Elía et al. 2007) offers evidence both of substantial intraregional divergence and of closely related sister species. For example, Feijoo et al. (2010) confirmed the distinction of *Abrothrix lanosa*, a species distributed in Tierra del Fuego and the southern tip of the continent, and its sister relationship to *A. longipilis*, which is widely distributed in Patagonia and neighboring areas in central Chile.

The long-clawed abrotrichines, currently assigned to the genera *Notiomys*, *Geoxus*, *Chelemys*, and *Pearsonomys*, are a clade with several adaptations to subterranean or semisubterranean life (Patterson, 1992; Pearson 1984; Rodríguez-Serrano, Palma, and Hernández 2008). *Notiomys* and *Pearsonomys* are restricted to Patagonia, whereas *Geoxus* and *Chelemys* extend farther north. The diversity of this group of long-clawed mice is not entirely understood (e.g., D'Elía et al. 2006), but available data suggests that much of its differentiation took place within Patagonia (Pardiñas et al. 2008; Feijoo et al. 2010; Lessa, D'Elía, and Pardiñas 2010).

In sum, the Patagonian-Fuegian region supports several endemic mammals associated with the Andean forest and the Patagonian steppe. Analyses of geographical and phylogenetic information suggest that some species have likely originated outside the region and secondarily invaded it, while others may have differentiated in Patagonia (e.g., Lessa, D'Elía, and Pardiñas 2010; Martínez et al. 2010; Parada 2007). Finally, there is convincing evidence of substantial in situ differentiation in two distinct groups of rodents, namely a Patagonian-Fuegian clade of tuco-tucos and the Abrotrichini (Feijoo et al. 2010; Lessa, D'Elía, and Pardiñas 2010; Parada 2007). Regrettably, the basic data required to assess the relative importance of these patterns (i.e., the number, limits, geographical distributions, and phylogenetic relationships of most Patagonian-Fuegian mammals) are not available in sufficient detail to adequately assess these patterns.

16.4 Intraspecific Variation: Phylogeography and Historical Demography

Three closely related developments have greatly increased our ability to make inferences about the geographical structure and demographic history of individual species: (1) direct sequencing of PCR-amplified segments of DNA, with a strong, but not exclusive, emphasis on the mitochondrial genome (reviewed by Avise 2004); (2) explicit genealogical approaches to the study of allelic variation in DNA sequences across geography (phylogeography, reviewed by Avise 2000; Hickerson et al. 2009); and (3) coalescent theory (reviewed by Wakeley 2007).

Within the framework of coalescent theory, it was quickly realized that the processes that generate genetic variation within populations are inherently highly variable as a result of both mutational and genealogical variation (Tajima 1983). Furthermore, patterns of variation are affected (sometimes in similar ways) both by demographic change and departures from neutrality (Tajima 1989a, 1989b). In principle, it is possible to extract substantial information about demographic history, as well as to distinguish the effects of demography from those of selection, by comparing and contrasting information from multiple, unlinked loci. If codistributed species show similar patterns (comparable phylogeographic breaks and similar indications of demographic change), inferences can be made about their shared history even with a single locus. Lessa, Cook, and Patton (2003) used this line of argument to uncover "genetic footprints" of demographic expansion shared by mammalian species of western North America, particularly at high latitudes. Phylogeographic studies of

European mammals have shown comparable paths of northward, postglacial colonization from glacial refugia at lower latitudes (Hewitt 2004).

Two important historical components of high-latitude species have been their retraction to refugia during glacial phases and subsequent postglacial recolonization associated with demographic expansion (Hewitt 2004; Lessa, Cook, and Patton 2003; Waltari et al. 2007). This raises the possibility that much of the current, intraspecific genetic variation within the Patagonian-Fuegian region results from its historical preservation in refugia and postglacial expansion into the current distribution of each species. In particular, the hypothesis of postglacial colonization from a single source, possibly located north of Patagonia, appears appealing. However, we have already reviewed evidence of intraregional differentiation and persistence of endemics at and above the species level, so alternative phylogeographic scenarios must be considered. We also provide an overview of available evidence, with an emphasis on data from sigmodontine mice.

The colilargo (*Oligoryzomys longicaudatus*) provides an example of a species widespread in the Patagonian-Fuegian region (including Isla Grande and Navarino in Tierra del Fuego) that lacks phylogeographic structure (Belmar-Lucero et al. 2009; Lessa, D'Elía, and Pardiñas 2010; Palma, Marquet, and Boric-Bargetto 2005). The following sigmodontine rodents from the Patagonian-Fuegian region are similarly characterized by a lack of phylogeographic subdivision in the region: *Akodon iniscatus*, *Calomys musculinus*, and *Graomys griseoflavus* from northern Patagonia, *Eligmodontia morgani*, *E. typus*, *Loxodontomys micropus*, *Phyllotis xanthopygus*, and *Chelemys macronyx* from continental areas in the region, and *Reithrodon auritus* from both the continent and Isla Grande (Lessa, D'Elía, and Pardiñas 2010). Increased sampling may reveal additional variation in some of these clades (e.g., Cañón et al. 2010).

Smith, Kelt, and Patton (2001) and Rodríguez-Serrano, Cancino, and Palma (2006) examined the phylogeography of *Abrothrix olivaceus* in Chile and neighboring areas of NW Patagonia in Argentina. A Patagonian mitochondrial DNA clade is differentiated from non-Patagonian Chilean areas to the north, and shows evidence of demographic expansion. Recent analyses (Lessa, D'Elía, and Pardiñas 2010) documented the presence of the Patagonian clade throughout the steppe in Argentina, thereby extending earlier results. However, a distinct clade was found on Isla Grande, which is as divergent from the Patagonian clade as either is from those in central and northern Chile. *Geoxus valdivianus*, *Abrothrix longipilis*, and *Euneomys chinchilloides* also are characterized by the presence of two phylogeographic clades, although in these cases the corresponding breaks separate northern from southern Patagonia (Lessa, D'Elía, and Pardiñas 2010).

In sum, sigmodontine mice provide numerous examples of species without phylogeographic subdivision in the Patagonian-Fuegian region, but also include four species that show breaks either within continental Patagonia or between the continent and Isla Grande (fig. 16.1). Interestingly, Himes, Gallardo, and Kenagy (2008) found substantial geographical fragmentation in the monito del monte (*Dromiciops gliroides*), a marsupial with a range restricted largely to the Valdivian forest that currently occupies areas that were both glaciated and unglaciated at the LGM. Remarkably, as with the sigmodontines, the breaks found in the monito del monte are somewhat latitudinally oriented, rather than following the Andes. In line with analyses of species relationships, these cases of positive documentation of phylogeographic structure indicate that the region has offered opportunities for geographical differentiation.

These assessments have several limitations that must be kept in mind. First, phylogeographic breaks represent an extreme form of geographic differentiation in the form of reciprocally monophyletic, allopatric, or parapatric mitochondrial clades. Examination of more subtle geographical differentiation is largely a pending issue in the region (but see Belmar-Lucero et al. 2009; Rodríguez-Serrano, Hernández, and Palma 2008) because it requires denser sampling and other types of analytical approaches. Second, informative as they are, mitochondrial DNA analyses must be supplemented with data on nuclear loci for independent corroboration of the suggested patterns. This is not uncommon at or above the species level (e.g., D'Elía 2003) but remains rare within species of the region (thus far, no published study has assessed nuclear DNA variation for a Patagonian small mammal). Finally, sampling of Patagonian-Fuegian species is still very limited, particularly in the steppe, in the numerous Chilean islands south of Chiloe, and in general at both high latitudes or elevations.

One significant finding of coalescent theory is that populations that have undergone demographic expansion deviate from the patterns of genetic variation resulting, under neutrality, from an equilibrium between mutation and drift (Tajima 1989b). Colonizing populations are also expected to carry less genetic variation than historically stable ones, but departures from equilibrial expectations offer greater power and are—with the caveat that neutrality is a requirement—less equivocal in their interpretation (Lessa, Cook, and Patton 2003). In the Northern Hemisphere, areas covered by major ice sheets during the LGM are reasonably assumed to have been colonized after the end of the last glaciations. As a result, evidence of demographic expansion is routinely tied to these relatively recent events (e.g., Lessa, Cook, and Patton 2003). This interpretation has been extended to cases in the Southern Hemisphere, including the Patagonian-Fuegian region (Belmar-Lucero et al. 2009; Smith, Kelt, and Patton 2001).

However, the timing of proposed demographic expansions should be addressed explicitly. There are already assessments of demographic events in Patagonian fishes that can be tied to the late Pleistocene but are likely older than the LGM (Ruzzante et al. 2008). We have recently addressed this issue in Patagonian-Fuegian sigmodontines by obtaining species-specific estimates of rates of evolution on the basis of relaxed molecular clocks (Drummond et al. 2006), and applying those rates to convert relative time, estimated using a model of spatial expansion (Excoffier 2004), to absolute time (Lessa, D'Elía, and Pardiñas 2010). Ten of the seventeen units of analysis (entire species or phylogeographic units) we considered showed evidence of demographic expansion. All of these events may be assigned to the Late Quaternary, but few can be reasonably established as having occurred subsequent to the LGM (table 16.3). With the caveats that these analyses assume strict neutrality, that estimates based on a single locus have large errors, and that the models of expansion used are relatively simple, these results suggest, minimally, that episodes predating the last glacial have left an historical imprint in the genetic makeup of Patagonian sigmodontines. The fact that some species or clades in the region do not show evidence of demographic expansion indicates a mixed history, including cases

Table 16.3 Estimates of times of expansion (T, in years BP) and corresponding lower and upper limits of the 95% confidence intervals of species (or phylogeographic clades within species) that have positive and significant estimates of expansion (from Lessa, D'Elía, and Pardiñas 2010).

Species/ phylogeographic unit	Area	Lower 95% CI	T (BP)	Upper 95% CI
Calomys musculinus	N. Patagonia	24,728	55,160	106,533
Graomys griseoflavus	N. Patagonia	26,869	64,739	92,091
Phyllotis xanthopygus	N. Patagonia	33,620	74,073	100,031
Chelemys macronyx	Patagonia	47,923	97,851	140,492
Eligmodontia typus	Patagonia	85,903	130,274	188,538
Abrothrix longipilis	S. Patagonia	78,709	153,224	211,812
A. olivaceus	Patagonia	118,848	165,163	216,330
A. olivaceus	Isla Grande	13,057	43,170	119,785
Oligoryzomys longicaudatus	Patagonia–Isla Grande	85,177	117,715	180,582
Reithrodon auritus	Patagonia–Isla Grande	209,645	343,304	552,788

in which stability and geographical structure has persisted through the last glacial cycles. This is not surprising, given the limited extent of late Neogene glaciations relative to those in the Northern Hemisphere. However, it reinforces the notion, already introduced in the discussion of species-level analyses, that the Patagonian-Fuegian region has been an area generating biological diversity, in addition to being a recipient of diversity generated elsewhere.

Detailed studies based on larger sample sizes and geographical coverage, as well as more than one locus per species, are required to refine these assessments, separate demographic effects from departures from neutrality, and, especially, reveal areas of population persistence.

16.5 Conclusions

Our current understanding of the historical biogeography of Patagonia and Tierra del Fuego is strongly conditioned by poor—and geographically biased—sampling of the mammalian fauna, and by limited data. However, in recent decades there has been a growing understanding of the levels of endemicity of the area, based on better systematic assessments and new discoveries. Although hampered by the same limitations, phylogenetic, phylogeographic, and population-genetic analyses have contributed important data that suggest that the Patagonian-Fuegian region has a biogeographical history that combines local differentiation (both within the region and relative to areas at lower latitudes) and colonization from more temperate areas. Ongoing and future efforts to document and analyze mammalian diversity will continue to contribute to our understanding of the biogeography of Patagonia and Tierra del Fuego.

Acknowledgments

We are grateful to Cecilia Da Silva for help in preparation of the manuscript, and to Leonora Costa and Bruce Patterson for encouraging us to produce this review. Douglas Kelt and two anonymous reviewers provided detailed comments that helped improve the manuscript substantially. Financial support was provided by Grant 7813-05 of the National Geographic Society, CSIC-Universidad de la República and PEDECIBA, Uruguay, CONICET PIP 6179, Agencia PICT 32405 and PICT 2008-0547, Argentina, and FONDECYT 1110737, Chile.

Literature Cited

Allnut, T. R., A. C. Newton, A. Lara, A. C. Premoli, J. J. Armesto, S. Vergara, and M. Gardner. 1999. "Genetic Variation in *Fitzroya cupressoides* (Alerce), a Threatened South American Conifer." *Molecular Biology* 8:975–87.

Ariztegui, D., M. M. Bianchi, J. Masaferro, E. LaFargue, and F. Niessan. 1997. "Interhemispheric Synchrony of Late-Glacial Climatic Instability as Recorded in Proglacial Lake Mascardi, Argentina." *Journal of Quaternary Science* 12:133–38.

Avise, J. C. 2000. *Phylogeography. The History and Formation of Species*. Cambridge: Harvard University Press.

———. 2004. *Molecular Markers, Natural History and Evolution*, 2nd ed. Sunderland, MA: Sinauer.

Bekessy, S. A., T. R. Allnutt, A. C. Premoli, A. Lara, R. A. Ennos, M. A. Burgman, M. Cortes, and A. C. Newton. 2002. "Genetic Variation in the Monkey Puzzle Tree (*Araucaria araucana* (Molina) K. Koch), Detected Using RAPDs." *Heredity* 88:243–49.

Belmar-Lucero, S., P. Godoy, M. Ferrés, P. Vial, and R. E. Palma. 2009. "Range Expansion of *Oligoryzomys longicaudatus* (Rodentia, Sigmodontinae) in Patagonian Chile, and First Record of Hantavirus in the Region." *Revista Chilena de Historia Natural* 82:265–75.

Cañón C., G. D'Elía, U. F. J. Pardiñas, and E. P. Lessa. 2010. "Phylogeography of *Loxodontomys micropus* with Comments on the Alpha Taxonomy of *Loxodontomys* (Cricetidae: Sigmodontinae)." *Journal of Mammalogy* 91:1449–58.

Cheviron, Z. A., A. P. Capparella, and F. Vuilleumier. 2005. "Molecular Phylogenetic Relationships Among the *Geositta* Miners (Furnariidae) and Biogeographic Implications for Avian Speciation in Fuego-Patagonia." *Auk* 122:158–74.

Clapperton, C. 1993. *Quaternary Geology and Geomorphology of South America*. Amsterdam: Elsevier.

D'Elía, G. 2003. "Phylogenetics of Sigmodontinae (Rodentia, Muroidea, Cricetidae), with Special Reference to the Akodont Group, and with Additional Comments on Historical Biogeography." *Cladistics* 19:307–23.

D'Elía, G., R. A. Ojeda, F. Mondaca, and M. H. Gallardo. 2006. "New Data of the Long-Clawed Mouse *Pearsonomys annectens* (Cricetidae, Sigmodontinae) and Additional Comments on the Distinctiveness of *Pearsonomys*." *Mammalian Biology* 71:39–51.

D'Elía, G., U. F. J. Pardiñas, P. Teta, and J. L. Patton. 2007. "Definition and Diagnosis of a New Tribe of Sigmodontine Rodents (Cricetidae: Sigmodontinae), and a Revised Classification of the Subfamily." *Gayana* 71:187–94.

D'Elía, G., P. Teta, and U. F. J. Pardiñas. 2006. "Sigmodontinae *incertae sedis*." In *Mamíferos de Argentina: Sistemática y Distribución*, edited by R. Barquez, M. Díaz, and R. Ojeda, 197–202. Mendoza: Sociedad Argentina para el Estudio de los Mamíferos.

Drummond, A. J., S. Ho, M. Phillips, and A. Rambaut. 2006. "Relaxed Phylogenetics and Dating with Confidence." *PLoS Biology* 4 (5): e88.

Excoffier, L. 2004. "Patterns of DNA Sequence Diversity and Genetic Structure After a Range Expansion: Lessons from the Infinite-Island Model." *Molecular Ecology* 13:853–64.

Feijoo, M., G. D'Elía, U. F. J. Pardiñas, and E. P. Lessa. 2010. "Systematics of the Southern Patagonian-Fueguian Endemic *Abrothrix lanosus* (Rodentia: Sigmodontinae): Phylogenetic Position, Karyotypic and Morphological Data." *Mammalian Biology* 75:122–37.

Fogwill, C. J., and P. W. Kubik. 2005. "A Glacial Stage Spanning the Antarctic Cold Reversal in Torres del Paine (51°S), Chile, Based on Preliminary Cosmogenic Exposure Ages." *Geografiska Annaler* 87:403–08.

Glasser, N. F., S. Harrison, V. Winchester, and M. Aniya. 2004. "Late Pleistocene and Holocene Palaeoclimate and Glacier Fluctuations in Patagonia." *Global and Planetary Change* 43:79–101.

Heusser, C. J., and L. Heusser. 2006. "Submillennial Palynology and Palaeoecology of the Last Glaciation at Taiquemó (50,000 cal yr, MIS 2–4) in Southern Chile." *Quaternary Science Reviews* 25:446–54.

Heusser, C. J., T. V. Lowell, L. E. Heusser, A. Hauser, B. G. Andersen, and G. H. Denton. 1996. "Full-Glacial–Late-Glacial Palaeoclimate of the Southern Andes: Evidence from Pollen, Beetle, and Glacial Records." *Journal of Quaternary Science* 11:173–84.

Heusser, C. J., and S. S. Streeter. 1980. "A Temperature and Precipitation Record of the Past 16,000 Years in Southern Chile." *Science* 210:1345–47.

Hewitt, G. M. 2004. "Genetic Consequences of Climatic Oscillations in the Quaternary." *Philosophical Transactions of the Royal Society of London, B* 359:183–95.

Hickerson, M. J., B. C. Carstens, J. Cavender-Bares, K. A. Crandall, C. H. Graham, J. B. Johnson, L. Rissler et al. 2009. "Phylogeography's Past, Present, and Future: 10 Years After Avise, 2000." *Molecular Phylogenetics and Evolution* 54:291–301.

Himes, C. M. T., M. H. Gallardo, and G. J. Kenagy. 2008. "Historical Biogeography and Post-Glacial Recolonization of South American Temperate Rain Forest by the Relictual Marsupial *Dromiciops gliroides*." *Journal of Biogeography* 35:1415–24.

Hulton, N. R. J., R. S. Purves, R. D. McCulloch, D. E. Sugden, and M. J. Bentley. 2002. "The Last Glacial Maximum and Deglaciation in Southern South America." *Quaternary Science Reviews* 21:233–41.

Jakob, S. S., E. Martinez-Meyer, and F. R. Blattner. 2009. "Phylogeographic Analyses and Paleodistribution Modeling Indicate Pleistocene In Situ Survival of *Hordeum* Species (Poaceae) in Southern Patagonia Without Genetic or Spatial Restriction." *Molecular Biology and Evolution* 26:907–23.

Lessa, E. P., J. A. Cook, and J. L. Patton. 2003. "Genetic Footprints of Demographic Expansion in North America, but not Amazonia, During the Late Quaternary." *Proceedings of the National Academy of Sciences, USA* 100:10331–34.

Lessa, E. P., G. D'Elía, and U. F. J. Pardiñas. 2010. "Genetic Footprints of Late Quaternary Climate Change in the Diversity of Patagonian-Fueguian Rodents." *Molecular Ecology* 19:3031–37.

Kirsch J. A., A. W. Dickerman, O. A. Reig, and M. S. Springer. 1991. "DNA Hybridization Evidence for the Australasian Affinity of the American Marsupial *Dromiciops australis*." *Proceedings of the National Academy of Sciences, USA* 88:10465–69.

Marchelli, P., and L. A. Gallo. 2004. "The Combined Role of Glaciation and Hybridization in Shaping the Distribution of Genetic Variation in a Patagonian Southern Beech." *Journal of Biogeography* 31:451–60.

———. 2006. "Multiple Ice-Age Refugia in a Southern Beech of South America As Evidenced by Chloroplast DNA Markers." *Conservation Genetics* 7:591–603.

Mares, M. A., J. K. Braun, B. S. Coyner, and R. A. Van Den Bussche. 2008. "Phylogenetic and Biogeographic Relationships of Gerbil Mice *Eligmodontia* (Rodentia, Cricetidae) in South America, with a Description of a New Species." *Zootaxa* 1753:1–33.

Markgraf, V., J. R. Dodson, A. P. Kershaw, M. S. McGlone, and N. Nicholls. 1992. "Evolution of Late Pleistocene and Holocene Climates in the Circum-South Pacific Land Areas." *Climate Dynamics* 6:193–211.

Markgraf, V., and M. McGlone. 2005. "Southern Temperate Ecosystem Responses." In *Climate Change and Biodiversity*, edited by T. E. Lovejoy and L. Hannah, 142–56. New Haven: Yale University Press.

Markgraf, V., R. S. Webb, K. H. Anderson, and L. Anderson. 2002. "Modern Pollen/Climate Calibration for Southern South America." *Palaeogeography, Palaeoclimatology, Palaeoecology* 182:375–97.

Martínez, J. J., R. E. González, G. R. Theiler, R. Ojeda, C. Lanzone, A. Ojeda, and C. N. Gardenal. 2010. "Patterns of Speciation in Two Sibling Species of *Graomys* (Rodentia, Cricetidae) Based on mtDNA Sequences." *Journal of Zoological Systematics and Evolutionary Research* 8:159–66.

Mathiasen, P., and A. Premoli. 2010. "Out in the Cold: Genetic Variation of *Nothofagus pumilio* (Nothofagaceae) Provides Evidence for Latitudinally Distinct Evolutionary Histories in Austral South America." *Molecular Ecology* 19:371–85.

McCulloch, R. D., M. J. Bentley, R. S. Purves, N. R. J. Hulton, D. E. Sugden, and C. M. Clapperton. 2000. "Climatic Inferences from Glacial and Palaeoecological Evidence at the Last Glacial Termination, Southern South America." *Journal of Quaternary Science* 15:409–17.

Morando, M., L. J. Avila, C. R. Turner, and J. W. Sites Jr. 2007. "Molecular Evidence for a Species Complex in the Patagonian Lizard *Liolaemus bibronii* and Phylogeography of the Closely Related *Liolaemus gracilis* (Squamata: Liolaemini)." *Molecular Phylogenetics and Evolution* 43:952–73.

Moreno, P. I. 1997. "Vegetation and Climate Near Lago Llanquihue in the Chilean Lake District between 20,200 and 9500 14C yr BP." *Journal of Quaternary Science* 121: 485–500.

Moreno, P. I., C. Villagrán, P. A. Marquet, and L. G. Marshall. 1994. "Quaternary Paleobiogeography of Northern and Central Chile." *Revista Chilena de Historia Natural* 67: 487–502.

Moreno, P. I., C. Villagrán, P. A. Marquet, and L. G. Marshall. 1994. "Quaternary Paleobiogeography of Northern and Central Chile." *Revista Chilena de Historia Natural* 67: 487–502.

Muellner, A. N., K. Tremetsberger, T. Stuessy, and C. M. Baeza. 2005. "Pleistocene Refugia and Recolonization Routes in the Southern Andes: Insights from *Hypochaeris palustris* (Asteraceae, Lactuceae)." *Molecular Ecology* 14:203–12.

Musser, G. M., and M. D. Carleton. 2005. "Superfamily Muroidea." In *Mammal Species of the World: A Taxonomic and Geographic Reference*, 3rd ed., edited by D. E. Wilson and D. M. Reeder, 984–1531. Baltimore: Johns Hopkins University Press.

Nilsson, M. A., A. Gullberg, A. E. Spotorno, U. Arnason, and A. Janke. 2003. "Radiation of Extant Marsupials After the K/T Boundary: Evidence from Complete Mitochondrial Genomes." *Journal of Molecular Evolution* 57:S3–S12.

Nuñez, L., M. Grosjean, and I. Cartajena. 2001. "Human Dimensions of Late Pleistocene/

Holocene Arid Events in Southern South America." In *Interhemispheric Climate Linkages*, edited by V. Markgraf, 105–17. San Diego: Academic Press.

Oesterheld, M., M. R. Aguiar, and J. M. Paruelo, eds. 1998. "Ecosistemas Patagónicos." *Ecología Austral* 8:75–308.

Osgood, W. H. 1943. "The Mammals of Chile." *Field Museum of Natural History, Zoological Series* 30:1–268.

Palma, R. E., P. A. Marquet, and D. Boric-Bargetto. 2005. "Inter- and Intraspecific Phylogeography of Small Mammals in the Atacama Desert and Adjacent Areas of Northern Chile." *Journal of Biogeography* 32:1931–41.

Palma, R. E., E. Rivera-Milla, J. Salazar-Bravo, F. Torres-Perez, U. F. J. Pardiñas, P. A. Marquet, A. E. Spotorno et al. 2005. "Phylogeography of *Oligoryzomys longicaudatus* (Rodentia: Sigmodontinae) in Temperate South America." *Journal of Mammalogy* 86:191–200.

Parada, A. 2007. "Sistemática Molecular de *Ctenomys* (Rodentia, Ctenomyidae): Límites y Grupos de Especies Abordados con un Muestreo Taxonómico y Geográfico Denso." Unpublished MSc thesis. Universidad de la República, Montevideo, Uruguay.

Pardiñas, U. F. J., P. Teta, S. Cirignoli, and D. Podesta. 2003. "Micromamíferos (Didelphimorphia y Rodentia) de norpatagonia extra andina, Argentina: Taxonomía Alfa y Biogeografía." *Mastozoología Neotropical* 10:69–113.

Pardiñas, U. F. J., D. E. Udrizar Sauthier, and P. Teta. 2009. "Roedores del Extremo Sudoriental Continental de Argentina." *Mastozoología Neotropical* 16:471–73.

Pardiñas, U. F. J., D. E. Udrizar Sauthier, P. Teta, and G. D'Elía. 2008. "New Data on the Endemic Patagonian Long-Clawed Mouse *Notiomys edwardsii* (Rodentia: Cricetidae)." *Mammalia* 72:273–85.

Pastorino, M. J., and L. A. Gallo. 2002. "Quaternary Evolutionary History of *Austrocedrus chilensis*, a Cypress Native to the Andean-Patagonian Forest." *Journal of Biogeography* 29:1167–78.

Pastorino, M. J., L. A. Gallo, and H. H. Hattemer. 2004. "Genetic Variation in Natural Populations of *Austrocedrus chilensis*, a Cypress of the Andean-Patagonian Forest." *Biochemical Systematics and Ecology* 32:993–1008.

Patterson, B. D. 1992. "A New Genus and Species of Long-Clawed Mouse (Rodentia: Muridae) from Temperate Rainforests of Chile." *Zoological Journal of the Linnean Society* 106:127–45.

———. 2010. "Climate Change and Faunal Dynamics in the Uttermost Part of the Earth." *Molecular Ecology* 19:3019–21.

Pearson, O. P. 1984. "Taxonomy and Natural History of Some Fossorial Rodents of Patagonia, Southern Argentina." *Journal of Zoology (London)* 202:225–37.

Premoli, A. C., T. Kitzberger, and T. T. Veblen. 2000. "Isozyme Variation and Recent Biogeographical History of the Long-Lived Conifer *Fitzroya cupressoides*." *Journal of Biogeography* 27:251–60.

Premoli, A. C., C. Souto, A. Rovere, T. T. Aunutt, and A. C. Newton. 2002. "Patterns of Isozyme Variation As Indicators of Biogeographic History in *Pilgerodendron uviferum* (D. Don) Florin." *Diversity and Distributions* 8:57–66.

Rabassa, J., ed. 2008. *The Late Cenozoic of Patagonia and Tierra del Fuego*. Amsterdam: Elsevier, *Developments in Quaternary Science* 11.

Rabassa, J., and C. M. Clapperton. 1990. "Quaternary Glaciations of the Southern Andes." *Quaternary Science Reviews* 9:153–74.

Rabassa, J., A. Coronato, G. Bujalesky, M. Salemme, C. Roig, A. Meglioli, C. Heusser, et al. 2000. "Quaternary of Tierra del Fuego, Southernmost South America: An Updated Review." *Quaternary International* 68:217–40.

Rabassa, J., A. M. Coronato, and M. Salemme. 2005. "Chronology of the Late Cenozoic Patagonian Glaciations and Their Correlation with Biostratigraphic Units of the Pampean Region (Argentina)." *Journal of South American Earth Sciences* 20:81–103.

Rodbell, D. T., J. A. Smith, and B. G. Mark. 2009. "Glaciation in the Andes During the Late Glacial and Holocene." *Quaternary Science Reviews* 28:2165–212.

Rodríguez-Serrano, E., R. A. Cancino, and R. E. Palma. 2006. "Molecular Phylogeography of *Abrothrix olivaceus* (Rodentia: Sigmodontinae) in Chile." *Journal of Mammalogy* 87:971–80.

Rodríguez-Serrano, E., C. E. Hernández, and R. E. Palma. 2008. "A New Record and an Evaluation of the Phylogenetic Relationships of *Abrothrix olivaceus markhami* (Rodentia: Sigmodontinae)." *Mammalian Biology* 73:309–17.

Rodríguez-Serrano, E., R. E. Palma, and C. E. Hernández. 2008. "The Evolution of Ecomorphological Traits within the Abrothrichini (Rodentia: Sigmodontinae): A Bayesian Phylogenetics Approach." *Molecular Phylogenetics and Evolution* 48:473–80.

Ruzzante, D. E., S. J. Walde, J. C. Gosse, and V. E. Cussac. 2008. "Climate Control on Ancestral Population Dynamics: Insight from Patagonian Fish Phylogeography." *Molecular Ecology* 17:2234–44.

Salazar-Bravo, J., J. W. Dragoo, D. S. Tinnin, and T. L. Yates. 2001. "Phylogeny and Evolution of the Neotropical Rodent Genus *Calomys*: Inferences from Mitochondrial DNA Sequence Data." *Molecular Phylogenetics and Evolution* 20:173–84.

Schaefer, J. M., G. H. Denton, D. J. A. Barrell, S. Ivy-Ochs, P. W. Kubik, B. G. Andersen, F. M. Phillips et al. 2006. "Near-Synchronous Interhemispheric Termination of the Last Glacial Maximum in Mid-Latitudes." *Science* 312:1510–13.

Schipper, J., J. S. Chanson, F. Chiozza, N. A. Cox, M. Hoffmann, V. Katariya, J. Lamoreux et al. 2008. "The Status of the World's Land and Marine Mammals: Diversity, Threat, and Knowledge." *Science* 322:225–30.

Smith, M. F., D. A. Kelt, and J. L. Patton. 2001. "Testing Models of Diversification in Mice in the *Abrothrix olivaceus/xanthorhinus* Complex in Chile and Argentina." *Molecular Ecology* 10:397–405.

Smith, M. F., and J. L. Patton. 2007. "Molecular Phylogenetics and Diversification of South American Grass Mice, Genus *Akodon*." In *The Quintessential Naturalist: Honoring the Life and Legacy of Oliver P. Pearson*, edited by D. A. Kelt, E. P. Lessa, J. A. Salazar-Bravo, and J. L. Patton, 927–59. Berkeley: University of California Press, *University of California Publications in Zoology* 134.

Soriano, A., W. Volkheimer, H. Walter, O. Box, A. Marcolin, J. M. Vallerini, C. P. Movia et al. 1983. "Deserts and Semi-Deserts of Patagonia." In *Temperate Deserts and Semi-Deserts*, edited by N. E. West, 423–60. Amsterdam: Elsevier.

Steppan, S. J., O. Ramirez, J. Banbury, D. Huchon, V. Pacheco, L. Walter, and A. O. Spotorno. 2007. "A Molecular Reappraisal of the Systematics of the Leaf-Eared Mice

Phyllotis and Their Relatives." In *The Quintessential Naturalist: Honoring the Life and Legacy of Oliver P. Pearson*, edited by D. A. Kelt, E. P. Lessa, J. A. Salazar-Bravo, and J. L. Patton, 799–826. Berkeley: University of California Press, *University of California Publications in Zoology* 134.

Stevens, R. D. 2006. "Historical Processes Enhance Patterns of Diversity Along Latitudinal Gradients." *Proceedings of the Royal Society of London, B, Biological Sciences* 273:2283–89.

Tajima, F. 1983. "Evolutionary Relationship of DNA Sequences in Finite Populations." *Genetics* 105:437–60.

———. 1989a. "Statistical Method for Testing the Neutral Mutation Hypothesis by DNA Polymorphism." *Genetics* 123:585–95.

———. 1989b. "The Effect of Change in Population Size on DNA Polymorphism." *Genetics* 123:597–601.

Wakeley, J. 2007. *Coalescent Theory: An Introduction*. Greenwood Village: Roberts and Co.

Waltari, E., E. P. Hoberg, E. P. Lessa, and J. A. Cook. 2007. "Eastward Ho: Phylogeographical Perspectives on Colonization of Hosts and Parasites Across the Beringian Nexus." *Journal of Biogeography* 34:561–74.

Wiens, J. J., and M. J. Donoghue. 2004. "Historical Biogeography, Ecology, and Species Richness." *Trends in Ecology and Evolution* 19:634–44.

Contributors

Ana Laura Almendra
Department of Biology
M. L. Bean Life Science Museum
Brigham Young University
Provo, UT 84602 US

M. Susana Bargo
División Paleontología Vertebrados
Museo de La Plata
B1900FWA La Plata, Argentina
& CIC

Cibele R. Bonvicino
Programa de Genética
Instituto Nacional de Câncer
Rio de Janeiro, RJ, Brazil

Ana Paula Carmignotto
Universidade Federal de São Carlos
Km 110, Bairro Itinga, 18052-780
Sorocaba, SP, Brazil

Guillermo H. Cassini
División Paleontología Vertebrados
Museo de La Plata
Paseo del Bosque s/n
B1900FWA La Plata, Argentina
& ANPCyT, Argentina

Reynaldo Charrier
Departamento de Geología
Universidad de Chile
Casilla 13518, Correo 21
Santiago, Chile

Laura Chornogubsky
Sección Paleontología Vertebrados
Museo Argentino de Ciencias Naturales
"Bernardino Rivadavia"
Buenos Aires, Argentina

Leonora P. Costa
Departamento de Ciências Biológicas
Universidade Federal do Espírito Santo
Avenida Marechal Campos 1468
Maruípe 29043-900 Vitória, ES, Brazil

Darin A. Croft
School of Medicine
Department of Anatomy
Case Western Reserve University
Cleveland, OH 44106 US

Liliana M. Dávalos
Department of Ecology & Evolution
SUNY Stony Brook
Stony Brook, NY 11794 US

Guillermo D'Elía
Instituto de Ciencias Ambientales y Evolutivas
Universidad Austral de Chile
casilla 567 Valdivia, Chile

Eduardo Eizirik
Faculdade de Biociências, PUCRS
Av. Ipiranga 6681, prédio 12
Porto Alegre, RS 90619-900
& Instituto Pró-Carnívoros
Av. Horácio Neto 1030
Atibaia, SP 12945-010, Brazil

John J. Flynn
Division of Paleontology and
Richard Gilder Graduate School
American Museum of Natural History
New York, NY 10024 US

Javier N. Gelfo
División Paleontología Vertebrados
Museo de La Plata
Paseo del Bosque s/n (B1900FWA)
La Plata, Argentina
& CONICET, Argentina

Francisco J. Goin
División Paleontología Vertebrados
Museo de La Plata
Paseo del Bosque s/n (B1900FWA)
La Plata, Argentina
& CONICET, Argentina

Alfredo Langguth
Departamento de Sistemática e Ecologia
Universidade Federal da Paraíba
João Pessoa, PE, Brazil

Yuri L. R. Leite
Departamento de Ciências Biológicas
Universidade Federal do Espírito Santo
Avenida Marechal Campos 1468
Maruípe 29043-900, Vitória, ES, Brazil

Enrique P. Lessa
Laboratorio de Evolución
Facultad de Ciencias
Universidad de la República
Iguá 4225
Montevideo 11400 Uruguay

Burton K. Lim
Department of Natural History
Royal Ontario Museum
100 Queen's Park
Ontario M5S 2C6 Canada

Thomas Martin
Steinmann-Institut für Geologie
Mineralogie und Paläontologie
Universität Bonn, Nussalle 8
53115 Bonn, Germany

Ulyses F. J. Pardiñas
Unidad de Investigación Diversidad
Sistemática y Evolución
Centro Nacional Patagónico
Puerto Madryn, Chubut, Argentina

Bruce D. Patterson
Department of Zoology
Field Museum of Natural History
1400 S. Lake Shore Drive
Chicago, IL 60605 US

Francisco J. Prevosti
Museo Argentino de Ciencias Naturales
"Bernardino Rivadavia"—CONICET
C1405DJR Buenos Aires, Argentina

Duke S. Rogers
Department of Biology
M. L. Bean Life Science Museum
Brigham Young University
Provo, UT 84602 US

Leopoldo H. Soibelzon
División Paleontología de Vertebrados
Museo de La Plata
Paseo del Bosque s/n
La Plata, Buenos Aires, Argentina

Sergio Solari
Instituto de Biología
Universidad de Antioquia
Calle 67 No. 53-108, A.A. 1226
Medellín, Colombia

Néstor Toledo
División Paleontología Vertebrados Museo de La Plata
Paseo del Bosque s/n
B1900FWA La Plata, Argentina
& CONICET

Samuel T. Turvey
Institute of Zoology
Zoological Society of London
Regent's Park
London, NW1 4RY UK

Paúl M. Velazco
Department of Mammalogy
American Museum of Natural History
Central Park West at 79th St.
New York, NY 10024 US

Mario de Vivo
Museu de Zoologia,
Universidade de São Paulo
Av. Nazaré 481
São Paulo, SP 04263-000, Brazil

Sergio F. Vizcaíno
División Paleontología Vertebrados Museo de La Plata
Paseo del Bosque s/n
B1900FWA La Plata, Argentina
& CONICET

Marcelo Weksler
Setor de Mastozoologia
Departamento de Vertebrados
Museu Nacional, UFRJ
Rio de Janeiro, RJ 20940-040, Brazil

Michael O. Woodburne
Museum of Northern Arizona
Flagstaff, AZ 86011 US

Andre R. Wyss
Department of Earth Science
University of California-Santa Barbara
Santa Barbara, CA 93106 US

Taxonomic Index

Subject Index